WIRELESS
AD HOC
NETWORKING

WIRELESS NETWORKS
AND MOBILE COMMUNICATIONS

Series Editor: Yan Zhang

Millimeter Wave Technology in Wireless PAN, LAN, and MAN
Shao-Qiu Xiao, Ming-Tuo Zhou and Yan Zhang
ISBN: 0-8493-8227-0

Security in Wireless Mesh Networks
Yan Zhang, Jun Zheng and Honglin Hu
ISBN: 0-8493-8250-5

**Resource, Mobility and Security Management in Wireless Networks
and Mobile Communications**
Yan Zhang, Honglin Hu, and Masayuki Fujise
ISBN: 0-8493-8036-7

Wireless Mesh Networking: Architectures, Protocols and Standards
Yan Zhang, Jijun Luo and Honglin Hu
ISBN: 0-8493-7399-9

**Wireless Ad Hoc Networking: Personal-Area, Local-Area, and
the Sensory-Area Networks**
Shih-Lin Wu and Yu-Chee Tseng
ISBN: 0-8493-9254-3

Wireless Metropolitan Area Networks: WiMAX and Beyond
Yan Zhang and Hsiao-Hwa Chen
ISBN: 0-8493-2624-9

**Distributed Antenna Systems: Open Architecture for Future
Wireless Communications**
Honglin Hu, Yan Zhang and Jijun Luo
ISBN:1-4200-4288-2

AUERBACH PUBLICATIONS
www.auerbach-publications.com
To Order Call: 1-800-272-7737 • Fax: 1-800-374-3401
E-mail: orders@crcpress.com

WIRELESS AD HOC NETWORKING

Personal-Area, Local-Area, and the Sensory-Area Networks

Edited by

Shih-Lin Wu ♦ Yu-Chee Tseng

CRC Press
Taylor & Francis Group
Boca Raton London New York

CRC Press is an imprint of the
Taylor & Francis Group, an **informa** business

CRC Press
Taylor & Francis Group
6000 Broken Sound Parkway NW, Suite 300
Boca Raton, FL 33487-2742

First issued in paperback 2019

© 2007 by Taylor & Francis Group, LLC
CRC Press is an imprint of Taylor & Francis Group, an Informa business

No claim to original U.S. Government works

ISBN-13: 978-0-8493-9254-2 (hbk)
ISBN-13: 978-0-367-38931-4 (pbk)

Library of Congress Cataloging-in-Publication Data

Wireless ad hoc networking : personal-area, local-area, and the sensory-area
 networks / editors, Shih-Lin Wu and Yu-Chee Tseng.
  p. cm. -- (Wireless networks and mobile communications)
 Includes bibliographical references and index.
 ISBN-13: 978-0-8493-9254-2 (alk. paper)
 ISBN-10: 0-8493-9254-3 (alk. paper)
 1. Wireless LANs. 2. Wireless communication systems. 3. Sensor networks.
I. Wu, Shih-Lin. II. Tseng, Yu-Chee. III. Title. IV. Series.

TK5105.78.W59 2007
004.6'8--dc22   2006031857

Visit the Taylor & Francis Web site at
http://www.taylorandfrancis.com

and the Auerbach Web site at
http://www.auerbach-publications.com

Contents

PART I: WIRELESS PERSONAL-AREA AND SENSORY-AREA NETWORKS

1 Coverage and Connectivity of Wireless Sensor Networks 3
1.1 Introduction 3
1.2 Computing Coverage of a Wireless Sensor Network 4
1.3 Coverage and Scheduling 14
1.4 Coverage and Connectivity 20
1.5 Conclusions 23

2 Communication Protocols 25
2.1 Introduction 25
2.2 Similarities and Differences between WSNs and MANETs 28
2.3 Communication Patterns in Wireless Sensor Networks 30
2.4 Routing Protocols in WSNs 39
2.5 Comparative Studies 53
2.6 Conclusions and Future Research Directions 56

3 FireFly: A Time-Synchronized Real-Time Sensor Networking Platform 65
3.1 Introduction 65
3.2 The FireFly Sensor Node 67
3.3 RT-Link: A TDMA Link Layer Protocol for Multihop Wireless Networks 71
3.4 Nano-RK: A Resource-Centric RTOS for Sensor Networks 87
3.5 Coal Mine Safety Application 98
3.6 Summary and Concluding Remarks 102

4 Energy Conservation in Sensor and Sensor-Actuator Networks 107
4.1 Introduction 107
4.2 Localized Algorithms Save Energy 110

4.3   Minimum-Energy Broadcasting and Multicasting   113
4.4   Power-Aware Routing   115
4.5   Controlled Mobility for Power-Aware Localized Routing   116
4.6   Power-Efficient Neighbor Communication and Discovery
       for Asymmetric Links   118
4.7   Challenges of Power-Aware Routing with a Realistic
       Physical Layer   119
4.8   A Localized Coordination Framework for Wireless Sensor
       and Actuator Networks   121
4.9   Localized Movement Control Algorithms for
       Realization of Fault Tolerant Sensor and Sensor-
       Actuator Networks   128
4.10   Conclusion   130

5   Security in Wireless Sensor Networks 135
5.1   Introduction   136
5.2   Physical Layer Security   136
5.3   Key Management   140
5.4   Link Layer Security   155
5.5   Network Layer Security   157
5.6   Application Layer Security   159

**6   Autonomous Swarm-Bot Systems for Wireless
     Sensor Networks** 167
6.1   Introduction   167
6.2   The System Architecture   168
6.3   Cooperative Localization Algorithm   169
6.4   Foraging and Gathering   172
6.5   Minimap Integration   178
6.6   The Collaborative Path Planning Algorithm   181
6.7   Conclusion   184

**7   A Smart Blind Alarm Surveillance and Blind
     Guide Network System on Wireless Optical
     Communication** 191
7.1   Introduction   191
7.2   The Manufacture of Wireless Optical Transceiver   193
7.3   The Design of Wireless Optical Network   196
7.4   Smart Wireless Optical Blind-Guidance Cane and
       Blind-Guidance Robot   199
7.5   The Design of a Smart Guide System with Wireless Optical
       Blind-Guidance Cane and a Blind-Guidance Robot   203
7.6   Smart Wireless Optical Communication of Blind Alarm
       Surveillance System   210
7.7   The Design and Implementation of a Smart Wireless
       Blind-Guidance Alarm Surveillance System   214

PART II: WIRELESS LOCAL-AREA NETWORKS

8 Opportunism in Wireless Networks: Principles and Techniques 223
8.1   Opportunism: Avenues and Basic Principles   223
8.2   Source Opportunism   227
8.3   Spatio-Temporal Opportunism over a Single Link   234
8.4   Spatio-Temporal Opportunism in Ad Hoc Networks   241
8.5   Spatiotemporal-Spectral Opportunism in Ad Hoc Networks   247
8.6   Conclusions   250

9 Localization Techniques for Wireless Local Area Networks 255
9.1   Introduction   255
9.2   Nondedicated Localization Techniques   256
9.3   Location Tracking   272
9.4   Conclusion   274

10 Channel Assignment in Wireless Local Area Networks 277
10.1   Introduction   277
10.2   Preliminaries   280
10.3   Rings   282
10.4   Grids   285
10.5   Interval Graphs   288
10.6   Trees   292
10.7   Conclusion   296

11 MultiChannel MAC Protocols for Mobile Ad Hoc Networks 301
11.1   Introduction   301
11.2   Design Issues of Multichannel Protocols   302
11.3   Multichannel Protocols   303
11.4   Comparison of Multichannel MAC Protocols   320
11.5   Open Issues   321
11.6   Conclusions   322

12 Enhancing Quality of Service for Wireless Ad Hoc Networks 325
12.1   Introduction   326
12.2   Background   327
12.3   The Proposed EDCF-DM Protocol   331
12.4   Performance Evaluation   334
12.5   Conclusions   340

13 QoS Routing Protocols for Mobile Ad Hoc Networks 343
13.1   Introduction   343
13.2   Reviews of the QoS Routing Protocols   345
13.3   Our QoS Routing Protocol   348

13.4 Simulation Results 360
13.5 Conclusions 367

14 Energy Conservation Protocols for Wireless Ad Hoc Networks 371
14.1 Introduction 371
14.2 Power Management 371
14.3 Power Control 382
14.4 Topology Control Protocols 387
14.5 Summary 397

15 Wireless LAN Security 399
15.1 WEP and Its Security Weaknesses 399
15.2 802.1X Security Measures 405
15.3 IEEE 802.11i Security 410
15.4 Summary 417

16 Temporal Key Integrity Protocol and Its Security Issues in IEEE 802.11i 419
16.1 Introduction 419
16.2 Wired Equivalent Privacy and Its Weakness 420
16.3 Wi-Fi Protected Access 421
16.4 Temporal Key Integrity Protocol 423
16.5 Fragility of Michael 430
16.6 TKIP Countermeasures 431
16.7 Key Handshake Procedure 433
16.8 Conclusions 434

PART III: INTEGRATED SYSTEMS

17 Wireless Mesh Networks: Design Principles 439
17.1 Introduction 439
17.2 Generic Architecture and Basic Requirements of Wireless Mesh Networks 439
17.3 Network-Planning Techniques 442
17.4 Self-Configuring Techniques 448
17.5 Conclusions 458

18 Wireless Mesh Networks: Multichannel Protocols and Standard Activities 461
18.1 Introduction 461
18.2 Multichannel MAC Protocols 462
18.3 Multichannel Routing Protocols 473
18.4 Standard Activities of Mesh Networks 478
18.5 Conclusions 481

19 Integrated Heterogeneous Wireless Networks 483
   19.1   Introduction   483
   19.2   Integration of Infrastructure-Based Heterogeneous
            Wireless Networks   485
   19.3   Heterogeneous Wireless Multihop Networks   492
   19.4   Research Issues for Heterogeneous Wireless Networks   500
   19.5   Conclusions   502

20 Intrusion Detection for Wireless Network 505
   20.1   Introduction   505
   20.2   Background on Intrusion Detection   505
   20.3   Intrusion Detection for Mobile Ad Hoc Networks   506
   20.4   Intrusion Detection for Wireless Sensor Networks   519
   20.5   Conclusion   530

**21 Security Issues in an Integrated Cellular
    Network—WLAN and MANET** 535
   21.1   Introduction   535
   21.2   Architecture of the Integrated Network   537
   21.3   Security Impacts from the Unique Network Characteristics   540
   21.4   Potential Security Threats   542
   21.5   An Investigation and Analysis of Security Protocols   554
   21.6   New Security Issues and Challenges   564
   21.7   Conclusion   566

22 Fieldbus for Distributed Control Applications 571
   22.1   Introduction   571
   22.2   Review on Distributed Control   577
   22.3   Fundamental Aspects of DCS   578
   22.4   Standards, Frequency Bands, and Issues   580
   22.5   Some of the Major Wireless Fieldbuses   583
   22.6   Selecting a Fieldbus   590
   22.7   Discussion and Conclusions   591

**23 Supporting Multimedia Communication in the
    Integrated WCDMA/WLAN/Ad Hoc Networks** 595
   23.1   Introduction   595
   23.2   Multiple Accesses in CDMA Uplink   599
   23.3   Multiple Accesses in CDMA Downlink   605
   23.4   Mobility Management   613
   23.5   Design Integration with Ad Hoc Networks   620

Index 629

Preface

The rapid progress of mobile/wireless communication and embedded microsensing MEMS technologies leads us toward the dream "Ubiquitous/Pervasive Computing." Wireless local-area networks (WLANs) have been widely deployed in many cities and have become a requisite tool to many people in their daily lives. Wireless personal-area networks (WPANs) provide a cableless personal working area by using wireless interconnection among devices, such as personal computer, printer, personal digital assistant, and digital camera. A wireless sensor node is an integrated device which consists chiefly of a sensor part, a wireless communication part, and an information processing unit. Wireless sensor nodes can be deployed in a sensing field to monitor events that we are interested in. In particular, the human body could be a potential sensing field. By backing events in the body area, wearable sensor networks may greatly improve our daily life. By integrating these multidiscipline technologies into a pervasive system, we can access information and acquire computing resources anytime, anywhere with any device. The pervasive system can be used in a wide range of applications such as home automation, health care, transportation, agriculture, scientific survey, industrial automation, and military applications.

This book aims to provide a wide coverage of key technologies in wireless ad hoc networks including networking architectures and protocols, cross-layer architectures, localization and location tracking, power management and energy-efficient design, power and topology control, time synchronization, coverage issues, middleware and software design, data gathering and processing, embedded network-oriented operating systems, mobility management, self-organization and governance, QoS and real-time issues, security and dependability issues, integrated wired/wireless/sensor networks and systems, applications, modeling and performance evaluation, implementation experience, and measurements. We believe that there is no other existing book that brings the concept—*ad hoc networking*—together with its key technologies and applications over the platforms—*personal-area, sensory-area,* and *local-area networks*—in a single work. The book contains three parts and each part has several self-contained chapters written by experts and researchers from around the world.

The potential audience for this book includes researchers and students who want to specialize in the field of computer science and electrical engineering, professionals and engineers who are designers or planners for wireless sensory-, personal-, and local-area networks, and those who are interested in the ubiquitous/pervasive environment.

This book is a collective work of many excellent experts and researchers. We would like to express our immense gratitude to these authors for their valuable contributions. It has been a pleasure to work with Richard O'Hanley of Auerbach Publications (Taylor & Francis Group) for their professional support and encouragement to publish this book. We would also like to thank Catherine Giacari of CRC Press (editorial project development) for her kindly help in this project. Finally, we appreciate our family for their great patience and enormous love throughout the publishing work.

The Authors

Shih-Lin Wu received his BS degree in Computer Science from Tamkang University, Taiwan, in June 1987 and his PhD degree in Computer Science and Information Engineering from National Central University, Taiwan, in May 2001. From August 2001 to July 2003, he was an Assistant Professor in the faculty of the Department of Electrical Engineering, Chang Gung University, Taiwan. Since August 2003, he has been with the Department of Computer Science and Information Engineering, Chang Gung University. Dr. Wu served as a Program Chair in  the *Mobile Computing Workshop*, 2005 and as a guest editor for the special issue of *Journal of Pervasive Computing and Communication* on "Key Technologies and Applications of Wireless Sensor and Ad Hoc Networks." Several of his papers have been chosen as best papers in international conferences. His current research interests include mobile computing, wireless networks, distributed robotics, and network security. Dr. Wu is a member of the IEEE and the Phi Tau Phi Society.

Yu-Chee Tseng received his BS and MS degrees in Computer Science from the National Taiwan University and the National Tsing-Hua University in 1985 and 1987, respectively. He obtained his PhD in Computer and Information Science from the Ohio State University in January 1994. He was an associate professor at the Chung-Hua University (1994–1996) and at the National Central University (1996–1999), and a professor at the National Central University (1999–2000). Since 2000, he has been a professor in the Department of Computer Science, National Chiao-Tung University, Taiwan, where he is currently its chairman.

Dr. Tseng served as a Program Chair in the *Wireless Networks and Mobile Computing Workshop*, 2000 and 2001, as a Vice-Program Chair in the *International Conference on Distributed Computing Systems (ICDCS)*, 2004, as well as of the *IEEE International Conference on Mobile Ad-hoc and Sensor Systems (MASS)*, 2004, as an associate editor for *The Computer Journal*, as a guest editor for the special issue of *ACM Wireless Networks* on "Advances in Mobile and Wireless Systems," as a guest editor for the special issue of *IEEE Transactions on Computers* on "Wireless Internet," as a guest editor for the special issue of *Journal of Internet Technology* on "Wireless Internet: Applications and Systems," as a guest editor for the special issue of *Wireless Communications and Mobile Computing* on "Research in Ad Hoc Networking, Smart Sensing, and Pervasive Computing," as an editor for *Journal of Information Science and Engineering*, as a guest editor for the special issue of *Telecommunication Systems* on "Wireless Sensor Networks," and as a guest editor for the special issue of *Journal of Information Science and Engineering* on "Mobile Computing."

Dr. Tseng received the Outstanding Research Award from the National Science Council, ROC, in both 2001–2002 and 2003–2005, the Best Paper Award at the *International Conference on Parallel Processing* in 2003, the Elite I. T. Award in 2004, and the Distinguished Alumnus Award from the Ohio State University in 2005. His research interests include mobile computing, wireless communication, network security, and parallel and distributed computing. Dr. Tseng is a member of ACM and a Senior Member of IEEE.

Contributors

Dharma P. Agrawal
OBR Center for Distributed and
  Mobile Computing
Department of ECECS
University of Cincinnati
Cincinnati, Ohio

Alan A. Bertossi
Department of Computer Science
University of Bologna
Bologna, Italy

Guohong Cao
Department of Computer Science and
  Engineering
The Pennsylvania State University
University Park, Pennsylvania

Dave Cavalcanti
Wireless Communications and
  Networking Department
Philips Research North America
Briarcliff Manor, New York

Chao-Hsin Chang
Department of Computer Science and
  Information Engineering
Chang Gung University
Taoyuan, Taiwan

Chih-Yung Chang
Department of Computer Science and
  Information Engineering
Tamkang University
Tamsui, Taiwan

Jenhui Chen
Department of Computer Science and
  Information Engineering
Chang Gung University
Taoyuan, Taiwan

Shu-Min Chen
Department of Computer Science
National Chiao Tung University
Hsinchu, Taiwan

Jen-Shiun Chiang
Department of Electrical Engineering
Tamkang University
Tamsui, Taiwan

Yung-Shan Chou
Department of Electrical Engineering
Tamkang University
Tamsui, Taiwan

Po-Jen Chuang
Department of Electrical Engineering
Tamkang University
Tamsui, Taiwan

Kamal Dass
Department of Computer Science
The University of Alabama
Tuscaloosa, Alabama

Cheng-Chang Ho
Department of Electrical Engineering
Tamkang University
Tamsui, Taiwan

Chih-Shun Hsu
Department of Computer Science and
 Information Engineering
Nanya Institute of Technology
Jhongli, Taiwan

Rong-Hong Jan
Department of Computer Science
National Chiao Tung University
Hsinchu, Taiwan

Yih-Guang Jan
Department of Electrical Engineering
Tamkang University
Tamsui, Taiwan

Huan-Chao Keh
Department of Computer Science and
 Information Engineering
Tamkang University
Tamsui, Taiwan

Cynthia Kersey
Department of Computer Science
University of Illinois
Chicago, Illinois

Ahmad Khoshnevis
Department of Electrical and Computer
 Engineering
Rice University
Houston, Texas

Kiseon Kim
Gwangju Institute of Science
 and Technology
Gwangju, South Korea

Anup Kumar
Department of Computer Engineering
 and Computer Science
University of Louisville
Louisville, Kentucky

Hsiao-Ju Kuo
Department of Computer Science
National Chiao Tung University
Hsinchu, Taiwan

Sheng-Po Kuo
Department of Computer Science
National Chiao Tung University
Hsinchu, Taiwan

Yang-Han Lee
Department of Electrical Engineering
Tamkang University
Tamsui, Taiwan

Chih-Yu Lin
Department of Computer Science
National Chiao Tung University
Hsinchu, Taiwan

Jheng-Yao Lin
Department of Electrical Engineering
Tamkang University
Tamsui, Taiwan

Jonathan C.L. Liu
Computer, Information Science
 and Engineering
University of Florida
Gainesville, Florida

Yunhuai Liu
Department of Computer Science
Hong Kong University of Science and
 Technology
Hong Kong, China

N.P. Mahalik
Gwangju Institute of Science and
 Technology
Gwangju, South Korea

Rahul Mangharam
Department of Electrical and Computer
 Engineering
Carnegie Mellon University
Pittsburgh, Pennsylvania

Lionel M. Ni
Computer Science Department
Hong Kong University of Science and
 Technology
Hong Kong, China

M. Cristina Pinotti
Department of Computer Science and
 Mathematics
University of Perugia
Perugia, Italy

Anthony Rowe
Department of Electrical and Computer
 Engineering
Carnegie Mellon University
Pittsburgh, Pennsylvania

Raj Rajkumar
Department of Electrical and Computer
 Engineering
Carnegie Mellon University
Pittsburgh, Pennsylvania

Ashutosh Sabharwal
Department of Electrical and Computer
 Engineering
Rice University
Houston, Texas

Loren Schwiebert
Department of Computer Science
Wayne State University
Detroit, Michigan

Jang-Ping Sheu
Department of Computer Science and
 Information Engineering
National Central University
Jhongli, Taiwan

Shiann-Tsong Sheu
Department of Communication
 Engineering
National Central University
Jhongli, Taiwan

Kuei-Ping Shih
Department of Computer Science and
 Information Engineering
Tamkang University
Tamsui, Taiwan

Ivan Stojmenovic
SITE
University of Ottawa
Ontario, Canada

Wai-Hong Tam
Department of Computer Science
National Chiao Tung University
Hsinchu, Taiwan

Jeffrey J.P. Tsai
Department of Computer Science
University of Illinois
Chicago, Illinois

Hsien-Wei Tseng
Department of Electrical Engineering
Tamkang University
Tamsui, Taiwan

Huei-Ru Tseng
Department of Computer Science
National Chiao Tung University
Hsinchu, Taiwan

Yu-Chee Tseng
Department of Computer Science
National Chiao Tung University
Hsinchu, Taiwan

Shen-Chien Tung
Department of Computer Science and
 Information Engineering
National Central University
Jhongli, Taiwan

Ju Wang
Department of Computer Science
Virginia Commonwealth University
Richmond, Virginia

Ching-Chang Wong
Department of Electrical
  Engineering
Tamkang University
Tamsui, Taiwan

Fang-Jing Wu
Department of Computer Science
National Chiao Tung University
Hsinchu, Taiwan

Kui Wu
Department of Computer Science
University of Victoria
Victoria, Canada

Rung-Hou Wu
Department of Computer Science and
  Communication Engineering
St. John's University
Tamsui, Taiwan

Shih-Lin Wu
Department of Computer Science and
  Information Engineering
Chang Gung University
Taoyuan, Taiwan

Yong Xi
Department of Computer Science
Wayne State University
Detroit, Michigan

Yang Xiao
Department of Computer Science
The University of Alabama
Tuscaloosa, Alabama

Bin Xie
Department of Computer Engineering
  and Computer Science
University of Louisville
Louisville, Kentucky

Jhen-Yu Yang
Department of Computer Science and
  Information Engineering
Chang Gung University
Taoyuan, Taiwan

Wuu Yang
Department of Computer Science
National Chiao Tung University
Hsinchu, Taiwan

Zhenwei Yu
Department of Computer Science
University of Illinois
Chicago, Illinois

Hao Zhu
Department of Electrical and
  Computer Engineering
Florida International University
Miami, Florida

Jun Zheng
School of Information Technology
  and Engineering
University of Ottawa
Ontario, Canada

WIRELESS PERSONAL-AREA AND SENSORY-AREA NETWORKS

Chapter 1

Coverage and Connectivity of Wireless Sensor Networks

Hsiao-Ju Kuo, Sheng-Po Kuo, Fang-Jing Wu, and
Yu-Chee Tseng

1.1 Introduction

In a wireless sensor network, one central issue is to evaluate how well
the sensing field is monitored by sensors, also known as the *coverage* of
the network. The coverage issue is directly related to the sensing capability of the network on monitoring phenomenons occurring in the sensing
area, such as intruders in a battlefield or fire events in a forest. After the
coverage is guaranteed, another important issue is to prolong the network
lifetime when sensors are more than necessary. More specifically, if sensors are more than the required number, some of them can be scheduled to
turn off their power. These powered-off sensors can be activated later when
some powered-on sensors run out of energy. The scheduling should always
guarantee the required level of coverage. Furthermore, ensuring communication connectivity is also essential when scheduling sensors' on-duty
time. *Connectivity* normally means that there is at least one path connecting any two on-duty sensors and the gateway. This chapter is dedicated
to reviewing the related coverage- and connectivity-maintaining protocols
and algorithms.

The organization of this chapter is as follows. Section 1.2 discusses how to compute the coverage of a wireless sensor network. Section 1.3 introduces scheduling protocols that can put some sensors into the sleep mode while maintaining the required level of coverage. Finally, some mechanisms for maintaining both sensing coverage and communication connectivity are presented in Section 1.4.

1.2 Computing Coverage of a Wireless Sensor Network

Coverage is an essential problem in wireless sensor networks. It is important to ensure that sensors provide sufficient coverage of the sensing field. However, one should use as few sensors as possible to cover the sensing field to reduce the hardware cost. Assuming that sensors are randomly deployed, this section discusses three general models to define the coverage problem and reviews some solutions to the coverage problem. The first one is the *binary model*, where each sensor's coverage area is modeled by a disk. Any location within the disk is well monitored by the sensor located at the center of the disk; otherwise, it is not monitored by the sensor. The second one is the *probabilistic model*. An event happening in the coverage of a sensor is either detected or not detected by the sensor depending on a probability distribution. Hence, even if an event is very close to a sensor, it may still be missed by the sensor. The last model considers the coverage problem by including the issue of how targets travel along the sensing field. The worst and best traveling paths of this model can be used to evaluate the sensing capability of the sensor network.

1.2.1 Coverage Solutions Based on the Binary Model

Under the binary model, the sensing range of a sensor is modeled by a disk. The radii of disks representing sensors' sensing ranges may be the same or different. When an event occurs within a sensor's sensing range, it is well monitored by the sensor (i.e., all locations within the sensing range of a sensor are under the same quality of surveillance). On the contrary, locations outside the sensing region of a sensor are considered undetectable by the sensor.

In Ref. [1], the coverage problem is defined as follows: we are given a set of sensors, $S = \{s_1, s_2, \ldots, s_n\}$, in a two-dimensional (2D) area A. Each sensor s_i, $i = 1, \ldots, n$, is located at a coordinate (x_i, y_i) inside A and has a sensing range of r_i, i.e., it can monitor any point that is within a distance of r_i from s_i.

Definition 1

A location in A is said to be *covered* by s_i if it is within s_i's sensing range. A location in A is said to be *j-covered* if it is within at least j sensors' sensing ranges.

Definition 2

A *subregion* in A is a set of points that are covered by exactly the same set of sensors.

Two versions of the coverage problem are defined as follows:

Definition 3

Given a natural number k, the *k-non-unit-disk coverage (k-NC) problem* is to determine whether all points in A are k-covered or not.

Definition 4

Given a natural number k, the *k-unit-disk coverage (k-UC) problem* is to determine whether all points in A are k-covered or not, subject to the constraint that $r_1 = r_2 = \cdots = r_n$.

There are several reasons for asking a wireless sensor network to be k-covered such that $k > 1$. First, it could be for the fault-tolerant reason. Second, some special applications, such as trilateration, may require each point in the sensing field to be at least 3-covered. Figure 1.1 illustrates how sub regions of a sensing field are covered.

The work[2] proposes a solution to the k-UC. To determine the coverage level, this work looks at how intersection points between sensors' sensing ranges are covered. It claims that a sensing field A is k-covered if all intersection points between each pair of sensors and between each sensor and the boundary of this field A are at least k-covered. Based on this property, a *coverage configuration protocol (CCP)* that can schedule sensors' on-duty time so as to maintain a given coverage level is presented in Ref. [2]. Initially, all sensors are in the *active* state. If some area are overly covered, redundant nodes may find themselves unnecessary and switch to the *sleep* state. More specifically, a sensor need not stay active if all intersection points within its sensing range are at least k-covered by other nodes in its neighborhood excluding itself. A sleeping node also periodically wakes up

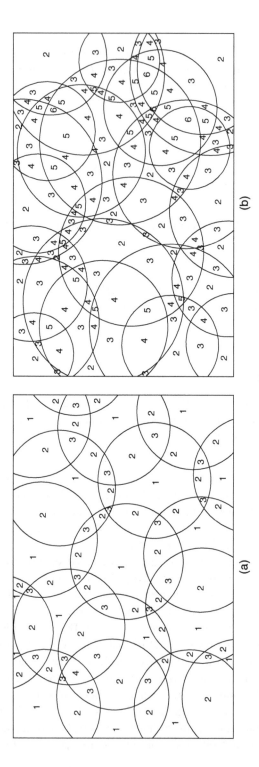

Figure 1.1 Examples of the coverage problem: (a) the sensing field is 1-covered by unit disks and (b) the sensing field is 2-covered by nonunit disks. The number in each subregion is its coverage level.

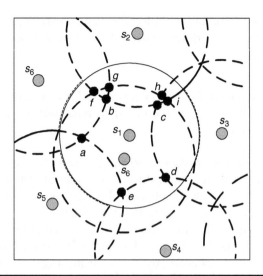

Figure 1.2 **A sensor network example where sensor s_1 locally decides itself to be redundant and thus goes to sleep based on the approach proposed in Ref. [2].**

and enters the *listen* state. While in the listen state, the sensor will evaluate whether it is necessary to return to the active state.

An example is shown in Figure 1.2. The objective is to achieve 1-coverage in this sensing field. Sensor s_1 has to check the intersection points, $a, b, \ldots, i$, in its sensing region. It finds that each point is covered by at least one other sensor and thus it can go to the sleep state. Note that when evaluating the coverage of an intersection point, the two sensors that intersect at this point are not considered. Also note that in the above example, since s_1 decides to go to the sleep state, sensor s_6 should not go to the sleep state. So some coordination mechanism among sensors is required.

In Ref. [1], an efficient approach to solving the k-UC and k-NC problems is proposed. Instead of checking the coverage level of intersection points, it looks at how the perimeter of each sensor's sensing range is covered. This leads to an efficient polynomial-time algorithm. Specifically, the algorithm tries to determine whether the perimeter of a sensor's sensing range under consideration is sufficiently covered. By collecting this information from all sensors, the coverage of the sensing field can be determined. The formal derivation is as follows:

Definition 5

Consider any sensor s_i. We say that s_i is *k-perimeter-covered* if all points on the perimeter of s_i's sensing range are perimeter-covered by at least k sensors other than s_i itself. Similarly, a segment of the perimeter is *k-perimeter-covered* if all points on the segment are perimeter-covered by at least *k* sensors other than s_i itself.

Theorem 1 *Suppose that no two sensors are located in the same location. The whole network area A is k-covered iff each sensor in the network is k-perimeter-covered.*

An example of determining the perimeter coverage of a sensor is shown in Figure 1.3(a). The sensor s_1 first determines which segments of its perimeter are covered by which neighboring nodes. For example, the segment of the perimeter from point *a* to point *b* is covered by s_2 and s_3. Thus, this segment is 2-perimeter-covered. A simplified representation of perimeter coverage is shown in Figure 1.3(b). The perimeter segment covered by s_2 can be denoted by an arc from angle α_l to angle α_r. To compute the perimeter coverage of each perimeter segment, we can mark angles α_l and α_r on the line segment $[0, 2\pi]$ as shown in Figure 1.3(c). By traversing this line segment, the perimeter coverage of the sensor can be easily determined. In this example, s_1 is 2-perimeter-covered. According to

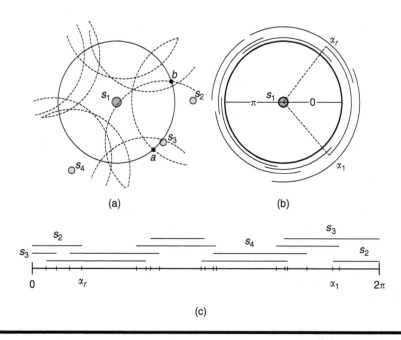

(a)　　　　　　(b)

(c)

Figure 1.3　An example for determining perimeter coverage: (a) sensors' intersection relationship with s_1, (b) a simplified representation of perimeter coverage, and (c) placing sensors' intersection relationship on a line segment $[0, 2\pi]$.

Theorem 1, if all other sensors are at least 2-perimeter-covered, the sensing field is 2-covered.

This algorithm can be easily translated to a distributed protocol, where each sensor only needs to collect local information to make its decision. The computational cost for each sensor is dominated by sorting the intersection angles. This incurs a time complexity of $O(d \log d)$, where d is the maximum number of sensors that are neighboring to another sensor. Compared to the algorithm in Ref. [2], which requires $O(d^2)$ for each sensor to observe the intersection points within its coverage, this algorithm incurs lower computational complexity.

Note that when a sensor s_i is k-perimeter-covered, it is not necessary that the sensing range of s_i is k-covered. In Figure 1.4, sensor s_1 is 2-perimeter-covered but its sensing range is not 2-covered since the shadow region contained in s_1 is only 1-covered. Since this algorithm looks at perimeter coverage, this can be explained by observing that s_2 is not 2-perimeter-covered (the dashed segment of s_2's perimeter is only 1-perimeter-covered). Although this property differs from the algorithm in Ref. [2], when gathering all sensors' local decisions, the coverage level of the whole field can be correctly determined.

Ref. [3] further proposes a new algorithm for determining the coverage level of a three-dimensional (3D) space. Given a set of sensors in a 3D sensing field, the goal is to determine if this field is sufficiently α-*covered*, where α is a given integer, in the sense that every point in the field is covered by at least α sensors. The sensing range of each sensor is modeled by a 3D ball. At the first glance, the 3D coverage problem seems very difficult since even determining the subspaces divided by the spheres of sensors' sensing ranges is very complicated. However, the authors show that tackling

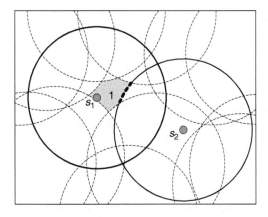

Figure 1.4 An example of the difference between 2-perimeter-coverage and 2-coverage. The sensor s_1 is 2-perimeter-covered but not 2-covered.

this problem is still feasible within polynomial time. The authors propose a novel solution by reducing the geometric problem from a 3D space to a 2D space, and further to a 1D space, thus leading to a very efficient solution. In essence, the solution tries to look at how the sphere of each sensor's sensing range is covered. As long as the spheres of all sensors are sufficiently covered, the whole sensing field is sufficiently covered. To determine whether each sensor's sphere is sufficiently covered, the authors in turn look at how each spherical cap and how each circle of the inter-section of two spheres is covered. By stretching each circle on a 1D line, the level of coverage can be easily determined.

1.2.2 Coverage Solutions Based on the Probabilistic Model

In the previous section, a sensor's sensing capability is modeled by binary values according to an artificial and uniform boundary. The binary sensing model is suitable for sensors to detect events whose quality of surveillance is only slightly affected by sensing distances, such as temperature, humidity, and light. However, this is sometimes oversimplified and may not be able to capture the stochastic nature of signals. In the real world, the sensing capabilities of sensor nodes are usually dependent on environmental factors and signal propagation characteristics. Therefore, the sensing model based on the probabilistic assumption may be more realistic to capture the actual sensing capability of sensor nodes.

The probabilistic sensing model uses probability distributions to express the quality of surveillance of events being sensed by sensors. Under this model, the sensing capability of a sensor i for a location u is expressed by a probability function $p_i(u)$. Suppose that there are n sensors in the sensor networks and an event appears at location u. The detection probability $P(u)$ contributed by these n sensors can be modeled by

$$P(u) = 1 - \prod_{i=1}^{n}(1 - p_i(u)) \tag{1.1}$$

Note that the sensing capability p_i depends on the signal propagation model. According to different propagational models, the evaluation of sensing capability could be different. Two common probabilistic sensing models of p_i are introduced as follows:

1. Signal decay model: Assume that these n sensors are deployed at locations l_i, $i = 1, 2, \ldots, n$. When a target at location u emits a signal, the energy (signal strength) sensed by sensor s_i is defined by

$$E_i(u) = \frac{K}{\|u - l_i\|^k} + N_i \tag{1.2}$$

where K is the energy emitted by the target, N_i the noise, and k the signal decaying coefficient, which typically ranges from 2 to 5. $\|u - l_i\|$ denotes the geometric distance between the target and sensor s_i. The noise effect is usually modeled by a zero-mean Gaussian distribution random variable.[4]

Based on this formulation, we can define the sensing capability of sensor s_i for target u as follows:

$$p_i(u) = \text{Prob}[E_i(u) \geq \eta] \tag{1.3}$$

where η means the receiver sensitivity, which is the minimum signal strength for a successful detection. The environmental noise may cause unstable event detection. Longer detection distance will decrease the sensing capability p_i.

2. Log-normal shadowing model: The log-normal shadowing model is a widely used propagational model to express the signal strength decay effect in an open space.[4–6] A path loss function PL is used to express the decay effect in terms of the propagation distance d as follows:

$$PL(d) = PL(d_0) + 10n \log \left(\frac{d}{d_0} \right) + X_\sigma \tag{1.4}$$

where n is a path loss exponent that indicates the decreasing rate of signal strength in an environment, d_0 a reference distance which is close to the transmitter, and X_σ a zero-mean Gaussian distribution random variable (in dB) with a variance σ to express the shadowing effect.

According to the path loss function, we can derive the received energy at the sensor s_i as

$$E_i(u) = K - PL(\|u - l_i\|) \tag{1.5}$$

where K is the initial transmission power from target u. Then, similar to the previous model, Eq. (1.3) can be used to indicate the sensing capability p_i.

Under such models, an event happening in the coverage of a sensor is either detected or not detected depending on some probability distribution. Therefore, even if an event is very close to a sensor, it may still be missed by the sensor. Ahmed et al.[7] propose that the coverage problem should be defined as follows: the whole region A is *sufficiently covered* by a set of sensors if the detection probability of each location in A is equal to or greater than a desired detection probability. An interesting problem is to determine whether the sensing region is sufficiently covered or not. Clearly, the geometric approaches used in Section 1.2.1 cannot be used here. One intuitive algorithm is to partition A into grids and to check the coverage of each grid point. For each grid point u, the nearest sensor is responsible for the computation of the detection probability $P(u)$. If all grid points are satisfied, we can conclude that the whole sensing field is sufficiently covered. This computation cost is based on the number of grid points. Hence, this approach is not very scalable.

1.2.3 Coverage Solutions Based on Exposure

A sensor's sensing capability normally decreases when its distance from the target event increases. As a result, a location that is farther from most sensors is considered harder to be monitored. In some works, the coverage problem is considered by assuming that targets may move through the sensing field and the objective is to evaluate the sensor network's capability of surveillance.

In Ref. [8], the authors try to find a path which is best or worst monitored by sensors when a mobile target traverses along the path. It is believed that such a path could reflect the best or worst sensing capability provided by the sensor network. This work defines the *maximal breach path* and the *maximal support path*, such that the distance from any point to the closest sensor is maximized and minimized, respectively. Polynomial-time algorithms are proposed to find such paths. The key idea is to use the Voronoi diagram and the Delaunay triangulation of sensor nodes to limit the search space for the optimal paths in each case. The Voronoi diagram is formed from the perpendicular bisectors of lines that connect two neighboring sensors, while the Delaunay triangulation is formed by connecting nodes that share a common edge in the Voronoi diagram. Examples of the Voronoi diagram and Delaunay triangulation are shown in Figure 1.5.

Because each point on the line segments of the Voronoi diagram has the maximal distance to its closest sensor, the maximal breach path must lie

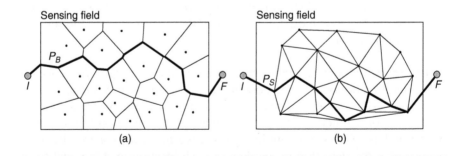

Figure 1.5 Examples of (a) the maximal breach path in the Voronoi diagram and (b) the maximal support path in the Delaunay triangulation. *I* and *F* denote the initial and final locations of the path.

on the line segments of the Voronoi diagram. To find the maximal breach path, each line segment is given a weight equal to its minimum distance to the closest sensor. The proposed algorithm then performs a binary search between the smallest and largest weights. In each step, a breadth-first search is used to check the existence of a path from the source point to the destination point using only line segments with weights that are larger than the search criterion, called *breach weight*. If a path exists, the criterion is increased to further restrict the lines considered in the next search iteration. Otherwise, the criterion is decreased. An example of the maximal breach path is shown in Figure 1.5(a). Similarly, since the Delaunay triangulation produces triangles that have minimal edge lengths among all possible triangulations of sensors, the maximal support path must lie on the lines of the Delaunay triangulation of sensors. To find the maximal support path, the weights of line segments of the Delaunay triangulation are set to their lengths. The rest of the search steps are similar as above. An example of the maximal support path is shown in Figure 1.5(b).

To further enhance the above results, the concept of traveling time should be included to reflect how a moving target is exposed to sensors when it moves along a path. Meguerdichian et al.[9] define the exposure problem and show how to determine the minimal exposure path given any sensor topology.

The exposure for a target in the sensing field during the interval $[t_1, t_2]$ along a path $p(t)$ is defined as

$$E(p(t), t_1, t_2) = \int_{t_1}^{t_2} I(F, p(t)) \left| \frac{dp(t)}{dt} \right| dt \qquad (1.6)$$

where $I(F, p(t))$ is the sensor field intensity measured at location $p(t)$ from the closest sensor or all sensors in the sensor field F, and $|dp(t)/dt|$ the element of arc length.

An example using Eq. (1.6) is illustrated in Figure 1.6. A sensor S is located in coordinate $(0,0)$ and a target moves from $A = (1, -1)$ to $B = (-1, 1)$. Since points A and B are at the same distance from S, Eq. (1.6) implies that the minimum exposure path will be the arc-centered S with a radius equal to the distance between S and A. Thus, path P_1 is the minimum exposure path. However, if the sensing field in Figure 1.6 is bounded by the given square, then the minimal exposure path will become P_2. Note that P_2 is the concatenation of two line segments and a quarter of the circle centered at S with a radius of 1. The reason that P_2 is the minimum exposure path can be explained by comparing P_2 to another arbitrary path, say, the dashed path P_3 in Figure 1.6. Since the minimal exposure path from A to B must be a symmetrical path, comparing a half of P_2 (the segment from A to q_4) to P_3 is sufficient. P_3 can be separated into two parts, the

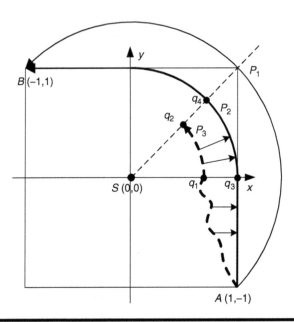

Figure 1.6 A mobile target moves from A to B in different paths: P_1 is the minimal exposure path on an unbounded field, P_2 the minimal exposure path on a square field, and P_3 half of an arbitrary path.

part from A to q_1 and the part from q_1 to q_2. The first part clearly has a higher exposure than the segment from A to q_3 since it is closer to S and is also longer in length. As to the second part, Eq. (1.6) implies that any two arcs lying on different concentric circles centered S with the same arc angle will have the same exposure value. So, the exposure value of this part in P_3 is equal to the corresponding segment in P_2. Therefore, we can conclude that P_2 is the minimal exposure path in this scenario.

In practice, if the target can move in arbitrary directions, the exposure problem will become an extremely difficult optimization problem. The approach proposed in Ref. [9] is to divide the sensing field into grids, and force the objects' roaming paths to only pass the edges of grids and/or diagonals of grids. Each line segment of a roaming path is assigned a weight equal to the exposure value of this segment. Then a single-source-shortest-path algorithm is used to find the minimal exposure path.

1.3 Coverage and Scheduling

Since sensors are usually supported by batteries, sensors' on-duty time should be properly scheduled to conserve energy. If some nodes share a common sensing region and task, then we can turn off some of them

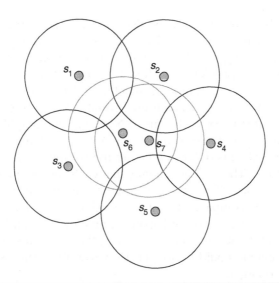

Figure 1.7   An example showing that one of the sensors s_6 and s_7 can be put into the sleep mode.

to conserve energy, and thus extend the lifetime of the network. This is feasible under the condition that turning off these nodes still provides the requested coverage level. An example is shown in Figure 1.7. Sensor s_6 can be put into the sleep mode since all its sensing area is covered by other nodes. Clearly, one of sensors s_6 and s_7 can be put into the sleep mode, but not both of them. As a result, we not only need to check if sensors satisfy certain eligibility rules but also need to schedule their sleeping time carefully. In this section, we will review some centralized and distributed scheduling algorithms.

1.3.1   Set Covering Solutions

One solution to extending network lifetime is to organize sensors into several sets, called *set cover*. The sensors of each set can fully cover the whole sensing field. At any time, only one active set is responsible for monitoring the sensing field, and the other sensors can be put into the sleep mode. Through a round-robin scheduling fashion, the network lifetime can be effectively prolonged.

The set covering problem has been modeled in two ways: *maximum disjoint set cover problem*[10] and *maximum set cover problem*.[11] Given a collection C of n sensors $s_1, s_2, \ldots, s_n$, the former problem is to find the maximum number of disjoint set covers for a finite target set. Here the target set is a set of target locations in the sensing field each of which has to be covered by at least one sensor. Every cover C_i is a subset of C, $C_i \subseteq C$, such that every target is covered by at least one sensor in C_i, and for any two covers C_i and C_j, $C_i \cap C_j = \emptyset$. Given a collection C of n sensors s_1, $s_2, \cdots, s_n$, the latter problem is to find a number of set covers $C_1, \ldots, C_p$ with time weights $t_1, \cdots, t_p$ in $[0, \zeta]$ such that $t_1 + t_2 + \cdots + t_p$ is maximized and for each sensor $s_i \in C$, its total time weight in $C_1, \cdots, C_p$ is at most ζ, where ζ is the lifetime of each sensor. Note that, for simplification, these problems only require sensors to cover a finite set of target locations.

These two problems differ in whether the computed set covers should be disjoint or not. The former requires that each sensor should belong to only one set. Hence, if there are n disjoint sets, the network lifetime is extended n times compared to the case without scheduling. Intuitively, the latter allows sensors to participate in multiple sets and each set can operate for a different time interval. Clearly, this is less restricted and thus the network lifetime could be longer. However, the search space for this problem is also larger.

For example, Figure 1.8 illustrates four sensors s_1, s_2, s_3, and s_4 deployed in a sensing field and there are four targets T_1, T_2, T_3, and T_4 to be monitored. Each sensor has the same sensing range. We can observe that sensor s_1 can cover the set of targets $\{T_1, T_2, T_4\}$, s_2 can cover $\{T_1, T_3, T_4\}$, s_3 can cover $\{T_1, T_2, T_3, T_4\}$, and s_4 can cover $\{T_2, T_3\}$. One solution of the disjoint set cover problem is to let $C_1 = \{s_1, s_2\}$ and $C_2 = \{s_3, s_4\}$. Assume that the remaining energy of all sensors is equal to an unit time ζ. Then, we can schedule sets C_1 and C_2 in a round-robin fashion, so the network lifetime

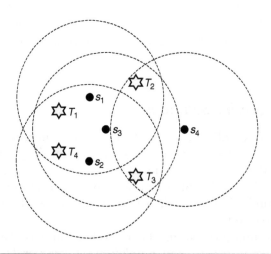

Figure 1.8 An example of the set cover problem. Sensors s_1, s_2, s_3, and s_4 cover the sets of targets $\{T_1, T_2, T_4\}$, $\{T_1, T_3, T_4\}$, $\{T_1, T_2, T_3, T_4\}$, and $\{T_2, T_3\}$, respectively.

is doubled. If we release the restriction that sets must be disjoint, it is possible to have a better scheduling. In the same example, one solution of the set cover problem is $C_1 = \{s_1, s_2\}$, $C_2 = \{s_2, s_4\}$, $C_3 = \{s_1, s_4\}$, and $C_4 = \{s_3\}$. Some sensors participate in multiple sets. Hence, it is necessary to make sure that the total active time of each sensor is less than or equal to the unit time ζ. One feasible weight assignment for this example is $t_1 = t_2 = t_3 = \frac{1}{2}\zeta$ and $t_4 = \zeta$. This scheduling gives total network lifetime of $2\frac{1}{2}\zeta$, which is longer than the scheduling result of the disjoint set solution.

These two problems have been proved to be NP-complete problems.[10,11] Hence, some heuristic algorithms are proposed. For example, the *Greedy-MSC Heuristic* algorithm builds set covers by a greedy strategy.[11] At each step, the most sparsely covered target will be considered. Then, from all available sensors, the sensor with the greatest contribution will be selected to cover this target. The sensor which can cover more uncovered targets is considered to have more contributions. After all targets have been covered, a new set cover is formed. Then, the final network lifetime is the product of the number of set covers and the active time of each set cover.

1.3.2 Distributed Scheduling Solutions

Another sensor scheduling scheme is proposed in Ref. [12]. In this scheme, each sensor decides its own on-duty time locally. The time axis is divided into a sequence of *frames*. In each frame, there are two phases, the *initialization phase* and the *sensing phase*. The sensing phase is further divided into *rounds* with equal duration. Figure 1.9(a) shows this idea. The whole sensing area is divided into grid points. Each grid point must be covered by at least one sensor in every moment during each round. In the initialization phase, each sensor node randomly generates a *reference time* and exchanges this time information with its neighboring sensors. A sensor has to join the schedule of each grid point covered by itself based on its reference time such that the grid point is covered by at least one sensor at any moment of a round. Since a sensor may cover multiple grid points, its on-duty time in each round is the union of schedules of all grid points covered by the sensor.

The detailed procedure is described as follows. In the initialization phase, sensor s_i will generate a reference time Ref_i in the interval $[0, T]$, where T is the length of a round. The time in each round for sensor s_i to keep awake for a grid point is written as $[Ref_i - T_{front}^i, Ref_i + T_{end}^i]$, where T_{front}^i and T_{end}^i are the time durations before and after Ref_i, respectively, for s_i to cover this grid point. For example, in Figure 1.9(b), sensor s_1 has to join the schedules of grid points A, B, C, and D. Assume that the duration of each round is $T = 20$ s, and that the selected reference times of s_1, s_2, s_3,

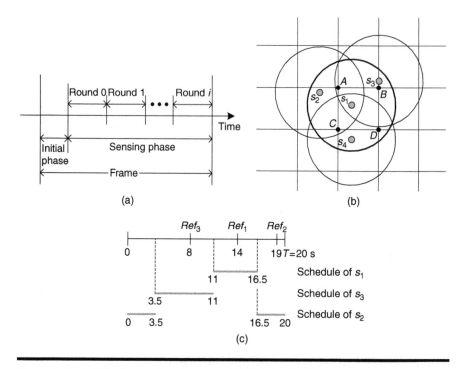

Figure 1.9 (a) The frame structure used in Ref. [12], (b) a sensor network configuration, and (c) the schedules of sensors s_1, s_2, and s_3 for the grid point A.

and s_4 are 14, 19, 8, and 11, respectively. Figure 1.9(c) shows the on-duty schedules of sensors s_1, s_2, and s_3 to cover point A. Note that T^i_{front} and T^i_{end} are always set to one half of the interval between two adjacent reference times. Therefore, the on-duty time period allocated for s_1 to cover point A is from 11 to 16.5. In this way, within a round, the point A is monitored by the three sensors in turn. Consider all grid points within the coverage of s_1. Its on-duty time in each round is the union of schedules to cover all grid points A, B, C, and D. Figure 1.10 illustrates the final integrated on-duty schedule of s_1, which is [11, 2.5]. Note that time in a round is considered in a wrap-around way.

The above grid approximation may lead to some small uncovered holes in the sensing field. To conquer this problem, we can reduce slightly the circle representing the coverage of each sensor. Also note that the reference time of each sensor in each frame will be recalculated to achieve a certain degree of randomness to balance sensors' energy expenditure. The recalculation is done in the initialization phase of each frame.

Ref. [13] proposes an enhanced scheme to reduce the computational complexity of Ref. [12] and to further balance sensors' energy consumption. The enhancements are achieved through several modifications. First,

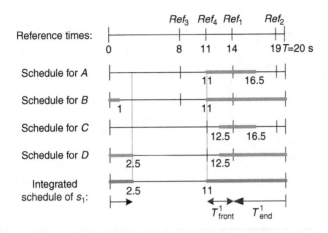

Figure 1.10 The schedule of s_1 for the example in Figure 1.9 after joining the schedules for all grid points covered by s_1.

instead of using grid points, this scheme utilizes the result in Ref. [2] by calculating sensors' schedules based on the intersection points of their sensing ranges. This will significantly reduce the computational complexity. Second, several optimization mechanisms are proposed to balance sensors' energy consumption by intelligently selecting sensors' reference times.

In Ref. [13], the first modification is called the *1-Coverage-Preserving (1-CP)* protocol. It adopts the same frame structure as in Ref. [12]. Instead of considering grid points, this algorithm considers the intersection points among the perimeters of sensors' sensing ranges. For example, to calculate s_1's on-duty schedule, points p, q, and r in Figure 1.11(a) are considered. The final integrated schedule of s_1 is the union of the schedules for these intersection points, as shown in Figure 1.11(b). Typically, the number of intersection points is much less than the number of grid points. So the computational cost is significantly reduced.

Ref. [13] also proposes an *energy-based 1-CP* protocol that can intelligently select sensors' reference times. To prolong network lifetime, this protocol utilizes sensors' remaining energies to balance their energy consumption. This is achieved in two steps: first, the reference times of sensors with more energies should be placed more sparsely than those with less energies in the duration of a round. For example, each round is logically divided into two zones with different lengths, $(0, 3T/4)$ and $(3T/4, T)$, where T is the time duration of a round. A sensor s_i with remaining energy E_i ranked at top 50% among its neighbors should choose its reference time randomly from the larger zone, while a sensor s_i with E_i ranked at the bottom 50% should choose from the smaller zone. Second, the parameters T^i_{front} and T^i_{end} of s_i are also chosen based on E_i. For any intersection point p,

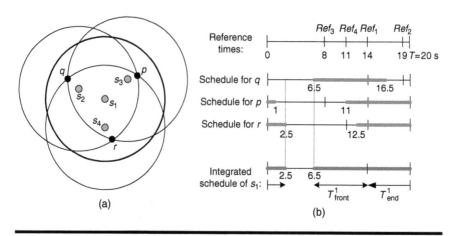

(a)

(b)

Figure 1.11   An example of the 1-CP protocol in Ref. [13]: (a) intersection points and (b) computation of s_1's integrated schedule.

the two parameters are decided according to the remaining energies of s_i and the two sensors with reference times before and after Ref_i. Specifically, the scheme defines

$$T_{front}^{p,i} = (Ref_i - prev(Ref_i)) \times \frac{E_i}{E_i + E_{i'}}$$

$$T_{end}^{p,i} = (next(Ref_i) - Ref_i) \times \frac{E_i}{E_i + E_{i''}}$$

where i' and i'' are the sensors which also cover point p and have reference times located before and after Ref_i (i.e., $prev(Ref_i)$ and $next(Ref_i)$) on the time axis, respectively. These designs may enforce sensors with more remaining energies to keep awake for longer time, and thus have the potential to extend the network lifetime. Simulation results are demonstrated in Ref. [13] to verify the advantages of these enhancements.

1.4   Coverage and Connectivity

For a wireless sensor network to work properly, it is necessary to maintain both sensing coverage and communication connectivity at the same time. When a sensor detects a target event, it often needs to send this information back to a sink node. Therefore, maintaining the network communication connectivity is another basic requirement in sensor deployment.

Some researchers discuss the connectivity issue combined with the coverage problem.[2,14,15] These algorithms are mainly based on an important fact, which claims that when the transmission range is greater than or

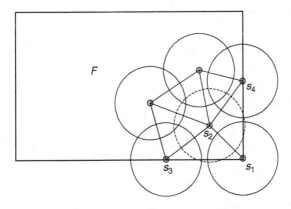

Figure 1.12 An example of a 1-covered network (only the sensors in the lower-right corner are shown). Circles are sensors' sensing ranges. The straight lines between sensors represent the communication links between them. When s_2 is removed, the network is disconnected.

equal to twice that of the sensing range, satisfying coverage implies satisfying connectivity as well. In other words, under this assumption, the goal of maintaining both coverage and connectivity can be reduced to simply preserving coverage of the sensing field.

Wang et al.[2] show that when the transmission range R_t of sensors is not smaller than twice the sensing range R_s ($R_t \geq 2R_s$), then if the sensor network is k-covered it implies that the sensor network is k-connected, where k-connected means that the network will remain connected after removing any arbitrary $k-1$ sensors from the network. This theorem can be explained by Figure 1.12. When $k=1$, it is possible to disconnect the network by removing one corner sensor. For example, in Figure 1.12, removing s_2 will disconnect the network. When $k > 1$, the sensing field can be regarded as covered by k disjoint sets of sensors each providing 1-coverage to the sensing field, and thus one must remove at least k sensors to disconnect the network.

Ref. [2] proposes a CCP that can determine a subset of sensors to be active to satisfy the required degree of coverage and connectivity. CCP is discussed under two different assumptions, $R_t \geq 2R_s$ and $R_t < 2R_s$. According to the above discussion, when $R_t \geq 2R_s$, maintaining k-coverage and k-connectivity is equivalent to maintaining only k-coverage. Therefore, for each sensor, it can simply check if its coverage area is already k-covered without itself. If so, it can put itself into the sleep mode and wake up in the next round. How to determine if a sensor's sensing area is at least k-covered has been discussed in Section 1.2.1.

However, when $R_t < 2R_s$, CCP cannot guarantee connectivity. Thus, it tries to integrate with an existing protocol, SPAN,[16] which is a power-saving

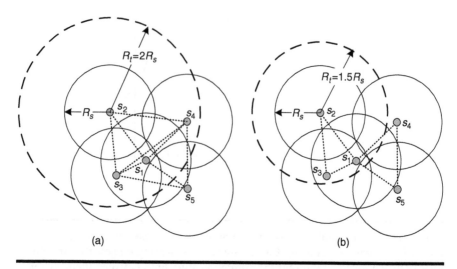

Figure 1.13 An example of the CCP + SPAN execution result. Solid circles mean sensing ranges R_s, and dotted circles mean transmission ranges R_t. The dotted lines between sensors represent communication links between sensors. (a) When $R_t = 2R_s$, 1-coverage implies 1-connectivity, so sensor s_1 can be turned off. (b) When $R_t = 1.5R_s$, sensor s_1 has to stay active to maintain the communication backbone.

connectivity maintenance protocol. SPAN tries to maintain a communication backbone to provide connectivity while sensors not on the backbone are kept in the sleep mode. The CCP-SPAN-integrated protocol decides a sensor to be active if it satisfies the active principle in CCP or SPAN. Otherwise, it can enter the sleep mode. For example, in Figure 1.13(b), the $R_t < 2R_s$ condition makes the network probably not connected after the coverage issue is satisfied. Although sensor s_1 is allowed to enter the sleep mode by CCP, it will be requested to stay active to connect with its neighbors to form a communication backbone according to SPAN. Note that the resulting network may remain k-covered but less than k-connected.

Ref. [15] discusses the coverage-connectivity-combined issue by adopting a probability-based coverage model and the concept of *connected dominating set (CDS)*. The coverage of any location depends not only on the sensing ranges of sensors but also on its distances from these sensors. This probability model is similar to the models discussed in Section 1.2.2. To compute the coverage of the network, the sensing field is divided into grids. For each grid point, we compute its detection probability contributed by nearby sensors. Then, each sensor decides if it should stay active by observing the *redundancy degrees* of grid points within its sensing area. The redundancy degree of a sensor at a grid point is the end result of comparing the detection probability contributed by this sensor with it by other sensors. The higher the latter probability is, the larger the redundancy degree will be. If the redundancy values of all grid points within the sensor's sensing area are higher than a predefined threshold, this sensor becomes a candidate that can be put into the sleep mode. Otherwise, it has to stay active.

After the above steps, each sensor which becomes a candidate should check if any two of its active neighbors are connected. If not, it has to remain in the active state to maintain network connectivity. In this way, a coverage-sufficient subset of sensors is identified to form a communication backbone.

1.5 Conclusions

In this chapter, we have surveyed some coverage-related solutions for wireless sensor networks. We first studied how to evaluate the coverage level of a specified sensing field under different models. Then, two important issues, the sensors-scheduling and network-connectivity problems, are discussed. Other related issues include localization algorithms[17–19] and identifying set problems,[20,21] which also deserve further studies.

References

1. C.-F. Huang and Y.-C. Tseng. The coverage problem in a wireless sensor network. In *WSNA '03: Proceedings of the 2nd ACM International Conference on Wireless Sensor Networks and Applications*, pp. 115–121, 2003.
2. X. Wang, G. Xing, Y. Zhang, C. Lu, R. Pless, and C. Gill. Integrated coverage and connectivity configuration in wireless sensor networks. In *SenSys '03: Proceedings of the 1st International Conference on Embedded Networked Sensor Systems*, pp. 28–39, New York, 2003.
3. C.-F. Huang, Y.-C. Tseng, and L.-C. Lo. The coverage problem in three-dimensional wireless sensor networks. In *IEEE GLOBECOM*, vol. 5, pp. 3182–3186, 2004.
4. T.S. Rappaport. *Wireless Communications: Principles and Practice*. Prentice Hall PTR, 1996.
5. S.-P. Kuo, Y.-C. Tseng, F.-J. Wu, and C.-Y. Lin. A probabilistic signal-strength based evaluation methodology for sensor network deployment. *International Journal of Ad Hoc and Ubiquotus Computing*, 1:3–12, 2005.
6. G. Xing, C. Lu, R. Pless, and J.A. O'Sullivan. Co-grid: An efficient coverage maintenance protocol for distributed sensor networks. In *Information Processing in Sensor Networks (IPSN)*, pp. 414–423, 2004.
7. N. Ahmed, S.S. Kanhere, and S. Jha. Probabilistic coverage in wireless sensor networks. In *The IEEE Conference on Local Computer Networks*, pp. 672–681, 2005.

8. S. Meguerdichian, F. Koushanfar, M. Potkonjak, and M.B. Srivastava. Coverage problems in wireless ad-hoc sensor networks. In *IEEE INFOCOM*, vol. 3, pp. 1380–1387, 2001.

9. S. Meguerdichian, F. Koushanfar, G. Qu, and M. Potkonjak. Exposure in wireless ad-hoc sensor networks. In *MobiCom '01: Proceedings of the 7th Annual International Conference on Mobile Computing and Networking*, pp. 139–150, New York, 2001.

10. M. Cardei and D.-Z. Du. Improving wireless sensor network lifetime through power aware organization. *Wireless Networks*, 11:333–340, 2005.

11. M. Cardei, M.T. Thai, Y. Li, and W. Wu. Energy-efficient target coverage in wireless sensor networks. In *IEEE INFOCOM*, vol. 3, pp. 1976–1984, 2005.

12. T. Yan, T. He, and J.A. Stankovic. Differentiated surveillance for sensor networks. In *SenSys '03: Proceedings of the 1st International Conference on Embedded Networked Sensor Systems*, pp. 51–62, New York, 2003.

13. C.-F. Huang, L.-C. Lo, Y.-C. Tseng, and W.-T. Chen. Decentralized energy-conserving and coverage-preserving protocols for wireless sensor networks. In *IEEE International Symposium on Circuits and Systems (ISCAS)*, vol. 1, pp. 640–643, 2005.

14. H. Zhang and J.C. Hou. Maintaining sensing coverage and connectivity in large sensor networks. Technical Report UIUCDCS-R-2003-2351, Department of Computer Science, University of Illinois at Urbana-Champaign, June 2003.

15. Y. Zou and K. Chakrabarty. A distributed coverage- and connectivity-centric technique for selecting active nodes in wireless sensor networks. *IEEE Transactions on Computers*, 54:978–991, 2005.

16. B. Chen, K. Jamieson, H. Balakrishnan, and R. Morris. Span: An energyefficient coordination algorithm for topology maintenance in ad hoc wireless networks. In *MobiCom '01: Proceedings of the 7th Annual International Conference on Mobile Computing and Networking*, pp. 85–96, New York, 2001.

17. P. Bahl and V.N. Padmanabhan. Radar: An in-building rf-based user location and tracking system. In *IEEE INFOCOM*, pp. 775–784, 2000,

18. T. Roos, P. Myllymaki, H. Tirri, P. Misikangas, and J. Sievanen. A probabilistic approach to wlan user location estimation. *International Journal of Wireless Information Networks*, 9:155–164, 2002.

19. Y.-C. Tseng, S.-P. Kuo, H.-W. Lee, and C.-F. Huang. Location tracking in a wireless sensor network by mobile agents and its data fusion strategies. *The Computer Journal*, 47:448–460, 2004.

20. S. Ray, D. Starobinski, A. Trachtenberg, and R. Ungrangsi. Robust location detection with sensor networks. *IEEE Journal on Selected Areas in Communications*, 22:1016–1025, 2004.

21. F.Y.S. Lin and P.L. Chiu. A near-optimal sensor placement algorithm to achieve complete coverage-discrimination in sensor networks. *IEEE Communications Letters*, 9:43–45, 2005.

Chapter 2

Communication Protocols

Lionel M. Ni and Yunhuai Liu

One of the most important issues in wireless sensor networks is the data delivery service between sensors and the data collection unit (the *sink*). Although sensor networks and mobile ad hoc networks (MANETS) are similar to some extent, they are radically distinct in many aspects. Sensor networks have many unique features, making them more challenging and in need of further development. Existing routing protocols for sensor networks can be classified as indicator-based or indicator-free. In this chapter, we make a comparative study of these protocols. Some open issues and research directions are pointed out as guidelines for future research work.

2.1 Introduction

In this chapter, we focus on the communication protocols in wireless sensor networks (WSNs).[1] As one of the most important parts of WSNs, the communication protocol provides data delivery services between entities. Apart from the traditional wired or mobile ad hoc networks, there is always one or a set of special data collection nodes (the *sink*) that functions as a gateway between the network and end users. The sink has reliable connections (e.g., wired or satellite) to the Internet, powerful processing capabilities, and adequate power supplies (see Figure 2.1). Due to this routing paradigm, the

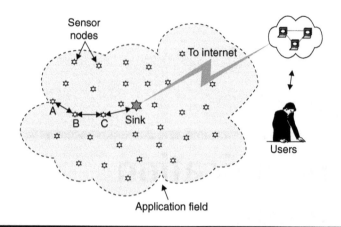

Figure 2.1   An paradigm of routing in WSN.

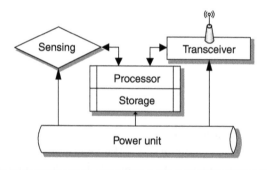

Figure 2.2   The basic components of a sensor node.

data request is often issued between the sink nodes and the sensor nodes. A typical sensor has four basic components, as shown in Figure 2.2: a sensing module, a processing module, a transceiver, and a power supply. There can be some other modules such as a location determining system, a power generator, and a mobilizer. Although sensor networks are similar to other networks to some extent, unique features make the routing problems in sensor networks significantly different from these of others.

■ Power issue is the largest concern.[1–4]
   Sensors are usually self-powered by batteries. The total energy of a sensor is on the order of 1 J.[4] Some may have solar cells that can scavenge energy at about 1 J per day outdoors and 1 mJ per day indoors. Since it is impractical to recharge a depleted battery, energy-aware protocols are preferred. Among all the components,

the transceiver for transmissions and receptions accounts for a major portion of the power usage.[5] This implies that power-efficient protocols have significant impacts on the network performance, and an energy-efficient routing protocol can dramatically elongate the network lifetime.

■ Resources for processing, storage, and transmitting are extremely limited.

Owing to the cost, sensors are made of simple circuits. All protocols and algorithms must be simple enough to fit into the sensor's onboard chipsets. At present, the Berkley MICA2 motes[6] have ATmega128l processors (8 MHz), 128K of programmable memory, 4K of configuration EEPROM and 512K of flash memory for storage. The packet is of fixed 36 bytes transmitted over 38.4 kbps (at most) wireless channels. Since the transmission range is around 10 m, it makes less sense for all the sensors to have directed connections to the sink. Alternatively, ad hoc mode by multihop routing is more likely to be appropriate. That is, sensors do not only function as data a generators, but also as routers to relay messages.

■ Global IDs or unique addresses are not always available.[1]

Owing to the nature of massive production and large numbered deployment, a globally unique ID needs a long bit. However, due to the lack of acknowledgement of transmission, the current platform of sensor networks supports short packets to increase the transmission success rate. In TinyOs,[7,8] all packets are of fixed 36 bytes. In ZigBee design,[9] given a 64-bit unique ID, nodes are preferred to employ a 16-bit local ID assigned by users or applications.

■ The positions of individual sensors do not need to be predetermined.[1]

WSNs allow the random deployment of sensors to terrains, e.g., by dropping from an airplane;[10] the application field could be a hostile environment which is inaccessible to humans (e.g., in forest fire detection and chemical pollution monitoring applications) and sensors are prone to failure. Thus, it is difficult to predetermine the precise locations of individual sensors. Under such scenarios, the topology control and management mechanisms must be flexible and robust in order to efficiently allow the network to operate efficiently. Protocols must be fault-tolerant.

■ Sensors are large-numbered deployed with high density.

Sensor nodes are large-numbered and densely deployed in a field to achieve some degree of redundancy. Here, the density is usually defined as the average number of neighbors[1] that a sensor can directly communicate with. Algorithms should be scalable and self-organized for the huge number of entities.

- Applications of sensor networks are more likely to be data-centric and content-oriented.[11,12]
  The sensing data is less relevant to the entity that generates the data. Instead, the data content is more important.

In the rest of this chapter, we first make a comprehensive comparison between WSNs and MANETs for their similarities and differences. Next, we give an overview of the applications of routing. We then present the current protocol designs by organizing them into two categories: indicator-based and indicator-free approaches. Afterwards, we compare these protocols for their applicability, performance evaluation, and limitations. Finally, we conclude the chapter with some existing problems and possible future research directions.

2.2   Similarities and Differences between WSNs and MANETs

Although MANETs and WSNs share some common features, they are different in many aspects, which make the routing protocols designed for MANETS unsuitable for sensor networks. In this section, we compare the similarities and differences of WSNs and MANETs.

2.2.1   Similarities

MANETs are characterized by multiple entities, a frequently changing network topology and the need for efficient dynamic routing protocols. WSNs are featured by tiny devices with extremely constrained resources, high-density deployment, and a data-centric routing paradigm.

WSNs and MANETs are similar in the following aspects:

- *Ad hoc mode.* No fixed infrastructure or base station is needed. Entities communicate with each other through multiple wireless links. Each node serves as a router to forward packets for others.
- *Resource constraint.* Nodes are resource-constrained. Devices of MANETs are typically hand-held devices with lower power capacities. Sensors are even smaller.
- *Power issue.* Nodes are usually powered by batteries (chargeable or nonchargeable). Power-aware and energy-efficient algorithms can significantly improve the performance of the systems.
- *Wireless communications.* Both MANETs and WSNs are built on top of wireless communication channels.

2.2.2 Differences

Despite the certain salient common characteristics, there are several critical differences between the two kinds of networks.

- *Node identifications.* Entities on MANETs always have globally unique IDs (e.g., MAC address or IP address), making the routing problems address-centric. In WSNs, IDs are not always available. Even distinguishing between neighbors becomes an issue.
- *Resources.* Although both MANETs and WSNs are resource-constrained computing environments, the critical levels are different. Resources of the entities in MANETs are several orders above those of the nodes in WSNs. For current popular hand-held devices, the typical components of MANETs, such as the processors often have a capacity of hundreds of megahertz, memories are of the order of dozens of megabytes, and rechargeable batteries can run for several hours. Using flash memory, storage spaces are pretty large so that complex statistics can be stored locally. WSNs are composed of devices with extremely limited resources. A typical sensor, for example, the Berkeley MICA2,[13] is equipped with an 8-bit ATmegal28 processor, 128K of programmable memory, 4K of EEPROM, and 512K flash storage spaces.
- *Communication paradigm.* Nodes in MANETs are peers. No node is given priority over another, and transmission could occur between any pair of nodes. The depletion energy of a single entity is unacceptable—no one likes to fail the first time. The "hot spot problem" must be prevented. In addition, the multicasts and global broadcasts are not so frequent, as is the case with the Internet. Transmissions on MANETs are address-centric. On the contrary, in WSNs the sink plays the most important role, serving as the only gateway between the network and users. Mostly, the sink is at one end of the transmissions, meaning that transmissions are always initialed from or destined to the sink. Sensor-to-sensor communications may be rare, but multicast and broadcast transmissions are so common that their impacts must be taken into consideration. In some cases, the failure of individual nodes can be ignored unless it is the sink, which has the problem.
- *Design objectives.* In MANETs, *QoS* and bandwidth usage are the most important concerns and first criteria of evaluation.[14] The power consumption is less important since the energy source can be replaced or recharged. WSNs consist of large numbers of unattended devices where battery replacement is much more difficult. The optimization goal of routing is to lower the power usage so as to enhance the network availability.

■ *Major research challenges.* Since the nodes in MANETs have great abilities to move, mobility becomes the greatest challenge.[15] Wireless links are up and down in an arbitrary fashion, and topologies of networks keep changing, making discovered routes prone to failure. In WSNs, although both the sink and the data source can move,[16] sensors always lack mobility and are usually stationary. The failure of sensor circuits and the depletion of batteries contribute the dominating factors influencing the change of topology.[17]

■ *Density.* The density μ is defined as the average number of neighbors per node.[1,18] It can be roughly calculated using[1] $\mu(R) = (N\pi R^2)/A$ with unit disk communication model, where N is the number of sensors in an area A, and R the radio transmission range. In MANETs, nodes are sparsely scattered while in WSNs they are densely deployed. Different levels of density lead to different solutions for robustness. In MANETs, the data-link layer protocols help to insure the reliability of one-hop transmissions. Single path is sufficient for end-to-end communications. The correctness of a routing protocol is the most important part of design. However, in WSNs, sensors are prone to failure, and local channels are broadcast without an acknowledgment mechanism. The one-hop transmission can be very unreliable. Under such situations, reliable transmission becomes a nontrivial issue.[19,20]

■ *Wireless multicast advantage (WMA).* Differing from traditional networks, the so-called *wireless multicast advantage* (WMA)[3] property is held in WSNs. When transmitting with WMA, all directed neighbors of the transmitter can the packet for free. It exploits the broadcast nature of the wireless medium, and allows the sender to reach multiple nodes simultaneously without penalty. Receivers outside the transmission range will not get enough signals to achieve a required signal-to-noise (SNR) threshold. They just regard the signal as noise. In MANETs however, transmissions are address-centric without WMA.

■ *Protocol design fashion.* Inherited from the traditional wired network, the layer concept is still valid in MANET design, and protocols are modulated and developed layer by layer. On the contrary, WSNs are on top of embedded systems. This all-in-one design is more appropriate.

2.3 Communication Patterns in Wireless Sensor Networks

In this section, we consider the routing problem from the application point of view. We first discuss the node identification problem, and then describe the communication patterns, which are the services that the routing protocols should provide.

2.3.1 ID Establishment

As application-dependent computing environments, WSNs are designed to collect and deliver valuable information depending on different types of applications. In many cases, these applications need to label the collected information to ensure correct processing. Otherwise, usage of the information is limited. For example, in the Glacier Bed Deformation Monitoring System,[21] sensors are required to report the temperatures of nodes periodically. Without knowledge of the data source, the collected temperatures make no sense to the data analyzer. Clearly, node identification is critical for its ability to distinguish and organize the information. A well-designed ID scheme is not only of value for applications, but also of paramount importance for the correct and efficient operations of the system itself. For instance, without knowing the source and destination of the information, routing the packets becomes such a challenging task that flooding becomes the only choice.

Having an ID in each of the sensors is a basic assumption in many applications.

Node ID can be either *global* or *local*. In the global scheme, each node has a globally unique ID, usually given during the manufacturing process. This is similar to the 48-bit MAC address used in the traditional network. However, the global ID is not feasible for WSNs due to its cost and related high overheads.[1] On the one hand, the large scale, high density, and disposable nodes require a node ID to be sufficiently long to avoid the ID conflict. On the other, embedded system cannot afford the large overhead introduced by a long ID. Owing to the harsh environments of wireless radios, most WSNs support short packets to reduce the probability of transmission error. For example, the Berkeley Motes[6] have fixed packet sizes of 36 bytes. Consider a transmission scenario assuming that the packet header carries no control information other than the source and destination IDs. Suppose the Zigbee standard[22] is applied with the unique ID of 64 bits. It results in an effective payload of, at most $(36 - 16)/36 < 60\%$ of packets. Thus, a 16-bit short address field is employed in Zigbee for packet transmission. If the IPv6 standard, with 128-bit addresses, is used, the effective payload is even less unless some header compression method is adopted. According to this observation, we have the argument that node ID schemes have a great impact on system efficiency and that the traditional global ID scheme is not suitable.

Instead of employing the static global ID scheme, a local ID scheme has more flexibility. Each node is assigned with an ID, either before or after the deployment. Ideally in an unattended environment, a node is able to automatically generate a unique local ID without having a centralized control and with a minimum number of ID bits. We define a local ID with *k*-hop uniqueness as having the feature that any two nodes of the same ID are at least *k* hops away. In general, the local ID has two extremes. One is $k = 3$, which is the least restrictive condition to ensure that none of nodes is confused by its direct neighbors.[1,23,24] The other extreme is that *k* is more than the diameter of the network, where the ID is distinctive within the field. The later ID scheme,[25] called *field ID*, is preferred due to its ability to help the data analyzer to uniquely identify a node without any other assistance. Examples of the local ID assignment with $k = 3$ and field ID scheme are shown in Figure 2.3 and Figure 2.4, respectively. Note that the current routing protocols can only operate with some form of IDs, except the blind flooding, For example, geography-based routing[26] uses the location information, which is a kind of local ID. Some other routing protocols, for example Refs. [20, 27, 28] have implicit node ID schemes.

An optimal ID establishment algorithm should capture several features. First, it must be a fully distributed algorithm. Second, it should generate an ID with acceptable establishment overhead, in terms of both the energy consumed and the time spent. Third, the ID should occupy minimum storage space in the packet header. Fourth, the ID should be useful for applications. ID establishment algorithms can be categorized as location-dependent ones, such as using the location information, and location-independent ones, such as randomly generated flat ID schemes. We argue that location-dependent IDs are more useful for efficient packet routing and data organizing, although location-dependent IDs very likely need more storage spaces than location-independent IDs.

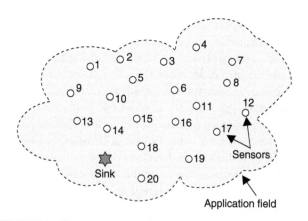

Figure 2.3   An example of field ID assignment.

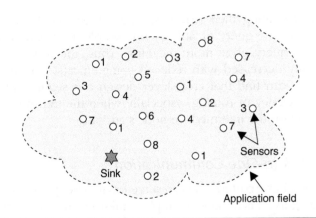

Figure 2.4   An example of local ID assignment with the 3-hop unique feature.

2.3.2   Route from Sink to Sensor: Sink-to-Sensor Communications

Interactions between sinks and sensors are always initialized by the sinks. It can be queries, events of interest, or commands. In this subsection, we consider the problem of routing a packet from the sink to one or some of the sensors.

2.3.2.1   A Sink-to-All Communication

Blind flooding is the foundation of message delivery from the sink to all sensors. The sink-to-all communication can be categorized by two forms.

The first form is that packets are just duplicated from one to another. Eventually, all sensors should receive a "same" copy. Most queries are of this form. Obviously, flooding is the most efficient way.

The other form is that the transmitted packet relies on some intermediate processing results. For example, in Ref. [27], it is shown that when building the minimal distance (e.g., hop counts) from a sensor to the sink, traditional blind flooding is a bad solution due to the unnecessary redundant traffic. In this scenario, data is processed by each of the receivers before it is sent out again. Otherwise, an intermediate (but not the final) processing result introduces unnecessary forwarding that is exponentially increased, proportional to the distance to the sink. Also in Ref. [29], it is shown that in-network data processing can dramatically decrease the communication overhead for event-driven applications. For example, given the query: "If more than 100 nodes have detected the temperature over 30°C, please report me.", instead of forwarding the query immediately after receiving it, sensors can have a back-off procedure to wait for the results from all the upstream nodes. After summing their readings, nodes then decide whether they should send out the query or not. By this simple aggregation technique, the forwarded query gives more meaning. Thus, the response time can be dramatically decreased with reduced communication overhead. From the above, we can find that cross-layer design can significantly improve the energy efficiency of routing, especially when the transmitted data are dynamic based on the intermediate node's attribute.

2.3.2.2 A Sink-to-One Communication

Sink-to-one is a basic pattern that can serve for all other patterns in the sink-to-sensor category. It is feasible to employ sink-to-all transmissions to serve for the sink-to-one transmissions. A more efficient option is to involve only those which are related (e.g., single-path transmissions, to save energy). For example, a sender broadcasts the packet in the local channel. A relevant receiver picks up the packet and forwards it. And all the other neighbors discard the overheard one. To obtain a desired level of reliability, the data delivering path can be single or multiple. Note that when there is a lack of a globally unique ID scheme, identifying the destination node becomes a critical issue.

2.3.2.3 A Sink-to-Region Communication

Definition. Given location information or a location-dependent ID scheme, a *region* is the set of nodes fulfilling some given conditions.

Many applications require the sink to reach one or a set of sensors according to their locations, for example, "node in Room1511, report your sensing data." While, in practice, all sensors in the vicinity may be valid due to the high density of node deployment, inaccurate locations of nodes may result mostly due to the loose requirements of applications. Under such scenarios, rather than communicating with a specific node at (x, y) position, all nodes within the region should be reached by a single specially designed scheme. By viewing the destination nodes as a single set and exploring the correlations between these nodes, some severe problems can be alleviated:

- Destinations are very close to each other. Their routes to the sink are potentially close, meshed or even interleaved with each other. Collisions and interference are severe and fatal—receivers hear nothing but noise. This is like the so-called *"response implosion problem."*[30] With the increase in density of sensors, the problem becomes more serious.
- Also because the destinations are near each other, intermediate nodes could be passed through several times. As transmission

accounts for the major portion of the total energy consumption, overutilization is likely to lead to the energy depletion of the nodes along the path. The network could be partitioned and the functionality of whole system affected. This is the so-called "*hot spot depletion*" problem.[31]

■ Multipath routing scheme can significantly increase the transmission reliability. However, power usage can be dramatically saved by applying some aggregation mechanism without sacrificing the reliability.[32,33]

A critical prerequisite of aggregations is that the sink-to-region transmissions must be synchronized. They happen at nearly the same time enabling the aggregation to be performed. A general technique to reduce the traffic is as follows: deliver a packet to any one of nodes falling within the region (or on the boundary of the region); the reception node then initials a restricted flooding within the destination region. Sink-to-region communications are also referred to as geo-cast.

2.3.2.4 A Sink-to-Subset Communication

In this scenario, the sink wants to send messages to sensors scattered through the whole field. Even given the global knowledge, it is still difficult to aggregate packets to save energy. More seriously, the sink may have no hint about which node to deliver the packet to. We believe that in such cases, blind flooding is the only choice. For example, a query is sent out: "nodes with temperatures, higher than 30°C, please report to me." There may be no response, several responses or responses from all the nodes.

Nevertheless, there could be some techniques to optimize such a transmission when the destinations are given by their locations or location-dependent IDs. A transmission tree could be constructed in advance to avoid the unnecessary traffic. This problem has many connections with the traditional multicast problem in the Internet.

Another variation is that some previous communications may have occurred, which may provide some useful information. Users want to send back further messages to the related nodes. This would constitute the dualoperation of communication, which we will discuss in later sections.

2.3.3 Routes from Sensor to Sink: Sensor-to-Sink

Let us suppose that the sensors have already received the stimulus, and thus have the data to be delivered to the sink. The stimulus could be a previous query or an event of interest.

2.3.3.1 A One-to-Sink Communication

One-to-sink communication is one of the most popular transmission patterns. A source node needs to send a message back to the sink. Most current routing protocols focus on this. It is much easier than the sink-to-one communication.

2.3.3.2 A Region-to-Sink Communication

A natural property of sensor networks is that sensors are placed with high density. One event may trigger several originators which want to report to the sink. They reside in a region too. Here, the region has the same meaning as defined previously.

Several one-to-sink services are still servable. However, the created problems are similar to those we listed in the sink-to-region transmissions. Aggregation techniques are preferred for reducing the unnecessary traffic.

■ *Combination.* Messages from different senders are combined to form a compressed one. This combination operation usually needs a local computation process. The aggregator must check the contents of the packets before it combines two packets as one. The appropriate operations of aggregation are SUM, MAX, MIN, TopK, and so on. A SUM example is:
  ● The sink sends a query: "Who detected a vehicle in last five minutes?"
  ● Node A replies: "Node A detected" and sends the packet to Node B for forwarding.
  ● Node B detected the vehicle too. After checking A's packet, B sends out the packet: "Both A and B detected a vehicle."
■ *Concatenation.* This is different from the combination. The receivers do not check the packet content. Instead, they encapsulate the received packet with their own to form a new packet. It is easier to implement, but the newly generated packet has approximately double the size of the original. It is believed that a single, larger packet consumes much less energy than the sum of many smaller ones.[2]
■ *Reduction.* In some cases, the sensing data are highly redundant. Another option to save energy is selecting only a subset of the data source to send back data instead of all of them, which is known as sample-cast.[30,34] A proposed approach is that samples are answered with a probability p that can be fixed or adaptively changed.

Again, an implicit prerequisite for aggregations is that the transmissions are synchronized.

2.3.3.3 A Subset-to-Sink Communication

Similar to region-to-sink, a subset-to-sink also involves several nodes which have messages to send back to the sink. This communication pattern can be employed for eliciting the information regarding a generic status of sensors. When the responses depend on some dynamic attributes of sensors, this pattern is preferred. For example, users need an energy map[35] (which nodes are low in power and how they are deployed). Nodes having power lower than a threshold should reply—they are likely to fail in the near future and some maintenance actions are needed to ensure the health of the system.

Another potential application field of this pattern is when the query is in the form of a sample-cast.[30,34] For example, in order to estimate a whole view of the network, the sink wants to achieve a 10% status of nodes. The sink does not care which 10% is functional, but scattered 10% nodes are preferred. It is not a bad idea to let all the nodes whose ID is divisible by 10 report their data.

Backoff-based strategies may help to dramatically decrease traffic and thus reduce the energy consumption, especially for those nodes closer to the sink which account for the more important roles in routing.

2.3.3.4 An All-to-Sink Communication

When the sink wants to obtain some information from the whole application field, an all-to-sink communication is appropriate. For example, in scientific research computing environments, sensors are designed to report the collected information periodically. At fixed intervals, each sensor is supposed to send back at least one message.

In this scenario, both "response implosion" and "hot spot depletion" are serious problems. However, this time the "hot spots" are not those nodes along some particular routes, but those that are nearer to the sink (e.g., nodes that are one-hop-away neighbors of sink) as they have more forwarding workload than others. It is one of the challenging jobs in literature.

2.3.4 Sensor-to-Sensor Communications

Researchers have noticed the importance of *in-network data processing* and aggregation in literature.[29,36–38] Sensor applications depend on the efficacy of being able to trace the needed data from the network. Often, there data consist of summaries or aggregations, that are more meaningful, rather than raw sensor readings. To save the communication overhead, the data should be moved from the originators to some intermediate nodes for in-network processing. To support this kind of transmission, sensor-to-sensor communications are needed. Also, Ratnasamy et al.[12] propose a *data-centric storage system* in literature to reduce the communication overhead of query. In data-centric storage, the data originator does not store the information locally or deliver it to the sink, but transmits it to other nodes. The whole data set of the system is managed in a structured way. When querying, the sink can look up and retrieve data without flooding the query. Both in-network data processing and data-centric storage systems need sensor-to-sensor communications—a sensor has to talk to other sensors.

Compared with other communication patterns, sensor-to-sensor routing experiences more overhead on energy, local storage, and control packets.

2.3.5 Dual-Operation Communications

It is interesting that since routing in sensor networks is data-centric, the data flow is bidirectional and transmission patterns are not isolated from each other. For example, a sink-to-region is very likely to be followed by a region-to-sink—sink queries the status of a region and all the involved sensors should give a response.

In general, there are two types of dual-operation communications:

- A sink-to-sensor followed by a sensor-to-sink. This happens when the sink sends out a query, and then the corresponding sensors send replies back. Some routing approaches[19,20,27,28,37] need to proactively build routes. For the latter sensor-to-sink operation, a previous sink-to-sensor communication can significantly decrease the route-building cost.
- A sensor-to-sink followed by a sink-to-sensor. This happens when the sink wants to have further interactions with sensors after a sensor-to-sink communication. For example, a sensor gives a daily report: "I detected a temperature higher than 80°C." The sink finds that it is a big event and the following actions are needed: "OK. Since this is so susceptible, please report to me every 5 minutes instead of daily." A previous sensor-to-sink can help identify the data source and the related path.

Dual operations can be explored to provide more energy-efficient routing services.

2.4   Routing Protocols in WSNs

In general, existing routing protocols can be divided into two categories: indicator-based and indicator-free. In the indicator-based, there is always an initialization phase where an indicator-generation algorithm is applied. Accordingly to the algorithm, every node generates an *indicator* to help determine the routes. In the indicator-free algorithms, routes are built on the fly.

2.4.1   Indicator-Based Approaches

In this category (Figure 2.5), indicators are built for sensors in a setup stage. They then follow those indicators to make decisions when routing. According to different types of indicators, we further categorize this kind of protocol into several subclasses: geography-, gradient-, and cluster-based.

2.4.1.1   Geography-Based

Some protocols assume that precise locations of sensors are available. Packets carry the source and destination locations, helping intermediate nodes to make routing decisions. In general, the forwarder selects the neighbor that is closer to the destination as the next hop. This greedy forwarding process is repeated until the packet eventually arrives at the destination.

■ Location-aided routing (LAR)[39] is a protocol originally designed for MANETs. Instead of flooding the whole network, LAR selects a small region (called *request zone*) to flood. In Ref. [39], two

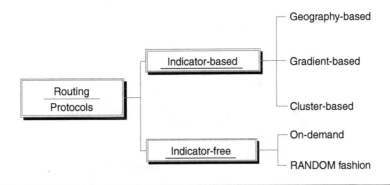

Figure 2.5   A taxonomy of routing protocols in WSNs.

shapes of request zone are proposed and evaluated. One is a rectangle bounded by the source and destination (as depicted in Figure 2.6). The other shape is a sector centered at the destination (depicted in Figure 2.7). DREAM[39] has a similar idea. The flooding region is also a sector-like shape, but the center is located at the source instead of at the destination (as shown in Figure 2.8). The two approaches were published in the same year at the same conference.

■ Greedy Perimeter Stateless Routing (GPSR)[26] is a single-path algorithm that avoids network-wide flooding. GPSR consists of two

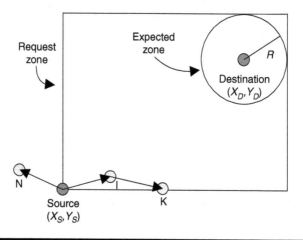

Figure 2.6 The rectangle shape of the request zone in LAR—scheme 1.

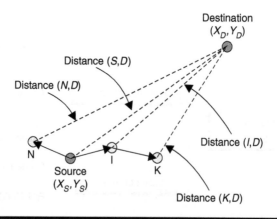

Figure 2.7 The sector shape of the request zone in LAR—scheme 2.

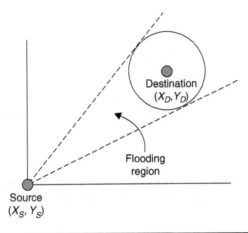

Figure 2.8   Sector shape of the flooding region in DREAM.

methods to determine a route: *greedy forwarding* when it is possible; and *perimeter forwarding* (also referred to *face routing*) when greedy forwarding fails. In GPSR, packets carry the source and destination locations. In the normal *greedy mode*, the forwarder selects the neighbor which is geometrically closest to the destination as the next hop. Accordingly, the destination is greedily approached hop by hop until it is reached. It is possible for an intermediate node to encounter *voids* where all its direct neighbors are farther from the destination than itself. Then the packet is transferred to the *perimeter mode* to route beyond the void. GPSR applies the right-hand rule to traverse a void (as shown in Figure 2.9). A perimeter-mode packet is marked with the location of the node from which the packet enters the perimeter mode. It can go further than the current forwarding node to traverse the perimeter of a void. The right-hand rule is followed until the packet arrives at a node that is closer to the destination than the node that makes the packet enter the perimeter mode. Perimeter forwarding is also referred to as face routing.

■ Geographical and Energy Aware Routing (GEAR)[40] is motivated to solve the problem of void in geography-based routing, inspired by the traditional distance-vector algorithms. The basic idea is that, for each transmitted packet, the forwarder will build a *learned cost H* as the estimated transmission cost to the destination. The learned cost H has an initial value and is adjusted adaptively in communications. Finally it reaches the optimal value which can reflect the real cost of transmissions to a particular destination. The detailed protocol is as follows. Given the destination location, the H of a

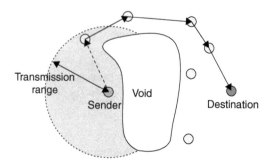

Figure 2.9   The void of forwarder and the right-hand rule to walk around it.

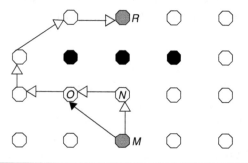

Figure 2.10   The right-hand rule of the GEAR.

forwarder is initialed as the physical distance from the forwarder to the destination. It is updated according to the following operations. Consider a scenario in which node M wants to send a packet to R, as shown in Figure 2.10. Suppose M selects N as the next hop. Let $H(N,R)$ denote the learned cost of N to R. H is then initialed as 2, the Euclidean distance from N to R. Black points represent failure nodes which are actually voids for N. N detects the voids and forwards the packet to its neighbor O which has the least H value. Note that currently, O has the H value $H(O,R) = $ Distance $(O,R) = \sqrt{5}$. N updates its own learned cost $H(N,R) = H(O,R) + \text{cost}(O,N)$ as the new estimated cost of N. From now on, M directly sends packets to O instead of N since O has a lower H value than N. By this feedback mechanism, learned cost can be calibrated closer and closer to the optimal value.

■ Fang et al. in BOUNDHOLE[41] gives a more restricted definition of a void. With the proved TENT rule, all voids can be predetermined. An example is shown in Figure 2.11. The solid circle is the transmission

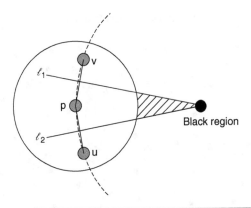

Figure 2.11 The Tent rule—black region is a void associated with *p*.

range of node p. u, v are two neighbors. l_1 and l_2 are perpendicular bisectors of segments joining up and vp. The black area is the region where p is closer than u and v but p cannot reach. Accordingly, this area is the void corresponding to node p no matter whether there is a destination or not. All such potential voids are precomputed and stored locally. With the right-hand rule, the strategies to route across those voids are precalculated and stored locally for future transmissions.

■ Two-tier data dissemination[16] (TTDD) is designed for scenarios of multiple mobile-sink applications. An example is illustrated in Figure 2.12. Each data source (assume they are static) in TTDD proactively builds a virtual grid structure. The grid is a $\alpha * \alpha$ square. The data source is in the origin of the grid. Nodes at the crossing point locations (or the nearest node to that position) serve as the router to forward data. Within the grid that sinks fall into, data are transmitted through flooding (called lower-tier transmission); in the higher tier, data is forwarded between crossing point nodes. TTDD is independent of low-layer geography routing algorithms. It does not care how data are delivered between crossing points through multihop. Neither does it show how the routes are found at the route setup stage.

■ There are also some other geography-based approaches, which share a similar idea, such as SPAN,[42] geographic hash table (GHT),[12] and so on.

Geography-based routing scheme is a promising one with good scalability since it needs less control states in both the packet header and the nodes. The route is guaranteed to be found when the destination exists;

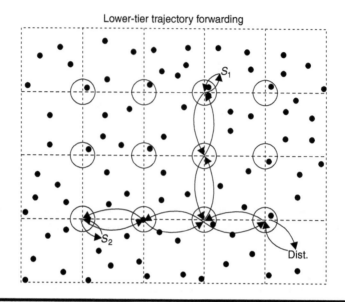

Figure 2.12 TTDD grids the application field to cells.

when it does not exist, the algorithm terminates at the node closest to the destination. It also provides omni-direction data delivery services by which messages can be sent between any pair of nodes. However, all these approaches need geography information—they are too extensive for simple sensors.

2.4.1.2 Gradient Based

Instead of employing the absolute geometric distance, another solution utilizes the relative distance between nodes. Such relative distance is the "gradient" indicating the direction of data flow to a particular originator. Some approaches select the sink as the originator, such as MCFN,[27] GRAB,[20] ARRIVE,[28] and GRAd.[43] Others utilize the data source as the origin, such as Direct Diffusion.[19,37]

■ MCFN[27] studies the problem of delivering the packet from the sensors to the sink. In MCFN, each node maintains a *cost field* as the minimum cost from the node to the sink. Once the cost field is established, packets flow to the sink along the cost field in the decreasing direction. Packets only carry the minimum cost from the source to the sink and the cost incurred till then from the source to the current node. The intermediate forwarding node simply broadcasts the packet to all its neighbors without assigning which one is the next hop. The neighborhood receiver decides to forward it if and only if the sum of consumed cost (contained in packet) and the cost of the node (stored in node) matches the cost of source node (contained in packet). That is, $Cost_{source} = Cost_{consumed} + Cost_{current\_node}$, implying that the current node is on the optimal path. In the example shown in Figure 2.13, the hop number is employed as the cost field. The main contribution of MCFN is that it presents a scheme to avoid the excessive advertisement packets that may overwhelm the network when establishing the cost field. It makes sure that every node only advertises its cost field once with the correct value. GRAB[20] (Figure 2.14) is a mesh version of MFCN to increase the transmission reliability. In GRAB, rather than using the single optimal path, a mesh between the source and the sink is formed, and all the nodes in that mesh are involved. The width of the mesh is proportional to the distance to the sink. To build the mesh, the concept of extra *credit* is introduced as the budget that packets could consume. GRAB greatly increases the reliability of MCFN without introducing too much overhead—only the packet header is a little longer. In practice, both hop counts and energy can be used as the cost field. We should note that MCFN and GRAB both need symmetric links so that the established cost field is valid when transmitting.

■ ARRIVE[28] applies the hop number as the gradient. Apart from GRAB, nodes in ARRIVE explicitly maintain the list of directed neighbors and assign the next hop. The main contribution of ARRIVE is the

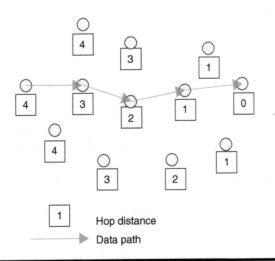

Figure 2.13   An example of transmission route in MCFN.

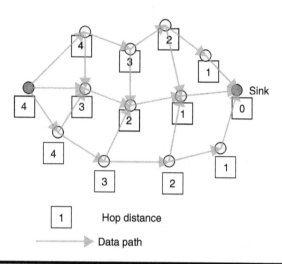

Figure 2.14   The transmission mesh of GRAB with 1 extra credit.

introduced random factor to increase the reliability instead of the deterministic one employed in GRAB.

◼ Different from the above destination-based gradient setup, Direct Diffusion[37] and its multipath version[19] establish the gradient originating from the data source but not the sink. An example is shown in Figure 2.15. First, the sink floods a query and every node should record from which node it receives the query. Thus, reverse paths back to the sink can be established. The data source node then follows these reverse paths to the sink. In the reply procedure, intermediate nodes again record the neighbor from which they receive the first data. After the data successfully reaches the sink, the sink then reinforces the best path from which the data first comes. The sink reinforces the data path by acknowledging to one of its neighbors: "you are on the optimal path." That neighbor then forwards this announcement to the neighbor from which it gets the very first data. This propagation of announcement never stops until it arrives at the data source. In Ref. [19], a braided version of Directed Diffusion is proposed. The alternate path has a performance similar to the optimal one in terms of hop numbers.

◼ There are also some other kinds of gradient-based protocols such as the GRAd.[43]

Gradient-based approaches do not need location information, but need symmetric links. They are scalable with optimal paths. However, they

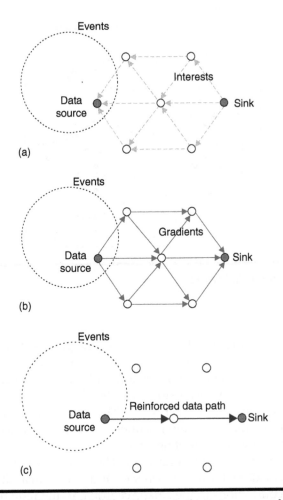

Figure 2.15 (a) Sink floods queries (interests). (b) Data are sent back to the sink. (c) One route is reinforced by the sink.

provide only a little portion of the data delivery services—only for the transmissions from others to the gradient zero.

2.4.1.3 Cluster-Based

Another kind of routing protocol is cluster-based. Some virtual clusters are formed and the cluster-heads are elected. When routing, sensors send the packet to a cluster-head first and the cluster-heads then take the responsibility of forwarding the packet to the sink.

■ The Cluster-head Gateway Switch Routing[44] (CGSR) selects the cluster-head by least ID or the highest connectivity. A cluster

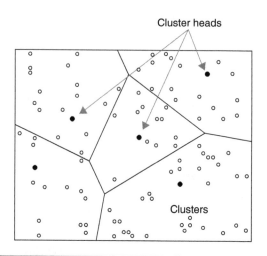

Figure 2.16 Clusters in LEACH.

contains a cluster-head and all its direct neighbors. Any two adjacent cluster-heads have at most two hops in between. Nodes between two heads belong to both clusters, and serve as the gateway to do the inter-cluster transmissions. Nonhead nodes communicate with others through their cluster-heads. The underlying inter-cluster routing utilizes the destination sequenced distance vector DSDV[45], a table-based wireless routing protocol.

■ Low Energy Adaptive Clustering Hierarchy(LEACH)[33] has a distributed cluster formation scheme. It assumes that all nodes have direct connections to the sink. According to the requirement of the application, sensors have a fixed probability to be a cluster-head. After being deployed to the sensing field, sensors independently determine whether they serve as a cluster-heads or not with a predefined probability. Cluster-heads then broadcasts an announcement to all others. After receiving the announcements, a nonhead node selects the closest one as its head and joins its cluster. In transmission, nodes send the data packets to the corresponding cluster-head which then forwards it to the sink. There is no inter-cluster operation, and all cluster-heads directly communicate with the sink. The shape of clusters in LEACH is like a voronoi diagram.[46] An example is depicted in Figure 2.16. Clusters are bounded by perpendicular bisectors between cluster-heads. Because of the huge amount of consumed energy for a cluster-head, LEACH proposed a rotation scheme that the members of a cluster serve as the cluster-head in turn. In Ref. [47], a variant of LEACH, called

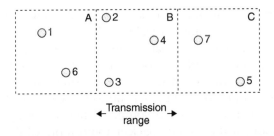

Figure 2.17   Clusters (grids) in GAF.

Power-Efficient Gathering in Sensor Information Systems (PEGASIS), was also proposed.

■ Geographic Adaptive Fidelity[48] (GAF) is a geography-based cluster scheme assuming that node location information is available. It is a density control mechanism more than a routing algorithm. As shown in Figure 2.17, the sensing field is divided into several fixed-size $r*r$-squared grids, where $r = R/\sqrt{5}$ and R is the transmission range. Sensors located in the same grid form a cluster and directly communicate with the sensors in adjacent grids. Nodes in a cluster take turns to be the cluster-head, which senses data and relays it. All the nonhead nodes go to sleep and are woken up by the head when needed (GAF assumed nodes can wake up). Being independent of the underlying routing protocols, GAF can help to lower the density of working nodes and hence save energy.

■ There are also some other cluster-based protocols, for example, cluster-based topology control (CLTC),[49] DIMENSION,[11] MECN,[50] and SMECN.[51]

Cluster-based protocols are always independent of the underlying routing algorithms. But they need extra communication overhead to maintain the cluster architecture.

2.4.2   Indicator-Free Approaches

Indicator-free algorithms have no initialization phases and the packets are transmitted by the on-demand or random fashion.

2.4.2.1   The On-Demand Fashion

In this fashion, route requests are sent out to find a route; the reply packet contains the route information.

■ Ad hoc On-demand Distance Vector Routing (AODV)[52] is an improvement algorithm of DSDV[45] designed for wireless networks. To send a message, the data source initiates a path-discovery process in order to find the route. The route request packet (REQ) is flooded to the network and the intermediate nodes record the neighbor from which they get the REQ first, in order to establish reverse paths back to the source. When the REQ arrives at the destination, it then sends back a route reply (REP) to the source following those reverse paths. AODV needs symmetric links; otherwise the REP may not be able to reach the source and AODV would fail.

■ Dynamic Source Routing (DSR)[53] eliminates the symmetric-link assumption held in AODV. When a sender has a message to send, a REQ is generated and flooded into the network. DSR is different from AODV, which records the route in the intermediate nodes, in that it holds all the route information in the REP packet. When the REQ arrive at the destination, the latter then has the whole route information from the source to the destination. The destination then floods another packet, the REP message, into the network. REP carries two bits of information, the REQ received by the destination and the route information thus far. When this REP arrives to the source node, the source will have both the whole route to the destination (carried by REQ) and the route from the destination back to the source (carried by REP).

■ As a part of PODS[54] project, Multipath On-demand Routing (MOR)[55] is a multipath version of AODV, The main contribution of MOR is that, many REPs are sent back instead of a single one as in AODV. Accordingly, multiple paths from the source to the destination are established. The source then employs all of these paths in a round-robin fashion.

On-demand fashions are always robust and reliable. However, they usually consume too much energy on individual routes.

2.4.2.2 The Random Fashion

To meet some optimal criteria, algorithms in this category introduce some random factors when making route decisions.

■ Rumor routing[56] is an algorithm intended to avoid flooding. In rumor routing, for each particular event, a subset of nodes (called witness of the event) is notified with the transmission route back to the event. The sink also employs a subset of nodes to query the event, hoping one of them was the corresponding witness. The

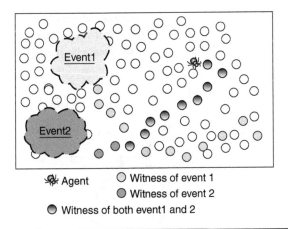

Figure 2.18   In Rumor routing, events send out agents and set witness.

detailed technique is as follows. *Agent* is a special packet carrying information about an event. It randomly travels the network with a certain lifetime (TTL) and notifies all the passing-by nodes about the event, making them the witnesses of the event. These witness nodes are then able to build reverse paths toward the event. When querying, the sink sends out some agents to scout for a witness of the event. Note that the interaction between an agent and a witness is two-way. If a node passed by an agent is already a witness of some other events, the agent then absorbs those events recorded in the node as well. So the further an agent travels, the more information it gathers until it dies (as shown in Figure 2.18). The novel point of[56] is that although the agent randomly traverses the network, a special technique is applied to try to make the travel path as straight as possible (but not guaranteed). An example of the technique is shown in Figure 2.19. The basic idea is that, agents not only contain event information, but also records of the neighbor list of the last passing node. When the agent arrives at a node (e.g., R in Figure 2.19), it discards the neighbor list of P but retains that of Q. After comparing this with the neighbor list of R, the agent finds that B and C are both neighbors of Q and R. Then the agent does not take them into consideration when selecting R's next hop—otherwise it is likely to make a curved path. Rumor routing avoids the flooding. However, rumor routing neither ensures success in finding a route in any sense, Nor in finding an optimal one.

■ Energy-aware routing[31] is intended to take the energy factor into consideration when selecting the next hop. First, nodes probe all

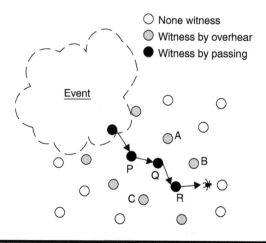

Figure 2.19   The paths of agents are random but relatively straight.

their upstream neighbors by hop number. They then calculate the probability of selecting a path from the estimated cost of delivering the message through the path. That is, for node N_j, the cost of passing by N_i is $C_{N_j,N_i} = Cost(N_i) + Metric(N_j, N_i)$. Suppose FT_j is the set of all the upstream neighbors of N_j. The probability that N_i is selected as the next hop is

$$P_{N_j,N_i} = \frac{1/C_{N_j,N_i}}{\sum_{k \in FT_j} 1/C_{N_j,N_k}}$$

N_j then calculates $Cost(N_j) = \sum_{i \in FT_j} P_{N_j,N_i} C_{N_j,N_i}$ as its estimated cost. It is for down-steam nodes of N_j. Energy-aware routing avoids overutilizing nodes in the optimal path to alleviate the "hot spot problem."

■ ReInForM[57] is another random multipath scheme. It proves that the average number of disjointed paths between a pair of nodes is close to the average number of neighbors in a large-scale sensor network. When the path is fairly long (e.g., hundreds of hops away), the difference between the optimal and the worst path is very small. So ReInForM randomly selects some of the available paths to deliver the packet to increase the reliability of transmissions. Load balancing between intermediate nodes is also achieved.

Random-fashion approaches always lack the stability of routing.

Table 2.1 The Usability of Routing Protocols for the Direction from Sink to Sensors

	Sink-to-One	Sink-to-Region	Sink-to-Subset	Sink-to-All
Blind flooding				Servable[a]
GPSR	Servable			
LAR,DREAM	Servable			
GEAR, Boundhold	Servable			
TTDD	Servable	Servable		
MCFN,GRAB				
ARRIVE				
Direct Diffusion	Servable			
CGSR	Servable	Servable		
LEACH, PEGASIS	Servable	Servable		
AODV, DSR	Servable			
Rumor routing	Servable			
Energy-aware	Servable			
ReInForM	Servable			

[a] "Servable" means that the protocol can provide appropriate services for that pattern.

2.5   Comparative Studies

In this section, we compare the performances of the protocols discussed above. The comparisons begin from the services, which we have categorized in Section 2.2, that there protocols support. We then give the performance metrics on which we base our comparisons.

2.5.1   Comparisons on Communication Pattern

In this subsection, we investigate the communication services that protocols can support. Owing to the lack of applications of sensor-to-sensor communications, we only list those for sink-to-sensor and sensor-to-sensor patterns. Table 2.1 shows the applicability of protocols for the patterns of sink-to-sensor communications; Table 2.2 shows for the patterns of sensor-to-sink communications.

From Tables 2.1 and 2.2, we can find that most protocols can be employed for transmissions between the sink and one sensor. However, a few algorithms are appropriate for grouping communications. Another observation is that the sink-to-sensors transmissions are given less attention.

Table 2.2   The Usability of Routing Protocols of Sensor-to-Sink

	One-to-Sink	Region-to-Sink	Subset-to-Sink	All-to-Sink
Blind flooding				
GPSR	Servable			
LAR, DREAM	Servable			
GEAR, Boundhole	Servable			
TTDD	Servable	Servable		Servable
MCFN, GRAB	Servable			
ARRIVE	Servable			
Direct Diffusion	Servable			
CGSR	Servable	Servable		
LEACH, PEGASIS	Servable	Servable		
AODV, DSR	Servable			
Rumor routing	Servable	Servable		
Energy-aware	Servable			
ReInForM	Servable			

Although geography-based solutions provide more services than average, they need location information, which seems too critical for simple sensors.

2.5.2   Performance Evaluation

In this subsection, we present some metrics to evaluate the performance of routing protocols. According to these metrics, we then comprehensively investigate the differences between different protocols.

2.5.2.1   Evaluation Metrics

The evaluation metrics can be categorized as:

1. Overhead of communications:
   - Number of packets for individual one-to-one communications: this is to show how much energy is consumed for a single transmission. Here we employ the metric of number of packets. Note that the consumed energy is also an option.
     - Max number of packets consumed by individual nodes (PKT-MAX): the maximum number of packets transmitted by individual nodes.
     - Total number of packets consumed by whole network (PKT-TOTAL): the sum of the total number of transmitted

packets over the whole network. Note that this metric has the same meaning as the average number of packets consumed by individual nodes.

- Packet header (PKT-HEADER): protocols need some control states stored in the packet header, e.g., node ID, the source locations and so on.

2. Overhead to maintain the routes:
   - Number of packets to establish routes in the setup phase (RTE-BUILD): this metric is to show the establishment overhead.
   - Number of packets to maintain routes (RTE-RECOVER): the number of packets to recover routes when the network topology changes.
   - Number of states in sensor (NODE-HEADER): the storage space in sensors to help routing is limited. Sensors only have around 100 K bytes of flash memory for storage. This is not only used for the protocol control states, but also for the application data. The control states may include but may not be restricted to the neighbor list, location information, reverse paths of individual transmissions.

3. Reliability and fault tolerance: Sensors may be deployed in a hostile environment. Channels may generate great noise and sensors are very prone to failure. Here are three kinds of errors and failures.
   - Channel errors (CH-ERR): transmissions temporarily fail due to collisions, interferences or circuit error; the sender, however, does not know that the packet has got lost during transmissions.
   - Random failure of nodes (RAND-FAIL): the failed nodes are scattered in the whole application area. Functioning nodes get to know the failure by systematic mechanism such as the periodic "Hello" beacon messages. No reaction has been recorded yet.
   - Pattern failure of nodes (PATN-FAIL): the failed nodes follow some patterns. For example, in a fire-detection application, an area of nodes fails due to the fire. No reaction has been recorded yet.

4. Route coverage (COVER): not all the protocols guarantee that they can find a route even if one exist. From the perspective of the sink, this metric is to show how much of the field is covered.

5. Cost to manufacture sensors (COST): some protocols require very expensive devices (e.g., GPS).

2.5.2.2 The Performance Evaluation of Protocols

From Table 2.3, a first observation is that most protocols can provide nearly optimal transmission overheads for the services they provide. LAR and DREAM need a degree of flooding that significantly increases the communication overhead. Rumor routing introduces random factors which also increase the transmission overhead. The inherent property of Directed Diffusion determined that a huge number of packets were needed to maintain routes even for simple transmissions. Second, except for TTDD, DSR, and rumor routing, which piggyback a great deal of information on packets, other protocols do not need a big packet header. Third, protocols with a good quality of route usually need great efforts on establishing and recovering routes. Finally, most of the protocols need great spaces in nodes to store the information they need. There are mainly two types of information: neighbor list and event list. Only MFCN/GRAB do not explicitly assign the next hop. So they do not need to maintain a neighbor list.

Table 2.4 shows that multipath versions of protocols outperform others in terms of reliability. But they can hardly deal with the pattern failure of nodes. It is because those multipaths are usually close to each other and even interleaved. When a region of nodes fails, all of these paths fail at the same time. Most of the protocols show good route coverage, except the rumor routing.

2.6 Conclusions and Future Research Directions

In this section, we conclude this chapter and point out some existing problems that may indicate possible directions of future research.

2.6.1 A Summary of the Current Protocols

Existing routing protocols provide some degree of desired data-delivering services. But all of them have their own drawbacks. Geography-based routing is scalable, and has relatively low system overhead, and can be employed for most of the communication patterns. Some multipath versions of geography-based routing also showed encouraging reliability and fault-tolerant capabilities. However, the assumption of precise location information significantly limits their application domain. Gradient-based protocols eliminate the need for location information. They also present attractive performances on overhead and reliability. However, they can only be employed in particular scenarios. Cluster-based routings always need extra control messages to maintain the structure of the cluster. How to efficiently form the cluster, how to assign the cluster head,

Table 2.3 Comparison of Overhead of Protocols

	PKT-MAX	PKT-TOTAL	PKT-HEADER	RTE-BUILD	RTE-RECOVER	NODE-HEADER
Blind flooding	1		Small	0	0	Small
GPSR	1	Optimal'	Small	0	0	
LAR	1		Small	0	0	Small
DREAM	1		Small	0	0	Small
GEAR	1	Optimal'	Small	0	0	
BOUNDHOLE	1	Optimal'	Small	0	0	
TTDD	1	Optimal'				
GHT(SPAN)	1	Optimal'	Small	0	0	
MCFN, GRAB	1	Optimal	Small			Small
ARRIVE	1	Optimal	Small			
Direct Diffusion	1		Small			
CGSR	1	Optimal'	Small			
LEACH, PEGASIS	1	Optimal'	Small			
GAF	1	Optimal'	Small			
AODV, MOR	1	Optimal	Small			
DSR	1	Optimal		0		
Rumor routing	1	Optimal'	Small		0	
Energy-aware	1	Optimal'	Small		0	
RelnForM	1	Optimal'	Small		0	

Note: Optimal' means semi-optimal.

Table 2.4   Comparison of Overhead of Protocols

	ERR-CHAN	FAIL-RAND	FAIL-PATN	COVER	COST
Blind flooding	Good	Good	Good	Good	Low
GPSR		Good	Good	Good	
LAR, DREAM	Good	Good		Good	
GEAR, Boundhole		Good	Good	Good	
TTDD		Good	Good	Good	
MCFN, GRAB	Bad, Good	Bad, Good		Good	Low
ARRIVE	Good	Good		Good	Low
Direct Diffusion		Good		Good	Low
CGSR			Good	Good	Low
LEACH, PEGASIS		Good	Good	Good	
AODV, DSR				Good	Low
Rumor routing	Good	Good	Good		Low
Energy-aware	Good	Good		Good	Low
ReInForM	Good	Good	Good	Good	Low

and efficiently transmit messages—intercluster as well as intracluster—are all challenging jobs. On-demand routing requires both the sensor and the packets to reserve enough spaces to contain routing control information. Although random fashion routing show great potential, a gigantic packet (and node) header should be decreased before it can be employed in practice.

2.6.2   Existing Problems

Although many efforts have been paid in this area, there are still many problems in literature.

First, node identification needs more developments. It is known that localizations are critical assumptions for WSNs. Even given a loose requirement of the accuracy and granularity of the locations of sensors, it is still a challenge job to do localizations. A first issue in routing is how to efficiently deliver the message without or with less location information. Whether some other information is helpful, for example, a globally unique ID or a local ID; how to establish such an ID; and how to let nodes cooperate with their local ID to provide better routing services are still open research issues. We believe that these problems are related to each other and need further research efforts.

Second, most of the routing protocols in WSNs are designed for sensor-to-sink communications. Few researchers have considered the sink-to-sensor and sensor-to-sensor transmissions for their applications, impacts and solutions. Recall that communication through WSNs is a process of interaction between users and the network where communication patterns are not isolated from each other. We are interested in how to harness such a connection to provide an efficient delivery of data, instead of fully disconnected, brute-force solutions. A more general expression of this problem is, how to do routing in a computing environment mainly composed of dual operations.

The third problem is the long-term connectivity maintenance of networks. Nodes are often homogeneous in terms of initial energy while their workloads are always unevenly distributed, causing some nodes to deplete their energy faster than others. More seriously, nodes closer to the sink always have more workload than more distant nodes. They not only transmit their own data, but also help to forward others' data. Consequently, they are prone to failure because of the depletion of the energy. This is the so-called "hot spot" problem mentioned in Refs. [28, 45]. But this has not been fully investigated. The routing hole is around the sink. In such cases, although most of the nodes still function, active nodes cannot communicate with the sink and thus lose their connection to the outside world. A larger application field makes the problem more serious. We call this a "sink-routing-hole." Routing protocols to avoid the existence of "sinkholes" can significantly increase the connectivity and lifetime of networks.

2.6.3 Conclusion

In this chapter, we presented a comprehensive survey on current routing techniques in wireless sensor networks. We discussed the communication patterns and classified existing approaches into two major categories. After that, we presented the performance-evaluation metrics and gave a comprehensive comparison between them. Finally, we highlighted some of the challenging jobs and pointed out future research directions.

References

1. I. F. Akyildiz, W. Su, Y. Sankarasubramaniam, and E. Cayirci, "A Survey on Sensor Networks," *IEEE Communications Magazine*, vol. 40, pp. 102–114, 2002.
2. Y. Yao and J. Gehrke, "Query Processing for Sensor Network," In *Proceedings of CIDR*, 2003.
3. Y. W. Hong and A. Scaglione, "Energy-efficient Broadcasting with Cooperative Transmission in Wireless Ad Hoc Networks," In *Proceedings of Allerton Conference*, July, 2003.

4. J. M. Kahn, R. H. Katz, and K. S. J. Pister, "Next Century Challenges: Mobile Networking for Smart Dust," In *Proceedings of ACM MobiCom*, pp. 271–278, 1999.

5. V. Raghunathan, C. Schurgers, S. Park, and M. B. Srivastava, "Energy-aware Wireless Microsensor Networks," *Signal Processing Magazine*, vol. 19, pp. 40–50, 2002.

6. J. Hill, R. Szewczyk, A. Woo, S. Hollar, D. Cullar, and K. Pister, "System Architecture Directions for Networked Sensors," In *Proceedings of ASPLOS-IX*, 2000.

7. Z. Alliance, "http://www.zigbee.org."

8. U. Berkeley, TinyOs, "http://www.tinyos.net."

9. E. M. Royer and C.-K. Toh, "A Review of Current Routing Protocols for Ad Hoc Mobile Wireless Networks," *IEEE Personal Communications*, vol. 6, pp. 46–55, 1999.

10. P. Corke, S. Hrabar, R. Peterson, D. Rus, S. Saripalli, and G. Sukhatme, "Autonomous Deployment and Repair of a Sensor Network using an Unmanned Aerial Vehicle," In *Proceedings of IEEE International Conference on Robotics and Automation*, pp. 3602–3608, May, 2004.

11. D. Ganesan, D. Estrin, and J. Heidemann, "DIMENSIONS: Why Do We Need a New Data Handling Architecture for Sensor Networks," *ACM Computer Communication Review*, vol. 33, pp. 143–148, 2003.

12. S. Ratnasamy, B. Karp, S. Shenker, D. Estrin, R. Govindan, L. Yin, and F. Yu, "Data-Centric Storage in Sensor Nets with GHT, a Geographic Hash Table," *Mobile Networks and Applications*, vol. 8, pp. 427–442, 2003.

13. J. Polastve, R. Szewczyk, C. Sharp, and D. Cullar, "The Mote Revolution: Low Power Wireless Sensor Network Devices," In *Proceedings of Symposium on High Performance Chips*, 2004.

14. J. N. Al-Karaki and A. E. Kamal, "A Taxonomy of Routing Techniques in Wireless Sensor Networks," In *Sensor Networks Handbook*, CRC Publishers, Boca Raton, FL, 2004.

15. S. R. Das, C. E. Perkins, and E. M. Royer, "Performance Comparison of Two On-demand Routing Protocols for Ad Hoc Networks," *IEEE Personal Communication Magazine*, vol. 8, pp. 16–28, 2000.

16. F. Ye, H. Luo, J. Cheng, S. Lu, and L. Zhang, "A TwoTier Data Dissemination Model for Largescale Wireless Sensor Networks," In *Proceedings of ACM MobiCom*, 2002.

17. K. Sohrabi, J. Gao, V. Ailawadhi, and G. J. Pottie, "Protocols for Self-organization of a Wireless Sensor Network," *IEEE Personal Communication Magazine*, vol. 7, pp. 16–27, 2000.

18. N. Bulusu, D. Estrin, L. Girod, and J. Heidemann, "Scalable Coordination for Wireless Sensor Networks:Self-configuring Localization System," In *Proceedings of ISCTA*, 2001.

19. D. Ganesan, R. Govindan, S. Shenker, and D. Estrin, "Highly-resilient, Energy-efficient Multipath Routing in Wireless Sensor Networks," In *Proceedings of ACM MobiCom*, 2001.

20. F. Ye, G. Zhong, S. Lu, and L. Zhang, "Gradient Broadcast: A Robust Data Delivery Protocol for Large Scale Sensor or Networks," *ACM Wireless Networks*, vol. 11, pp. 285–298, 2003.
21. J. K. Hart and J. Rose, "Approaches to the Study of Glacier Bed Deformation," *Quaternary International*, vol. 86, pp. 45–58, 2001.
22. ZigBee, Alliance, "http://www.zigbee.org."
23. D. Culler, D. Estrin, and M. Srivastava, "Overview of Sensor Network," *IEEE Computer*, vol. 37, pp. 41–78, 2004.
24. C. Schurgers, G. Kulkarni, and M. B. Srivastava, "Distributed On-Demand Address Assignment in Wireless Sensor Networks," *TPDS*, vol. 13, pp. 1056–1065, 2002.
25. Y. Liu, L. M. Ni, and M. Li, "A Geography-free Routing Protocol for Wireless Sensor Networks," In *Proceedings of International Workshop on High Performance Switching and Routing (HPSR)*, pp. 145–150, 2005.
26. B. Karp and H. T. Kung, "GPSR: Greedy Perimeter Stateless Routing for Wireless Networks," In *Proceedings of Mobicom*, 2000.
27. F. Ye, A. Chen, S. Lu, and L. Zhang, "A Scalable Solution to Minimum Cost Forwarding in Large Sensor Network," In *Proceedings of ICCCN*, pp. 304–309, 2001.
28. C. Karlof, Y. Li, and J. Polastre, "ARRIVE: An Architecture for Robust Routing in Volatile Environments," University of California at Berkeley, UCB//CSD-03-1233, 2003.
29. S. Madden, M. J. Franklin, J. Hellerstein, and W. Hong, "TAG: A Tiny Aggregation Service for Ad Hoc Sensor Networks," In *Proceedings of OSDI*, 2002.
30. C. Jaikaeo, C. Srisathapornphat, and C.-C. Shen, "Diagnosis of Sensor Networks," In *Proceedings of ICC*, 2001.
31. R. C. Shah and J. M. Rabaey, "Energy Aware Routing for Low Energy Ad Hoc Sensor Networks," In *Proceedings of WCNC*, pp. 350–355, 2002.
32. W. R. Heinzelman, J. Kulik, and H. Balakrishnan, "Adaptive Protocols for Information Dissemination in Wireless Sensor Networks," In *Proceedings of ACM MobiCom*, 1999.
33. W. R. Heinzelman, A. Chandrakasan, and H. Balakrishnan, "LEACH: Energy-efficient Communication Protocol for Wireless Microsensor Networks," In Proceedings of Hawaii International Conference on System sciences, 2000.
34. T. Imielinski and S. Goel, "DataSpace: Querying and Monitoring Deeply Networked Collections in Physical Space," *IEEE Wireless Communications*, vol. 7, pp. 4–9, 2000.
35. R. A. F. Mini, A. A. F. Loureiro, and B. Nath, "The Best Energy Map of a Wireless Sensor Network," In *Proceedings of SBRC*, 2004.
36. C. Intanagonwiwat, D. Estrin, R. Govindan, and J. Heidemann, "Impact of Network Density on Data Aggregation in Wireless Sensor Networks," In *Proceedings of ICDCS*, 2001.
37. C. Intanagonwiwat, R. Govindan, and D. Estrin, "Directed Diffusion: A Scalable and Robust Communication Paradigm for Sensor Networks," In *Proceedings of ACM MobiCom*, 2000.

38. J. Heidemann, F. Silva, C. Intanagonwiwat, R. Govindan, D. Estrin, and D. Ganesan, "Building Efficient Wireless Sensor Networks with Low-level Naming," In *Proceedings of SOSP*, 2001.

39. Y.-B. Ko and N. H. Vaidya, "Location-aided Routing (LAR) in Mobile Ad Hoc Networks," In *Proceedings of ACM MobiCom*, 1998.

40. Y. Yu, R. Govindan, and D. Estrin, "Geographical and Energy Aware Routing: A Recursive Data Dissemination Protocol for Wireless Sensor Networks," UCLA Computer Science Department, UCLA/CSD-TR-01-0023, 2001.

41. Q. Fang, J. Gao, and L. J. Guibas, "Locating and Bypassing Routing Holes in Sensor Networks," In *Proceedings of INFOCOM*, pp. 2458–2468, 2004.

42. B. Chen, K. Jamieson, H. Balakrishnan, and R. Morris, "Span: An Energy-Efficient Coordination Algorithm for Topology Maintenance in Ad Hoc Wireless Networks," *ACM Wireless Networks*, vol. 8, 2002.

43. R. D. Poor, "Gradient Routing in Ad Hoc Networks," Media Laboratory, Massachusetts Institute of Technology.

44. P. Bahl and V. N. Padmanabhan, "RADAR: An In-building RFbased User Location and Tracking System," In *Proceedings of the Conference on Computer Communications*, 2000.

45. C. E. Perkins and P. Bhagwat, "Highly Dynamic Destination Sequenced Distance Vector Routing (DSDV) for Mobile Computers," In *Proceedings of SIGCOMM*, 1994.

46. F. Aurenhammer, "Voronoi Diagrams—A Survey of a Fundamental Geometric Data Structure," *ACM Computing Surveys*, vol. 23, pp. 345–405, 1991.

47. S. Lindeey and C. Raghavendra, "PEGASIS: Power-efficient Gathering in Sensor Information Systems," In *Proceedings of IEEE Aerospace Conference*, 2002.

48. Y. Xu, J. Heidemann, and D. Estrin, "Geography-informed Energy Conservation for Ad Hoc Routing," In *Proceedings of ACM MobiCom*, pp. 70–84, 2001.

49. C.-C. Shen, C. Srisathapornphat, R. Liu, Z. Huang, C. Jaikaeo, and E. L. Lloyd, "CLTC: A Cluster-based Topology Control Framework for Ad hoc Networks," In *Proceedings of ACM MobCom*, 2004.

50. V. Rodoplu and T. H. Meng, "Minimum Energy Mobile Wireless Networks," *IEEE Journal on Selected Areas in Communications*, vol. 17, pp. 1333–1344, 1999.

51. L. Li and J. Y. Halpern, "Minimum-Energy Mobile Wireless Networks Revisited," In *Proceedings of ICC*, pp. 278–283, 2001.

52. C. E. Perkins and E. M. Royer, "Ad hoc On-Demand Distance Vector Routing," In *Proceedings of IEEE Workshop on Mobile Computing Systems and Applications*, 1999.

53. D. B. Johnson and D. A. Maltz, "Dynamic Source Routing in Ad Hoc Wireless Networks," In *Proceedings of ACM MobiCom*, 1996.

54. D. B. Johnson, D. A. Maltz, and J. Broch, "DSR: The Dynamic Source Routing Protocol for Multi-Hop Wireless Ad Hoc Networks," in *Ad Hoc Networking*, C. E. Perkins, Ed.: Addison-Wesley, 2001, pp. 139–172.

55. S. h. Chen, "Multipath On-Demand Routing in Sensor Network Topologies," Master of Science Thesis, Information and Computer Sciences Department, University of Hawaii in MANOA, 2003.

56. D. Braginsky and D. Estrin, "Rumor Routing Algorithm for Sensor Networks," In *Proceedings of ACM MobiCom*, pp. 562–567, 2002.

57. B. Deb, S. Bhatnagar, and B. Nath, "ReInForM: Reliable information Forwarding using Multiple Paths in Sensor Networks," In *Proceedings of ACM MobiCom*, pp. 406–415, 2001.

Chapter 3

FireFly: A Time-Synchronized Real-Time Sensor Networking Platform

Anthony Rowe, Rahul Mangharam, and Raj Rajkumar

3.1 Introduction

The rapid proliferation of sensor networks has placed increasing demands upon the system infrastructure for supporting scalable and energy-efficient sensor applications. As applications for sensors in areas as diverse as security surveillance, traffic monitoring, smart spaces, and smart buildings continue to grow, infrastructural support for sensor network applications in the form of comprehensive hardware, software, and networking support is becoming increasingly important.

The Real-Time Multimedia Systems Laboratory in Carnegie Mellon University has developed a complete wireless sensor networking platform called FireFly comprising of FireFly sensor nodes (SNs), a real-time multitasking operating system (OS) called *nano-RK* and a time-division-multiplexed-access protocol stack. The platform design has been directly driven by the following goals:

1. Low cost: Each node must be relatively inexpensive to enable cost-effective acquisition and deployment.

2. Energy efficiency: The installation cost of wireless SNs is greatly facilitated by the lack of wiring installation, but this (often) requires that a SN be battery-operated. Battery lifetimes must be acceptable to current and future application domains of interest, and this can only be achieved by using an energy-efficient design and implementation.

3. Scalability: No aspect of the system including hardware, system software support, and networking must introduce any bottlenecks in deploying very large-scale FireFly SNs. For example, while the range of any single radio will be relatively short, communications must be feasible across large distances necessitating multihop communications.

4. Extensibility: Since different applications will likely require different sensors, each node must have the capability to interface with new sensors.

5. Ease of programmability: Each node must be programmed using a well-known programming language like C, and not be restricted to any specialized language subsets (or supersets). The programming model available to application developers must be compatible and consistent with the popular models such as multitasking used very commonly in both embedded real-time and desktop environments. The interface between the development environment and target nodes must also be simple and convenient.

6. Self-configuring: The wireless sensor network (WSN) must be self-configuring when initially deployed; and when the network topology changes due to the introduction, removal, or death of SNs, the network must be capable of healing itself.

7. Support for time-sensitive applications: Many application domains for WSNs will stream time-sensitive data, such as events of interest from security, privacy, or emergency viewpoints, while media streaming such as voice (or even video) will become important soon in many contexts.

8. Hierarchical architecture: Many practical WSN will have a multitiered architecture where the SNs will be at the leaves, intermediate gateway nodes will collect and coordinate information at the next higher tier, and one or more control/command centers (or gateways) will interface with human operators and/or external systems. It must therefore be possible to configure wireless SNs to form network topologies of interest to a particular application's needs.

The rest of this chapter is organized as follows: Section 3.2 provides a detailed description of the FireFly sensor nodes. Section 3.3 presents

RT-Link, our TDMA-based link layer that supports multihop wireless networks. Section 3.4 describes nano-RK, our multitasking energy-aware real-time OS running on FireFly SNs. In Section 3.5, we summarize some of our early experiences with deploying FireFly sensor networks in a coal mine that enables miner tracking and two-way communications. Finally, we make some concluding remarks in Section 3.6.

3.2 The FireFly Sensor Node

The low-cost low-power hardware SN called FireFly that we have developed is shown in Figure 3.1. The board uses an Atmel Atmega32L[1] 8-bit microcontroller and a Chipcon CC2420[2] IEEE 802.15.4 wireless transceiver. The microcontroller operates at 8 MHz, and has 32 kB of ROM and 2 kB of RAM. The FireFly board includes light, temperature, audio, dual-axis acceleration, and passive infrared motion sensors. We have also developed a lower-cost version of the board called the FireFly Jr. that does not include sensors, and is used for forwarding packets in the network or can be used as a module inside other devices such as actuators that do not require sensing. Any FireFly board can interface with a computer using an external USB dongle.

Table 3.1 shows a breakdown of the typical energy consumption of the different components on the FireFly board. Since the transmit energy on the boards is quite low (1 mW), the analog components in the radio's power amplifier are not as dominant as they would be in other forms of radio like 802.11. This accounts for why the radio receive energy is greater than the transmit energy and means that nodes should not only try to minimize packet transmission, but should also minimize listening time.

Figure 3.1 **FireFly and FireFly Jr. board with AM synchronization module.**

Table 3.1   Energy Statistics for Current Hardware Setup

	Power (mW)	Energy
CPU		$(0.05\,\text{mW} * t_{idle}) + (24.0\,\text{mW} * t_{active})$
Idle	0.05	$0.05\,\text{mW} * t_{idle}$
Active	24.0	$24.0\,\text{mW} * t_{active}$
Network		$(0.06\,\text{mW} * t_{idle}) + (1.8\,\mu\text{J} * N_{rx\_bytes})$
		$+ (1.6\,\mu\text{J} * N_{tx\_bytes})$
RX	59.1	$1.8\,\mu\text{J}$ per byte
TX	52.1	$1.6\,\mu\text{J}$ per byte
Idle	0.06	$00.06\,\text{mW} * t_{idle}$
Sensor		
Light, temperature	0.09	11.25 nJ per reading
Microphone	2.34	$2.87\,\mu\text{J}$ per reading
PIR	5.09	$1\,\mu\text{J}$ per reading
Acceleration	1.8	11.25 nJ per reading

3.2.1   Hardware-Assisted Time Synchronization

3.2.1.1   Benefits of Time Synchronization

Tight and energy-efficient time synchronization in the wireless multihop domain yields the following benefits:

1. Energy-efficient communication. An effective approach to energy-efficient service for applications with either periodic or aperiodic flows is to operate all nodes at low duty cycles so as to maximize the shutdown intervals between packet exchanges. Time synchronization is crucial to tightly pack the activity of all nodes so that they may maximize a common sleep interval between activities. Contention-based MAC protocols suffer from nearby nodes overhearing packets that are not addressed to them. In a time-synchronized network, nodes can be scheduled such that even though they could physically communicate with each other, the topology can be logically pruned.

2. Bounded message latency. Time synchronization allows messages to be scheduled such that they are collision-free. This provides guarantees on timeliness, eliminating ambiguity about whether or not a message was dropped or simply delayed. This is important for latency-sensitive applications such as control automation systems or interactive media streaming applications. Furthermore, collision-free transmissions save energy that may otherwise be wasted on retransmissions, caused by packet loss arising from collisions.

3. High throughput. A tightly scheduled collision-free environment allows for higher throughput than a system using a contention-based scheme. In the wireless sensor network setting this accommodates ondemand bulk transfers of data such as firmware updates, logged sensor data, or streaming of high data rate sensors.

4. Deterministic lifetime. The energy required to power sensor network radios is typically 10 to 20 times more than the underlying CPU power. Since all communication is scheduled in advance, time synchronization enables the vast majority of energy consumption to be allocated in advance. We will discuss later how resource reservations can further refine the deterministic lifetime of these systems by managing other energy-consuming resources.

5. Total event ordering. Many applications such as localization and tracking require ordering of event samples taken from different nodes at different times. Network-wide time synchronization greatly simplifies this process by providing absolute time stamps that can be compared against each other.

3.2.1.2 Out-Of-Band Time Synchronization in FireFly Nodes

We make available two out-of-band time-synchronization sources to achieve highly accurate time synchronization even over very large-scale geographical areas. One uses the WWVB atomic clock broadcast, and the other relies on a carrier-current AM transmitter. In general, the synchronization device should be low-power, inexpensive, and consist of a simple receiver. The time-synchronization transmitter must be capable of covering a large area.

The WWVB atomic broadcast is a pulse-width modulated (PWM) signal with a bit starting each second. Our system uses an off-the-shelf WWVB receiver (Figure 3.2) to detect these rising edges, and does not need to decode the entire time string. When active, the board draws 0.6 mA at 3 V and requires less than 5 μA when powered off.

Inside buildings, atomic clock receivers are typically unable to receive any signal, so we use a carrier-current AM broadcast. Carrier-current uses a building's power infrastructure as an antenna to radiate the time-synchronization pulse. We used an off-the-shelf low-power AM transmitter and power coupler[3] that adhere to the United States Federal Communications Commission (FCC) Part 15 regulations without requiring a license. The transmitter provides time synchronization to two five-story campus buildings which operate on two AC phases. Figure 3.2 shows an add-on AM receiver module capable of decoding our AM time-synchronization pulse. We use a commercial AM receiver module in a custom-designed add-on board which performs threshold detection of the demodulated signal to

Figure 3.2   Left to right: WWVB atomic clock receiver, AM receiver, and USB interface board.

decode the presence of a pulse. The supporting AM board is also capable of controlling the power to the AM receiver.

The energy required to activate the AM receiver module and to receive a pulse is equivalent to sending one and a half 802.15.4 packets. The use of a more advanced custom radio solution would reduce these values and allow for a more compact design. We estimate that by using a single-chip AM radio receiver, the synchronization energy cost would be less than sending a single in-band packet.

To maintain scalability across a wide campus area including the interior of multiple buildings, our AM transmitter locally rebroadcasts the atomic clock time signal. The synchronization pulse for the AM transmitter is a line-balanced $50\,\mu s$ square wave, generated by a modified FireFly node capable of atomic clock synchronization.

Evaluation: We evaluated the effectiveness of the synchronization by placing five nodes at different points inside an eight-story building. Each node was connected to a data collection board using several hundred feet of cables. The data collection board timed the difference between the generation synchronization pulse and acknowledgment of the pulse by each node. This test was performed while the MAC protocol was active to get an accurate idea of the possible jitter including MAC-related processing overhead. Figure 3.3 shows a histogram with the distribution of each node's synchronization time jitter. An AM pulse was sent once per every second for 24 hours during normal operation of a classroom building. As seen in the graph, we found that the jitter is bounded to within $200\,\mu s$ in the absolute worst case. Furthermore, 99.6% of the synchronization pulses were correctly detected. Based on our experiences, we anticipate that with more

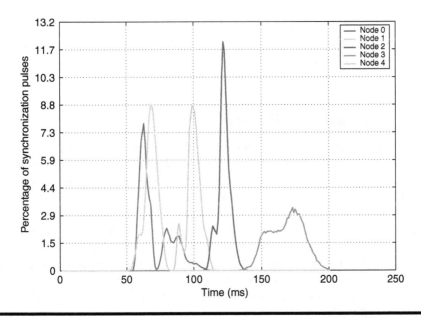

Figure 3.3   Distributions of AM carrier-current time-synchronization jitter over a 24-h period.

refined tuning of the AM radios, the jitter could be bounded to be well within $50\,\mu$s.

To maintain synchronization over an entire TDMA cycle duration, it is necessary to measure the drift associated with the clock crystal on the processor. We observed that the worst of our clocks was drifting by $10\,\mu$s/s giving it a drift rate of 10^{-5}. Our previous experiment illustrates that the jitter from AM radio was at worst $100\,\mu$s, indicating that the drift would not become a problem for at least 10 s. The drift due to the clock crystal was also relatively consistent, and hence could be accounted for in software by timing the difference between synchronization pulses and performing a clock-rate adjustment. In our final implementation, we are able to maintain global time synchronization to within $20\,\mu$s.

3.3   RT-Link: A TDMA Link Layer Protocol for Multihop Wireless Networks

RT-Link is the FireFly TDMA-based link layer protocol designed to provide predictability in throughput, latency, and energy consumption. All packet exchanges occur in well-defined time slots. Global time synchronization is provided to all fixed nodes by a robust and low-cost out-of-band channel. We now describe in detail our RT-Link protocol, packet types, supported node types, and the different modes of protocol operation.

3.3.1   Current MAC Protocols

Several MAC protocols have been proposed for low-power and distributed operation for single- and multihop wireless mesh networks. Such protocols may be categorized by their use of time synchronization as asynchronous, loosely synchronous, and fully synchronized protocols. In general, with a greater degree of synchronization between nodes, packet delivery is more energy-efficient due to the minimization of idle listening when there is no communication, better collision avoidance, and elimination of overhearing of neighbor conversations. We now briefly review key low-power link protocols based on their support for low-power listening (LPL), multihop operation with hidden terminal avoidance, scalability with node degree, and offered load.

3.3.1.1   Asynchronous Link Protocols

The Berkeley MAC (B-MAC)[4] protocol performs the best in terms of energy conservation and design simplicity. B-MAC supports carrier sense multiple access (CSMA) with LPL where each node periodically wakes up after a checking interval and checks the channel for activity for a short duration of 2.5 ms. If the channel is found to be active, the node stays awake to receive the payload following an extended preamble. Using this scheme, nodes may efficiently check for neighbor activity. For each transmission instance, the transmitter must remain active for the duration of the receiver's channel check interval. This creates a major drawback since it forces the receiver to check the channel very often (in milliseconds) even when the event sampling interval spans several seconds or minutes. For example, if an event occurs every 20 min, all B-MAC receivers check the channel for activity approximately every 80 ms to limit the transmitter's burst duration to 80 ms.[4] This coupling of the receiver's sampling interval and the duration of the transmitter's preamble severely restricts the scalability of B-MAC when operating in dense networks and across multiple hops. B-MAC does not inherently support collision avoidance due to the hidden terminal problem and the use of request to send–clear to send (RTS–CTS) handshaking with LPL is inefficient because the RTS requires the use of the extended preamble. Furthermore, upon wakeup, B-MAC employs CSMA which is prone to wasting energy and adds nondeterministic latency due to packet collisions. We argue that in a multihop network, topology awareness can be exploited even in practice to schedule packets without collision leading to both predictable end-to-end timing behavior and additional energy savings.

3.3.1.2 Loosely Synchronous Link Protocols

Protocols such as S-MAC[5] and T-MAC[6] employ local sleep–wake schedules know as *virtual clustering* between node pairs to coordinate packet exchanges while reducing idle operation. Both schemes exchange synchronizing packets to inform their neighbors of the interval until their next activity and use CSMA prior to transmissions. As all the neighbors of a node cannot hear each other, each node must set multiple wakeup schedules for different groups of neighbors. The use of CSMA and loose synchronization trades energy consumption for simplicity. WiseMAC[7] is an iteration on Aloha designed for downlink communication from infrastructure nodes and has been shown to outperform 802.15.4 for low traffic loads. WiseMAC does not support multihop communications. Both T-MAC and WiseMAC use preamble sampling to minimize receive energy consumption during channel sampling. The use of CSMA in either scheme can degrade performance severely with increasing node degree and traffic.

3.3.1.3 Fully Synchronized Link Protocols

TDMA protocols such as TRAMA[8] and LMAC[9] are able to communicate between node pairs in dedicated time slots. TRAMA supports both scheduled slots and CSMA-based contention slots for node admission and network management. LMAC is a lightweight bit-mask schedule reservation scheme; it establishes collision-free operation by negotiating nonoverlapping slot across all nodes within the two-hop radius. Both protocols assume the provision of global time synchronization but consider it an orthogonal problem. RT-Link has similar support for contention slots but employs slotted-Aloha[10] instead of CSMA as it is more energy-efficient with LPL. Furthermore, RT-Link integrates time synchronization within the protocol and also in the hardware specification.

RT-Link has been inspired by dual-radio systems such as those described in Refs. [11,12] used for low-power wakeup. However, neither system described in Refs. [11,12] has been used in practice for time-synchronized operation. Several in-band software-based time-synchronization schemes such as RBS,[13] TPSN,[14] and FTSP[15] have been proposed and provide good accuracy. In Ref. [16], Zhao provides experimental evidence showing that over one third of the population of immobile nodes in an indoor environment routinely suffer a link error rate of over 50%, even when the receive signal strength is above the sensitivity threshold. This severely limits the diffusion of in-band time-synchronization updates and hence reduces the network performance.

3.3.1.4 Description of RT-Link

RT-Link employs an out-of-band time-synchronization mechanism, which also globally synchronizes all nodes and is less vulnerable than the above schemes. We believe that hardware-based time synchronization adds new enabling capabilities to wireless sensor networks and warrants extensive exploration and usage in practical large-scale environments.

RT-Link supports two node types: fixed and mobile. Both node types include a microcontroller, 802.15.4 transceiver, and multiple sensors as described in Section 3.2. The fixed nodes have an add-on time-synchronization module, which is normally a low-power radio receiver designed to detect a periodic out-of-band global signal. In our implementation, we use an AM/FM time-synchronization module for indoor operation and an atomic clock receiver for outdoors. For indoors, we use a carrier-current AM transmitter,[3] which plugs into the power outlet in a building and uses the building's power grid as an AM antenna to radiate the time-synchronization pulse. We can also feed an atomic clock pulse as the input to the AM transmitter to provide the same synchronization regime for use both indoors and outdoors across a wide coverage area.

The time-synchronization module detects the periodic synchronization pulse and triggers an input pin in the microcontroller which updates the local time. As shown in Figure 3.4, this marks the beginning of a finely slotted data communication period. The communication period is defined as a fixed-length cycle and is composed of multiple frames. The synchronization pulse serves as an indicator of the beginning of the cycle and the first frame. Each frame is divided into multiple slots, where a slot duration is the time required to transmit a maximum-sized packet. RT-Link supports two kinds of slots:

- *scheduled slots* (SS) within which nodes are assigned specific transmit and receive time slots, and
- a series of unscheduled or *contention slots* (CS) where nodes, which are not assigned slots in the SS, select a transmit slot at random as in slotted-Aloha.

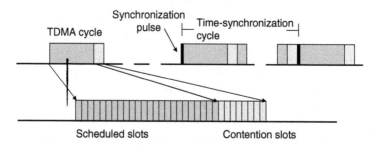

Figure 3.4   RT-Link time slot allocation with out-of-band synchronization pulses.

Nodes operating in SS are provided timeliness guarantees as they are granted exclusive access to the shared channel, consequently enjoy the privilege of interference- and collision-free communication. While the support for SS and CS is similar to that available in 802.15.4, RT-Link is designed for operation across synchronized multihop networks. After an active slot is complete, the node schedules its timer to wake up just before the (expected) next active slot and promptly switches to sleep mode. In our default implementation, each cycle consists of 32 frames and each frame consists of thirty-two 5 ms slots. Thus, the cycle duration is 5.12 s and nodes can choose one or more slots per frame up to a maximum of 1024 slots every cycle. The common packet header includes a 32-bit transmit and 32-bit receive bit-mask to indicate which slots of a node is active. RT-Link supports five packet types including HELLO, SCHEDULE, DATA, ROUTE, and ERROR. The packet types and their formats are described in detail in Ref. [17].

3.3.2 Network Operation Procedures

RT-Link operates on a simple three-state machine as shown in Figure 3.5. In general, nodes operating in the CS are considered *Guests*, while nodes with SS are considered *Members* of the network. When a fixed node is powered on, it is first initialized as a *Guest* and operates in the CS. It initially keeps its synchronization radio receiver on until it receives a synchronization pulse. Following this, it waits for a set number of slots (spanning the SS) and then randomly selects a slot among the CS to send a HELLO message with its node ID. This message is then forwarded (via flooding if explicit routes are not present) to the gateway and the node is eventually scheduled a slot in the SS.

However, when a *Mobile* node needs to transmit, it first stays on until it overhears a neighboring *Member* operate in an SS. The mobile node

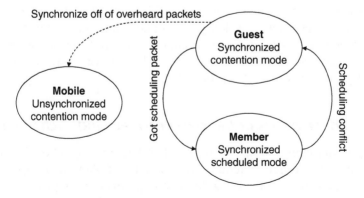

Figure 3.5   RT-Link node state machine.

achieves synchronization by observing the *Member*'s slot number and computes the time until the start of the CS. *Mobile* nodes are never made members because their neighborhood can change frequently and hence remain silent until a *Member* provides it with a time reference. All nodes with scheduled slots listen to every slot in the CS using LPL. When a node chooses to leave the network, it ceases broadcasting HELLO packets and is gracefully evicted from each of its neighbors' list. The gateway eventually detects the absence of the departed node from each of the neighbors' HELLO updates, and may reschedule the network if necessary.

For fixed nodes that are unable to receive the global time beacon and for mobile nodes, RT-Link provides software-based in-band time synchronization. Nodes can implicitly pass time synchronization onto another node using the current slot in the packet header. This implicit time synchronization can cascade across multiple hops.

3.3.3   Logical Topology Control

RT-Link schedules communication based on the global network topology. This requires a topology-gathering phase followed by a scheduling phase. To acquire the network connectivity graph, we aggregate the neighbors lists from each node at the gateway. We then construct connectivity and interference graphs and schedule nodes based on k-hop coloring, such that two nodes with the same slot schedules are mutually separated by at least $k+1$ hops. Figure 3.6 shows the impact of node degree on lifetime. As the number of neighbors with which a node communicates increases, the number of transmit and receive slots correspondingly increases, consuming more energy. Figure 3.7(a) shows the connectivity graph of a randomly generated topology with 100 nodes. The connectivity graph is colored using the two-hop constraint to ensure that it is free of collisions. Links can then be removed by instructing an adjacent node to no longer wake up to listen on that particular time slot. Using this principle we can reduce the degree of nodes while checking to maintain network connectivity. The reduced degree topology shown in Figure 3.7(b) reduces the average network energy and simplifies routing. Such logical topology control is not possible with random access protocols.

3.3.4   Effectiveness of Interference-Free Node Scheduling

A major underlying assumption in RT-link is that two-hop scheduling results in an interference-free schedule. Traditionally, TDMA multihop wireless scheduling has been solved as a distance-k graph coloring problem, where k is set to 2. To validate this assumption, we tested the interference range of a node with respect to its transmission range. We placed a set of nodes

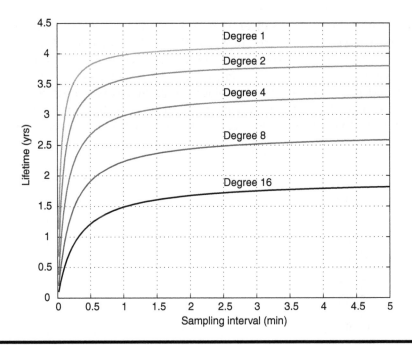

Figure 3.6   Life given increasing node degree between 1 and 16. Node degree one yields the longest life.

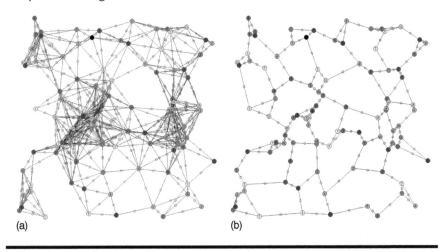

Figure 3.7   TDMA topology control by not activating receiver for selected time slots. (a) Physical connectivity graph. (b) Pruned logical connectivity.

along a line in an open field and measured the packet loss between a transmitter and receiver as the transmitter's distance was varied. Once the stable communication distance between a transmitter and receiver was determined, we evaluated the effect of a constantly transmitting node (i.e., a jammer) on the receiver. Our experimental results for stable transmission (power level 8) are shown in Figure 3.8. We noticed that 100% of the packets are received up to a transmitter–receiver distance of 10 m. Following this, we placed the transmitter at a distance of 2, 4, 6, 8, and 10 m and for each transmitter position, a jammer was placed at various distances from the receiver. At each point, the transmitter sent one packet every cycle to the receiver for 2000 cycles. We measured the impact of the

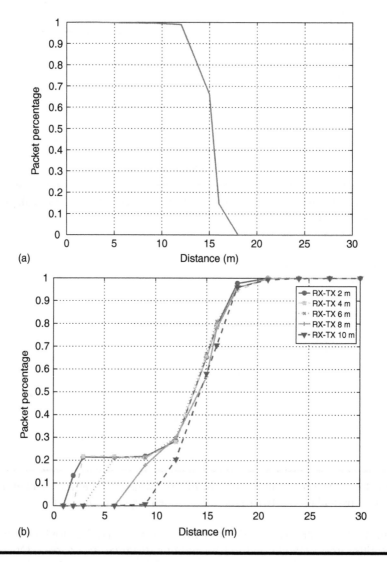

Figure 3.8   Packet success rate while transmitting in a collision domain. (a) RX to TX distance with no jammer. (b) Fixed RX to TX distance and varying jammer.

jammer by observing the percentage of packets from the transmitter that are received successfully by the receiver.

We observe two effects of the jammer: first, the effect of the jammer is largely a function of the distance of the jammer from the receiver and not of the transmitter from the receiver. Between 12 and 18 m, the impact of the jammer is similar across all transmitter–receiver distance pairs. Second, when the transmitter and jammer are close to the receiver (i.e., under 8 m), the transmitter demonstrates a "capture effect" and maintains an approximately 20% packet reception rate.

The above results show that the jammer has no impact beyond twice the stable reception distance (i.e., 20 m) and a concurrent transmitter may be placed at thrice the stable reception distance (i.e., 30 m). Such parameters are incorporated into the node coloring algorithm in the gateway to determine collision-free slot schedules.

3.3.5 Network Scheduling

In multihop wireless networks, the goal of scheduling has often been to maximize the set of concurrent transmitters in the network.[18] This is achieved either by scheduling nodes or links such that there are no collisions. In the FireFly networks we study in this paper, the applications generate steady or low data rate flows but require low energy consumption and low end-to-end delay. Figure 3.9 presents two schedules: one with the minimal number of time slots corresponding to the minimal use of colors in the graph-coloring step, and the other containing extra slots but provisioned such that leaf nodes deliver data to the gateway in a single TDMA cycle. The minimal color schedule maximizes concurrent transmissions, but

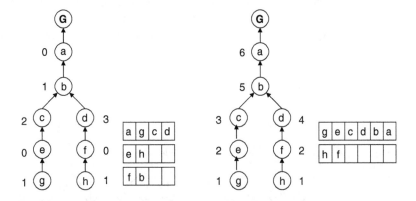

Figure 3.9 Maximal concurrency schedule (left) compared to a delay-sensitive schedule (right). Note that the maximal concurrency schedule needs two frames to deliver all data.

causes queueing delays and hence does *not* minimize the upstream latency of all nodes. By assigning the time slots appropriately in preference to faster uplink and downlink routes, we emphasize that for networks with delay-sensitive data, the ordering of slots should take priority over maximizing spatial reuse (and increased concurrent transmissions).

The generation of minimum delay schedules is similar to the distance-two graph coloring problem that is known to be NP-complete.[19] Fortunately, many heuristics work well in practice yielding a small constant deviation from the optimal.[19] To illustrate the minimum-delay capability of RT-Link, we briefly discuss one such heuristic to schedule a FireFly network where the traffic consists of small packets being routed up a tree to a single gateway. The heuristic consists of four steps. The first step builds a spanning tree over the network rooted at the gateway. Using Dijkstra's shortest-path algorithm, any connected graph can be converted into a spanning tree. As can be seen in Figure 3.10(b), the spanning tree must maintain "hidden" links that are not used when iterating through the tree to ensure that the two-hop constraint is still satisfied in the original graph. Once a spanning tree is constructed, a breadth-first search is performed starting from the root of the tree. The heuristic begins with an initially empty set of colors. As each node is traversed by the breadth-first search, it is assigned the lowest value in the color set that is unique from any 1- or 2-hop neighbors. If there are no free colors, a new color must be added into the current set. The next step in the heuristic tries to eliminate redundant slots that lie deeper in the tree by replacing them with larger valued slots. This manipulation allows data from the leaves of the tree to move as far as possible toward the gateway in a single TDMA cycle, and will be easily seen in the next and final step. Figure 3.10(c) shows how the previous three nodes are given larger values to minimize packet latencies. The final step in the heuristic inverts all of the slot assignments such that lower slot values are toward the edge of the tree allowing information to be propagated and aggregated in a cascading manner toward the gateway.

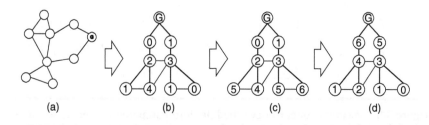

Figure 3.10 Ordered coloring to minimize upstream end-to-end delay.

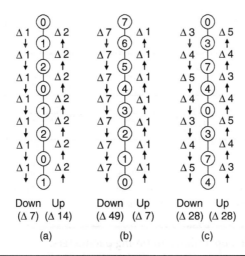

Figure 3.11   Different schedules change latency based on direction of the flow.

3.3.6   Explicit Rate Control

RT-Link allows explicit rate control by specifying a 4-bit rate index r, in the schedule assigned to each node (Figure 3.11). A flow's rate is defined by the number of active frames that it transmits specified by 2^{r-1}. For example, rate 1 transmits on every frame while rate 3 transmits on every 4th frame. Using this scheme, we can vary a flow's rate by controlling the number of slots and the rate index assigned to it.

3.3.7   TDMA Slot Mechanics

When a node is first powered on, it activates the AM receiver and waits for the first synchronization pulse. Figure 3.12 shows the actual timing associated with our TDMA frames. Once the node detects a pulse, it resets the TDMA frame counter maintained in the microcontroller, which then powers down the AM receiver. When the node receives its synchronization pulse, it begins the active TDMA time cycle. After checking its receive and transmit masks, the node determines which slots it should transmit and receive on. During a receive time slot, the node immediately turns on the receiver. The receiver will wait for a packet, or it will time out if no preamble is detected.

   The received packet is read from the CC2420 radio chip into a memory address that was allocated to that particular slot. We employ a zero-copy buffer scheme to move packets from the receive to the transmit queue. In the case of automatic packet aggregation, the payload information from a packet is explicitly copied to the end of the transmit buffer. When the

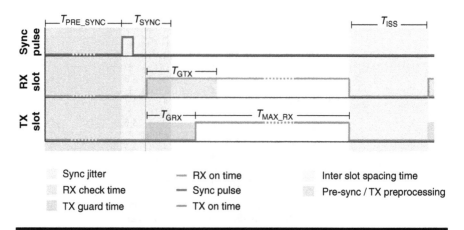

Figure 3.12 **RT-Link operation and timing parameters.**

Table 3.2 **Power Consumption of the Main Components**

Power Parameters	Symbol	I (ma)	Power (mW)
Radio transmitter	P_{radio_TX}	17.4	52.2
Radio receiver	P_{radio_RX}	19.7	59.1
Radio idle	P_{radio_idle}	0.426	1.28
Radio sleep	P_{radio_sleep}	$1e^{-3}$	$3e^{-3}$
CPU active	P_{CPU_active}	1.1	3.3
CPU sleep	P_{CPU_sleep}	$1e^{-3}$	$3e^{-3}$
AM sync active	P_{sync}	5	15

node reaches a transmit time slot, it must wait for a guard time to elapse before sending data. Accounting for the possibility that the receiver has drifted ahead or behind the transmitter, the transmitter has a guard time before sending and the receiver preamble-check has a guard time extending beyond the expected packet. Tables 3.2 and 3.3 in the next section shows the different time-out values that work well for our hardware configuration. Once the time slot is complete, there needs to be an additional guard time before the next slot. We provide this guard time plus a configurable interslot processing time that allows the MAC to do the minimal processing required for inter-slot packet forwarding.

This feature is motivated by the need to satisfy stringent memory limitations and the desire to minimize network queue sizes.

Figure 3.13 shows a sample trace of two nodes communicating with each other. The rapid receiver checks at the end of the cycle show the contention period with LPL for the duration of a preamble.

Table 3.3   Energy of Components with Respect to Power and Time

Energy Parameters	Symbol	Energy (mW)
Synchronization	E_{sync}	$P_{sync}* (T_{sync} + T_{sync\_setup})$
Active CPU	E_{CPU_active}	$P_{CPU\_active} * T_{active}$
Sleep CPU	E_{CPU_sleep}	$P_{CPU\_sleep} * T_{idle}$
TX radio	E_{radio_tx}	$P_{radio\_tx} * (T_{max\_payload} + T_{GTX})$
RX radio	E_{radio_rx}	$P_{radio\_rx} * T_{max\_payload}$
Idle radio	E_{radio_idle}	$P_{radio\_idle} * T_{active}$
Sleep radio	E_{radio_sleep}	$P_{radio\_sleep} * T_{idle}$
RX radio check	E_{GRX}	$P_{radio\_rx} * T_{GRX}$

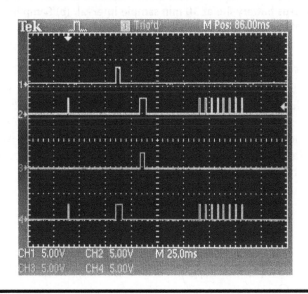

Figure 3.13   Channels 1 and 2 show transmit and receiver activity for one node. Channels 3 and 4 show radio activity for a second node that receives a packet from the first node and transmits a response a few slots later. The small pulses represent RX checks that timed out. Longer pulses show packets of data being transmitted. The group of pulses toward the right side show the contention slots.

3.3.8   Lifetime

Two major factors that control node lifetime in sensor networks are the topology and event sampling rate. We have already shown how RT-Link allows for logical pruning of topology to conserve energy. We will now investigate the lifetime with respect to event sampling rate. A typical LPL-CSMA approach must balance long preamble transmit times with the

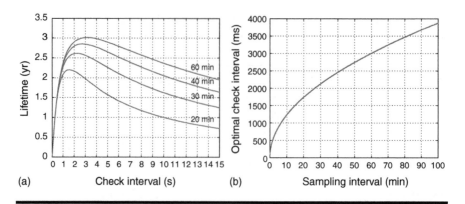

(a) Check interval (s) (b) Sampling interval (min)

Figure 3.14 **Effect of sample interval on LPL CSMA check rate. (a) LPL CSMA check rate versus battery life at 30 min sample interval. (b) Sampling interval versus optimal check rate.**

frequency of channel activity checks. As described in Ref. [4] we observe a curve similar to Figure 3.14(a) where at a given sampling rate, there is an optimal lifetime produced by a particular check rate. The authors in Ref. [4] have accidentally neglected to include the voltage when calculating power and hence the node battery lifetimes are exaggerated. We show the corrected graph using power values based on our hardware. The battery lifetime can be computed using Eq. (3.2).

$$E_{\text{idle}} = P_{\text{sleep}} * \left(T_s - \left(T_{\text{cca}} * \frac{T_s}{T_c} \right) \right) \tag{3.1}$$

$$L = C_{\text{bat}} \frac{\left(\frac{T_s}{T_c} * E_{\text{sample}} \right) + (T_c * P_{tx}) + E_{rx} + E_{\text{cpu}} + E_{\text{idle}}}{T_s * V * 24 * 365} \tag{3.2}$$

Table 3.4 describes the above values where L is the node lifetime in years. For a given sampling rate, checking the channel more or less frequently can be quite inefficient. In a multihop environment, this means that for a particular event rate of interest, the end-to-end latency is a function of the system check rate which must be fixed to achieve the optimal lifetime. This implies that without time synchronization, large sampling intervals will lead to longer latencies. Figure 3.14(b) shows the optimal check rate as a function of the sampling rate. This is determined by the zero-crossing of the derivative of Eq. (3.2) for every sampling rate. The dots represent the optimal check rate at the 30-min sampling rate obtained from the previous graph. We see that even as the event rate approaches 100 min, the check rate must still be less than 4 s to achieve the best lifetime. In this 100 min

Table 3.4   LPL-CSMA Parameters

Parameter	Symbol	Value
Sleep power	P_{sleep}	90 mW
Sample time	T_s	
Check interval	T_c	
Channel check time	T_{cca}	2.5 ms
Sample energy	E_{sample}	150 mJ
Battery capacity	C_{bat}	2500 mAh
Voltage	V	3.0

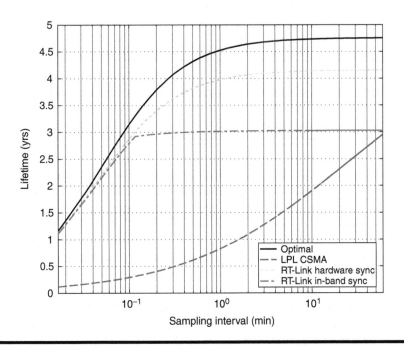

Figure 3.15   Sampling interval versus lifetime for both CSMA and TDMA.

interval, approximately 1500 checks would have been wasted with a single neighbor alone. Figure 3.15 shows the sampling rate with respect to lifetime for RT-link (with and without hardware time synchronization), the optimal node lifetime and the optimal LPL-CSMA lifetime. The overall optimal lifetime assumes perfect node synchronization meaning that the only energy to be consumed is the minimum number of perfectly coordinated packet transmit and receives, and the system idle energy. The LPL-CSMA line represents the lifetime, given the optimal check rate. We see that for fast sampling rates, hardware time synchronization makes less of a difference.

This is because synchronization can be achieved by timing the arrival of normal data messages that already contain slot information. As the sampling rate increases, extra messages must be sent to maintain in-band time synchronization. We see that across the range of a few seconds to nearly 2 h, RT-Link with hardware synchronization is quite close to the optimal lifetime and outperforms the LPL-CSMA mac protocol by a significant margin.

3.3.9 End-to-End Latency

To investigate the performance of RT-Link, we simulated its operation to compare the end-to-end latency with asynchronous and loosely synchronized protocols across various topologies. To study the latency in a multi-hop scenario we focused on the impact of the hidden terminal problem on the performance of B-MAC and S-MAC. All the tests in Ref. [4] were designed to avoid the hidden terminal problem and essentially focused on extremely low-load and one-hop scenarios. We simulated a network topology of two "backbone" nodes connected to a gateway. One or more leaf nodes were connected to the lower backbone node as shown in Figure 3.16. Only the leaf nodes generated traffic to the gateway. The total traffic issued by all nodes was fixed to 1000 1-byte packets. At each hop, if a node received multiple packets before its next transmission, it was able to aggregate them up to 100-byte fragments. The tested topology is the base case for the hidden terminal problem as the transmission opportunity of the backbone nodes is directly affected by the degree of the lower backbone node.

We compare the performance of RT-Link with a 100- and 300-ms cycle duration with RTS–CTS enabled B-MAC operating with 25 and 100 ms check times. The RTS–CTS capability was implemented as outlined in Ref. [4]. When a node wakes up and detects the channel to be clear, RTS and CTS with long preambles are exchanged followed by a data packet with a short preamble. We assume B-MAC is capable of perfect clear channel assessment, zero packet loss transmissions, and zero cost acknowledgement of packet reception. We observed that as the node degree increases (Figure 3.17), B-MAC suffers a linear increase in collisions, leading to an

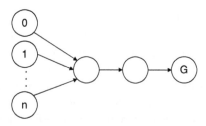

Figure 3.16 Multihop network topology with hidden terminal problem.

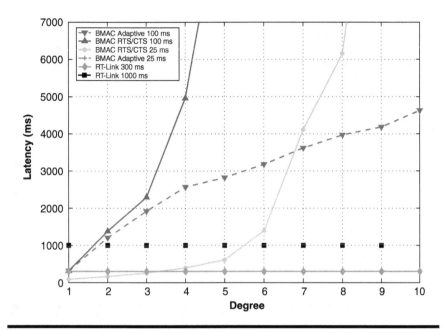

Figure 3.17 Impact of latency with node degree.

exponential increase in latency. With a check time of 100 ms, B-MAC saturates at a degree of 4. Increasing the check time to 25 ms, pushes the saturation point up to a degree of 8. Using the schedule generated by the heuristic in Section 3.5, RT-Link demonstrates a flat end-to-end latency.

The clear drawback to a basic B-MAC with RTS–CTS is that upon hidden terminal collisions, the nodes immediately retry after a small random backoff. To alleviate this problem, we provided nodes with topology information such that a node's contention window size is proportionate to the product of the degree and the time to transmit a packet. As can be seen in Figure 3.18, this allows for a relatively constant number of collisions since each node shares the channel more efficiently. This extra backoff, in turn increases latency linearly with the node degree. We see that RT-Link suffers zero collisions and maintains a constant latency.

3.4 Nano-RK: A Resource-Centric RTOS for Sensor Networks

The push provided by the scaling of technology and the need to support increasingly complicated and diverse applications has resulted in the need for traditional multitasking OS abstractions and programming paradigms. The case for small-footprint real-time OS support in sensor networks is

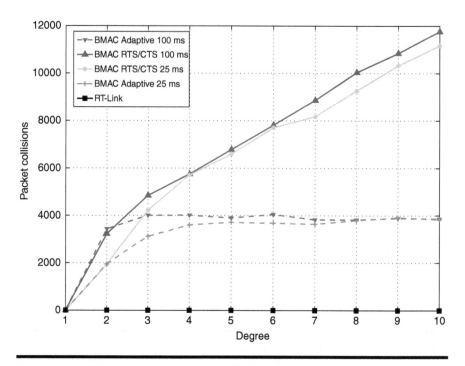

Figure 3.18 Effect of node degree on collisions for B-MAC.

strengthened by the fact that many sensor networking applications are time-sensitive in nature, i.e., the data must be delivered from the source to the destination within a timing constraint. For example, in a surveillance application, data relayed by a task which is responsible for detecting intruders and subsequently alerting the gateway nodes of the system should be able to reach the gateway on a timely basis. In this section, we present Nano-RK, a small-footprint embedded real-time OS with networking support.

Nano-RK supports the classical OS multitasking abstractions allowing sensor application developers to work in a familiar paradigm resulting in small learning curves, quicker application development times and improved productivity. We show that an efficient implementation of such a paradigm is practical. We associate tasks with priorities and support priority-based preemption, i.e., a task can always be preempted by a higher-priority task that becomes eligible to run. For timing-sensitive applications, we use priority-based preemptive scheduling to implement the rate-monotonic paradigm[20] of real-time scheduling so that a periodic sensor task set with timing deadlines can be scheduled such that their timing guarantees are honored. Since modern sensor networks use ad hoc multihop wireless networking for packet relaying, we provide port-based socket abstractions that can be used by sensing tasks for sending and receiving data.

Since sensor nodes are resource- and energy-constrained, we provide functionality, whereby the OS can enforce limits on the resource usage of individual applications and on the energy budget used by individual applications and the system as a whole. In particular, we implement CPU reservations and network bandwidth reservations wherein dedicated access of individual application to system resources is guaranteed by the OS. The OS also implements sensor reservations to enforce usage on the number of accesses to individual sensors. Since the energy used by each task is the sum total of energy consumed by the CPU, the radio interface, and the individual sensors, a particular setting for each of these leads to an energy reservation. Since we use a static design-time approach for admission control, we provide tools for estimating the energy budget of each application and (hence) the system lifetime. The CPU, network, and sensor reservation values of tasks can be iteratively modified by the system designer until the lifetime requirements of the node are satisfied.

3.4.1 Current Sensor Network Operating Systems

Infrastructural software support for sensor networks was introduced by Hill et al.[21] They proposed TinyOS, a low footprint OS that supports modularity and concurrency using an event-driven approach. TinyOS supports a cyclic-executive model, wherein interrupts can register events, which can then be acted upon by other nonblocking functions. We believe that there are several drawbacks to this approach. The TinyOS design paradigm is a significant departure from the traditional programming paradigm involving threads, making it less intuitive for application developers. In contrast, we support a traditional multitasking paradigm retaining task abstractions and multitasking. Unlike TinyOS, where tasks cannot be interrupted, we support priority-based preemption. Nano-RK provides timeliness guarantees for tasks with real-time requirements. We provide task management, task synchronization, and high-level networking primitives for the developer's use. While our footprint size and RAM requirements are larger than that of TinyOS, our requirements are consistent with current embedded microcontrollers. Sensor network hardware typically has ROM requirements of 32–64 kB and RAM requirements of 4–8 kB. Nano-RK is optimized primarily for memory and secondarily for ROM. SOS[22] is architecturally similar to TinyOS with the additional capability for loading dynamic run-time modules. In contrast to SOS, we propose a static, multitasking paradigm with timeliness and resource reservation support.

The Mantis OS[23] is the most closely related work to ours in the existing literature. In comparison to Mantis, we provide explicit support for periodic task scheduling that naturally captures the duty cycles of multiple sensor tasks. We support real-time task sets that have deadlines associated with their data delivery. We use the novel mechanisms of CPU and network reservations to enforce limits on the resource usage of individual tasks. With respect to networking we provide a rich API set for socket-like abstractions, and a generic system support for network scheduling and routing. Nano-RK provides a host of built-in interfaces for extensive power management. Several power management interfaces are used by the system with some APIs available for use (with prudence) by applications.

While low-footprint OSs such as uCOS and Emeralds[24] support real-time scheduling, they do not have support for wireless networking. Our networking stack is significantly smaller in terms of footprint as compared to existing implementation of wireless protocols like Zigbee (around 25 kB ROM and 1.5 kB RAM) and Bluetooth (around 50 kB ROM). We also provide high-level socket-type abstractions, and hooks for users to develop or modify custom MAC protocols.

Our system infrastructure can be used to complement distributed sensor applications such as an energy-efficient surveillance system.[25,26] Our contributions are orthogonal to the literature on real-time networking/resource allocation protocols,[27,28] energy-efficient routing/scheduling schemes,[5,29] data aggregation schemes,[30] energy-efficient topology control,[31] and localization schemes.[32,33] Our OS can be used as a software platform for building higher-layer middleware abstractions like in Ref. [34]. Our energy reservation mechanism can be used to prevent the type of energy DoS attacks described in Ref. [35].

Finally, our work complements Ref. [36] in extending the resource kernel paradigm to energy-limited resource-constrained environments like sensor networks.

3.4.2 Design Goals for a Sensor Networking RTOS

We present the following design goals for an RTOS targeting WSNs:

- Multitasking: The OS should provide a simple and intuitive programming paradigm for easy use by application developers. It is desirable to retain the traditional multitasking paradigm familiar to both desktop and embedded system programmers. Application developers should be able to concentrate on application logic rather than low-level system issues such as scheduling and networking.
- Networking Stack Support: The OS should support multihop networking, routing, and simple user-level networking abstractions similar to sockets. In particular, low-level networking details such as reliable packet transmission, multicasting, queue management, etc. should be handled by the OS.
- Support for Priority-Based Preemption: Node battery lifetime continues to be a major challenge in sensor networks. Hence, given that

energy consumed by processing per bit is significantly less than the per-bit energy consumed by the radio interface, there is a trend toward increased local processing (such as embedded vision and sound processing). This typically results in increased task execution times. In such situations where task run-times are large, there is a need for priority-based preemption to give precedence to higher priority events.

True preemptive multitasking becomes necessary in a system where multiple inputs to the system must be serviced at different rates within a required period. For instance, imagine a sensing platform consisting of a microphone, light sensor, radio interface, a GPS for position information or time synchronization, and a smart camera system. Table 3.5 shows typical period and execution times for each of these devices. Manually scheduling such a task set can become daunting using timer interrupts.

A nonpreemptive scheme might handle the radio with an external interrupt, the light and microphone with two priority-based timers, and leave the GPS and camera processing for the main program loop. Even in this situation, the developer may encounter difficulties because the camera servicing time is longer than the period of the GPS. Given that many low-end microcontrollers have limited timer interrupts, it can become difficult to schedule such a task set. Developers may need to resort to manual time splicing of their functions, thus making future modifications difficult. With a preemptive priority-based system, each of these sensing functions would be supported by a prioritized periodic task.

■ Timeliness and Schedulability: Most sensor applications such as surveillance tend to be time-sensitive in nature where packets must be relayed and forwarded on a timely basis. While routing and network link scheduling are important components in ensuring that packets meet their end-to-end delay bounds, timing support

Table 3.5 A Sample Task Set with Update Periods and Execution Times

	Period	Execution Time
Radio	Sporadic	10 ms
Microphone	200 Hz	10 μs
Light sensor	166 Hz	10 ms
Smart camera	1 Hz	300 ms
GPS	5 Hz	10 ms

on each node in the network is also essential. To honor end-to-end deadlines, local tasks on each node have deadlines associated with the completion of their local data relaying and processing. Managing the deadlines of these tasks requires support of a *real-time* OS.

■ Battery Lifetime Requirements: Guaranteeing sensor node battery lifetimes of 3 to 5 years continues to be a major challenge in sensor networks. If limits on the usage of energy can be enforced, lifetime guarantee requirements of the system as a whole can likely be provided (under reasonable assumptions about operating conditions such as network connectivity). The OS can also ensure that the system energy is apportioned in a manner commensurate to the importance of the tasks so that critical tasks are guaranteed their energy budget.

■ Enforcement of Resource Usage Limits: Since sensor nodes are resource-constrained, precious CPU cycles, network buffers, and bandwidth should be apportioned to application needs. OS support for guaranteed, timely, and limited access to system resources is necessary for supporting application deadlines and balanced apportioning of system slack (residual unused resources). This mechanism can also be used to place some limits on the impact of faulty or malicious tasks on system operation.

■ Unified Sensor Interface Abstraction: Providing a unified and simple abstraction for accessing sensor readings and actuating responses would greatly benefit the end user. In particular, low-level details associated with sensor/actuator configurations should be abstracted away from the user. Sensors should be supported using device drivers that can return real-world units as well as raw ADC values.

■ Small Footprint: The current trend of low-end embedded processors is toward larger ROM sizes (64 to 128 kB) and smaller RAM sizes (2 to 8 kB). The OS architecture should be compliant with this trend by optimizing for RAM with a higher priority than ROM and optimizing for run-time efficiency. This memory constraint also implies that when the choice exists, one prefers a static configuration to a dynamic decision that requires additional data storage and run-time manipulations.

3.4.3   The Architecture of Nano-RK

In this section, we describe the architecture of Nano-RK, its constructs and capabilities. The overall system architecture of Nano-RK is shown in Figure 3.19.

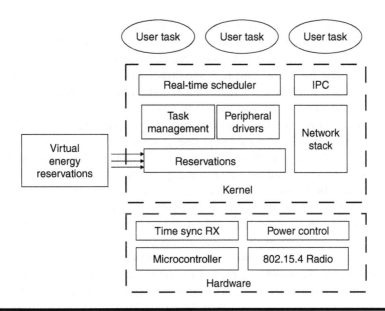

Figure 3.19 Nano-RK architecture diagram.

3.4.3.1 Static Approach

Given the stringent memory constraints of embedded sensor OSs, Nano-RK uses a static design-time framework. Specifically, tasks under nano-RK are *not* spawned dynamically, and applications are co-located with the OS in a single address space. In addition, admission control and real-time schedulability analyses are carried out offline. We emphasize that a static approach does *not* necessarily mean that task properties and configuration parameters cannot be reconfigured during run-time. Data-(or control-) dependent modifications to the task code such as changing task periods, resource usage limits, resource priorities, and configuration of various parameters such as the network buffer sizes and stack sizes of each task can be changed to accommodate mode changes. With current energy and memory constraints, these run-time configurations will need to be verified offline at design-time.* This results in a lightweight real-time multitasking OS with a small memory footprint while retaining the rich set of functionality found in conventional RTOSs.

3.4.3.2 The Reservation Paradigm

The reservation paradigm, as implemented in a *resource kernel,*[36] is a simple practical paradigm for guaranteeing timeliness and enforcing temporal

---

* Future revisions of Nano-RK may relax this constraint.

isolation in real-time OSs. A resource kernel allows applications to specify and obtain guarantees on their resource and timeliness requirements. Once a guaranteed reservation is obtained after appropriate admission control, the kernel schedules and enforces guaranteed access to system resources such that each application's reservation parameters are honored. In the past, the resource kernel abstraction has been used only in dynamic run-time settings. However, the resource reservation paradigm, particularly its enforcement aspect, is desirable for use in static settings as well. A sensor application task can specify its requirement of CPU cycles, network band-width, and network buffers over fixed periods, which will be enforced by the Nano-RK kernel. Only tasks that have not depleted their reserva-tion quota rates are eligible for scheduling. In deference to the stringent constraints of sensor nodes, exactly a single task can be associated with a reservation. In contrast, the classical resource kernel concept permitted zero, one or more tasks to be bound to a reservation.

In summary, Nano-RK supports CPU reservations, sender/receiver net-work bandwidth reservations and sensor/actuator reservations. All of these reservations can be combined to enforce a virtual node-wide (and even system-wide) energy reservation.

3.4.3.3 Energy Management Support

Since maximizing battery lifetime continues to be a major objective (and challenge) in sensor networks, there is a need for aggressive power savings, which is obtained in practice by operating at low duty cycles. Nano-RK enforces this in the form of *virtual energy reservations*, by leveraging the fact that the energy consumed by a task is the total sum of the CPU energy, radio interface energy, and the energy associated with the use of sensors and actuators. The CPU and radio energy consumed by a task can be controlled by changing the CPU and network reservation sizes. To bind the energy consumed by sensors, Nano-RK provides *sensor reservations*, by which a sensor/actuator can be used only for a finite (specified) number of times over successive periodic intervals. For example, a sensor reservation may impose a maximum of two accesses within a period of 100 ms. Our unified sensor interface provides functionality wherein sensors are turned off (gated) by default and any access to a sensor is an atomic operation that consists of the sensor being turned on, its value being read, and then being turned off again. This makes it possible for the OS to set an upper limit on the number of accesses made to a sensor over a particular period.

As a consequence, it is possible to map a resource tuple of (*CPU, net-work, sensor*) reservations to a particular level of energy consumption. Given periodic tasks, one can calculate the mean power used by all tasks over a hyperperiod, giving a reasonably accurate estimate of the node life-time. By modifying the values of the (*CPU, network, sensor*) reservation tuple, the mean energy consumed by each task can be controlled. This can be used for either controlling the battery lifetime of a node or for modifying the proportion of system energy alloted to each task (e.g., certain mission-critical tasks can be allocated a higher energy budget). We again note that energy reservations are implemented by controlling the (*CPU, network, sensor*) reservation tuple at a predeployment stage. This is consistent with the predominantly static approach that Nano-RK adopts.

Nano-RK employs a novel energy-efficient time-management scheme using one-shot timer interrupts instead of a polling interrupt. Traditionally, an OS will periodically call the scheduler to check if a context swap should occur and to update the time-of-day (TOD) counter. This consumes excess energy especially during long periods of sleep. Nano-RK operates the main hardware timer in a one-shot mode wherein the next timing interrupt is triggered when a time-out has occurred or a scheduled event occurs. This is inherently more energy-efficient than schemes that needlessly call the scheduler on a periodic basis. There are two conflicting requirements at this point. One, applications may need a high granularity of timers and time-of-day clocks when they are in an active mode of operation. Conversely, when the system is in mostly sleep mode, the processor and the node should not be awakened frequently just to update timers and the time-of-day clock. This conflict is resolved by Nano-RK using hardware timers that internally count at a high clock rate (several MHz when available) while the rest of the node can be asleep. The OS or application tasks can read the fine-grained internal timer value to obtain high-granularity timing and time-of-day information. This internal timer hardware counter is set to accumulate relatively large values (of the order of minutes or more) and a hardware interrupt is generated when this accumulator overflows. In the case of extended idle intervals that last several tens of minutes, or even hours or days, this timer overflow maintenance has an associated energy penalty (which is significantly less than the penalty that would be incurred if the periodic OS timer wakes up the CPU at MHz rates).

Figure 3.20 shows how timer granularity affects extended idle periods. The top curve shows the average power consumption when the OS periodically calls the scheduler. The bottom curve shows Nano-RK's power consumption using our one-shot timer method. The arrow marker shows Nano-RK's default operating point, which allows for better than 1 ms timer resolution for use by tasks.

Nano-RK's representation for time is based on the portable operating system interface (POSIX) time structure timeval. This consists of two 32-bit numbers to represent the (`seconds, nanoseconds`) fields. The OS TOD counterfield is incremented as needed, and overflows will not occur for practically foreseeable intervals of time. This allows Nano-RK to support the fine-granularity timing requirements of real-time applications, while maintaining a (practically) nonoverflowing notion of absolute time.

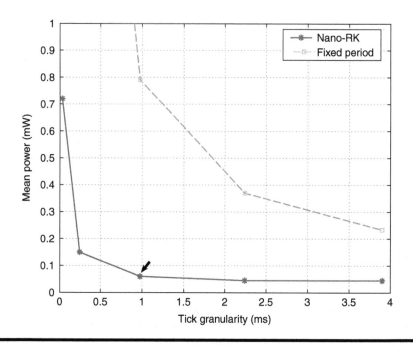

Figure 3.20 **Power consumption versus timer granularity.**

Table 3.6 **Breakdown of Nano-RK's Resource Requirements**

Component	Resource
Context swap time	45 μs
Mutex structure overhead	5 bytes per resource
Task stack overhead	32–128 (typically 64) bytes per task
Task overhead	30 bytes per task
Link layer overhead	200 bytes
2 tasks, 2 mutexes,	
2 128-byte network buffers	1 kB RAM, 16 kB ROM

3.4.3.4 Task Management and Scheduling

Nano-RK task control block (TCB) structures are populated during initialization.[†] They store the register context of all task (registers and stack), the task's priority, period of recurrence, (*CPU, network, sensor*) reservation sizes, port identifiers, etc (Table 3.6). Two linked lists of TCB pointers are

---

[†] While we provide API support to modify fields in the TCB during run-time, we encourage static configuration for footprint reduction of the ROM image.

used to order the set of active and suspended tasks respectively, based on period of recurrence. Tasks can block on certain events (such as being woken up at a certain point of time or the arrival of a network packet) and can be unsuspended and enqueued in the OS active list when the events occur. We suspend tasks that have pending events rather than using a polling-based implementation of Nano-RK system calls. This is done for energy-efficiency reasons because if there are no tasks eligible to run, the system can be powered down to sleep.

Our system uses priority-based preemptive scheduling, and while we provide explicit support for periodic tasks, we also support aperiodic and sporadic tasks in our framework. The highest priority task that is eligible to run in the system is always scheduled by the OS. A periodic task can suspend itself after the completion of its current instance using the `wait_until_next_period` () system call.

We implement the Priority Ceiling Protocol Emulation (also known as the "Highest Locker Priority") protocol to bound the duration of priority inversion encountered by a higher-priority process due to the phenomenon of priority inversion (wherein a shared resource needed by the higher-priority process is currently being used by a lower-priority process). In particular, each mutex is associated with a priority ceiling, defined to be the priority of the highest priority task that will ever access the mutex. When a mutex is acquired (using `lock_mutex` ()), the priority of the task is elevated to the priority ceiling of the mutex. Once the mutex is released (using the `unlock_mutex` () system call), the priority of the task reverts to its original level. This results in bounded priority inversion which can be accounted for in the offline schedulability test. Thus, real-time synchronization is supported in Nano-RK.

Instead of providing explicit message box support, we provide system support for conventional semaphores that can be used by tasks to manipulate application buffers in a controlled manner for facilitating interprocess communication. This obviates the necessity for OS buffer space for storing message data and allows efficient *zero copying*[37] mechanisms to facilitate information sharing among tasks.

3.4.3.5 Nano-RK Integration with RT-Link

An effective approach to energy-efficient communication is to operate all nodes at low communication duty cycles so as to maximize the shutdown intervals between packet exchanges. The two fundamental challenges in delivering delay-bounded service in such networks are (a) coordinating transmissions so that all active nodes communicate in a tightly synchronized manner and (b) ensuring that all transmissions are collision-free. In Nano-RK, we exploit time synchronization to tightly pack the activity of all nodes so that they may maximize a common sleep interval between activities. Furthermore, in the absence of dropped packets, it provides guarantees on timeliness, throughput, and network lifetime for end-to-end communications.

Using global time synchronization, Nano-RK provides a TDMA-based link layer protocol that schedules communication based on the global topology. This requires a topology-gathering phase followed by a scheduling phase. Once the network has been scheduled, communication transactions occur during assigned slots that fit within a periodic set of frames. Thus, a TDMA cycle consists of a finite set of frames each of which repeat a pattern of active and inactive slots.

One challenge associated with this scheme is ensuring that communication still fits within Nano-RK's real-time framework. We model the communication in our system in such a way that it fits within the rate-monotonic scheduling paradigm.[20] An individual slot inside each TDMA frame is represented as a periodic task with an execution time of a single slot size and a period equal to the frame interval. The link layer schedule can then be considered to be an aggregate of these tasks with a single common period. The behavior of this aggregate task corresponds to that of a classical periodic task with the worst-case execution time of the sum of its active slots, and a period equal to the frame size (except that the execution is split into multiple segments that are separated from one another by exactly one period). Nano-RK implements the link layer schedule as a single *compound task* running at the highest priority, with the reservation parameters of the aggregate link layer schedule. In addition, the standard period-based reserve at the individual task level still exists and enforces that an application task transmits for no more than its allocated time. In summary, the TDMA communication is modeled as a single compound task that aggregates multiple fixed period tasks, but in actuality, it executes as a single task with all of the transmission slots composed together.

3.5 Coal Mine Safety Application

Over the past decade, there has been a surge of accidents in coal mines across the world. In many of these cases, miners are trapped several thousand feet below the surface for several hours or even days, while rescuers try to find their location and attempt to communicate with them. As a tragic example, in January 2006, 13 miners were trapped for nearly two days in the Sago Coal Mine in West Virginia, USA. The miners were less than few hundred feet from an escape route but were not aware of it. Similarly, in February 2006, in the Pasta de Conchos coal mine in Mexico, 65 miners were trapped more than 1 mile below the ground level. After more than 100 h of rescue attempts, the authorities were still unable to locate or establish communication with the miners. In both cases, the prevalent wired communication systems were destroyed when a portion of the mine collapsed and there was no way to reestablish connection to the affected areas.

The normal practice to check the status of the trapped miners is to drill a narrow hole (of 1–2 inch radius) from the surface to a mine tunnel and drop a microphone, camera, and air quality sensors at different locations around the disaster area (Figure 3.21). This method provides limited access to the affected region as medium-sized mines may span several miles across. Another method of communicating to the miners is by installing a loop antenna that is several miles long, over the surface of the mine. This scheme uses a low-frequency transmitter on the surface to send one-way broadcasts of short text messages, but is unable to get feedback about the status or location from the miners below.

We propose the establishment of a self-healing wireless network in such mine-like environments to maintain communication in the remaining connected network. If a wireless node was lowered through the drill-hole, it could re-establish communication with the remaining network and initiate two-way communication with the miners. In addition, the miners would be able to leave broadcast voicemail-type messages and allow it to propagate to all nodes in the remaining network.

We were invited to investigate the use of wireless sensor nodes to track miners and to evaluate the viability as an end-to-end rescue communication network for miners during an incident. It is important to note that during normal operation, the network's primary task is to track miners and record

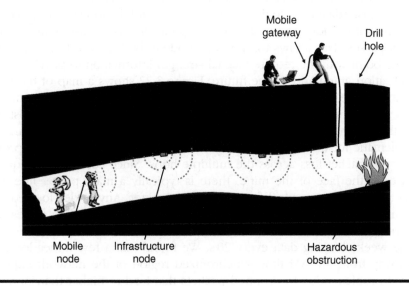

Figure 3.21 Rescue sensor network in coal mine.

environmental-quality data. To keep normal network maintenance costs low, it is necessary to meet the following design goals:

1. All nodes are to be battery-powered.
2. Nodes must have a predictable lifetime of 1 year or more under normal operation.
3. Nodes must provide continuous two-way voice communication for several days with fully charged batteries.
4. Voice communication must include two-way interactive calling, one-way "push-to-talk" voice messaging and support for store-and-broadcast voicemail messaging.
5. The network must be able to tolerate topology changes and self-heal to maintain connectivity after a network partition.

We performed two sets of experiments at the National Institute for Occupational Safety and Health (NIOSH) experimental coal mine facility. The goal of the first deployment was to track the location of mobile nodes carried by miners as well as monitor an assortment of sensors over a period of 3 weeks. The second deployment was to test an emergency mode of operation where interactive two-way voice is streamed over the network.

3.5.1 Location Tracking

For our location tracking tests, we deployed 42 nodes across almost 2 miles of underground corridors. Every 20 s each infrastructure node would send a neighbor list with signal strength values for each link as well as the most recent sensor data back to the gateway. As the mobile node moves through the mine, it is added to the different neighbor lists of infrastructure nodes that it passes. This allows for a mobile node to be localized to the nearest access point with the required signal strength information to provide finer localization granularity in the future. Figure 3.22 shows a map of the coal mine with the overlaid network topology. We see that due to the remaining coal pillars, the degree of the network graph is quite low (at most 5), but the depth is quite large (over 15 hops). Long, linear chains can be problematic for in-band time-synchronization due to the increasing probability of packet loss across the multiple hops. Since coal miners require power at the face of the mine, there is typically a main power line fed into the mine that is ideal for our AM transmitter. Any nodes located on the main corridor can use the AM time synchronization while nodes on the periphery can use in-band time synchronization. We left the nodes for three weeks logging data every 20 s. We found that a few nodes located far away from the AM time synchronized region of the network experienced problems due to dropped packets that lead to higher than normal power consumption. During the network setup, we saw that all nodes had

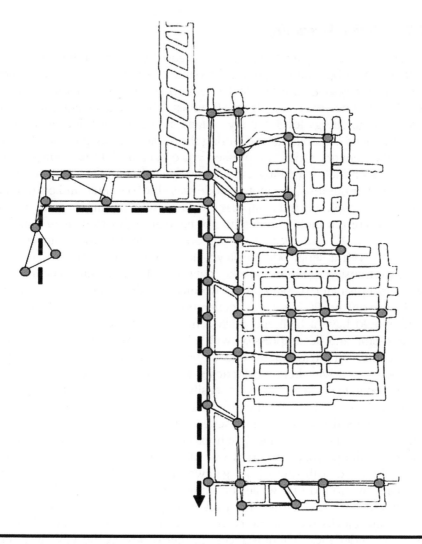

Figure 3.22   Coal mine map with network topology. Dotted line shows leaky feeder time-synchronization cable.

reliable links. Narrow passageways, miners, and machinery increase packet loss by blocking line-of-sight communication. We have learned two valuable lessons from this deployment. First, even in controlled environments link quality can change due to motion in the environment and unforeseen perturbations over time however the topology will return to a steady state. Second, as hop length increases, reliability decreases, which causes time synchronization degradation and increased energy consumption in the form of extended synchronization wait times. This indicates that we should further explore how to address link faults in an energy-efficient manner.

3.5.2 Voice Streaming

In our second deployment, we wanted to add the ability for the network to switch into a high rate of operation capable of streaming interactive voice from a gateway to a mobile node. Under normal circumstances, sensor data is collected once every 20 s from all nodes. This includes light, temperature, average energy level, battery voltage, and the SNR values associated with any nearby mobile nodes. During the audio streaming mode of operation, the network sends compressed voice data at 13 kBps. Our primary focus was on the networking aspects and evaluating the feasibility of such a system. For our tests, the mobile node was able to sample audio from the onboard microphone and compress the data while running the networking task. Our current mobile nodes do not have an onboard digital-to-analog converter (DAC) and speaker output, so we used a laptop connected to the node with a serial port to playback the received audio. To simplify tests, we transferred the raw packet data over the universal asynchronous reciever/transmitter (UART) and performed the decompression and playback live on the PC. In a next-generation device, the handheld mobile node would have a high-end microcontroller capable of doing both the compression and decompression with a built-in DAC and speaker system.

In controlled environments outside of the mine, we found that the system performed with below 3% packet loss per hop. Sending redundant data in separate packets allowed for easily understandable end-to-end voice transfers. Figure 3.23 shows the distribution of packet loss clustering at four different hops along an eight-hop series of nodes inside the coal mine. The end-to-end latency across the eight hops between when audio was sampled and when the playback occurred was just under 200 ms. Each hop along a prescheduled path toward the gateway maintained an average latency of 24 ms. We found that while the mine corridor is clear of obstructions the wireless channel shows few packet drops. In some situations when a machine blocks the narrow corridor we see packet loss rates as high as 50%. Under these circumstances, packet drops are heavily clustering making error concealment or recovery difficult. Since occupancy inside a coal mine is relatively sparse (usually less than five groups) compared to the mine's size, clear paths are quite common. Future work will investigate protocols that use the mesh nature of sensor networks to ameliorate broken links by using redundant paths.

3.6 Summary and Concluding Remarks

We presented in this chapter a detailed description of the FireFly WSN platform, developed by the Real-Time and Multimedia Systems Laboratory at Carnegie Mellon University. This platform consists of extensible FireFly sensor nodes, a time-driven link layer multihop networking protocol called

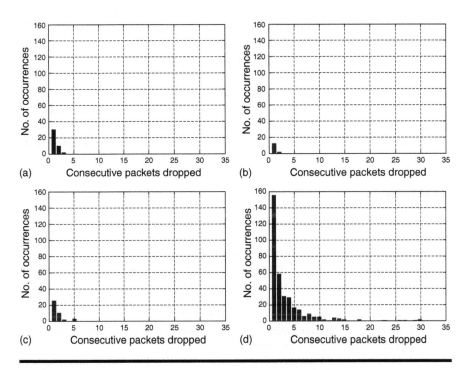

Figure 3.23 Packet loss clustering at four points in a multihop chain of nodes streaming audio. (a) 1.5% loss. (b) 0.04% loss. (c) 2.1% loss. (d) 52.3% loss.

RT-Link, and a multitasking power-aware real-time OS called nano-RK. The primary objectives of FireFly are to support large-scale ad hoc wireless sensor nodes and long battery life at each node.

Each FireFly sensor node has a low-power 802.15.4 radio and built-in support for hardware-assisted time synchronization. Tight time synchronization of a few microseconds is obtained by each node receiving periodic timing pulses from an AM/FM beacon indoors or the atomic clock outdoors. Software time synchronization can also be used in lieu of out-of-band synchronization, if desired. Using the common time reference, nodes can be preassigned slots during which they can transmit or receive packets such that collisions are guaranteed not to occur. A time-driven link layer protocol named RT-Link utilizes this approach across all nodes to support a repeatable frame-based schedule. The schedule is comprised of scheduled slots (for preallocated transmission and reception of packets), and contention slots that allow new nodes to use channel contention to become members of the network. Mobile nodes are also similarly supported. The nano-RK real-time sensor OS supports a classical multitasking paradigm along with novel support for the resource kernel paradigm using enforced resource reservation on CPU cycles, network packets, and sensor accesses. A real-time synchronization scheme for minimizing priority inversion is also included. Multiple power management techniques to facilitate low duty cycle operation are used to extend battery life.

We also described our experiences with deploying a 42-node WSN within a coal mine that can be used to track the locations of miners in real time, and for two-way continuous voice streaming in case of emergency situations. Our future work will extend self-configuring and self-healing capabilities of the FireFly platform. We also plan to deploy large-scale versions of the FireFly network for other applications of practical interest.

References

1. Atmel Corporation, atmega32 data sheet, 2005.
2. Chipcon Inc., chipcon cc2420 data sheet, 2003.
3. Radio Systems 30w tr-6000 am transmitter data sheet, 2001.
4. J. Polastre, J. Hill, and D. Culler. Versatile low power media access for wireless sensor networks. *SenSys*, November 2005.
5. W. Ye, J. Heidemann, and D. Estrin. An energy-efficient mac protocol for wireless sensor networks. *INFOCOM*, June 2002.
6. T. Dam and K. Langendoen. An adaptive energy-efficient mac protocol for wireless sensor networks. *SenSys*, November 2003.
7. A. El-Hoiydi and J. Decotignie. Wisemac: An ultra low power mac protocol for the downlink of infrastructure wireless sensor networks. *ISCC*, 2004.
8. V. Rajendran, K. Obraczka, and J.J. Garcia-Luna-Aceves. Energy-efficient, collision-free medium access control for wireless sensor networks. *Sensys*, 2003.
9. L.F.W. van Hoesel and P.J.M. Havinga. A lightweight medium access protocol for wireless sensor networks. *INSS*, 2004.
10. L.G. Roberts. Aloha packet system with and without slots and capture. *SIGCOMM*, 5(2):28–42, 1975.
11. C. Schurgers, V. Tsiatsis, S. Ganeriwal, and M. Srivastava. Topology management for sensor networks: Exploiting latency and density. *MobiHoc*, 2002.
12. C. Guo, L.C. Zhong, and J. Rabaey. Low power distributed mac for ad hoc sensor radio networks. *Globecom*, 2001.
13. J. Elson, L. Girod, and D. Estrin. Fine-grained network time synchronization using reference broadcast. *USENIX OSDI*, 2002.
14. S. Ganeriwal, R. Kumar, and M.B. Srivastava. Timing-sync protocol for sensor networks. *Proceedings of the ACM Sensys*, 2003.
15. M. Maroti, B. Kusy, G. Simon, and A. Ledeczi. The flooding time synchronization protocol. *Proceedings of the ACM Sensys*, 2004.
16. J. Zhao and R. Govindan. Understanding packet delivery performance in dense wireless sensor networks. *Proceedings of the ACM Sensys*, 2003.
17. A. Rowe, R. Mangharam, and R. Rajkumar. RT-Link: A time-synchronized link protocol for energy-constrained multi-hop wireless networks. *CMU Tech Report TR05-08*, 2005.

18. R.D. Nelson and L. Kleinrock. Maximum probability of successful transmission in a random planar packet radio network. *INFOCOM*, pp. 365–370, 1983.

19. H. Balakrishnan, C.L. Barrett, V.S.A. Kumar, M.V. Marathe, and S. Thite. The distance-2 matching problem and its relationship to the mac-layer capacity of ad hoc wireless networks. *IEEE Journal on Selected Areas in Communications*, 22(6):1069–1079, 2004.

20. C.L. Liu and J.W. Layland. Scheduling algorithms for multiprogramming in a hard real-time environment. *Journal of ACM*, 20(1):46–61, 1973.

21. A. Woo, S. Hollar, D. Culler, K. Pister, J. Hill, and R. Szewczyk. System architecture directions for network sensors. *ASPLOS 2000*, November 2000.

22. R.S. Shea, E. Kohler, M.B. Srivastava, S. Han, and R. Rengaswamy. Sos: A dynamic operating system for sensor nodes. *Mobisys*, June 2005.

23. J. Carlson, H. Dai, J. Rose, A. Sheth, B. Shucker, J. Deng, R. Han, H. Abrach, and S. Bhatti. Mantis: System support for multimodal networks of in-situ sensors. *2nd ACM International Workshop on Wireless Sensor Networks and Applications (WSNA)*, 2003.

24. P. Pillai, K.M. Zuberi, and K.G. Sin. Emeralds: A small-memory real-time microkernel. *In Proceedings of the 17th ACM Symposium on Operating System Principles*, June 1999.

25. J.A. Stankovic, T. Abdelzaher, L. Luo, R. Stoleru, T. Yan, L. Gu, J. Hui, B. Krogh, T. He, and S. Krishnamurthy. Efficient surveillance system using wireless sensor networks. *MobiSys*, June 2004.

26. Y. Wang, M. Martonosi, L.-S. Peh, D. Rubenstein, P. Juang, and H. Oki, Energy-efficient computing for wildlife tracking: Design tradeoffs and early experiences with zebranet. *ACM ASPLOS*, 2002.

27. C. Lu, L. Sha, J. Hou, J.A. Stankovic, and T. Abdelzaher. Real-time communication and coordination in embedded sensor networks. *Proceedings of the IEEE*, 91(7), July 2003.

28. C.-S. Shih, S. Giannecchini, and M. Caccamo. Collaborative resource allocation in wireless sensor networks. *ECRTS*, pp. 35–44, 2004.

29. Wook Hyun Kwon Hyung Seok Kim, Tarek Abdelzaher.Minimum-energy asynchronous dissemination to mobile sinks in wireless sensor networks. *ACM SenSys*, November 2003.

30. R. Govindan, C. Intanagonwiwat, and D. Estrin. Directed diffusion: A scalable and robust communication paradigm for sensor networks. *Mobicom*, August 2000.

31. J. C. Hou, N. Li, and L. Sha. Design and analysis of a mst-based distributed topology control algorithm for wireless ad-hoc networks. *IEEE Transactions on Wireless Communications*, 4(3):1195–1207, May 2005.

32. B. Blum, J. Stankovic, T. Abdelzaher, T. He, and C. Huang. Range-free localization schemes for large scale sensor networks. *Mobicom*, September 2003.

33. J.C. Hou, W.-P. Chen, and L. Sha. Dynamic clustering for acoustic target tracking in wireless sensor networks. *Mobicom*, 3(3), July 2004.

34. T. Abdelzaher, B. Blum, Q. Cao, Y. Chen, D. Evans, J. George, S. George, L. Gu, T. He, S. Krishnamurthy, L. Luo, S. Son, J. Stankovic, R. Stoleru, and

A. Wood. Envirotrack: Towards an environmental computing paradigm for distributed sensor networks. *IEEE International Conference on Distributed Computing Systems*, March 2004.

35. A.D. Wood and J. Stankovic. Denial of service in sensor networks. *IEEE Computer*, 35(10):54–62, 2002.

36. A. Molano, R. Rajkumar, K. Juvva, and S. Oikawa. Resource kernels: A resource-centric approach to real-time and multimedia systems. *Proceedings of the SPIE/ACM Conference on Multimedia Computing and Networking.*, January 1998.

37. H. Keng and J. Chu. Zero-copy tcp in solaris. *USENIX*, 1996.

Chapter 4

Energy Conservation in Sensor and Sensor-Actuator Networks

Ivan Stojmenovic

4.1 Introduction

4.1.1 Homogeneous Sensor Networks

Recent technological advances have enabled the development of low-cost, low-power, and multifunctional sensor devices. Large collections of these devices organized in multihop sensor networks have the potential to revolutionize the way humans can "sense" the physical world. Owing to their theoretical challenges and myriads of practical applications, wireless sensor networks are emerging as one of the priority research and development areas. Applications of sensor networks are envisioned primarily for monitoring the environment (e.g., motion, target tracking, fire detection, chemicals, temperature, etc.) or as embedded systems (e.g., biomedical sensor engineering, and smart homes). Sensors may measure distance, direction, speed, humidity, wind speed, soil makeup, temperature, chemicals, light, vibrations, motion, seismic data, acoustic data, strain, torque, load, pressure, and so on.

Sensors provide radically new communication and networking paradigms. They have small size, low battery capacity, nonrenewable power supply, small processing power, limited buffer capacity (thus routing tables, if used at all, must be small), and a low-power radio, and lack unique identifiers. They can be static and placed to monitor an area, or mobile (e.g., when attached to robots, soldiers, or vehicles). These nodes are autonomous devices with integrated sensing, processing, and communication capabilities. Sensor networks consist of a large number of sensor nodes that collaborate together using wireless communications with an asymmetric many-to-one data transfer model—sensors will typically send their data to a specific node called the sink node or monitoring station, which collects the requested information. Nodes in a sensor network are generally densely deployed. Thousands of sensors may be placed, mostly at random, either very close to or inside a phenomenon to be studied. Once deployed, the sensors are expected to self-configure into an operational wireless network, and must work unattended. The limited energy budget at the individual sensor level implies that to ensure longevity, the transmission range of individual sensors should be restricted. In turn, this implies that wireless sensor networks must be multihop.

To provide monitoring capability, it is important to place or select sensors so that as much of the monitored area is covered as possible. The problem of deciding which sensors need to remain active so that a certain area is minimally but fully monitored by the active sensors is known as the sensor area coverage problem. A survey of sensor area coverage solutions is given in Ref. [1], and it will not be covered here. There are also a variety of relevant issues like object location, path exposure, data dissemination, gathering, fusion, and so on.

The sensors, in most cases, use position information in their decisions. The availability of position information was widely recognized in the research community as highly desirable for the proper operation of the sensor network; however, it is a nontrivial problem and the precision of the location information may impact the performance of communication protocols. There exists a variety of position determination (or localization) protocols,[2] with a variety of message complexities and position accuracies.

4.1.2 Heterogeneous Sensor Networks

Current research and implementation efforts are mostly oriented toward a traditional scenario with stationary sensors and a single static sink that collects information from sensors. However, latest research has considered generalized scenarios, such as multiple stationary sinks, single mobile sink, or multiple mobile sinks. Future sensor network systems will be more heterogeneous and radically distributed, potentially with millions of nodes. They will respond to multiple tasks, to multiple and potentially mobile sinks, and multiple sensor networks will be integrated into a single network. There are algorithmic challenges in the rapidly emerging field of future heterogeneous supernetworks where ad hoc and sensor wireless networks will be integrated into wired and/or wireless fixed infrastructure. Challenges of such wide-area sensor systems include scalability, robustness/manageability, actuation, and privacy. This chapter will consider this broader view of sensor networks in selecting particular protocols focus on. From a wider perspective, communication is not the only energy consumer. Mobile sensors, mobile sinks, and mobile actuators also spend energy for moving around the network, and there is a tradeoff between energy spent on mobility and energy spent for communication. Further, energy needs to be saved not only by sensors, but potentially by sinks and actuators. Some examples presented in this chapter also consider the energy consumption due to mobility, in addition to the reducing communication overhead.

Sensor-actuator networks are heterogeneous networks that comprise networked sensor and actuator nodes that communicate among each other using wireless links to perform distributed sensing and actuation tasks. Actuators (called also *actors*) are resource-rich, usually mobile, and are involved in taking decisions and performing appropriate actions. Such networks are expected to operate autonomously in unattended environments. Nonconventional algorithmic techniques more suited to the characteristics of sensor-actuator networks need to be employed to resolve the many complicating issues. An example of an algorithmic challenge is in sensor-area coverage. After sensors are randomly dropped over an area, there might be areas with many sensors, and an algorithm to decide which of them may sleep, and which of them should remain active, can be applied. However, there might be areas which are not covered at all or are insufficiently covered by sensors. Actuator nodes may be given the task of moving some of the sensors from one position to another so that coverage is improved, with small cost.

Protocols designed for sensor networks must consider energy conservation as the prime objective because of power limitations in tiny sensors and the need to promote system longevity. There are several possible ways in which sensor network longevity can be extended. Many researchers propose self-organization and clustering protocols that will arguably extend the useful lifetime of sensor networks. Sensor organization includes scheduling active and sleep activity periods. In the sensor area coverage problem, for example, a set of sensors is selected for monitoring fully a given area, while the rest of the sensors are allowed to sleep, to conserve energy. But the process of deciding active sensors for coverage is also energy-consuming. Similarly, the process of organizing into clusters is energy-consuming. What then is a general principle for energy conservation? We argue in this chapter that the main source of energy (in)efficiency comes from the amount of messages sent in any given protocol. Protocols requiring fewer messages are normally those that conserve more energy. Messages are roughly divided into data traffic and control traffic. Data traffic includes messages that are functional to the stated purpose of having sensors, such as reporting to a sink. Control traffic includes other messages involved in established protocols, and can also be referred to as the communication overhead. The question then is how to reduce this communication overhead. Reduction of communication overhead leads to energy savings in two ways: First, fewer messages spend less energy. Next, increased messaging leads to increased collision probability, with the need for retransmissions (thus more than linear increase in energy consumption) or to the detrimental performance of protocols.

Broadly speaking, communication overhead is reduced, and longevity is promoted by the use of *localized* protocols. The basic characteristics of sensor networks, including distributed operation and a highly dynamic topology, make it imperative to design protocols that are *localized* rather than *centralized*. In the next section, we will elaborate on the notion of localized algorithms.

4.2   Localized Algorithms Save Energy

Because of the dynamic nature of ad hoc networks, topology changes are frequent and unpredictable. The main paradigm shift is to apply localized schemes as opposed to existing protocols requiring global information. The local information must suffice for a sensor node to make protocol decisions; otherwise, the increased communication overhead could offset the energy savings and increase latency. Localized algorithms are distributed algorithms where simple local node behavior achieves a desired global objective. Localized protocols provide *scalable* solutions, that is, solutions for wireless networks with an arbitrary number of nodes. Sensor networks with hundreds or thousands of nodes are envisioned.

The community does not have a unique understanding of what a localized algorithm is. We would like to provide here some further explanations, examples, and some further relevant criteria for the design of such protocols. Localized protocols offer the best prospect for achieving energy efficiency. In a localized protocol, each node makes protocol decisions based solely on some local knowledge available (to be more precise, based on the information from neighbors within k hops for certain small k, such as $k = 1$ or $k = 2$), without resorting to global network information. Localized protocols only require local information which, as a rule, is readily available to individual sensors by virtue of data collection, data fusion, and strictly local communication with immediate neighbors. Some (called beaconless) protocols do not even need the knowledge of neighbors. For example, in a beaconless routing protocol, a node currently holding a packet may send a request to all its neighbors asking them to forward it to the destination and a neighbor closest to the destination may respond first, offering help.

Centralized protocols require global information at each sensor. These protocols can perform well only for small networks. In a *centralized* (or *globalized*) algorithm, one or more nodes (or a central entity like a base station) need to learn the global node and/or edge structure. This can be either the whole graph (for instance, to find the route using the shortest path algorithm), or a global structure derived from the graph (such as a minimal spanning tree [MST]). Because of the huge communication overhead involved in gathering such information in dynamically changing ad hoc and/or sensor networks, globalized protocols cannot be energy-efficient solutions in dynamic networks with large numbers of nodes.

We further classify localized protocols according to the amount of information required and the overheads in the construction and maintenance phases. The amount of information required is related to the *message complexity*, which can be defined as the average number of transmitted messages per node in a protocol. Although some protocols appear to be localized, extensive message exchanges among neighbors amount to the collection and use of global information. In a *strictly localized protocol*, all information processed by a node is either local or global in nature, but obtainable in short (bounded by a constant) time by querying only the its own neighbors or itself. In other words, only a small constant number of message exchanges with one's neighbors is allowed. Strictly localized protocols may need some information that is part of their input (such as destination position in a routing protocol) but cannot use structures that are global in nature (e.g., information on which of the outgoing links belongs to the MST). Emergent algorithms are an interesting similar notion. An *emergent* algorithm is any computation that achieves formally or stochastically predictable global effects, by communicating directly with only a bounded number of immediate neighbors and without the use of central control or global visibility.

The definitions given so far still leave room for discussion and misinterpretation. For example, it is possible to construct an MST by a "localized" protocol which is a straightforward adaptation of a well-known centralized algorithm. One node can simply play the role of central processor, issue requests for necessary information, and make and distribute decisions. For instance, consider a protocol that involves building a spanning tree by doing a distributed breath-first search involving only local communications. Such a protocol would be a localized protocol but not strictly localized since it takes time proportional to the diameter of the network and the entire network must be traversed before the spanning tree can be constructed.

Thus one needs to consider other criteria while deciding whether a protocol is truly localized. For example, one needs to consider both the construction and the maintenance phases. A change in one part of the network may require message propagation and recomputation of the entire structure, such as that of an MST or a typical cluster organization. While clustering has an elegant construction phase, "chain effects" do occur in the maintenance phase. One can consider also the *time complexity* of a protocol. Depth-first search (DFS)- based routing, for example, may have unbounded delay (more precisely, delay proportional to the number of nodes in the network) between two visits to the same node. It also requires nodes to *memorize* some information. But the network is dynamic, and the node keeping stored information may not be available when DFS search returns to that area. Thus it is preferable to have a solution that avoids memorization, controls decision delays, minimizes messages between nodes, and is able to make instant decisions.

It may be desirable to design a formal definition of localized algorithms. Such a definition would allow a clear classification of any protocol with respect to locality. This, however, might be a difficult task, and other factors may come into play. Consider, for instance, the consensus protocol[3] in a sensor network. Each sensor has a binary value (0 or 1). At the end of the protocol, all sensors should have the same value, corresponding to the majority "decision." The protocol[3] consists of iterations. In a given iteration, each sensor takes the value of the majority of its neighbors. The process stops when all the sensors have the same value. Is this protocol localized? The number of iterations may be constant for some inputs (although sensors may need some time to be convinced that there will be no further changes), such as in the cases of rare "opposing" values. However, the number of iterations may be larger than the diameter of the network, since the final value at each sensor may depend on the values in all the sensors. If the consensus is reached within a number of iterations comparable to network diameter then this is the best possible protocol and may be argued to be localized. However, the protocol may require many more iterations to converge, and it is not clear whether it will always converge.

The remaining sections in this chapter will describe a number of localized protocols for sensor networks, which are all energy-conserving compared to other existing protocols. To reduce the amount of references in this chapter, we will only refer to the key articles, and additional chapters by this author that give a comprehensive list of references in the field. The problems studied include broadcasting, routing, anycasting, and connectivity issues.

Most protocols are based on a commonly assumed *unit disk graph* (UDG) model, where nodes communicate directly if and only if the distance between them is at the most R, where R is the transmission radius, equal for all nodes.

4.3 Minimum-Energy Broadcasting and Multicasting

4.3.1 Minimum-Energy Broadcasting

Suppose that nodes in an ad hoc network can adjust their transmission radii, and that they are aware of their own geographic position and of their neighbors. The problem is to broadcast a packet to all the nodes in the network so that the sum of all the power used for each transmission is minimized. The power consumption for two nodes at distance r is assumed to be proportional to $r^{\alpha} + c$, where $\alpha \geq 2$ and c is a constant that includes signal-processing cost and minimal reception power. It is shown[4] that, for $c > 0$, it is not optimal to minimize the transmission range. Furthermore, it was demonstrated that there exists an optimal radius—computed with a hexagonal tiling of the network area—that minimizes the power consumption. For $\alpha > 2$ and $c > 0$, the optimal radius is $r = (2c/(\alpha - 2))^{1/\alpha}$, which is derived theoretically, and confirmed experimentally. A localized broadcast algorithm, called TR-LBOP is proposed,[4] which takes this optimal radius into account. This protocol is experimentally shown to have limited energy overhead with respect to globalized algorithms for all network densities.

A survey of minimal-energy broadcasting protocols is given in Ref. [5]. We will describe here a competitive, localized protocol[6] that is not mentioned there. It is based on a topology control protocol called the local shortest path tree (LSPT), proposed by Wang, Wei, and Kuo[7] for power-aware routing and broadcasting in ad hoc and sensor networks. Each node u applies Dijkstra's algorithm to find the shortest weighted (weight of an edge of length r is $r^{\alpha} + c$) path to each of its neighboring nodes, using only its local 1-hop information (the concept can be extended to the k-hop knowledge). Node u then keeps only outgoing edges in this structure (toward neighbors with direct link requiring less power than the sum total of the power on any path between them), and removes the others. The result is sent to its neighbors. Each node then removes unidirectional links and adjusts its transmission radius according to the remaining logical links. The experimental data show that this tree has an average degree of about 2.4. LSPT is a connected structure, which can be used to describe an efficient minimum-energy broadcasting protocol as follows:[6]

Broadcasting starts from a node, which transmits with a radius equal to its longest edge in LSPT. After receiving the first copy of packet that is being disseminated, each node A applies neighbor elimination scheme (NES) on the set of its neighbors in LSPT. It sets a timeout that can be defined in a variety of ways (e.g., a random number, or a number corresponding to the number of neighbors in LSPT not eliminated by NES, or to distance to the furthest such neighbor). For every further copy of packet received, NES is similarly applied. Note that packet does not need to arrive from a neighbor in LSPT. Knowing the position of the sender and transmission radius applied, A may determine which of its neighbors in LSPT have received it. At the end of the timeout, node A verifies whether or not its list of LSPT neighbors became empty after all the NES applications. If so, retransmission is canceled. Otherwise the transmission radius is set to the furthest edge leading to a neighbor from LSPT which has not been eliminated yet. Note that this method corresponds to several others that used other connected structures instead of LSPT, such as (local) MST, or relative neighborhood graph.[5] It is conceptually much simpler than TR-LBOP[4] but is very competitive with it. LSPT targets routing optimization rather than broadcasting optimization. Optimal forwarding distance is $(c/(\alpha - 1))^{1/\alpha}$,[8] while optimal broadcasting radius is $(2c/(\alpha - 2))^{1/\alpha}$.[4] To get a broadcasting optimization radius rather than a routing one, one can apply LSPT with this new metric: $d^{\alpha-1} + 2c$.[6]

4.3.2 Minimum-Energy Multicasting

In *minimum-energy multicasting* problem, a packet is sent to certain destinations instead of all the nodes in the network. This problem was studied in Ref. [9]. The model considered in Ref. [9] has $c = 0$ in energy consumption $d^{\alpha} + c$ between two nodes at distance d, and assumes that nodes know the distances to all their neighbors, but not their positions. Flooding is used for preprocessing to find all the destinations; then destinations report back to source, creating reporting trees. The source takes a centralized decision on the power optimization of these reporting trees.

This protocol can be improved by restricting all paths from source and destinations to use only edges from a locally defined topology, such as LSPT.[10] This can be used in both the flooding steps and in reporting back to the source. If other localized structures are applied for $c > 0$, then the target radius idea (similar to Ref. [4]) needs to be applied. This solution promotes energy-efficient paths to all destinations, although overall it is suboptimal, since Steiner tree philosophy is not in the solution. If the positions of all the destinations are known to the source then no flooding step is needed at the beginning, and Steiner tree approach may be applied from the source. Hop-count-based multicasting protocol[11] can be adapted here, by modifying cost metrics appropriately. Suppose that node S has received the responsibility to forward the message to destinations $D_1, D_2, \ldots, D_k$. The progress made by selecting neighbor A to "cover" destination D_i is $T_i = |SD_i - AD_i|$. One neighbor is selected for each destination. The overall progress is $T = T_1 + \cdots + T_k$. The cost of forwarding packet to selected neighbors is $d^{\alpha} + c$, where d is distance of S from the furthest selected neighbor A. The cost metric should also include the term L, where L is the number of selected neighbors, because each of them, with its retransmission, will require some minimum energy. The objective is to select neighbors so that $L(d^\alpha + c)/T$ is minimized. The selection of the best forwarding neighbor set can be made by testing all set partitions of neighbors, or by applying a faster greedy method described in Ref. [11].

4.4 Power-Aware Routing

In power-aware geographic routing, the task is to route a message from a source to the destination node such that the sum of all the used powers on the path is minimized. This problem can be solved by a centralized (Dijkstra's) shortest weighted path algorithm by placing weight $d^\alpha + c$ on the edge connecting two nodes at distance d. Localized routing algorithms, using a variety of metrics, are surveyed in Refs. [12, 13]. They are mostly based on a simple framework[13] for designing network layer protocols for sensor networks including localized routing, broadcasting, area coverage, and so on. The framework is general enough and is applicable to a variety of problems, network assumptions, and optimality criteria. It is based on optimizing the ratio of the cost of making certain decisions, e.g., selecting forwarding neighbor for routing, and the progress made in doing so, e.g., reduction in the distance to the destination. This general guideline is applied (cf. Ref. [13]) for the design of hop count, power-aware, maximal lifetime, beaconless and physical layer-based routing, minimal-energy broadcasting, sensor area coverage, and multicasting protocols.

This general framework for localized routing works as follows. Suppose that each edge has a cost measure. The *cost* measure depends on the assumptions and metrics (such as hop count, power, reluctance, delay, and expected hop count) used, while *progress* measures the advance toward the destination (such as the reduction in distance to destination of current node to the distance of its neighbor). The cost of the selected link should be evaluated against the progress made toward the destination, and the minimal one is selected. The rationale for the method[13] is that it attempts to minimize the total cost by favoring neighbors closer to the destination (thus reducing the number of hops) and with smaller costs, with a suitable simple formula to choose a "winner" at each step.

Consider the case of power-aware routing. The power needed to send a packet from the current node C to its neighbor A is proportional to $r^\alpha + c$, where $r = |CA|$. This power measure can be used as a cost measure. Therefore the neighbor that minimizes $(r^\alpha + c)/(d - a)$ will be selected,[13,14] where $d = |SD|$ and $a = |AD|$. This means that the selected neighbor minimizes the power spent per unit of progress made, in terms of getting closer to the destination.

In Ref. [8], it is shown that, if additional nodes can be placed at desired locations, the optimal forwarding distance is $(c/(\alpha - 1))^{1/\alpha}$. This is used to derive a formula for minimal power $v(a)$ for routing between two nodes at distance *a*. A neighbor that minimizes $r^\alpha + c + v(a)$ is then selected in the so-called *ideal* routing protocol.[8]

The described position-based power-aware routing schemes assumed that each node knows the position of all its neighbors. This is not necessary, because of an adjustment that can be done if neighbor knowledge is not available. One can apply the idea of beaconless routing, proposed earlier for the hop count metric (cf. Ref. [12]). Current node *S* sends the (ready To send) (RTS) signal instead of the message, and waits for a node to respond with a clear to send (CTS) signal. Each neighbor, closer to the destination than *S*, sets a timeout duration proportional to the function being minimized, such as $(r^\alpha + c)/(d - a)$. When its timeout expires, neighbor will offer services (unless it hears another neighbor offering it). If several responses are received, *S* selects the first one that arrived.

Durresi, Paruchuri, and Barolli[15] described a delay-energy-aware routing protocol for sensor and actuator networks. Energy is saved by random wakeup scheme which allows sensors to be active during a randomly chosen fixed interval in each time frame. A packet is forwarded to any of the active neighbors that are closer to the destination than current node by at least the threshold distance *Th*. Sensor's active duration is based on its queue size. Energy needed to change between sleep and active states is not considered. At wakeup time, the sensor is not aware of already active sensors unless there is a hello message exchange with them. Shorter active periods also mean their lesser density and the increased failure of described greedy routing.

We considered here only the problem of localized routing for minimizing the total power for a given routing task. The problem of localized routing which has a goal of maximizing the network lifetime is also considered in Refs. [8, 12, 13].

4.5 Controlled Mobility for Power-Aware Localized Routing

We will consider, in this section, power-aware routing in networks with mobile sensors or sensor networks with mobile actuators. It is assumed that a path is to be created between two nodes, such as between a sensor node and a sink. Routing then uses either mobile sensors or mobile actuators as intermediate nodes. The motivation could be an application where a sensor continuously monitors and reports an event to a sink, with the other sensors being mobile and moving to positions that will reduce the energy need for reporting.

Goldenberg et al.[16] studied controlled node mobility to improve communication performance. In their localized protocols, intermediate nodes move to desirable locations to improve power efficiency for one unicast flow, multiple unicast flows, and many-to-one concast flows. Here, we only consider unicast flow, which corresponds to the power-aware routing. The routing scheme[16] works as follows:

1. Apply greedy routing or nearest forward progress (NFP) routing to establish initial path;
2. Iteratively, each node (except source and destination nodes) moves to the midpoint of the previous and the next node on the path.

Each node knows its position of itself, its neighbors, and destination. In *greedy* algorithm (proposed by Finn, cf. Ref. [12]), source or current node S forwards the packet to neighbor A that is closest to destination D. Only neighbors closer to destination than S are considered. If there is no such node, consider routing a failure. NFP is similar to NP (nearest with progress) method,[8] where a message is sent to the neighbor that is closest to S, among those that are closer to D than S.

Several movement scenarios are considered in Ref. [16]. Nodes may move in rounds. All nodes finish the current movement before the next one starts. At the end of each round, they exchange their new positions with neighbors on the path, and move toward the next position according to the protocol. In the continuous-node mode movement, each node adjusts its movement objective whenever it learns current positions of two neighboring nodes (which also move) on the path. In both cases, unconstrained and constrained versions are considered. In the latter case, nodes must preserve connectivity with other neighbors (not on current route) while moving. In their protocols,[16] the number of hops on the route does not change, but each hop reduces the distance between communicating nodes and therefore requires less communication power when transmission powers can be adjusted to the actual distances.

We describe here another algorithm[17] for routing from source S to destination D, $|SD| = d$. It works in the following way:

■ Find $d((\alpha - 1)/c)^{1/\alpha}$, and round it to the nearest integer $n + 1$ (n is the desired number of intermediate nodes; $x = d/(n + 1)$ is the ideal hop length if all nodes are on a straight line);
■ If $n \leq 0$ then S transmits directly to D;
■ Otherwise, route from S to D as follows. Node S currently holding the packet will forward it to neighbor A that minimizes:
■ $||AD| - |d - x||$ (version 1: neighbor is expected to reduce the distance D by $\approx x$), or
■ $(r^\alpha + c)/(d - y)$, where $r = |SA|, y = |AD|$ (version 2: the best cost over progress ratio).

Only neighbors closer to destination than the current node are considered. If no such neighbor exists then routing fails. Alternatively, face routing[18] can be applied to recover from void areas. Note that the actual number of hops by both versions may not be equal to the expected one which is precalculated. After the destination is reached, the number of actual hops is calculated. All nodes on the constructed route may learn it by backward routing from the destination to the source. The proposed method has multiple advantages over those described in Ref. [16]. The energy gain of the constructed route for long-term reporting is optimized, while for Ref. [16] it is limited by fixing the number of hops to a suboptimal value. Instead of moving zig-zag in rounds or continuous movements, nodes may move straight to their final positions. This reduces to zero further communication overhead for learning intermediate moving destinations, and also optimizes the length of moving route. In all the methods, the number of packets sent over final route should be sufficiently high to compensate for the energy needed for controlled mobility to better forwarding positions.

4.6 Power-Efficient Neighbor Communication and Discovery for Asymmetric Links

Protocol designers almost exclusively assume a bidirectional link between two sensors, although experimental measurements show that this assumption does not hold. However, dealing with asymmetric links is very difficult, starting from their very recognition, and power adjustment for mutual communication, which is briefly discussed here.

Srivastava et al.[19] described three algorithms for learning neighbor topology in case of realistic physical layers and asymmetric links. One algorithm selects minimal possible power to reach all neighbors that are reachable with maximal power. The second algorithm finds bidirectional neighbors. The third algorithm converts unidirectional links into bidirectional links. However, this protocol is questionable. A node receiving signal from a neighbor may not be able to reach that neighbor back, with its maximal possible power.

It may be desirable, for protocols making use of asymmetric links, to learn which sensor can hear transmissions from a given sensor A. After receiving a packet from A, neighbor B may not be able to acknowledge it directly, or to merely inform once about its existence. To respond, an alternate path is needed, via some other neighbors. Such an alternate path can be found in several ways.[20] Sensor B can initiate limited flooding up to k hop distance from itself, for fixed k, or by expanded ring search, starting with small k and increasing it if no confirmation arrives. If reaching A is necessary, full flooding can be applied, or it can make an attempt to optimize it, if position information is available. In the latter case, any position-based routing can be applied, including a DFS-based one (cf. Ref. [12]).

If the search finishes without finding the neighbor then the link is truly directional, that is, there is no reverse path.

4.7 Challenges of Power-Aware Routing with a Realistic Physical Layer

Current data communication protocols have serious drawbacks. They are designed for the UDG model, where two nodes communicate if and only if the distance between them is up to the transmission radius, equal for all nodes. Protocols are normally evaluated on simulators that implement realistic physical layers and scenarios. Because of the discrepancy between the model used to design protocol and the model used to evaluate it, simulators have frequently attacked some good properties of new protocols. For example, in UDG, long hops are good choices to create short paths toward the destination. However, in simulators, such hops may receive low packet reception probabilities, being distant nodes, and consequently declared mostly as packet failures. However, existing greedy routing could be modified by introducing appropriate new models, metrics, and modifications in routing protocols to address a more realistic physical layer. One suggested approach is to evaluate proposed protocols on simulators that exactly match the assumptions used for their design, and to pose new challenges by introducing new features from a realistic physical layer. Evaluations with realistic physical layers and existing simulators still have merits, but more as indicators of performance rather than for a final judgement. The goal is to design routing and other data communication protocols that are meant for using in real equipment, with much better performance than existing solutions.

Log-normal shadowing is one of the frequently used realistic physical layer models which accounts for slow fading effects in signal strengths. It was used in current studies searching for new routing schemes such as Refs. [21, 22]. However, there exist different interpretations of this model. Bit error rate (*BER*) is a function of signal-to-noise ratio (*SNR*). For instance, Ref. [22] uses formula $BER = e^{-SNR/2}/2$ referring to noncoherent FSK channel modulation. *SNR* depends on transmitting power P_T, distance d between sender and receiver, attenuation factor α, current thermal noise N, a random variable X_σ, and some constant C. For example, Ref. [22] uses formula $SNR = P_T/(d^\alpha C X_\sigma N)$, where $X_\sigma^{dB} = 10\log_{10} X_\sigma$ is a Gaussian random variable with mean zero and variance σ^2. Note that dB means "decibels," which is a measure of signal strength (and not an exponent in the formula). Since X_σ cannot be known before message transmission, exact *SNR* cannot be predicted even if other variables in the expression are fixed. Indirectly, *BER* depends on X_σ. To find the expected *SNR* and *BER*, integrals need to be applied over given formulas. Exact computation therefore is time-consuming, and unacceptable for application in a tiny sensor.

More energy can be wasted on computing the exact formula based already on some approximate values of terms involved. For this reason, Ref. [21] advocates the use of simple approximate formulas with sufficient accuracy, and gives such approximations for certain cases. Packet reception rate *prr* depends on *BER* and selected encoding. For example, in NRZ encoding, each bit is received independently of other bits, and the formula is then $prr = (1 - BER)^F$, where F is packet length in bits (cf. Ref. [22]). A number of localized greedy routing protocols based on log-normal shadowing model, fixed transmission power, and acknowledgments and data packets having the same lengths are proposed in Ref. [21]. In this case the expected number of packets sent by sender is $1/prr^2$ while the expected number of acknowledgments sent is $1/prr$. If data packets are assumed to be very long and the size of the acknowledgment very short with negligible impact on the expected hop count, then the sender is expected to send $1/prr$ packets (cf. Ref. [22]). If acknowledgments are not sent at all, then the goal is to maximize the probability of receiving the packet at the destination, and this was considered in Ref. [23].

Consider now the problem of power-aware routing with realistic physical layer. The goal is to minimize the total power needed for routing a packet, assuming that nodes can adjust their transmission radii. In Ref. [21] it is shown that an earlier proposed solution to this problem was incorrect. Another attempt to solve this problem was made by Li, et al.[22] They derive the optimal power needed for transmitting between two nodes at distance d, by maximizing $prr/(P_T+P_E)$, where P_T is adjustable while P_E is the fixed cost of using circuits in sender and receiver nodes. The optimality criterion $A - FA \ln(A) + 4FAP_E/(d^\alpha CX_\sigma N = 1$ is then derived, where $A = e^{-SNR/2}$. They claim that the optimal transmission power can be "easily calculated by numerical approaches" from this equation, and, when $P_E = 0$, it "only depends on the packet size." However, this equation depends on X_σ even when $P_E = 0$ since A depends on *SNR*, while *SNR* depends on X_σ. In turn, X_σ is a random variable and cannot be simply replaced by a constant value (e.g., its expected value 1) and still retain the optimality claim. The optimality "road" is more complex. Eventually the optimal power indeed depends on packet size but for the arbitrary value of P_E. The problem of finding the optimal transmitting power for minimizing overall energy for packet transmission between two nodes remains unsolved.

In experimental design, Refs. [21, 23] assumed equal probability of packet reception for every packet, by using a simple approximation formula with sufficient accuracy. Alternatively, packet reception probability can be based on an exact formula for a log-normal shadowing model by following a two-step randomization approach. First, randomly decide X_σ based on its distribution. This then decides *SNR* and *BER*. Based on such *BER*, find prr for that packet. The assumption is that each bit in a packet has equal *BER*, but different packets have different *BERs* and ultimately different *prr*s. Single randomization with an approximate formula has been replaced with double randomization with exact formula, while preserving the computation speed.

Using such modeling, assume that $p(d) = prr$ is the packet reception probability for two nodes at a distance d when maximum possible power $P_{\max}$ is used. The transmission energy is then proportional to d^α. Let a reference distance be R such that $p(R) = 0.5$ (other reference distances can be used). Suppose that the actual transmission power P_T is adjusted so that $P_T/P_{\max} = r^\alpha/d^\alpha$ for some r. Then it can be shown[24] that the corresponding packet reception probability is $p(Rd/r)$. Assuming a negligible cost of acknowledgment, the function to be optimized is then $(2E + r^\alpha)/p(Rd/r)$, where E is the term accounting for the energy cost of running the circuitry. This is a function of one variable that can be solved by numerical means. If the acknowledgment packets are considered then the optimization function has two variables, even when packet and acknowledgments are of the same length (it may not be optimal for the sender and the receiver to use the same powers). In all scenarios, localized power-aware routing protocols[24] then follow cost-to-progress ratio paradigm.[13] Assuming the positions of the neighboring nodes are known, the optimal power to reach the candidate neighbor is calculated, which is then divided by the progress which that neighbor generates.

If the positions of neighboring sensors are not known, beaconless position-based routing with realistic physical layers may be applied. Sensors receiving request for data forwarding will calculate the appropriate cost, and select timeout duration proportional to the ratio of that cost and the progress they achieve. If no message cancelling the service request is received before the timeout expiration, the packet is forwarded. The protocol[25] has details regarding nontrivial acknowledgments. The cost corresponds to the expected hop count in case of fixed transmission radius, or optimal energy consumption for a given distance in case of adjusted transmitting powers.

4.8 A Localized Coordination Framework for Wireless Sensor and Actuator Networks

After the sensors detect an event occurring in the environment, the event data is locally processed and transmitted to the actuators, which gather, process, and eventually reconstruct the event data. We refer to the process of establishing data paths between sensors and actuators as *sensor-actuator coordination*. Once an event has been detected, the actuators coordinate with each other to make a decision on the most appropriate way to perform the action. We refer to this process as *actuator–actuator coordination*. In this section, actuators are considered to be static, and protocols are

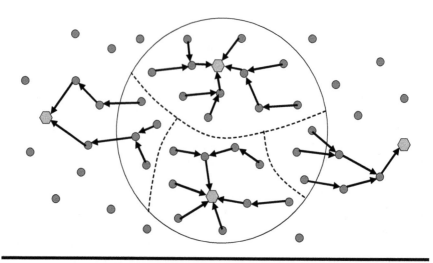

Figure 4.1   Anycasting in a sensor-actuator coordination framework.

designed based on the UDG model. The scenario is illustrated in Figure 4.1, where sensors located inside the event area (circle) report to their nearest actuators among four actuators (shown with hexagonal shape). Such a sensor-actuator scenario is a special case of hybrid networks considered in Ref. [26].

4.8.1   Sensor-Actuator Coordination

The monitoring area is divided into regions, one per each actuator. All sensors within a region, when event occurs, are reporting, each one to its nearest actuator. The first question is how the sensors can learn the position of the nearest actuator, or positions of all actuators. Positions of all the actuators can be learned by broadcasting from each one of them. Actuators normally have larger transmission radii than sensors, which enables them to communicate among themselves, either via a common sink or in a multihop fashion. They may transmit therefore with larger transmission radii than the one available to sensors, possibly even large enough to reach all the sensors with one transmission. Only sensors on the border of transmission radius need to retransmit: these are exactly the sensors that have neighbors which did not receive the message directly from the actuator. The protocol then may follow any existing broadcasting algorithm to reach all sensors. Other actuators may help in the process, by retransmitting the positions of all the actuators before the sensors start retransmitting among themselves. In these cases, the sensors learn the positions of actuators, but not how to reach them.

Actuators may also initially transmit using the same transmission radius as the one used by sensors (such an assumption is made in Ref. [26] for all the protocols described there). The motivation for such restriction is that sensors otherwise may not learn their hop distance for reporting to the nearest actuator because of asymmetric links. Worse, sensors may actually be unable to report to their physically nearest actuators because of some void areas between them. Therefore, it may be "safer" to restrict actuators to the same transmission radius that the sensors are using. This also allows for immediate construction of backward paths from sensors to actuators, also including alternative neighbors with the same hop count distance in case the first choice fails at reporting time.

Instead of the described periodic and synchronous flooding, sensors and actuators may also use "hello" messages at their own pace. They attach their believed hop distance to the (communicationally) nearest actuators to their "hello" messages, together with their own position. Actuators send their own "hello" messages with distance 0. Each sensor is able to recursively determine its hop distance to the nearest actuator based on recent "hello" messages received from other sensors. Each sensor associates itself with a "parent" sensor, and forwards the field reports to it. The parent sensor may collect reports from several associated sensors. It may perform data aggregation (which may include its own measurement) before forwarding the report to its own parent sensor.

Melodia, et al.[27] discussed the construction of data-aggregation trees using a metric different from hop count, by using energy model $u(d) = d^\alpha + c$ for cost of communicating at distance d. It is assumed that, initially, sensor nodes know their own positions, positions of their neighbors, and positions of actuators. It is also assumed that there are no void areas. Thus there exists an advance from a given sensor to any actuator, if direct transmission is not possible. They formulate an integer linear program to construct a data-aggregation tree to minimize overall energy spent for reporting. More precisely, a set of actuators is first selected, and then a minimum-energy tree toward these actuators that meet the required event reliability is constructed. Basically, several trees are created inside the geocasting region, and the paths/trees are then constructed toward each actuator (see Figure 4.1), including those outside that region. Event reliability is the ratio of unexpired packets to all received packets (data and notifications). The delay of a packet can be measured as the largest delay on a reporting aggregated tree (each sensor must wait for data from its "children" sensors before aggregating and forwarding the result to its parent sensor).

Since integer linear programming-based solutions are centralized algorithms with high time complexities, they are highly impractical. Authors[27] then consider some distributed solutions using some localized heuristics and state-based behavior of sensors. Each sensor alternates among four different states, namely *idle, startup, speedup,* and *aggregation state.* The main objective of state transitions is to reduce the number of hops when the reliability requirement is violated and to save energy when the reliability requirement is met. In the startup state, each sensor node selects its next hop based on the two-hop rule. Let S be a sensor, N one of its neighboring nodes, and A one of the actuators. Sensor S selects neighbor N for which $u(|SN|) + u(|NA|)$ is minimized, over all neighbors and over all actuators. We briefly analyze this choice and propose some alternatives.[28] Consider a candidate neighbor N so that $y = |SN|$ and $x = |NA|$, and candidate neighbor B so that $y + \Delta = |SB|$ and $x + \nabla = |BA|$. It can be shown that the difference between the two neighbors in the considered criterion is $u(|SN|) + u(|NA|) - u(|SB|) - u(|BA|) \approx \alpha y^{\alpha-1}\Delta + \alpha x^{\alpha-1}\nabla$.

When the sensor is far from the actuator, $y \ll x$ and the difference is $\approx \alpha x^{\alpha-1}\nabla$. This means that the selection is biased toward the neighbor that is closest to A, that is, the selection is close to the one made in greedy routing. This is not really a power-efficient selection. In Ref. [8], a different criterion is applied for power-aware routing toward the nearest actuator, as described in an earlier section here. $u(|NA|)$ is replaced by ideal cost $v(|NA|)$, where function v is linear, and $u(|SN|) + v(|NA|)$ is minimized. In Ref. [14], this can be further marginally improved and simplified by considering cost-to-progress ratio. Thus $u(|SN|)/(|SA| - |NA|)$ is minimized over all neighbors N and all actuators A.

There exists another option for creating power-efficient reporting trees in a localized manner. Each sensor may first decide which of its edges belong to LSPT,[7] discussed earlier. A tree is then constructed from these edges, starting from an actuator. Sensors will retransmit only to neighbors on LSPT. Those that are not already attached to the tree will respond to their new "parents" and then will search for possible "children" in the tree. This process will in fact delete all loops from LSPT.

The objective of heuristics[27] is to converge to a solution with reliability close to the event reliability threshold and minimal energy consumption. Sensor nodes associated with an actuator base their state transitions on the reliability observed by the actuator, which is broadcast at the end of each decision interval. The objective of the collaborative operation of nodes in the speedup state is to minimize the number of hops between sources and actuators. This is achieved by applying the greedy routing scheme by Finn (cf. Ref. [12]), which minimizes hop count. However, the details on who calls the speedup, how and when it stops are quite sophisticated. If all the nodes apply speedup procedure, too much advance can be made with excessive cost in energy. Authors[27] then discuss an aggregation state where nodes select the nearest neighbor with advance toward an actuator as the next hop. Details of this protocol remain unclear to the author of this chapter.

The main idea in Ref. [27] is to construct any path from a sensor to an actuator, and then improve this path by optimizing energy while any delay is controlled. We describe an alternative[28] to the transition-state-based heuristics of Ref. [27]. The goal would be to construct a path from a given sensor to one of the actuators based only on the position of itself, its neighboring sensors, and actuators, without the need to "repair" it. The optimization criteria could be to minimize the hop count, or minimal total power on the path to an actuator, or some other criterion. We will refer to this problem as the *anycasting* problem, with a variety of optimization metrics.

Assume first that the greedy approach will work, and therefore that the path length to each actuator is proportional to the distance to it. Then each sensor will naturally report to its geographically closest actuator. Anycasting is then equivalent to routing from a given sensor to its nearest actuator, and this case will be considered first. Source nodes are informed about tolerable delay, which is translated into a desired number of hops toward a particular actuator. The objective is to find a path with approximately that many hops, and optimize total power on such candidate paths. We will first consider the problem without the power-optimization part. Let k be the desired number of hops from a current sensor S toward actuator A, $|SA| = d$. Let N be a candidate neighbor (closer to A than S), $|SN| = r$, $|NA| = a$. The desired progress toward the destination A is d/k. Therefore neighbor N that minimizes $|d - a - d/k|$ is selected. The report is sent to N, which reduces desired k and proceeds similarly, until either A is reached or no neighbor closer to A exists. Now consider the power minimization requirement. The minimization formula is then $|u(r) - u(d/k)||d - a - d/k|$, which attempts to move a selected node closer to the straight line SA. Note that the minimization formula $u(r)|d - a - d/k|$ may favor neighbors very close to S. The power is minimized by reducing the lengths of selected hops subject to desired progress.

This algorithm may be also applied in the context of improving the current path by increasing or decreasing the observed hop count to achieve the desired delay. It is the very same algorithm where candidate neighbors are exactly nodes on the already established paths, communicated to a given node by either transmitting from neighbors with maximal power, or attaching a few last neighbors to the message originated from an actuator. Note that, up to a certain optimal routing distance,[8] the decrease in hop count corresponds to the decrease in total power on the path. The optimization discussed above is therefore applied for forwarding distances that exceed the power optimal one.

In more general scenarios, void areas may exist, leading to routing failures, and the algorithm needs to be modified. In the next section, we look at such a case, and design a more general algorithm. That algorithm will reduce to the described one when there are no void areas.

4.8.2 Anycasting with Guaranteed Delivery

Let us suppose that each node knows the position of all actuators, itself, and its neighbors. However, either the paths toward the actuators are not recorded, or are considered unreliable because of the failure of some sensors on possibly recorded paths. The anycasting task is to find a path toward one of the actuators, optimizing one of metrics (we will consider hop count for simplicity). Greedy anycasting is a straightforward generalization of greedy routing. The distance of a node to the actuators is its distance to the nearest of them. When neighboring nodes are considered, advance is measured with respect to that distance. Thus a given node and its neighbor may not use the same actuator to measure their distances.[27] In greedy mode, select a neighbor that offers the best advance.[27] This can be measured in terms of hop count, or power consumption. In the latter case, cost-to-progress ratio-based selection[13,14] can be applied. In greedy mode, if advance can be made toward any of the actuators, it is made. Greedy routing advances toward the closest actuator as much as possible. If a sensor without a closer neighbor to any of the actuators is encountered, greedy routing fails.

We consider now the problem of anycasting with guaranteed delivery.[29] One option is to initiate separate routing tasks toward each actuator, and apply greedy-face-greedy (GFG) routing algorithm[18] for each of them. All connected actuators will be encountered. GFG algorithm applies greedy routing as long as advance is possible. When a void area is reached (no advance is possible), it records the node C which faced the void area, and is switched to the face mode routing. It remains in the face mode until a node closer to the destination than C is encountered. Greedy routing then continues. Details of the face mode routing are given in Refs. [12, 18].

We now describe an anycasting algorithm that is path-based, that is, a single path from source node is followed until one of the actuators is reached. It generally follows GFG approach, where distance to a single destination has been replaced by distance to the nearest actuator, and face routing is applied toward a single selected actuator until recovery. Let S be source node, and $A_1, A_2, \ldots, A_k$ be actuators. We describe the algorithm with a general metric, using cost-to-progress ratio paradigm. For hop count metric, the cost is 1 (meaning one hop). Progress from current node S to candidate neighbor N is equal to the $|SA_i - NA_j|$, where A_i is the nearest actuator to S, and A_j is the nearest actuator to N.

Source node S may already have a dilemma on how to proceed. Suppose that actuator A_1 is geographically farther than actuator A_2. However, neighbors leading closer to A_1 exist, while there is no neighbor to advance toward A_2. Should S apply greedy mode toward A_1, or face mode toward A_2? Various decision criteria (including parameter-based) are possible. We decide to apply this one: if there exists a neighbor N so that $|NA_1| < |SA_2|$ then apply greedy mode, otherwise apply face mode.

Source node S verifies the distances from the neighbors to all the actuators, and its own distance to all the actuators, and decides whether to begin with the greedy or with the face mode. If the decision is to proceed with the greedy mode, it selects the neighbor that minimizes the selected cost-over-progress ratio. The same decision is made by the receiving nodes, until the report is either delivered to an actuator, or transition to the face mode is deemed more appropriate. Node S that decides to initiate the face-mode traversal will record its distance f from the selected actuator A_i in the forwarding packet. The packet is forwarded to the neighbor according to the face routing part of GFG, that is, to a neighbor along an edge of the Gabriel graph (see Ref. [12] for details). Face traversal continues until either the destination actuator is reached, or a loop is detected, or a neighbor of a current node is detected at distance $< f$ from its closest actuator. Loop detection criterion is the same as in GFG protocol.[18] If a loop is detected, this means that actuator A_i is unreachable, and is deleted from the list of actuators. In this case, a brand new anycasting protocol is set on a smaller set of actuators. If a neighbor is at a distance $< f$ from its closest actuator then the protocol returns to the greedy mode. Algorithm is loop-free because face routing either delivers or reports a loop, and, between two eliminated actuators, advances are made at each step (either by a greedy step or by several hops of a face mode). When a single actuator remains, the protocol behaves as GFG. Delivery is guaranteed since at each greedy step distances to selected destination are decreasing, which is also achieved by face routing.

If power-aware routing is followed, it is followed only in the greedy mode. Face routing forces the use of edges from Gabriel graph, which are generally short, thus supporting naturally power awareness. If limited delay considerations are added then this algorithm first needs to consider all the delay imposed by the face mode. The remaining delay is then "distributed" to greedy portions of the route. The receiving actuator will learn the total length or hop count of these portions, and total delay, and may decide what the optimal number of hops is. The greedy portions of the route are then adjusted on a backward scan from the actuator to the source.

4.8.3 Actuator–Actuator Coordination

In the actuator–actuator coordination step, report from one actuator is distributed to others, to decide which of them will perform a required action. A mixed centralized integer nonlinear program is proposed as a solution in Ref. [27]. Another solution[27] is localized, and is based on "auction protocol," where each actuator reports back to the originating actuator the offer to provide service and the cost of it. Therefore the request for service is flooded from the actuator node that collected the report, and each actuator responds back to it.

The communication overhead of this algorithm can be reduced by performing an "auction aggregation" protocol.[35] The collecting actuator A may itself have a low cost of performing needed action. It includes its own cost in the message that is flooded to other actuators. Actuators receiving the service request will compare their own cost, and attach to the message they retransmit only the lower cost (between incoming one and their own). The specific protocol has tree "expansion" and tree "contraction" phases. Tree expansion starts from collecting actuator A, creating a tree rooted at A. Each remaining actuator B, when the request was received for the first time, retransmits it with the lower cost appended to the packet. Note that an efficient broadcasting protocol (among those surveyed in Ref. [1]), can be applied to stop this retransmission if B has a higher cost than the cost recorded in the incoming packet and B should not retransmit by the protocol. These retransmissions create a response tree. Each node, with retransmission, includes its parent actuator in the message, so that actuators can decide whether or not they are leaves in the created tree. They become leaves if they do not hear any other actuator listing them as their parent. Leaf nodes start responding back to actuators, with the best cost they are aware of. Each intermediate node waits to hear from all the neighbors which declared it as a parent. After hearing, they select the best cost and report further toward the collector. The collector then decides which is the best node to perform the required action and routes the task to that node. Note that expansion tree may be terminated at some branches if a node estimates that better response or timely response will be impossible to provide by its children. This may be a consequence of accumulated delay or accumulated cost of moving to perform certain action.

4.9 Localized Movement Control Algorithms for Realization of Fault Tolerant Sensor and Sensor-Actuator Networks

We consider here the scenario of mobile sensors in a sensor network, or mobile actuators in a sensor-actuator network. They are assumed to know their geographic positions. A single sensor or single actuator failure may become fatal for the ability of the network to collect data from the field or perform certain actions on sensors. Fault tolerance can be greatly improved if the network is made *biconnected*. This means that any two nodes (by a node we mean a sensor or an actuator) may be connected by two disjoint paths. In other words, the network remains connected if one of the nodes (arbitrary one) is removed. We study here the problem of minimizing the total distance traveled by mobile sensors (or mobile actuators) so that the network becomes biconnected. One solution to make the network biconnected is the contraction algorithm, where nodes move toward their centroid, until biconnectivity is established. This algorithm in some scenarios may even minimize the total movement. However, this may lead to impractical solutions, since such movement can leave a good portion of the sensor network without coverage. Therefore, the formulated problem has, in reality, additional constraints related to preserving a reasonable area coverage.

Basu and Redi[30] consider a network of mobile robots/actuators (the protocol is valid also for mobile sensors), which are all aware of global network information, and which want to move so that the network becomes biconnected, therefore more fault-tolerant. In a one-dimensional case, authors reduce the total movement optimization problem to a linear programming formulation, which has a solution in polynomial time. For a two-dimensional case, authors conjecture that the problem is NP-complete and give an approximation algorithm. The network is decomposed into biconnected components, and the corresponding block tree is constructed. The block with the maximum number of nodes is the root. Other blocks are assigned to their parent blocks in the tree. Each leaf block is then translated toward its parent. It is translated toward the nearest node, or the nearest cutvertex if the parent block is empty. New connectivity is then calculated, blocks are recalculated, and the process repeats until the whole graph becomes biconnected.

We now describe a localized algorithm,[31] which is preferred for large-scale networks. The algorithm is a localized adaptation of block movement algorithm.[30] Each actuator decides locally whether or not it is critical. An actuator is critical if the subgraph of p-hop neighbors of this actuator (when the actuator itself is removed) is disconnected.[32] In practice, only small values of p (1, 2, and perhaps 3) may be considered. Critical actuators inform neighbors about their critical status. If an actuator has no neighbor which is also critical then it serves as the local center for actuator (block) movement. If two critical actuators are connected, and none has more critical neighbors, then this pair together leads the block robot movement. Otherwise, each critical actuator that has only one neighbor that is also a critical actuator will lead the process. These are endpoints of chains of critical nodes. Thus, each of the selected critical actuators (or pair of selected critical actuators) will constitute a local decision center for making network locally biconnected. This is done using p-hop information. Details of the protocol will be elaborated in Ref. [31]. The local decision center will first test whether the movement of just one of the actuators within p-hop distance (for small value of p such as 1, 2, or 3) will create a locally biconnected network. If so, such movement is made, and the algorithm advances to the next iteration, following the same block movement algorithm. However, blocks can be large, while restricted p-hop information is used. If a single node movement cannot biconnect the network, then the movement center will apply, for its p-hop neighbors' subgraph (if movement center has two nodes, it includes p-hop neighbors of any of them), the centralized algorithm.[30] However, this may destroy some good parts of the network. The process therefore continues with such iterations, until no critical actuator is discovered. An alternate solution may attempt both options and take the result that has less overall movement length.

Protocols can be further developed to add other important criteria. In addition to overall movement distance, coverage preservation is also important. It may be evaluated by considering the cost metric for sensor communication to the nearest actuator, and can be measured by, for example, average- or worst-case hop distance from a sensor to its nearest actuator, or average- or worst-case energy needed to send a message from a sensor to the nearest actuator. Alternatively, actuators can be assigned certain coverage areas and the total area covered by them measures how well they cover the network after becoming biconnected. There exists one more cost measure for the protocol: the number of messages sent in the process. This can be further subdivided into messages to maintain k-hop information, and messages to decide what the next move is.

4.10 Conclusion

In addition to the traditional network layer problems such as routing, broadcasting, and multicasting, sensor networks impose their own specific challenges. Because of their low power and low traffic demands, activity scheduling (deciding which sensors are active and which are sleeping, and the rules of status changes) is a major issue for prolonged network life. To save energy, sensors are in sleep mode most of the time, waking up at random or at predetermined time for short intervals under the control of an internal timer. It is however known that each transition from sleep to wake state consumes a certain fixed amount of energy. Therefore, certain sleep–awake patterns proposed in literature are, in fact, detrimental to longevity.[1] This is especially the case with the schemes that attempt to place the sensors in sleep mode between any two actions, often neglecting the need for time synchronization. We mention here two proposals that are worth considering, in addition to sleep schedules coming from sensor area coverage issues.[1]

In Ref. [33], the selected transmission time is included in the node's *intent* message and is specified relative to the *intent* message transmission time. When an *intent* message with a relative data message transmission time is received, it is converted into local time. The receiving node later wakes up at the specified time to receive the data message.

The recent IEEE 802.15.4 standard for low-rate wireless networks is widely considered as one of the technology candidates for wireless sensor networks. Two network topologies are allowed by the standard, both relying on the presence of a central coordinator. In peer-to-peer topology, sensors may communicate directly, while in star-shaped topology, they must communicate through a coordinator. Sensors are time-synchronized, and follow a joint sleep–active schedule. That is, they are active at the same time, followed by longer sleep periods. At the beginning of active periods, they compete for upcoming slots to send messages. Cross-layer activity management in IEEE 802.15.4 sensor network has been discussed in Ref. [34]. In their distributed scheme, sensors decide to sleep after sensing their reports if the network is dense and event reliability is therefore high. They remain active all the time if event reliability is very low. Otherwise they sleep with a certain probability or stay active and send more reports, depending on the reported event reliability.[34]

Energy conservation in sensor networks remains a prime consideration when designing any protocol. This chapter presented some examples of energy conservation following a wider view and more general aspects. We emphasized localized protocol design, reduction of message overhead, and adjusting transmission radii as ways to save energy. We also considered reducing mobility in some scenarios as a factor in power conservation. We expect more research on the design of power-efficient protocols, especially with respect to cross layer design.

References

1. D. Simplot-Ryl, I. Stojmenovic, and J. Wu, Energy efficient backbone construction, broadcasting, and area coverage in sensor networks, in *Handbook of Sensor Networks: Algorithms and Architectures* (I. Stojmenovic, ed.), Wiley, New York 2005, 343–379.

2. J. Bachrach and C. Taylor, Localization in sensor networks, in New York, *Handbook of Sensor Networks: Algorithms and Architectures* (I. Stojmenovic, ed.), Wiley, New York, 2005, pp. 277–310.

3. K.H. Jones, K.N. Lodding, S. Olariu, L. Wilson, and C. Xin, Biology-inspired distributed consensus in massively-deployed sensor networks. *ADHOC-NOW*, 2005, pp. 99–112.

4. F. Ingelrest, D. Simplot-Ryl, and I. Stojmenovic, Optimal transmission radius for energy efficient broadcasting protocols in ad hoc and sensor networks, *IEEE Transactions on Parallel and Distributed Systems*, Vol. 17, No. 6, June 2006, 536–547.

5. F. Ingelrest, D. Simplot-Ryl, and I. Stojmenovic, Energy-efficient broadcasting in wireless mobile ad hoc networks, in *Resource Management in Wireless Networking* (M. Cardei, I. Cardei, and D.-Z. Du, eds.), Springer, Berlin, 2005, pp. 543–582.

6. I. Stojmenovic, Minimum energy broadcast algorithm based on local shortest path tree in sensor networks, in preparation.

7. S.-C. Wang, D.S.L. Wei, and S.-Y. Kuo, SPT-based topology algorithm for constructing power efficient wireless ad hoc networks, *ACM International World Wide Web Conference*, 2004.

8. I. Stojmenovic and X. Lin, Power-aware localized routing in wireless networks, *IEEE Transactions on Parallel and Distributed Systems*, 12, 1122–1133, 2001.

9. S. Guo and O. Yang, Localized operations for distributed minimum energy multicast algorithm in mobile ad hoc networks, *IEEE Transactions on Parallel and Distributed. Systems*, to appear.

10. I. Stojmenovic, Minimum energy multicasting in sensor networks, in preparation.

11. P.M. Ruiz and I. Stojmenovic, Cost-efficient multicast routing in ad hoc and sensor networks, in *Handbook on Approximation Algorithms and Metaheuristics*, Chapman & Hall/CRC (T. Gonzalez, ed.), to appear.

12. H. Frey and I. Stojmenovic, Geographic and energy aware routing in sensor networks, in *Handbook of Sensor Networks: Algorithms and Architectures* (I. Stojmenovic, ed.), Wiley, New York 2005, pp. 381–415.

13. I. Stojmenovic, Localized network layer protocols in sensor networks based on optimizing cost over progress ratio, *IEEE Network*, 20(1), 21–27, 2006.

14. J. Kuruvila, A. Nayak, and I. Stojmenovic, Progress based localized power and cost aware routing algorithms for ad hoc and sensor wireless networks, *International Journal of Distributed Sensor Networks*, 2(2), 147–159, 2006.

15. A. Durresi, V. Paruchuri, and L. Barolli, Delay-energy aware routing protocol for sensor and actor networks, *IEEE International Conference on Parallel and Distributed Systems ICPADS*, 2005.

16. D.K. Goldenberg, J. Lin, A.S. Morse, B.E. Rosen, and Y.R. Yang, Towards mobility as a network control primitive, *ACM Mobihoc*, Japan, May 2004, pp. 163–174.

17. I. Stojmenovic, Controlled mobility for power aware localized routing in ad hoc, sensor and sensor-actuator networks, in preparation.

18. P. Bose, P. Morin, I. Stojmenovic, and J. Urrutia, Routing with guaranteed delivery in ad hoc wireless networks, *ACM Wireless Networks*, 7, 6, November 2001, pp. 609–616.

19. G. Srivastava, J.F. Chicharo, and P. Boustead, Power efficient connected topologies in ad hoc networks, *IEEE ISCC*, 2005, pp. 22–27.

20. I. Stojmenovic, Which sensors can hear me? in preparation.

21. J. Kuruvila, A. Nayak, and I. Stojmenovic, Hop count optimal position based packet routing algorithms in ad hoc wireless networks with a realistic physical layer, *IEEE Journal of Selected Areas in Communications*, 23, 1267–1275, 2005.

22. C.P. Li, W.J. Hsu, B. Krishnamachari, and A. Helmy, A local metric for geographic routing with power control in wireless networks, *IEEE SECON*, 2005.

23. J. Kuruvila, A. Nayak, and I. Stojmenovic, Greedy localized routing for maximizing probability of delivery in wireless ad hoc networks with a realistic physical layer, *Journal of Parallel and Distributed Computing*, 66(4), 499–506, 2006.

24. I. Stojmenovic, Power aware routing in sensor networks with a realistic physical layer, in preparation.

25. I. Stojmenovic, Beaconless routing in sensor and ad hoc networks with realistic physical layers, in preparation.

26. F. Ingelrest, D. Simplot-Ryl, and I. Stojmenovic. Routing and broadcasting in hybrid ad hoc and sensor networks, in *Theoretical and Algorithmic Aspects of Sensor, Ad Hoc Wireless and Peer-to-Peer Networks* (Jie Wu, ed.), Auerbach Publications (Taylor & Francis Group), Boston, MA, 2006, pp. 415–426.

27. T. Melodia, D. Pompili, V.C. Gungor, and I.F. Akyildiz, A distributed coordination framework for wireless sensor and actor networks, *ACM Mobihoc*, May 2005, pp. 99–110.

28. I. Stojmenovic, A delay constrained and power efficient data aggregation for wireless sensor and actuator networks, in preparation.

29. I. Stojmenovic, Anycasting with guaranteed delivery in sensor actuator networks, in preparation.

30. P. Basu, and J. Redi, Movement control algorithms for realization of fault tolerant ad hoc robot networks, *IEEE Network*, 18, 4, July/August 2004, pp. 36–44.

31. I. Stojmenovic, Localized movement control algorithms for realization of fault tolerant sensor and sensor actuator networks, in preparation.

32. M. Jorgic, I. Stojmenovic, M. Hauspie, and D. Simplot-Ryl, Localized algorithms for detection of critical nodes and links for connectivity in ad hoc networks, *The Third Annual Mediterranean Ad Hoc Networking Workshop Med-Hoc-Net*, Bodrum, Turkey, June 27–30, 2004, pp. 360–371.

33. W.S. Conner, J. Chhabra, M. Yarvis, and L. Krishnamurthy, Experimental evaluation of topology control and synchronization for in-building sensor network applications, *Mobile Networks and Applications* 10, 545–562, 2005.

34. J. Misic, S. Shafi, and V. Misic, Cross-layer activity management in an 802.15.4 sensor network, *IEEE Communications Magazine*, January 2006, pp. 131–136.

35. I. Stojmenovic, Auction aggregation for actuator-actuator coordinated response, in preparation.

Chapter 5

Security in Wireless Sensor Networks

Yong Xi and Loren Schwiebert

Security in wireless sensor networks (WSNs) have strong relationship with those in wireless ad hoc networks, due to the similar system models for both types of networks. However, due to different application scenarios of WSNs, security issues in WSNs are considerably different from those in wireless ad hoc networks. Specifically, most of the services in a sensor network are designed to be distributed. Distributed services are susceptible to various attacks.

Besides that, the implementation constraints of a WSN are considerably different from those of a wireless ad hoc network. Because of the tight integration of sensor nodes with the physical environment, sensors are smaller than nodes in a wireless ad hoc network, often having less computation capability, storage, confined communication range, and limited power.

This chapter mainly views the security issues from the above two facets. Traditional security issues such as availability, confidentiality, integrity, and authentication also have to be addressed in WSNs. However, nonrepudiation and access control are not part of the issues any more since one WSN deployment generally falls under one administrative domain.

The attacking model on sensor networks is quite different from that on the Internet. On the Internet, attackers form a part of the Internet. A sensor network is generally under a single administrative domain. Sensors work cooperatively. Attackers are from outside the network. In this sense, sensor network security is closer to computer security since each sensor network can be collectively seen as a single entity whose security should be protected.

Throughout this chapter, we use the following notations:

s_i	The ith sensor in the network. Assuming that every sensor has a unique identity, we interchangeably use s_i to denote the sensor and its identity
$E_k(m)$	Symmetric encryption of message m using a key k
$m\vert v$	Concatenation of message m and message v
$MAC_k(m)$	Message authentication code for message m using a symmetric key k

5.1 Introduction

Although security is a requirement from the application layer, the network implementation generally follows an open systems interconnection (OSI) model. In an OSI model, functions on the higher layers depend on functions on the lower layers. To provide security services in sensor networks, we also have to work from the ground up to the application layer. We discuss how security issues are addressed in each layer.

First, we address physical layer security. In sensor networks, all the attacks originate from the physical layer because the attacker has to have physical access to a sensor network. Attack through remote access is possible but can be dealt with in the context of Internet security. Since the techniques used for recovering digital information from the radio waves are open standards, there is no confidentiality, integrity, and authentication provided on the physical layer. On the physical layer, the security mainly deals with availability problems.

Then, we discuss how keys are managed in a sensor network. Keys are crucial for implementing a secure sensor network. Individual keys not only identify individual sensors, but also enable establishment of common session keys between different sensors.

Finally, we discuss how availability, confidentiality, integrity, and authentication are supported at each layer above the physical layer.

5.2 Physical Layer Security

WSNs are tightly integrated in the physical environment. It is no longer physically inaccessible from attackers, which poses serious threats on a WSN as sensors can be extracted and compromised. The goal of physical layer security is to guarantee that sensors are located at the designated locations, and to provide designated functions individually, i.e., physical layer security defends against those attacks through physical access to the sensors.

The types of physical layer attacks include

- Topology change. The sensors can be physically removed from the network. It may reduce coverage and connectivity of the network. In the worst case, if there is a single point of failure in the network, it can render the whole network useless.
- Sensor compromise. The removed nodes may be compromised. A compromised sensor may lose its secrecy and allow the attacker to access the network. In the worst case, a compromised sensor can camouflage as multiple sensors, introduce false data into the network, or redirect the traffic.

Physical security can certainly be addressed through human patrolling and inspection. However, this is inefficient for a sensor network deployed over a wide area. To protect physical security, a sensor network has to be aware of states of the sensors. This can be realized through a periodic polling mechanism. To conserve more energy, a hardware-based approach can also be used. For example, the sensors can be made tamperproof. Any tampering attempt can be detected and possibly reported.

To enable the polling mechanism, each sensor should be individually identified. To prevent the identity from being forged, membership management should be enforced. Cryptography-based approaches can be used here. Sensors can authenticate each other based on either key predistribution or public-key cryptography.

In a general sense, there is no absolute physical security guarantee in an open environment. It is difficult to provide physical security without involving social policies. However, from a technical point of view, physical security can still be addressed in terms of the cost of attack. For example, the propagation speed of radio waves is beyond the moving speed of any man-made object. An ideal network can always report the incident to remote stations at the time of the incident. (By ideal we mean that there is no other delay associated with the communication.)

5.2.1 Topology Change

5.2.1.1 Jamming Attack

An attacker can use radio waves as the method of attack. This type of attack is identified as jamming attack. In the next section, we will also discuss jamming attacks on the link layer. We differentiate jamming attacks on the link layer from those on the physical layer by looking at whether or not link layer information is used for launching the attack. A physical layer jamming attack can be seen as a brutal force jamming attack. For example, constant jamming, deceptive jamming, and random jamming[1] are brutal force attacks. A constant jammer sends out a continuous random signal. The signal either prevents CSMA-based protocols from sending out a packet or damages legitimate packets. A deceptive jammer sends out legitimate packets, which in turn puts cooperative MAC layer protocols into the reception state. A random jammer alternates between sleeping state and jamming state, making a tradeoff between jamming efficiency and energy conservation. Most of the time, the sleeping interval is shorter than the packet interval, rendering the channel unusable.

Jamming attacks can be defended through active means. Active mechanisms to defend jamming-based attacks include spread-spectrum communication, fault-tolerant coding, and alternative communication techniques. The effectiveness of spread-spectrum communication relies on the assumption that an attacker can have only limited energy on a specific frequency band. A powerful attacker equipped with a broadband jamming device is not in our discussion scope, such as a hostile party in electronic warfare. Fault-tolerant coding can be used to defend random jamming, in that it can be modeled as bursty errors. Sensors can also use other means of communication when the normal radio channel is being jammed, such as infrared or optical.[2] Other means of communication include strategically placed wired pairs of sensors, a secondary frequency-hopping radio with some sensors, or uncoordinated channel hopping.[3]

In the case where active means are not applicable, such as constant jamming, passive means can be used. Also, low-cost sensors are less likely to have sophisticated components to efficiently detect jamming attacks. Thus, it is preferable for them to use passive defenses. In passive defenses, sensors in the jammed area sleep to avoid unnecessary energy consumption.

The jammed area is most likely to be confined since a large jammed area is easy to be differentiated from natural noise, exposing the attacker. Then sensors outside the jammed area can identify the jammed area and take proper actions. Since the power from a single source decays at a rate of d^2, where d is the distance from the energy source to the observation point, sensors close to the boundary of the jammed area can use a relatively high transmission power to supersede the decayed jamming power. Possible actions include routing traffic around the area so that its effects on the rest of the network are minimized.

5.2.1.2 Physical Removal and Relocation

Single-point sensor removal exposes a model similar to that of jamming attack. Instead of receiving the removal message from the sensor, neighboring sensors actively probe each other for possible removal. Assume that the attacker removes a group of neighboring sensors, the identification of the affected area can be made similarly.

Another type of physical relocation attack is called the wormhole attack. In this attack, the attacker can move a sensor to a distant location. Instead of leaving the original spot empty, the attacker places a bridging device at the spot. There is another bridging device traveling along with the sensor. Any traffic between the sensor and its original neighbors is forwarded back and forth on this bridging link. The bridging link can use another dedicated frequency to avoid interfering with the sensor network. In this way, the sensor being relocated can still respond to its neighbors. However, the designated area is not covered by the sensor.

To defend against wormhole attacks, topological information has to be somehow embedded in the packet. This is generally done by embedding distance information into the packet. For example, temporal leashes[4] can be used to detect a wormhole attack. In this approach, neighboring sensors are tightly synchronized. Since the distance between two neighbors is bounded, the time spent on packet propagation is also bounded. By carrying an encrypted timestamp with each packet, a sensor can effectively tell whether extra delays are involved in the packet propagation. If so, it is most likely caused by a wormhole attack.

Wormhole attacks can also be addressed through positioning techniques. Secure positioning techniques can be used to verify the location of each sensor. A relocated sensor cannot pass this test. The details of this defense will be discussed in the secure positioning section.

Another way to identify each neighbor is to associate each sensor with its physical characteristics, specifically its radio characteristics, i.e., its radio fingerprint. Although radio fingerprinting has been widely adopted in cellular communications to verify the authenticity of each transmission source, it may not be applicable to WSNs due to physical limitations of each sensor.

5.2.2 Node Compromise

If a sensor is compromised, the attacker can have full control of the sensor. There is really no defense against this attack. Compromised sensors can function normally under regular situations and inject false data only when needed. Since a sensor network is intended to detect random events, there is no predefined model to verify the sensed data by each sensor. If the sensed events at different sensors are independent, there is no way to tell whether each event is authentic or not.

Cooperation among legitimate sensors has been proposed to solve this problem. In this approach, each event has to be signed by a group of sensors. The effectiveness of this approach depends on applications where the fidelity of events can be independently verified by different sensors.

It also depends on the assumption that an attacker cannot compromise a group of sensors in a cost-efficient way.

Node compromise can affect protocols on all the upper layers. For example, on the link layer, a compromised sensor can selectively drop packets. On the network layer, it can drop packets based on network information or disrupt routing protocols by broadcasting fake routing information. On the application layer, it can insert fake data or drop critical application data. As discussed in the above paragraph, node compromise can be best defended using application-specific information. Thus, we will discuss the defenses against node compromise in corresponding sections.

To summarize, on the physical layer, node removal and relocation can be detected. Jamming attacks through node addition cannot be prevented. However, effective passive defense by identifying the jammed area is available. Certain social policy and appropriate actions should be planned based on such services in case of attacks. Random, elusive attacks may not be located, but it only degrades the sensor network performance. A limited number of attackers can degrade the performance to only a certain degree, and cannot completely stop the application.

5.3 Key Management

In this section, we discuss how keys are predistributed for authentication, how session keys are established, and how keys are managed in a sensor network.

There are various types of traffic in a sensor network to support a wide range of applications. Sensor-to-sensor traffic is common since cooperative detections are necessary for detection reliability, and energy efficiency. Broadcasting and multicasting are common for base stations to initiate certain detections.

We first discuss pairwise key establishment. Pairwise key establishment is to establish a key between an arbitrary pair of sensors. It is the basic mechanism for key management in sensor networks. Group-based key management and network-wide key management can be based on pairwise key management.

Then, we discuss system aspects of key management, including group key management and broadcasting key management.

5.3.1 General Key Management Concepts

Public key cryptography is widely used for session key agreement protocols. In public key cryptography, the private key is well protected while the public key can be openly distributed. The signing and verification are separated. The separation of secret storages and processes enables highly flexible secure system designs. This flexibility is highly desirable in a fully distributed environment such as WSNs.

However, public key cryptography has higher computation cost than symmetric key cryptography. A longer key is needed for the same security strength, which requires more storage and communication bandwidth. The increased computation, storage, and communication cost may not be met by tiny sensors.

Symmetric key predistribution has been proposed to address this problem. To enable two sensors to establish a secure link, they have to share a common secret. Two classes of predistributing the common secrets are being proposed. The first class uses a shared key space. The second class uses time-dependent secrets.

A key space is a set of different keys. In the shared key space approach, the key space is projected on the set of sensors so that there are intersections between different sensors. An ideal key space having only one key has minimum overhead. However, one compromised sensor leads to the whole network being compromised. Another extreme approach is to project a different key for every pair of sensors. However, this approach is not scalable since each sensor needs to store a key for every other sensor. Various approaches have been proposed between those two ends.

A time-dependent secret depends on time. It can be a sequence of secrets used at different times. It can also be the sequence of values computed from a common secret through a one-way function.

In the following sections, we first introduce some key space-based approaches. Then we discuss some time-dependent secret-based approaches.

5.3.2 General Model about Key Space-Based Predistribution

To enable pairwise key establishment, a pair of sensors needs some common knowledge for them to establish a secure link later. This sharing of common knowledge can be modeled as an edge between the pair of sensors. The whole network can be modeled as a connectivity graph. Any predistribution scheme will thus generate a predefined connectivity graph.

However, the development of a sensor network generally follows a process of in-lab programming, and then field deployment. Various deployment methods may be used depending on different applications. For example, sensors may be dropped out of a plane in a military application. In this situation, there is no precise control over the placement of individual sensors. Even in a more controlled deployment method, such as strategically placing sensors for environmental monitoring, it is better not to fix a specific sensor at a specific location due to possible human error in a large-scale deployment.

The predefined connectivity graph should be able to handle this deployment randomness. Certain graphs can achieve this goal. For example, in the previously discussed case where each sensor stores a different key for every other sensor, the connectivity graph is a complete graph. No matter how random the deployment is, a complete graph can always guarantee the connectivity between two neighbors.

A complete graph is certainly not scalable. If we relax the requirement of every sensor being able to establish a secure link with every other sensor directly, instead we only need the graph to be connected so that any pair of sensor can establish a secure link directly or with the help of other sensors.

Random graph theory provides some insight on this. According to Erdös,[5] a random graph can be connected with high probability if the average degree of each vertex is above a certain value. This can reduce the storage requirement at each sensor greatly compared with the complete graph approach.

5.3.3 Random Key Predistribution

The most basic key predistribution is the random key predistribution scheme. Random key predistribution uses a pool of keys. For each sensor, a group of keys are randomly drawn from the pool without replacement. This group of keys is called the key ring of each sensor. The size of the key pool, and that of the key ring determine the probability of two sensors sharing a common key.

Assume that the size of the key pool is P and the size of the key ring is k. Each key ring is randomly drawn without replacement of the keys to prevent duplicate keys. Then the probability of two key rings sharing a common key p' is:[6]

$$p' = 1 - \frac{(P - k)!}{(P - 2k)!P!} \tag{5.1}$$

Given a proper key ring size, two sensors share a key with high probability. However, this is based on a complete graph, where every sensor is possibly communicating with every other sensor in the network. In a real WSN, every sensor can only communicate with those sensors within their radio range. In this scenario, every sensor should have a high probability of sharing a key with its neighbors. Effectively, the key ring size should be larger than that in the complete graph.

An immediate question here is that the increased key ring size also leads to an even higher probability of sharing a key globally, which is not needed in a real network. In the following sections, we will discuss some other approaches that utilize deployment knowledge to avoid unnecessary overhead. This over-deployment also creates problems for an extremely nonuniform network. In the case where a sensor has only a few neighbors, it may not be able to establish secure connections with the latter, thus creating a disconnected network.

When choosing the parameters for this approach, the following factors should be considered: the cost of establishing secure links and resilience to node capture. Given a fixed size network with n sensors, assuming random deployment, the probability of the whole network being connected can be computed from the probability of each sensor being connected with another sensor according to the random graph theory by Erdös and Renyi.[5] In order for a random graph to be connected with probability P_c, the expected degree of each sensor is:

$$d = \left(\frac{n-1}{n}\right)(\ln(n) - \ln(-\ln(P_c))) \tag{5.2}$$

Given a WSN, each sensor should be able to establish secure links with the expected number of its neighbors. Since the neighbors are randomly chosen, given a pool size P and a key ring size k, every sensor has a probability p' of sharing a key with one of its neighbors. The required p' can be derived from the expected number of its neighbors to the number of neighbors in its radio range. The pool size and key ring size can be chosen according to Eq. (5.1). In case there are not enough neighbors, a sensor can extend its radio range by increasing the transmission power.

There are still some neighbors that a sensor does not share a key with. To enable direct communication with those sensors, it can establish a secure connection through other sensors. Since the random graph of secure links is almost certainly connected, a path can be found between any two sensors with high probability. Also, the probability of not sharing a key decreases exponentially as the hop length of the path increases. This means that most connections can be established through a limited length path.

If a sensor is compromised, assuming the worst scenario, all the keys in the sensor's key ring are known to the adversary. In such a case, the adversary may use those keys to gain access to the system, or worse, to launch attacks. While the method to detect attacks depends highly on the specific application, it is necessary to have a method to revoke the compromised sensor once it is detected. To prevent the adversary from accessing the network, all keys in the key ring have to be revoked. Since the keys are shared among sensors, revoking even one key disables some other links in the network. However, revoking one compromised sensor can affect the connectivity of the network only slightly. This is due to two reasons: the probability of another sensor sharing a significant number of keys with the compromised sensor is small; the revoked keys take only a very small part of the whole key pool.

The revocation process is simple. The list of keys being revoked is distributed by the base station. The list is signed by the key distribution center

(KDC). Every sensor verifies the authenticity of the list using the KDC's public key. It then removes keys in the revocation list from its key ring. The revocation does not take extra storage. If the number of revoked sensors is big, the connectivity of the network may degrade significantly. In this case, the central controller can initiate rekeying and restore the connectivity of the network. Since sensors being compromised is a rare event, it can be expected that only after a long time, a network needs rekeying.

Another type of key predistribution is the key space approach. Instead of relying on independent key sets, a key space approach uses some mathematical space definition, which we will discuss later. Two types of key spaces have been proposed. The first type is based on a matrix. The second type is based on a bivariate polynomial. Both key space schemes exhibit a threshold property. There exists a positive constant λ. When there are less than λ compromised sensors, the rest of the network is still perfectly secure. But once λ sensors are compromised, all the links in the network are no longer secure.

The storage and computation overhead of key space approaches are proportional to λ. In a large sensor network, using the key space scheme is not scalable due to a possible larger number of sensors being compromised. To make the key space approach scalable and still preserve the threshold property, both schemes propose to use multiple key spaces. The multiple key spaces approach is the combination of the key space approach and random key predistribution. In the multiple key spaces approach, each sensor is randomly assigned a set of key spaces. So each sensor can establish a secure connection with any other sensor sharing the same key space.

Since any two sensors within the same key space can establish a secure connection, the connectivity of the multiple key spaces approach is on the same level with that of random key predistribution. However, because of the threshold property, compromising a few sensors no longer affects any other part of the network.

The matrix-based key space is based on the following idea. First, a primitive element s from a finite field $GF(q)$ is chosen, where q is the smallest prime larger than the key size. The following matrix G of size $(\lambda + 1) \times N$ is generated:

$$G = \begin{bmatrix} 1 & 1 & 1 & \cdots & 1 \\ s & s^2 & s^3 & \cdots & s^N \\ s^2 & \left(s^2\right)^2 & \left(s^3\right)^2 & \cdots & \left(s^N\right)^2 \\ & & \vdots & & \\ s^\lambda & \left(s^2\right)^\lambda & \left(s^3\right)^\lambda & \cdots & \left(s^N\right)^\lambda \end{bmatrix}$$

The jth column of G, which is denoted by $G(j)$, is distributed to sensor j. The key distribution center also generates ω random symmetric matrices $D_1, D_2, \ldots, D_\omega$ of size $(\lambda + 1) \times (\lambda + 1)$. Then each tuple

$S_i = (D_i, G), i = 1, \ldots, \omega$, is a key space. ω matrices are computed as $A_i = (D_i G)^T$.

Then, for each sensor j, τ distinct key spaces are randomly chosen. For each space S_i selected for sensor j, the jth row of A_i, which is denoted as $A_i(j)$, is stored in this sensor. Those rows are the private keys kept at each sensor and never revealed.

When two sensors i and j are trying to establish a secure link, they first exchange information about the common key space. If they share a common key space S_k, sensor i sends $G(i)$ to sensor j and sensor j sends $G(j)$ to sensor i. Then a common key can be calculated at sensor i as $A_k(i)G(j)$ and at sensor j as $A_k(j)G(i)$.

The mathematical proof is briefly described as follows:

$$AG = (DG)^T G = G^T D^T G = G^T DG = (AG)^T$$

A bivariate polynomial key space is defined with a bivariate polynomial $f(x, y) = \sum_{i,j=0}^{\lambda} a_{ij} x^i y^j$ over a finite field $GF(q)$, where q is a prime number that is large enough to accommodate a cryptographic key and λ is the threshold. The bivariate polynomial has the property $f(x, y) = f(y, x)$. For each sensor i, the KDC computes a polynomial secret $f(i, y)$. By evaluating $f(i, y)$ at the point $y = j$, sensor i can generate a key for sensor j as $f(i, j)$. Similarly, sensor j can generate a key for sensor i as $f(j, i)$ with its own secret $f(j, y)$. Since $f(i, j) = f(j, i)$, a common key can be established between sensors i and j.

Note that both approaches use the same order of storage on each sensor. The storage is $O(t \log q)$.

The basic random key predistribution can also be modified in the following way. After the network is bootstrapped, the sensor can destroy all the key material it has, thus eliminating the risk of being compromised. Although it does not provide the ability to add sensors later, this scheme is acceptable for a small scale disposable sensor network such as a battlefield sensor network in a hostile area.

The pairwise random key predistribution has some natural optimizations. There is no inherent difference between each sensor holding a set of keys and a group of sensors holding a bigger set of keys. It is natural to introduce a cooperative key establishment procedure based on the basic key establishment procedure. In the cooperative key establishment scheme, a sensor will ask another t sensors to distribute different secrets to the sensor being authenticated. So a pair of sensors can establish a secure link only if both of them can establish secure links with the other t sensors. To eliminate a single point of failure, the paths used for establishing the key should be independent. Finding independent 2-hop paths is relatively easy and does not involve too much overhead.[7] For an arbitrary pair of sensors to establish a secure key through s independent paths, Wacker et al.[8] proposed a recursive way to construct s such multihop paths.

Another optimization is to use q common keys instead of one common key for authentication.[7] The attacker has less chance to compromise other links by compromising a limited set of sensors, because it is more difficult to have a combination of q common keys than having one common key. This scheme increases the difficulty of small-scale attacks. However, it also favors a large-scale attack since the overall key pool is smaller than that of the one common key scheme to guarantee the same connectivity.

As we pointed out before, certain graphs, such as a complete graph, can guarantee connectivity. However, a complete graph may not be necessary at all times. A strongly regular graph has better performance than random graphs.[9–11] For example, a complete bipartite graph is shown to have better connectivity than a random graph with the same number of keys in each sensor.[9] This regular graph approach can be combined with any key predistribution scheme. It works well for a small-scale network. For a large-scale network, each sensor can have only a limited number of neighbors. In this case, a random graph approximates a regular graph asymptotically. PIKE[12], which uses a virtual grid to derive pairs of sensors sharing common keys, can also be viewed as a special case of this type.

5.3.4 Group-Based Key Predistribution and Management

Another way to improve the basic random predistribution schemes is to reduce the randomness. Fortunately, a network deployment generally exhibits some locality. For example, a group of sensors are deployed around an area of interest. Several group key predistribution schemes have been proposed to utilize three advantages of a group-based approach. First, since a local group is generally much smaller than the whole network, a intra-group key predistribution can be more flexible. Second, the connectivity between groups is improved due to a larger number of neighboring sensors. Third, since the number of groups is generally limited, a complete connectivity graph on the group level is acceptable.

A group key predistribution scheme also has better scalability. In the original random key predistribution, the whole network is considered as a single entity. This is not reasonable for any large-scale deployment. A large-scale sensor network is likely to be divided into many small administrative domains. Dividing it into smaller groups allows more flexible network management. For example, each group can have different hardware.

Depending on how groups are defined, different key predistribution strategies can be used.[13–15] In the following, we briefly discuss two typical schemes.

5.3.4.1 Grid-Based Group Key Predistribution

Grid-based group key predistribution[13] uses two types of groups for key predistribution. Each sensor belongs to a deployment group. In each deployment group, any key predistribution scheme can be used. Several cross groups are also formed. Assume that there are n sensors and sensors have IDs from 1 to n. Each deployment group should have m sensors. The deployment groups and cross groups are formed in the following way. Each deployment group G_i and cross group G_i' are defined in the following way:

$$G_i = s_k : \quad k = (i-1) \times m + j, \quad j = 1, \ldots, m$$

$$G_i' = s_k : \quad k = i + (j-1) \times m, \quad j = 1, \ldots, m$$

Any key predistribution scheme can be used for each cross group.

5.3.4.2 Cross Function-Based Group Key Predistribution

Cross function-based group key predistribution[14] also defines two types of groups. The deployment groups are similarly defined. However, the cross groups are defined differently. For a deployment group G_u, t sensors can be chosen to have shared keys with another group G_v and vice versa. The ith sensor being chosen is calculated in the following way:

$$F_i(G_u, G_v) = (t(v-1) + i) \bmod m + (u-1)m$$

Those t sensors chosen from G_u are paired with the t sensors chosen from G_v. Each pair of sensors is installed with a common key.

5.3.5 Time-Dependent Key Predistribution

In a key space-based approach, a shared secret is randomly scattered around the network. This creates a challenge for an attacker to compromise the network. Essentially, a shared secret is hidden over space. Similarly, in a time-dependent key predistribution scheme, a shared secret is hidden over time. Generally a one-way function $f_k(v)$ is used to guarantee perfect security of the original secret, where k is the secret and v some random value. For example, f can be a symmetric encryption function.

A basic time-dependent key predistribution scheme assumes that the network is physically safe during its boot-up time so that only one master key, K_b, is required for the sensors to establish pairwise secure connections. After the network is booted up, K_b is erased from the memory. This approach is typically represented by localized encryption and authentication protocol (LEAP)[16] and opaque transitory master key (OTMK) establishment.[17] The difference between LEAP and OTMK is how they use the master key K_b.

In LEAP, a pairwise key is generated in the following way:

1. Key predistribution. Each sensor u computes its own master key $K_u = f_{K_b}(u)$.

2. Neighbor discovery. Each sensor u broadcasts a message with its identity and a nonce r and waits for each neighbor v to reply with the following message: $v|MAC_{K_v}(v|r)$. Since sensor u can compute K_v from K_b, it can verify v's identity.

3. Pairwise key generation. Sensor u computes the pairwise key between u and v as $K_{uv} = f_{K_v}(u)$. Sensor v can also compute this key. The identity of u is implicitly authenticated by subsequent message exchanges using key K_{uv}.

In OTMK, a pairwise key is generated in the following way:

1. Challenge. Sensor u chooses a random value r and broadcasts a challenge message $JOIN|E_{K_b}(u|r)$.

2. Response with pairwise key. Sensor v verifies u's identity. It then generates a random key K_{vu} and replies with the following message: $REPLY|E_{K_b}(v|r+1|K_{vu})$. Sensor u then decrypts the message with K_b and verifies its authenticity. K_{vu} is then used as the pairwise key between sensor u and v.

In LEAP, if the master key K_b is compromised during the bootstrapping process, all the pairwise keys in the network can be computed from it with the sensors' identities. OTMK improves LEAP by using the master key only for authentication and generation of the pairwise key independently from the master key and sensors' identities. The disadvantage of OTMK is that once the initial master key is erased, the network can no longer accept new sensors because there is only an ad hoc trust relationship between neighboring sensors left and the pairwise key does not reflect any information on the original master key.

In a time-dependent key predistribution approach, the ability to admit new sensors into the network without the risk of losing the original master key is the key challenge. Not being able to admit new sensors may be acceptable for a small-scale network. For a large network, not only should we have the option to add new sensors into the network later, it is also very hard to have a tight time schedule for the network to boot up synchronously.

LEAP addresses this problem by using the master key K_b to generate a sensor key K_u for each sensor u. So later when a new sensor v, which possesses K_b, joins the network, v can authenticate u by verifying u, which possesses K_u. Vice versa, u authenticates v implicitly by assuming that v holds K_b, thus being able to generate K_u and finish the pairwise key establishment procedure.

The basic OTMK scheme lacks such a mechanism since no information is left about a sensor previously possessing the master key. In an improved scheme, OTMK Scheme II, two pieces of information are generated before the master is being erased. First, a sensor u uses the master key K_b to generate a signature of its identity $MAC_{K_b}(u)$. Then, u generates several verifiers. Each verifier is composed of a random value r and its verification $y = f_{K_b}(r)$. When a new sensor v with K_b joins the network, v can verify u's identity by requesting u to show $MAC_{K_b}(u)$. Vice versa, u can challenge v with one of the random values and expect v to show $f_{K_b}(r)$. To prevent replay attacks or being cheated to disclose the keys, neither $MAC_{K_b}(u)$ nor $f_{K_b}(r)$ is disclosed directly. Instead, they are used as keys to encrypt a nonce. This enables legitimate sensors, which already have the keys, to verify whether the other party possesses the keys without risking the keys being intercepted by a third party.

In a time-dependent key predistribution scheme, the master key K_b is a single point of failure. To mitigate this problem, a time-dependent key predistribution scheme should put constraints on the master key K_b. LEAP uses a time limit for the master keys. Instead of one single master key K_b, the key predistribution center generates m master keys $K_1, K_2, \ldots, K_m$ for m consecutive deployment slots. Each sensor is loaded with the network master key K_i when it is deployed in the ith slot. It is also loaded with its own master keys for all the slots after the ith slot, i.e., for the jth slot $(j > i)$, $f_{K_j}(u)$ is loaded in sensor u. Essentially, each master key K_i is only useful for an attacker to gain access to the network during the ith slot.

OTKM uses a μTESLA[18] like protocol. Instead of one master key K_b, there are also m master keys $K_1, K_2, \ldots, K_m$ for m consecutive deployment slots. However, those m keys are not independent. They are on a one-way hash chain $K_m, K_{m-1}, \ldots, K_1$, i.e., $K_{i-1} = H(K_i)$. A new sensor deployed in a later slot can generate a previous master key, thus gaining access to the network. An old master key cannot be reused in a later slot since a new master key is expected. The freshness of the new master key can be verified by hashing it several times to get the old master key.

Compared with random-key predistribution schemes, a time-dependent key predistribution scheme has the advantage of guaranteeing security if the boot strapping can be protected against attackers. The disadvantage is that it requires network-wide loose synchronization.

5.3.6 Public Key-Based Key Establishment

Compared with random key predistribution and time-dependent key predistribution, public key-based key establishment relies on the inherent difficulty of cracking public key cryptography. However, a sensor has only highly constrained physical resources. As against symmetric key cryptography, public key cryptographic computation is considerably slower, which consumes extra energy and also may not satisfy the real-time requirement of quickly responding to detected events.

Table 5.1   Comparable Security Strengths for RSA and ECC

Security Lifetimes	Bits of Security	RSA Key Size (bits)	ECC Key Size (bits)
Through 2010	80	1024	160
Through 2030	112	2048	224
Beyond 2030	128	3072	256
N/A	192	7680	384
N/A	256	15,360	512

Table 5.1 gives the key sizes suggested by NIST for certain levels of security between symmetric cryptography, RSA public key cryptography, and elliptic-curve cryptography (ECC) public key cryptography.

It is easy to see that at least the 112 bits security level, is desired for a sensor network. If RSA is used, a 2048-bit key is needed. However, a sensor network has only a limited packet size. For example, a typical TinyOS packet has only 29 bytes. The shorter packet is needed because of the low transmission rate of the low-power, energy-efficient radio in a sensor node. Transmitting a 2048-bit key has too much overhead in this scenario. An ECC-cryptography key is much preferred in a sensor network. To provide the same level of security, only a 224-bit key is needed, which is about nine times less transmission overhead than an RSA key. Also, we will discuss in the following sections that practical research on implementing both RSA and ECC in an embedded environment has shown that a limited data path and memory storage in a microcontroller severely degrade the performance of large integer operations.

Although it is widely accepted that ECC signing is faster than RSA signing, it is also widely accepted that ECC verification is slower than RSA verification. This looks like a disadvantage to adopt ECC in sensor networks. However, in practical uses, peer authentication involves both signing and verification. To enable the ad hoc networking model of sensor networks, peer authentication is required. ECC holds significant advantages over RSA in the combined operations, due to the large overhead of RSA signing.[19]

To use public key-based cryptography, the public key of each sensor also needs to be authenticated. While public keys can be signed by the key predistribution center using public key cryptography, certain optimizations can be done in sensor networks to avoid high computational cost. The optimization uses the characteristic of a sensor network that sensors have relatively stable neighbors and hence can cooperate with each other. In the following, we first briefly discuss some optimizations on applying ECC in sensor networks. Then we discuss how public keys can be authenticated. For simplicity, we limit our discussions on ECC over prime field $GF(p)$.

ECC is built on top of finite field algebra over elliptic curves. Before two parties begin communicating, they must agree on a common set of parameters. The parameters include the curve to be used, the finite prime field $GF(p)$ on which the curve is defined, and the generator G of a cyclic group with order n defined on the curve. On top of ECC, the elliptic curve digital signature algorithm (ECDSA) is used for authentication and integrity. In ECDSA, the signer first chooses a private number d on the prime field $GF(p)$. The corresponding point $D = dG$ on the curve is sent out as the public key. The number d is kept as the private key. Denote the user as D. When D wants to sign a message, he first generates an ephemeral point $K = kG$ for a random number k in $GF(p)$. For the message m to be signed, its hash $H(m)$ is generated. Then the signature (r, s) is generated in the following way:

$$r = x_{\text{coord}}(K) \bmod n \tag{5.3}$$

$$s = k^{-1}(H(m) + dr) \tag{5.4}$$

Using its public key D, the signature can be verified in the following way:

$$R = (s^{-1}H(m))G + (s^{-1}r)D \tag{5.5}$$

$$r' = x_{\text{coord}}(R) \tag{5.6}$$

The signature is accepted if $r = r'$.

In ECC, the computation is dominated by point multiplications. Various optimizations have been proposed to reduce the computation on point multiplications. Cohen et al.,[20] proposed a mixed coordinate system to offer the best performance. Their method is implemented by Gura et al.,[21] on two 8-bit CPUs, one of which is the processor used in Crossbow Mica sensor nodes. The implementation of the algorithm on 8-bit CPUs is not trivial. Modular multiplication and squaring of large integers are expensive in those CPUs due to a limited-size data path and RAM. Gura et al. optimized the implementation through assembly language procedures. The evaluation shows that the relative performance of ECC point multiplication over RSA modular exponentiation increases as the word size of the processor decreases and the key size increases. This clearly favors using ECC in small, embedded devices like sensor nodes. Furthermore, ECC-160 point multiplication outperforms the RSA-1024 private-key operation by an order of magnitude and is within a factor of 2 of the RSA-1024 public-key operation. The performance results are shown in Table 5.2.

It is also necessary to point out that other implementations of ECC in sensor networks are also emerging while this chapter is being written. For example, Liu and Ning[22] wrote a package for TinyOS called TinyECC.

To further improve the performance of ECC, other optimizations target reducing the number and computational cost of point multiplications.

Table 5.2   ECC and RSA Performance on Two 8-Bit CPUs

Algorithm	Atmega128 at 8 MHz			CC1010 at 14.7456 MHz		
	Time (s)	Data (bytes)	Code (bytes)	Time (s)	Data ext + int (bytes)	Code (bytes)
ECC secp160r1	0.81	282	3682	4.85	180 + 86	2166
ECC secp192r1	1.24	336	3979	7.56	216 + 102	2152
ECC secp224r1	2.19	422	4812	11.98	259 + 114	2214
RSA-1024 public-key	0.43	542	1073	>4.48		
RSA-1024 private-key	10.99	930	6292	106.66		
RSA-2048 public-key	1.94	1332	2854			
RSA-2048 private-key	83.26	1853	7736			

Antipa et al.[23] improve the speed of ECDSA verification by up to 40%. Their method works by supplementing the original ECDSA signature with a few extra bits. From those bits, the original ephemeral point can be efficiently reconstructed, thus the computational cost on point multiplications is reduced. It is also necessary to point out that while this chapter is being written, improving ECC efficiency is still a very active research field.

Another way to reduce the computational cost of point multiplications is to use an asymmetric protocol, in which sensors have different capabilities, thus high-cost computation can be moved to a more powerful sensor. Huang et al.[24] proposed two roles in a sensor network, security managers and sensors. To reduce the overall protocol overhead, they use implicit certificates for authentication and key establishment. Their protocol also employs a hybrid key establishment procedure, in which symmetric key cryptography is combined with ECC to offload the computational cost to the more powerful security manager. This method is useful for a heterogeneous deployment, where the physical security of security managers can be guaranteed. For example, the security manager can be a mobile hand-held device that is used when the sensors are being deployed. Sensors can then establish secure connections with each other with the help of the security manager.

To authenticate the public key, each public key can be signed by the key predistribution center. The public key of the key predistribution center is stored in each sensor so that the authenticity of any certificate issued by it can be verified. Another way to authenticate a public key is to use the general assumption that sensors in a sensor network belong to the same group, thus group authentication methods may be used. A Merkle tree-based authentication protocol for sensor networks is proposed by Du et al.[25] In their protocol, the hash value of the concatenation of a sensor's ID and its public key is put at a leaf of the tree. An internal node stores the hash value of the concatenation of its two children. The procedure continues until the root is generated. The value in the root, essentially the signature of the whole group, is stored in each sensor. To authenticate another sensor, all the hash values on the path from the leaf, where the sensor is located to the root, have to be known by the authenticator. A basic scheme is to allow each sensor to store this "proof" and send them out for authentication. However, since two sensors within the same subtree can use the root of the subtree for authentication, in practice just a few hash values need to be sent. A further optimization is to put neighboring sensors into a smaller subtree by utilizing deployment knowledge. In this authentication method, only hashing is required. Although the communication overhead is higher, the total amount of energy used for the authentication is still less than that of the public key-based certificate method, due to the much faster hashing than public key cryptography operations.

5.3.7 Group Key Management in Sensor Networks

The key establishment between two sensors enables point-to-point secure communication. In sensor networks, cooperative processing is often required. For example, to determine the location of a signal source, triangulation is necessary, which, in practical scenarios, generally involves more than four sensors. In another example, TinyDB, a distributed dynamic database interface of sensor networks, multicasts a query into a sensor network depending on its scope.

One may argue that since ad hoc secure connections already exist between sensors, multicast can simply be realized through unicast connections. However, this is neither energy nor bandwidth-efficient. In this subsection, we limit our discussion to group key management. We first discuss local broadcasting key establishment, we then discuss network-wide multicasting key establishment. Broadcasting can be viewed as a special case of multicasting. Thus similar approaches can be used.

Local broadcasting is the basis for many distributed mechanisms to work efficiently in sensor networks. Thus it is necessary to protect the local broadcasting with cryptographic keys. A local broadcasting key can be securely distributed through existing secure point to point connections. In case of one sensor being revoked, a new key can be generated and redistributed.

A network-wide multicasting key is more difficult to manage due to a larger number of sensors in the group. To manage the multicasting tree scalably, a multiple-level scheme can be used.[26] Instead of a single key for the whole multicasting tree, a multiple-level scheme uses multiple keys, each subtree using a different key. In this way, joining and leaving can be handled locally.

The above scheme trusts intermediate sensors to deliver the message without modification. If one sensor is compromised, it may modify a message, resulting in its children receiving the wrong message. To prevent such attacks, in addition to the confidentiality provided by a multicasting tree, the message should also be signed by the base station. The signature can be generated through public key cryptography. However, if the multicasting is frequent, public key cryptography involves too much overhead. In this case, μTESLA[18] can be used.

5.3.8   Key Revocation

Sensors are susceptible to physical attacks. The probability of a sensor being compromised is thus much higher than a physically protected system. If a sensor is compromised, we need to remove it from the network so that the attacker cannot gain access to the network through this sensor. Intuitively, a sensor can be revoked by removing its identity from the network, canceling its membership in any multicasting group. A naive way to remove its identity is to refresh the network with new identities for legitimate sensors. However, this involves too much overhead. So, in practice, the identity of the sensor being compromised has to be distributed to the whole network so that individual sensors can handle the revoked sensor directly.

The detection and revocation of compromised sensors can be done centrally. Raw detection has to be done in a distributed way owing to the distributed nature of sensor networks. Raw detection should be delivered to the base station reliably. The revocation announcement should also be delivered to all sensors using reliable broadcasting. The detection and the announcement should be signed by a sensor and the base station, respectively.

Centralized detection and revocation involve much overhead. Also, the base station is a single point of failure. On the contrary, distributed detection and revocation utilize the network itself. Communication can be local and the single point of failure is eliminated. However, the distributed revocation protocol itself is susceptible to attacks from compromised sensors. To work correctly, a distributed revocation protocol relies on the existence of a majority of uncompromised nodes.

For symmetric key predistribution schemes, a sensor can be revoked if all the other sensors sharing some keys with it remove their shared keys. We do not consider the transient keys generated. Transient keys only reflect temporary relationships. These temporary relationships are highly localized while the predistributed keys are equivalent to identities. For asymmetric key schemes, its identity has to be revoked network-wide due to the global trust on its certificate.

Distributed revocation relies on distributed voting. For distributed voting to work correctly, the propagation of the votes should not be susceptible to denial-of-service attacks, including selectively dropping and delaying votes. The voting should be loosely synchronized so that a majority can be established within a stipulated time. Chan et al.[27] used this approach. For each sensor, a group of participants are designated as its voting members. The assignment of the membership is mainly based on relationships. In a symmetric key predistribution scheme, the members include all the sensors that share common keys with it. The vote by each member is authenticated with a Merkle tree-based authentication scheme so that all the other members can verify it. A maximum propagation bound is assumed so that each vote can be reliably received by all the other members. Consecutive revocation sessions are used so that a majority can be received in a set time.

5.4   Link Layer Security

The basic function of the link layer in sensor networks is to discover neighbors and establish radio connections with them. Security on the link layer is to guarantee that functions are executed correctly and confidentially. We assume that all link layer packets are encrypted. Otherwise, an attacker can have full access to the link layer. The set up of encrypted links can be achieved through key establishment protocols described in the previous section.

The discovery of neighbors is based on local broadcasting. The announcement can be signed by using different key management techniques introduced in the previous section. The announcement may be jammed and forwarded to a remote location. The announcement being forwarded to a remote location is a worm-hole attack and can be defended as we discussed in the physical layer security section.

Defenses against brute force jamming attacks have been discussed in our physical layer security section. In this section, we defend against intelligent jamming attacks that utilize link layer information.

The cooperation among neighbors on the link layer may be taken advantage of by an attacker. For example, in S-MAC,[28] a sensor switches between the listen state and the sleep state periodically to save the energy spent on unnecessary idle listening. Schedules among neighbors are synchronized to facilitate communications among them. However, packet transmission only occurs when sensors are awake. This synchronized schedule can easily be observed by an attacker.[29] An attacker can sleep to save energy when sensors are sleeping and reactively jam the channel when sensors are awake.

To hide the sleeping schedule, S-MAC can be modified in the following way. Each time sensors are awake, they set up a new random schedule. In this way, an attacker can no longer predict when the sensors are awake. An attacker has to listen to the channel continuously to jam it. This method is essentially a time-hopping protocol. The disadvantage is that sensors may be out of synchronization if the channel is very unreliable. To improve the protocol, sensors may set up a pseudo-random schedule in which the sleeping schedule is derived from a seed by using a pseudo-random function.

To support confidentiality, a link layer security protocol should be used. The link layer security protocol should possess the following properties: access control, message integrity, and message confidentiality. Message integrity can be protected with a message authentication code (MAC). A MAC is generated by encrypting the hash value of the message content. Since only legitimate sensors can have the correct keys and generate legitimate MACs, access control can be enforced through session key establishment and message integrity. Message confidentiality in sensor networks should be carefully designed. A typical packet in sensors networks is only 30 bytes long, which means that there are at most 2^{240} different packets. The security strength of these packets is much weaker than that of the packets in traditional networks.

However, a closer look at the sensor networks revealed that the low duty property of a sensor network helps mitigate the above drawback. Karlof et al.[30] showed that in a typical sensor network with a 19.2 kb/s channel, it will take an attacker over 20 months to try all 2^{32} possible packets for a given 4-byte MAC. To reduce the overhead associated with authentication and encryption, TinySec[30] takes advantage of this property.

TinySec provides two security modes: authenticated encryption (TinySec-AE) and authentication only (TinySec-Auth). TinySec-AE encrypts the data and authenticates the entire packet with a MAC. To further reduce overhead, TinySec-Auth only authenticates the entire packet with a MAC and leaves the data intact. The default mode in TinySec is TinySec-Auth.

Each TinySec packet has the format shown in Figure 5.1.

TinySec uses cipher block chaining (CBC) as the encryption scheme. To use CBC, a random initial vector (IV) is needed for each packet to

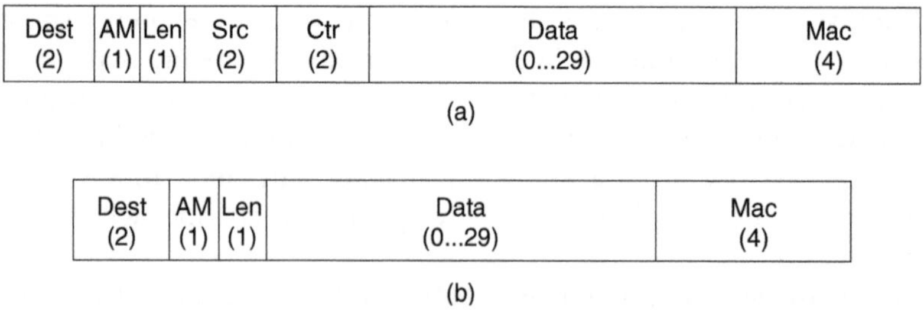

Dest (2)	AM (1)	Len (1)	Src (2)	Ctr (2)	Data (0...29)	Mac (4)

(a)

Dest (2)	AM (1)	Len (1)	Data (0...29)	Mac (4)

(b)

Figure 5.1 TinySec packet format. The number below each field name is the size of the field in number of bytes. (a) TinySec-AE packet format. (b) TinySec-AUTH packet format.

achieve semantic security. TinySec uses an 8-byte IV. To reduce space overhead, TinySec uses packet destination (Dest), active message type (AM), packet length (Len), and packet source (Src) as part of IV. A counter (Ctr) is used to guarantee that each packet has a different IV. Although a 2-byte counter seems small, the same value will only be reused after 45 days if the data rate is one packet per minute per sensor. During such a period, the cryptographic key can be updated to avoid IV reuse.

TinySec-AE encrypts the data payload. Both TinySec-AE and TinySec-Auth use MAC to protect the whole packet. The MAC is generated through CBC-MAC. The default block cipher in TinySec is 64-bit Skipjack. For simplicity, TinySec uses network-wide keys. Specifically, two network-wide keys are defined for all the sensors in a network: an encryption key and an authentication key. The encryption key is used by TinySec-AE to encrypt the data payload. The authentication key is used by both TinySec-AE and TinySec-Auth to generate the MAC.

5.5 Network Layer Security

A sensor network is generally a multihop network. Each sensor has a limited radio range. The deployment of sensor nodes into the area of interest requires an ad hoc network to be formed for collaborative processing and for communicating to a limited number of base stations.

Most traffic in sensor networks can be classified into three categories:

- Many-to-one. This type of traffic can be network-wide. For example, sensors report to the base station. It can also be local. For example, sensors report to a cluster head.
- One-to-many. This is generally the broadcasting from the base station.
- Local communication. This type of traffic is typically represented by local collaborative processing. The communication is denoted as local in which the observed physical phenomena is generally localized.

To accommodate these traffic types, several routing protocols have been proposed. TinyOS beaconing is a one-to-many routing protocol. It also builds reverse paths and serves as a many-to-one routing protocol. Directed diffusion[31] is a data-centric routing protocol. It uses flooding to propagate data interests across the network. Reverse paths are used for reporting events to the interest source. A preferred path can be reinforced. Geographic routing is also a natural choice for routing in sensor networks. This is because most sensor network applications already use a geographic model. For example, many-to-one and one-to-many traffic are generally intended to cover an area of interest. Direction information is also reflected in mobile tracking applications.

The goal of the security on the network layer in sensor networks is to protect the above traffic efficiently. In traditional networks, most traffic is end to end. Thus an end-to-end security solution is more efficient. The network layer instead focuses on providing high availability. In a sensor network, the network itself is the application. To provide sufficient protection for the application, security has to be put into the network layer.

The basic mechanism to deliver a message scalably in sensor networks is flooding. TinyOS beaconing and directed diffusion are typical examples of flooding. A nonauthenticated broadcasting protocol is easily defeated by an attacker. Thus we do not consider such a case. There are secure broadcasting protocols that are already in the literature.[18,32] We assume that the authenticity of each packet can be verified.

Intuitively, flooding seems the most resilient against attacks since every sensor is forwarding the packet, thus great redundancy has been provided. This is true for one-to-many traffic. However, in most one-to-many protocols, a reverse many-to-one routing tree is built to facilitate data reporting. The construction of the tree purely relies on the last hop from which a sensor hears the packet. In other words, it depends on the order in which each sensor receives the packet. However, this order is not protected. Karlof and Wagner[33] have shown that by using worm holes, an attacker can launch a sinkhole attack, redirecting traffic through a compromised sensor. Then it can efficiently launch other types of attacks, such as selective forwarding.

To defend against this attack, information about the source in the structure of the network has to be associated with each packet. An ideal solution is to have each sensor possess the full knowledge of the network. This is of course not practical. A more practical solution is to use geographic routing. Then, the base station can be conveniently represented with its geographic position. The geographic positions of sensors can be calculated and verified using different positioning techniques.[34–51] This type of calculation and verification often involves sensors in the whole neighborhood. Thus, it is difficult for an attacker to launch a successful attack with only one compromised sensor.

The basic flooding can be combined with other metrics. For example, in minimum cost forwarding, a cost is associated with each link in the network. The cost is application-dependent: energy, latency, loss, etc. It can be dynamically changed to reflect the changing condition of the network. A shortest path tree originating from the base station is constructed through flooding. Minimum cost forwarding can be attacked since there is no verification of the cost advertised by a sensor. A compromised sensor can advertise a zero cost to attract all neighboring traffic. An immediate fix to limit the attack is to have every sensor watch for each other's advertisement. The advertised values should be above the minimum value among all the values received by every sensor.

To reduce the overhead associated with flooding, fixed paths can be built once the relationship among different sensors has been revealed. However, fixed paths are susceptible to attacks. A path is a chain of trust. Any compromised sensor on the path can hijack the whole trust chain. To counteract this attack, multiple paths can be constructed.[52,53]

5.6 Application Layer Security

5.6.1 Secure Localization

Position information in sensor networks has wide applications. It is thus important to guarantee the correctness of this position information. Note that due to imprecise measurement of the relative distance between two sensors, the precision of the position information is only limited. Secure localization is not intended to guarantee the precision of localization. Instead, its main purpose is to make sure that the error in the calculated position cannot be over a limit. Effectively, any sensor cannot lie about its position in a significant way. Any application using the position information should be aware of this.

All localization techniques are based on solving a group of equations in a distributed way. A group of reference points are distributed in the network to help calibrate the real positions of sensors. To improve the estimation precision, multiple rounds of calculations are involved. In those calculations, it is difficult for a single compromised sensor to lie about its position. For example, neighboring sensors surrounding a sensor can form triangles to verify its position claim.[34,54] A range-independent position verification scheme based on intersections among beaconing sites is proposed.[42] Computation methods that can tolerate a few outliers are proposed.[43] Those mechanisms can be built into the network localization process. Their correctness relies on a majority of honest sensors.

While the reference points are generally base stations and are assumed well protected physically, there should still be some way to detect a malicious reference point due to its critical role on the whole network. Liu et al.[57], proposed a method to filter out malicious beacon signals as long as the majority of beacon signals are correct.

5.6.2 Secure Time Synchronization

Time information also has wide applications in sensor networks, especially for collaborative signal processing among sensors. For example, to identify the location of a signal source, sensors compare their received signals with each other and agree on a maximum-likely correlation. Then the time of arrival (ToA) differences among those signals can be used by the sensors to estimate the location of the signal source.

To use the time information, a common reference needs to be established. This is done through synchronization. A time synchronization protocol should address three factors causing time differences among different sensors: initialization differences, skew errors, and drift errors. Initialization differences are introduced when different sensors boot up at different times. Skew errors are caused by slight differences of the hardware specifications of different sensors. Drift errors are caused by ambient environmental factors.

Time synchronization between two sensors is called sender–receiver time synchronization.[55,56]. It can be realized by exchanging the clock information and measuring the round-trip time to estimate the communication delay. The protocol packets can be protected through cryptographic keys. However, a packet may be delayed by an attacker through caching the packet and jamming the receiver. A sensor can only limit the attack by bounding the round-trip time. Apparently, the tighter the upper bound of the round-trip time, the better the precision of the time synchronization. For this reason, time information in each packet is often stamped by media access control (MAC) layer programs when the packet is actually sent out.

Group time synchronization can be achieved by using one sensor as a common reference. Besides the defense against external attackers mentioned above, attacks from compromised sensors can also be detected through collaborations among sensors in the group. The detection is based on a triangle principle,[55] that is, the summation of the pairwise time offsets along a triangle of sensors should be zero. Deviations from zero caused by drift and skew errors should be bounded, and can be tolerated with a threshold-based policy. However, this method of detecting internal attackers involves message exchanges among different neighboring sensors, thus incurring higher costs.

Multihop time synchronization possesses some similar features as those in secure routing. If it is based on a chain of sender–receiver synchronizations, a single compromised sensor can affect the whole chain. Either the above defense proposed in the pairwise sender–receiver time synchronization or a multipath solution can be used.

Network-wide time synchronization is a generalization of multihop time synchronization. TPSN[57], and FTSP[58] are two typical protocols. TPSN uses a fixed spanning tree for time synchronization, in which a child synchronizes with its parent. So a compromised subtree root can affect all its descendants. FTSP uses flooding for time synchronization, in which a random sensor is elected as the flooding source. Although flooding is more resilient against attacks than a fixed spanning tree, the flooding source is a single point of failure. To protect the root in TPSN and the flooding source in FTSP, a set of roots can be used and each one of them should be able to set up authenticated broadcasting. To protect the routing structure in the network-wide time synchronization, a sensor should use time information from multiple neighbors.[56]

5.6.3 Secure Aggregation

Aggregation is another important distributed function provided by sensor networks. Owing to the locality of events, raw data collected by different sensors often has great redundancy. To improve energy efficiency, the raw data should be aggregated, with redundancy removed before it is transmitted to the base station. However, since the aggregation is a distributed function, it is susceptible to attacks by compromised sensors. The compromised sensors can either be sensors reporting raw data to aggregators or aggregators themselves.

Similar to the localization mentioned in the previous section, it is impossible to guarantee an exact aggregation result due to the noisy nature of the raw data. Instead, secure aggregation is intended to bind the errors caused by the compromised sensors.

The aggregation can be modeled as a function f of raw data $x_1, \ldots, x_n$ from n different sensors, i.e., $y = f(x_1, \ldots, x_n)$. A compromised sensor can contribute an arbitrary value, thus affecting the aggregation result. Depending on how the function f is defined, the effects of a corrupted input on the aggregation result varies. For example, if f is the average of the inputs, a highly deviated value can arbitrarily change the aggregation result. Wagner[59] points out that *average, sum, minimum,* and *maximum* are not secure while *count* can be secure. A general model for the function f is to adopt the statistical estimation theory. Some general methods can be applied to improve the robustness of the estimator. For example, in truncation, a sensor reading outside the valid range $[l, u]$ is ignored. In trimming, the highest 5% and lowest 5% readings are ignored, and only 90% readings in the middle are used to compute the aggregation.

To defend against compromised aggregators, Przydatek et al. proposed an aggregate-commit-prove[60] framework so that cheating can be detected by the base station. When the aggregator does the computation, it will commit to the set of raw sensor readings. The commitment is generated through a Merkle hash tree over the set of raw data. When the base station receives the aggregation result and the commitment, it does a random sampling of the raw data and checks that the random sample is consistent with the aggregation result and the commitment.

References

1. W. Xu, W. Trappe, Y. Zhang, and T. Wood. The feasibility of launching and detecting jamming attacks in wireless networks. In *Proceedings of the 6th ACM International Symposium on Mobile Ad Hoc Networking and Computing*, May 2005.
2. I.F. Akyildiz, W. Su, Y. Sankarasubramaniam, and E. Cayirci. Wireless sensor networks: A survey. *Computer Networks*, 38(4):393–422, 2002.
3. M. Cagalj, S. Čapkun, and J.P. Hubaux. Wormhole-based anti-jamming techniques in sensor networks. In *IEEE Transactions on Mobile Computing*, 6(1): 100–114, 2007.
4. Y.C. Hu, A. Perrig, and D.B. Johnson. Packet leashes: A defense against wormhole attacks in wireless networks. In *Proceedings of IEEE Conference on Computer Communications (INFOCOM'03)*, 2003.
5. J.H. Spencer. *The Strange Logic of Random Graphs* Berlin: Springer, c2001.
6. L. Eschenauer and V.D. Gligor. A key management scheme for distributed sensor networks. In *Proceedings of ACM CCS'02*, October 2002.
7. H. Chan, A. Perrig, and D. Song. Random key predistribution schemes for sensor networks. In *Proceedings of ACM CCS'03*, October 2003.
8. A. Wacker, M. Knoll, T. Heiber, and K. Rothermel. A new approach for establishing pairwise keys for securing wireless sensor networks. In *Proceedings of ACM SenSys '03*, November 2005.
9. J. Lee and D.R. Stinson. Deterministic key predistribution schemes for distributed sensor networks. In *11th International Workshop on Selected Areas in Cryptography (SAC 2004)*, 2004.
10. D. Sánchez and H. Baldus. A deterministic pairwise key pre-distribution scheme for mobile sensor networks. In *Proceedings of SecureComm 2005*, 2005.
11. J. Lee and D.R. Stinson. A combinatorial approach to key predistribution for distributed sensor networks. In *Proceedings of the IEEE Wireless Communications and Networking Conference (WCNC'05)*, 2005.
12. H. Chan and A. Perrig. Pike: Peer intermediaries for key establishment in sensor networks. In *Proceedings of IEEE Conference on Computer Communications (INFOCOM'05)*, 2005.
13. D. Liu, P. Ning, and W. Du. Group-based key pre-distribution in wireless sensor networks. In *Proceedings of the 2005 ACM Workshop on Wireless Security (WiSe 2005)*, 2005.
14. L. Zhou, J. Ni, and C.V. Ravishankar. Efficient key establishment for group-based wireless sensor deployments. In *Proceedings of the 2005 ACM Workshop on Wireless Security (WiSe 2005)*, 2005.
15. Y. Zhou, Y. Zhang, and Y. Fang. Llk: A link-layer key establishment scheme in wireless sensor networks. In *Proceedings of the IEEE Wireless Communications and Networking Conference (WCNC'05)*, 2005.
16. S. Zhu, S. Setia, and S. Jajodia. Leap: Efficient security mechanisms for large-scale distributed sensor networks. In *Proceedings of the 10th ACM Conference on Computer and Communications Security (CCS 2003)*, 2003.

17. J. Deng, C. Hartung, R. Han, and S. Mishra. A practical study of transitory master key establishment for wireless sensor networks. In *Proceedings of SecureComm 2005*, 2005.

18. A. Perrig, R. Szewczyk, V. Wen, D. Culler, and J.D. Tygar. Spins: Security protocols for sensor networks. *Wireless Networks Journal (WINET)*, 8(5):521–534, 2002.

19. A.S. Wander, N. Gura, H. Eberle, V. Gupta, and S.C. Shantz. Energy analysis of public-key cryptography for wireless sensor networks. In *the 3rd IEEE International Conference on Pervasive Computing and Communications (PerCom 2005)*, 2005.

20. H. Cohen, A. Miyaji, and T. Ono. *Efficient elliptic curve exponentiation using mixed coordinates*, pp. 51–65. London, UK, Springer-Verlag, 1998.

21. N. Gura, A. Patel, A. Wander, H. Eberle, and S.C. Shantz. Comparing elliptic curve cryptography and rsa on 8-bit cpus. In *Proceedings of the 2004 Workshop on Cryptographic Hardware and Embedded Systems (CHES 2004)*, August 2004.

22. A. Liu and P. Ning. Tinyecc: Elliptic curve cryptography for sensor networks (version 0.2), http://discovery.csc.ncsu.edu/software/TinyECC/, September 2006.

23. A. Antipa, D.R.L. Brown, R.P. Gallant, R.J. Lambert, R. Struik, and S.A. Vanstone. Accelerated verification of ecdsa signatures. In *Selected Areas in Cryptography*, pp. 307–318, Lecture Notes in Computer Science 3897 Springer Berlin 2006, ISBN 3-540-33108-5.

24. Q. Huang, J. Cukier, H. Kobayashi, B. Liu, and J. Zhang. Fast authenticated key establishment protocols for self-organizing sensor networks. In *Proceedings of the 2nd ACM International Conference on Wireless Sensor Networks and Applications (WSNA'03)*, 2003.

25. W. Du, R. Wang, and P. Ning. An efficient scheme for authenticating public keys in sensor networks. In *Proceedings of the 6th ACM International Symposium on Mobile Ad Hoc Networking and Computing*, May 2005.

26. J. Huang, J. Buckingham, and R. Han. A level key infrastructure for secure and efficient group communication in wireless sensor networks. In *Proceedings of SecureComm 2005*, 2005.

27. H. Chan, V.D. Gligor, A. Perrig, and G. Muralidharan. On the distribution and revocation of cryptographic keys in sensor networks. *IEEE Transactions on Dependable and Secure Computing*, 2(3):233–247, 2005.

28. W. Ye, J. Heidemann, and D. Estrin. An energy-efficient mac protocol for wireless sensor networks. In *Proceedings of IEEE Infocom'02*, New York, June 2002.

29. Y.W. Law, L. van Hoesel, J. Doumen, P. Hartel, and P. Havinga. Energy-efficient link-layer jamming attacks against wireless sensor network mac protocols. In *Proceedings of the 3rd ACM Workshop on Security of Ad Hoc and Sensor Networks (SASN 2005)*, November 2005.

30. C. Karlof, N. Sastry, and D. Wagner. Tinysec: A link layer security architecture for wireless sensor networks. In *Proceedings of ACM SenSys 2004*, November 2004.

31. C. Intanagonwiwat, R. Govindan, and D. Estrin. Directed diffusion: A scalable and robust communication paradigm for sensor networks. In

Proceedings of the 6th Annual ACM/IEEE International Conference on Mobile Computing and Networking (MobiCom'00), August 2000.

32. L. Lazos and R. Poovendran. Secure broadcast in energy-aware wireless sensor networks. In *IEEE International Symposium on Advances in Wireless Communications (ISWC'02)*, 2002.

33. C. Karlof and D. Wagner. Secure routing in sensor networks: Attacks and countermeasures. *Ad Hoc Networks*, 1(2–3):293–315, 2003.

34. S. Čapkun and J.P. Hubaux. Secure positioning of wireless devices with application to sensor networks. In *Proceedings of IEEE Conference on Computer Communications (INFOCOM'05)*, 2005.

35. S. Čapkun, M. Hamdi, and J.P. Hubaux. Gps-free positioning in mobile ad-hoc networks. In *Proceedings of International Symposium on Communication Theory and Applications*, July 2001.

36. L. Doherty, K.J. Pister, and L. Ghaoui. Convex position esitination in wireless sensor network. In *Proceedings of IEEE Infocom'01*, May 2001.

37. X. Ji and H. Zha. Sensor positioning in wireless ad-hoc sensor networks using multidimensional scaling. In *Proceedings of IEEE Conference on Computer Communications (INFOCOM'04)*, March 2004.

38. D. Niculescu and B. Nath. Ad-hoc positioning system (aps). In *Proceedings of the IEEE GlobeCom'01*, November 2001.

39. C. Savarese, K. Langendoen, and J. Rabaey. Robust positioning algorithms for distributed ad-hoc wireless sensor networks. In *Proceedings of 2002 USENIX Annual Technical Conference*, pp. 317–328, June 2002.

40. T. He, C. Huang, B.M. Blum, J.A. Stankovic, and T. Abdelzaher. Range-free localization schemes for large sensor networks. In *Proceedings of the 9th Annual ACM/IEEE International Conference on Mobile Computing and Networking (MobiCom'03)*, September 2003.

41. L. Hu and D. Evans. Localization for mobile sensor networks. In *Proceedings of the 10th Annual ACM/IEEE International Conference on Mobile Computing and Networking (MobiCom'04)*, September 2004.

42. L. Lazos and R. Poovendran. Serloc: Secure range-independent localization for wireless sensor networks. In *Proceedings of the 2004 ACM Workshop on Wireless Security (WiSe 2004)*, 2004.

43. Z. Li, W. Trappe, Y. Zhang, and B. Nath. Robust statistical methods for securing wireless localization in sensor networks. In *Proceedings of Information Processing in Sensor Networks (IPSN'05)*, 2005.

44. D. Moore, J. Leonard, D. Rus, and S. Teller. Robust distributed network localization with noisy range measurements. In *Proceedings of ACM SenSys 2004*, November 2004.

45. N. Patwari and A.O. Hero III. Using proximity and quantized rss for sensor localization in wireless networks. In *Proceedings of the 2nd ACM International Conference on Wireless Sensor Networks and Applications(WSNA'03)*, September 2003.

46. A. Savvides, C. Han, and M.B. Strivastava. Dynamic fine-grained localization in adhoc networks of sensors. In *Proceedings of the 7th Annual ACM/IEEE International Conference on Mobile Computing and Networking(MobiCom'01)*, July 2001.

47. A. Savvides, H. Park, and M. Srivastava. The bits and flops of the n-hop multilateration primitive for node localization problems. In *Proceedings of 1st ACM International Workshop on Wireless Sensor Networks and Applications (WSNA)*, September 2002.
48. Y. Shang and W. Ruml. Improved mds-based localization. In *Proceedings of IEEE Conference on Computer Communications (INFOCOM'04)*, March 2004.
49. Y. Shang, W. Ruml, Y. Zhang, and M. Fromherz. Localization from mere connectivity. In *Proceedings of the 4th ACM International Symposium on Mobile Ad Hoc Networking and Computing*, June 2003.
50. Z. Zhu and W. Shi. A practical, energy-efficient, robust localization algorithm in wireless sensor networks. Technical Report MIST-TR-2004-019, Wayne State University, November 2004.
51. D. Liu, P. Ning, and W. Du. Attack-resistant location estimation in sensor networks. In *Proceedings of Information Processing in Sensor Networks (IPSN'05)*, 2005.
52. D. Ganesan, R. Govindan, S. Shenker, and D. Estrin. Highly resilient, energy efficient multipath routing in wireless sensor networks. *Mobile Computing and Communications Review (MC2R)*, 5(4):11–25, 2001.
53. F. Ye, S. Lu, and L. Zhang. Gradient broadcast: A robust, long-live large sensor network. Technical Report, University of California at Los Angeles Computer Science Department, 2001.
54. Y. Zhang, W. Liu, Y. Fang, and D. Wu. Secure localization and authentication in ultra-wideband sensor networks. *IEEE Journal on Selected Areas in Communications*, 24(4):829–835, 2006.
55. S. Ganeriwal, S. Čapkun, C. Han, and M.B. Srivastava. Secure time synchronization service for sensor networks. In *Proceedings of the 2005 ACM Workshop on Wireless Security (WiSe 2005)*, 2005.
56. M. Manzo, T. Roosta, and S. Sastry. Time synchronization attacks in sensor networks. In *Proceedings of the 3rd ACM Workshop on Security of Ad Hoc and Sensor Networks (SASN 2005)*, November 2005.
57. S. Ganeriwal, R. Kumar, and M. Srivastava. Timing-sync protocol for sensor networks. In *Proceedings of the 1st ACM SenSys'03*, November 2003.
58. M. Maroti, B. Kusy, G. Simor, and A. Ledeczi. The flooding time synchronization protocol. In *Proceedings of ACM SenSys 2004*, November 2004.
59. D. Wagner. Resilient aggregation in sensor networks. In *Proceedings of the 2nd ACM Workshop on Security of Ad Hoc and Sensor Networks (SASN 2004)*, 2004.
60. B. Przydatek, D. Song, and A. Perrig. Sia: Secure information aggregation in sensor networks. In *Proceedings of ACM SenSys 2003*, November 2003.

Chapter 6

Autonomous Swarm-Bot Systems for Wireless Sensor Networks

Shiann-Tsong Sheu, Jenhui Chen, and Shih-Lin Wu

6.1 Introduction

This chapter introduces a practical model, which is a distributed and autonomous swarm-bot system, for wireless sensor networks. This work is achieved with the cooperation of three universities, the National Central University, the Chang Gung University, and the Tamkang University, and sponsored by National Science Council (NSC) inter-institutional collaboration project, Taiwan.

Recent advances in electronics and wireless communications have enabled the development of low cost, low power, and multifunctional sensor nodes that are small in size and communicate with others in short range. These tiny sensor nodes consisting of sensing, data processing, and communicating components can be used to form wireless sensor networks further. The sensor networks can be used for various application areas, e.g., health care, monitoring environment, military usage, and home security. The application of wireless sensor networks needs to satisfy the constraints introduced by factors such as fault tolerance, scalability, cost, hardware, topology change, environment, and power consumption. Since the above

Figure 6.1 The sixth-generation ant-bot.

constraints are highly stringent and specific, new wireless ad hoc networking techniques are required, e.g., the hardware and software architectures, communication protocols, and application programs.

Moreover, the study of the distributed coordination function of robots has become increasingly feasible due to the development of wireless communications, distributed systems, artificial intelligence, sensor networks, and system-on-chip (SoC). The goal of the project is to take inspiration from ant self-assembling mechanisms and structures to translate some features into a robotic system. The project includes a design, hardware implementation, test, and use of this type of self-assembling, self-organizing, metamorphic robotic systems. An important part of the project consists in the physical construction of at least one *swarm-bot system*, that is, a self-assembling and self-organizing robot colony composed of a number (30–35) of smaller devices, called ant-bots as shown in Figure 6.1.

Our focus is not the mechanical aspects of mobile sensor nodes: the sensor, power, motor, and wheels. Instead, this integrated project focuses on designing and implementing the communication medium access control (MAC) protocols, communication strategies, embedded systems, positioning, and cooperative behaviors, that are the core technologies of the wireless sensor network system.

6.2 The System Architecture

The autonomous swarm-bot system consists of many small and movable ant-bots, which comprise several different functional sensors such as a magnetic compass sensor, ultrasonic sensors, the infrared sensors, and a

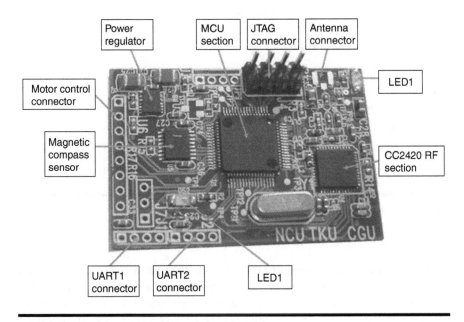

Figure 6.2 The embedded wireless sensor module.

light sensor. Each of them is also equipped with a wireless communication component for message exchange, and indoor positing usage. Aside from the sensing system of the sensor device, it can move according to commands issued by the administrator. It can imitate the behaviors of other sensor nodes to move or coordinate their actions by adopting a predesigned algorithm to archive some missions.

The hardware platform shown in Figure 6.2 is based on the Atmega128L microcontroller of Atmel Corporation and the cc2430 chip, which is the real SoC ZigBee/IEEE 802.15.4[15] designed communication chip developed by Chipcon company. Besides the central processing unit and communication component, the ant-bot comprising the mobile system and the power system, and the block diagram of the system is also shown in Figure 6.3. Owing to the battery energy and processing system constrain, the computing ability and memory capacity of the ant-bots are limited. Since these small ants can accept commands to achieve a given mission, several algorithms such as path planning, positing, and inter-ant collision prevention algorithms should be investigated.

6.3 Cooperative Localization Algorithm

The design of localization algorithm is one of the most important research issues for mobile sensor robots. Although many localization algorithms

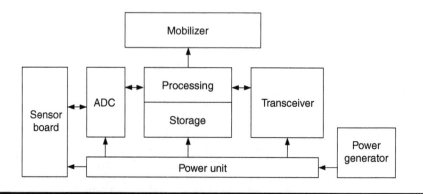

Figure 6.3   The hardware structure block diagram.

have been proposed, they usually assume that an infrastructure, which can provide a sign of location information such as landmarks, has been established in an operated environment or assume that some expensive and power-consumed devices, such as laser range finder, are mounted on these mobile sensor robots. As mentioned before, we have implemented a lot of cheap and autonomous ant-bots and each of them is equipped with a compute-limited processor plus a tiny memory, a wireless transceiver, and an electric compass. These ant-bots are organized as a team and one of them plays a coordinators' role. The others serve as members and communicate with one another by way of wireless transceivers. The coordinator is like a leader to guide its members to get to a target. The communication protocol within the team members is IEEE 802.15.4,[2] which is a low cost, low data-transfer rates (up to 250 kbps in normal), and low-energy consumption equipment for short range (≈ 10 m in indoor environment) wireless communications. Although our ant-bots are made up of these resource-constrained components, we have the most important unique character, i.e., *a lot of ant-bots* for an assigned mission. Thus we develop a novel and efficient localization algorithm by taking the advantage of the unique character.

A coordinate used in robot system can be cataloged into two types: *absolute coordinate* and *relative coordinate*. In absolute coordinate system, each robot calculates its coordinate with respect to an outside common coordinate system. In relative coordinate system, each robot determines its coordinate with respect to a reference point within the operated environment. The reference point can be the start coordinate of the coordinator of a team or a specific point at the place of a mission start, such as an entrance of a building needed to perform a mission. The localization of absolute coordinate system, using global positioning system (GPS) or preestablished landmarks, has been well studied in many research works[5,7–9] and will be not addressed here. For a cooperative robot team operated in an indoor or underground environment, there is no need to incur an extra cost to maintain a global absolute coordinate system. It is only necessary to know the relative coordinate to the original reference point for the robot team.

Our localization scheme is based on the relative coordinate system, and the reference point that is the start position of the selected coordinator of a team. How to elect the coordinator is stated in the previous section. Owing to the limited space, we only give an introduction to present the main idea of our localization scheme. Our algorithm includes two main phrases: an *initial* phrase and an *iterative* phrase. Following is an example combined with a figure to illustrate our algorithm, as shown in Figure 6.4.

At the initial phrase, we let the selected coordinator move around the circumference of a circle, and broadcast a sequence of beacons with its maximum transmitted power. Each beacon contains the coordinate of the selected coordinator, the coordinator's identification, and the transmitted power level. As shown in Figure 6.4, the coordinator (the ant-bot with blue color) turns around along the circumference of blue dotted circle with the radius of r_0. Every other ant-bot within the transmission range can determine its coordinate if it has received at least three different positions' beacons transmitted from the coordinator. This is because the trilateration scheme is employed in our localization algorithm. Figure 6.4 shows the green ant-bots within the second outer circle from the center point, that can determine their locations at the initial phrase.

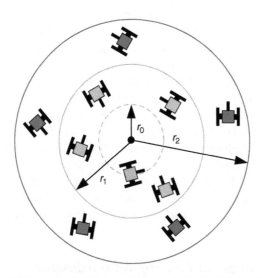

Figure 6.4   An illustration of the localization algorithm for our mobile sensor robots.

As shown in Figure 6.4, a condition may exist in which, some red ant-bots located in the two outermost circles cannot complete their localization process because, they are located far from the coordinator due to the limited transmission range. If this condition occurs, we perform the iterative phrase of our algorithm. At the iterative phrase, each ant-bot that has a coordinate at the end of the previous phrase will broadcast a sequence of beacons to its neighbors. The idea is similar to the initial phrase, that any red ant-bot can determine its own coordinate, if it has received at least three different green ant-bots' beacons. At the end of this phrase, we mark the red ant-bots with green color if the coordinate has been determined. The iterative phrase is performed until all ant-bots are marked with green color.

6.4 Foraging and Gathering

Another problem of the swarm-bot system is to avoid collisions among robots. The ant-bot adopts the received signal strength indication (RSSI) of wireless communications as the main tool for avoiding collisions. Meanwhile, we made use of the information about mutual orientations of ant-bots, using a two-stage threshold to achieve collision-free results. Furthermore, for power-saving, we propose a brand-new distributed approach that determines dynamic routes for each ant-bot instead of the *optimum routing search method*. Our plan includes integrating an algorithm with the system developed during the previous work, and testing simultaneously for its compatibility with cooperative project. To estimate effectively both the performance of the system and the feasibility, we have been devising a text/graphics arrangement (TGA) game as the end-product of this plan. It uses a sink node employed for announcing commands, to send out instructions regarding the required arrangement to ant-bots, which will then decide by itself as to how to do the team work, i.e., a process executed by distributed ant-bots. The results will determine whether the method is correct for controlling the behavior of group ant-bots. We plan to use the simulator, already under development, to do some predefined tests, and try to decrease the development time before proceeding to next stage of implementation. Since the plan first evolved 2 years ago, we have been executing the design of a TGA algorithm that focuses on instilling the ant-bots with both autonomy and coordination. Such use of coordination communications avoids collisions, and also saves time when executing the commands.

6.4.1 The Process of Electing a Coordinator

In our definition of sensor networks: the information detected by the ant-bots will make use of existing networks to send back to sink node.

In the field of sensor networks, each sensorial ant-bot is so individual, that they can be hardly controlled in the distributed systems environment. Thus, we turn the architecture of distributed systems into group computing systems.

A main ant-bot is elected from a group as the coordinator, which becomes in change of the entire operation of networks. The elected coordinator must communicate with other ant-bot members within the transmission scope of a sink node. Based on these requirements, we present a method that elects a coordinator, being in the light of values of RSSI. First, a sink node broadcasts a Coordinator Election message (see Figure 6.5a). After the ant-bots receive a Coordinator Election message, they back off for a short period based on the degree of RSSI (shorter the back-off period, farther the distance from the sink node). They separately broadcast a Coordinator Election message again when their back-off time exhausts. The other ant-bots will then stop the present back-off counting to execute the next back-off upon hearing another Coordinator Election messages during the computing back-off period. The ant-bots that receive more than two Coordinator Election messages, compute back-off period based on RSSI first (larger the back-off period, nearer the distance from the sender), as shown in Figure 6.5b. Then the ant-bot announces that it is the elector of the group, by broadcasting Coordinator message when the back-off period exhausts (see Figure 6.5c). The ant-bots that receive Coordinator message send an ACK Coordinator Message, and broadcast Stop Electing message. As soon as other ant-bots receive a Stop Electing message, they stop the process of Coordinator Election. This is the way a coordinator of an ant-bot group is elected, and located in the center of a group. Thus orientations and communications can benefit from this method. Further, the ID assignment for ant-bot is based on the MAC short address in each robot, and in this manner, we can save memory spaces to avoid building table for each ant-bot, so that the extra memory space is utilized effectively.

6.4.2   Position Measurement

The swarm-bot systems can be used to search for a target in an unknown environment with many obstacles, and need a collision-free path planning algorithm to reach the target. These mini ant-bots are equipped with several sensors, and wireless communication equipment. The sensor ant-bots can be equipped with the GPS. If there is no GPS (GPS-less), since all sensor ant-bots are in indoors then they need indoor position estimation methods to indicate the position. Most existing localization algorithms make use of the trilateration or the multilateration, based on range measurements obtained from the RSSI,[1] time of arrival (ToA),[9] time difference of arrival

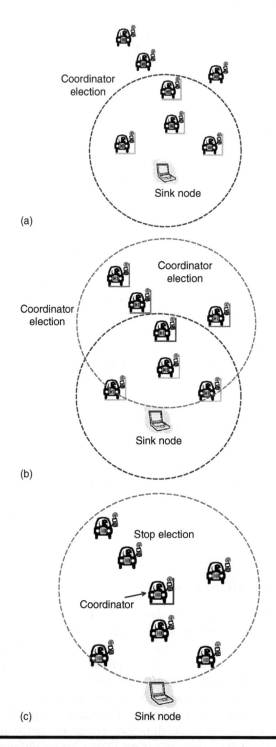

Figure 6.5 The process of the elected coordinator in multiple ant-bots. (a) An elected coordinator. (b) A second electing coordinator. (c) Stop election coordinator.

(TDoA),[8] and the angle of arrival (AoA).[7] The RSSI technique is employed to measure the power of the signal at the receiver. Because of its simplicity, it has been considered in our work.

6.4.3 Direction Measurement

Each ant-bot is equipped with an electronic compass, which can support a global direction of an ant-bot, to indicate its current direction.[10] With the electronic compass, the ant-bots can know which direction they are facing. The direction of the ant-bot d is represented as an angular magnitude by an electronic compass, and is denoted by ψ. The direction of the ant-bot i is defined as $d_i(\psi)$. The ant-bot can turn to any direction by giving an angle ϑ plus; it can set its own direction $d_i(\vartheta)$.

All ant-bots have to determine how to plan a route to the target, if the target's coordinate is given. First, the coordinator has to determine which is direction of the target. Assume that the coordinates of both the coordinator, and the target are (x_c, y_c) and (x_t, y_t), respectively. We denote the vector from the coordinator to the target by v, where

$$v = (x_t - x_c, y_t - y_c) \tag{6.1}$$

The length of the vector v is defined as

$$|v| := \sqrt{(x_t - x_c)^2 + (y_t - y_c)^2} \tag{6.2}$$

According to the vector v, we can determine in which quadrant the target is. The conditions are given as follows:

$$\begin{cases} \text{if } x_t - x_c \geq 0 \text{ and } y_t - y_c \geq 0, & \text{then target is in (I)} \\ \text{if } x_t - x_c < 0 \text{ and } y_t - y_c > 0, & \text{then target is in (II)} \\ \text{if } x_t - x_c \leq 0 \text{ and } y_t - y_c \leq 0, & \text{then target is in (III)} \\ \text{if } x_t - x_c > 0 \text{ and } y_t - y_c < 0, & \text{then target is in (IV)} \end{cases} \tag{6.3}$$

Figure 6.6 shows the relationship between the current direction of a coordinator $d_c(\psi)$, and the vector v of the coordinator to the target, where the angle ϕ is defined as the angular magnitude of v, and the θ is defined as the angular magnitude to the X-axis. θ can be obtained by using tangent theory given by

$$\theta = \tan^{-1}\left(\frac{y_t - y_c}{x_t - x_c}\right) \tag{6.4}$$

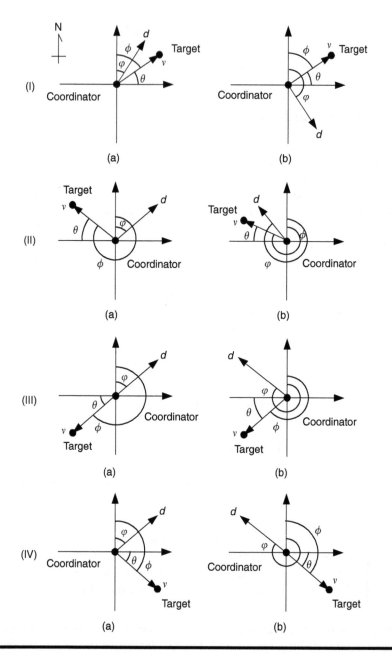

Figure 6.6 **The relationship between the *v* and the *d* in the four quadrants. All angles *ψ*, *φ*, and *θ* can be obtained by a electronic compass.**

According to (1.3) and (1.4), the turning angle $\Delta\gamma$ is defined as the angular magnitude, and the ant-bot has to turn to the target shown as below:

$$\Delta\gamma = \phi - \psi \qquad (6.5)$$

where

$$\phi = \begin{cases} \pi/2 - \theta, & \text{when } v \in \text{(I)} \\ \pi/2 + \theta, & \text{when } v \in \text{(II)} \\ 3\pi/2 - \theta, & \text{when } v \in \text{(III)} \\ 3\pi/2 + \theta, & \text{when } v \in \text{(IV)} \end{cases} \qquad (6.6)$$

When $\Delta\gamma > 0$, it represents turning clockwise, and $\Delta\gamma < 0$ represents turning counterclockwise.

6.4.4 Inter-Distance Maintenance

The inter-distance of ant-bots in a cluster should be maintained to avoid collisions or wandering away. We consider a cluster consisting of n ant-bots as $R_1, R_2, \ldots, R_n$, and the D_{ij} represents the distance between two ant-bots,[4] R_i and R_j. The ant-bot collision avoidance problem can be considered similar to the obstacle avoidance by using several techniques such as infrared[4] and ultrasound.[6] These techniques are designed to overcome the collision problems by maintaining their minimaps, but not wandering away problem. We use the RSSI values η to estimate the distance between two robots.[3] The RSSI values can be obtained from the communication packets received by the ant-bots. The estimated distance from a coordinator R_c, and an ant-bot R_i is given by

$$E_{ci}(\eta) = D_{ci} + n_i \qquad (6.7)$$

where the n_i represents the distance measurement error, and is modeled as zero-mean-independent white Gaussian processes with variance σ_n^2.

If the measured signal strength η_i from the received packets exceeds that of an ant-bot R_i by a threshold h, we say that the ant-bot R_i is in the lower boundary range. Otherwise, the ant-bot R_i wanders away from a coordinator, and then it has to pulled back. To maintain the distances among the coordinator and its members, the coordinator periodically broadcasts a beacon B with its own coordinate to the members. To avoid the *ping-pong effect* of using the threshold, we set two thresholds—the upper bound h_u, and the lower bound h_l. Each ant-bot, after receiving that beacon, judges the η_i, whether it is smaller than the h_l or not. If the $\eta_i < h_l$, the ant-bot R_i will turn it direction through using (1.5) and (1.6) to set the $d_c(\phi)$, and go forward the coordinator until the η_i is larger than or equal to the h_u.

6.5 Minimap Integration

We have known that the coordinator tries to collect the information about the environments from its own members for maintaining the global map information to carry out the path planning. So the members should periodically maintain the minimap in their memory for retaining the information of the environments beside its own. Because the coordinator have to maintain the global map built up by the minimap, it should use more memory than its members for holding the global map. If the coordinator runs out of its energy or breaks, then the backup coordinator replaces the original coordinator part to lead the cluster. Therefore, we should define the minimap held within the members, and know the packet format of the data communications among the coordinator and the members. In this section, we try to make the sensing data of environments a fixed packet format. There are two packet formats of communications for the coordinator and the members. One packet format is for the members to send information of environments to its coordinator, while the other is for the coordinator to request the members to gather the information of their environments. With these formats, the ant-bots can easily communicate with the others in a common protocol.

First, we define a fixed 8 × 8 minimap model to represent the environmental data of an ant-bot in its own coverage. In Figure 6.7, an ant-bot uses the distance measurements of the infrared or the ultrasound to fill the grids with color (environment data = 1) if it senses an obstacle in the

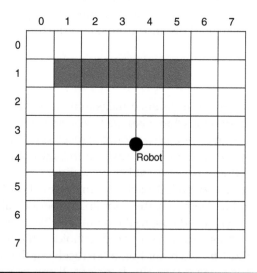

Figure 6.7 The example of 8 × 8 fixed structural minimap with obstacles for an ant-bot in an unknown environment.

environment. The ant-bot can adjust the range of the minimap according to the range of the sensors within itself. The size of the $m \times m$ minimap is decided by the number of sensors. Therefore, we make the information of the minimap into a fixed packet of 64 (8×8) bit streams. For example, the 64 bits environmental data of the ant-bot in Figure 6.7 are "00000000, 01111100, 00000000, 00000000, 00000000, 01000000, 01000000, and 00000000," respectively. So we can make the sensing data into a packet format as in Figure 6.8a. There are five columns in the packet of a member:

- Source_ID_Num. There are 4 bits in this column. This column records its ID number (member).
- Destination_ID_Num. There are 4 bits in this column. This column records the ID number of its coordinator.
- X-Coordinate. There are 10 bits in this column. This column records its current coordinate of the x-axis.

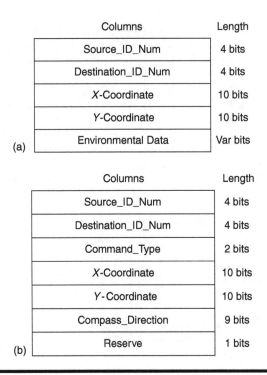

(a)

Columns	Length
Source_ID_Num	4 bits
Destination_ID_Num	4 bits
X-Coordinate	10 bits
Y-Coordinate	10 bits
Environmental Data	Var bits

(b)

Columns	Length
Source_ID_Num	4 bits
Destination_ID_Num	4 bits
Command_Type	2 bits
X-Coordinate	10 bits
Y-Coordinate	10 bits
Compass_Direction	9 bits
Reserve	1 bits

Figure 6.8 **The two packet formats of the minimap sending from the members to their coordinator and the request sending from the coordinator to its members. (a) Packet format of the members. (b) Packet format of the coordinator.**

- *Y*-Coordinate. There are 10 bits in this column. This column records its current coordinate of the *y*-axis.
- Environmental Data. In this column there are variable bits decided by the size of the minimap. This column records the information of its minimap. Take the Figure 6.7 for example, there are 64 bits in this column.

After defining the minimap model and packet formats in Figures 6.7 and 6.8, the ant-bots can easily send out the packets, and the coordinator can correctly recognize its sensing information by the standardized packet formats. Figure 6.9 is the sketch of the environmental shape integrated by the coordinator in the path planning. The coordinator receives all standardized minimaps of its members which are interested in building the partial view of the environment, and check all the

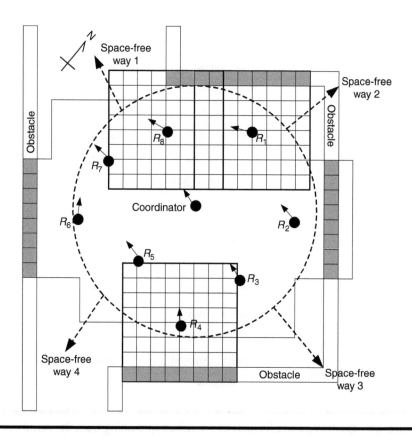

Figure 6.9 The members sense the information of environments integrated by the coordinator, and decide all the free-space ways.

space-free ways for decision-making. The packet format of the coordinator is shown in Figure 6.8b. There are seven columns in the packet of the coordinator:

- Source_ID_Num. There are 4 bits in this column. This column records its ID number (coordinator).
- Destination_ID_Num. There are 4 bits in this column. This column records the ID number of its members. If this message is a broadcast packet, we fill this column with "1111" (reservation for broadcasting).
- Command_Type. There are 2 bits in this column. This column records the passive action when the members receive the command. There are three commands in the Command_Type column of the packet that we can categorize with the help of its command actions. The Command_Type $= 00$ represents the broadcast message (or beacon) with an unchanged direction of its cluster, and the Command_Type $= 01$ is also the broadcast message, but a changed direction of its cluster. The last, the Command_Type $= 11$, is used by the coordinator to gather the information on environments from the specified ID ant-bots in the column of the Destination_ID_Num.
- X-Coordinate. There are 10 bits in this column. This column records its current coordinate of the x-axis.
- Y-Coordinate. There are 10 bits in this column. This column records its current coordinate of the y-axis.
- Compass_Direction. There are 9 bits in this column. This column is used to contain the degrees of the coordinator's angle by its electronic compass.
- Reserve. There are 9 bits in this column. This column is reserved for the future.

6.6   The Collaborative Path Planning Algorithm

By the integration of the minimaps, we can clearly construct the communication models between the coordinator and the members. Therefore, in this section, we describe the proposed collaborative path planning algorithm (CPPA) in detail. Consider Figure 6.9 for example; there are nine ant-bots ($n = 9$) gathered as a cluster cooperating with one another to reach a given target. First, the R_c receives the (x_t, y_t) of the target, and calculates the angular magnitude of the target ϕ to turn its direction $d_c(\phi)$ toward the target. The R_c then broadcasts a request message M_R, with the packet format of the coordinator including the target's coordinate (x_t, y_t), its coordinate (x_c, y_c), and the $d_c(\phi)$ to request its members for collecting relative map information. The member $R_i, i \in \{1, \ldots, n-1\}$, after receiving the M_R, collects the relative environments' information, and delivers a message M_i with the packet formats of the members, including the shapes of obstacles, if any, its coordinate (x_i, y_i), and its current direction $d_i(\psi)$ back to the R_c.

Using the information of the (x_i, y_i), and the R_c's direction $d_c(\psi)$, the coordinator R_c can determine in which locality the R_i is (quadrant). The R_c waits for a while to collect all the messages from its members, combines various information from both its members, and the members' corresponding quadrants; and can build a temporary map as shown in Figure 6.9. According to this global map, the R_c search as for the v and several possible space-free ways, and chooses the nearest space-free way for the target to make a decision using the priority of avoiding the obstacle in coordinator. The sketch of the priority in coordinator is shown in Figure 6.10. Assume that the angular magnitude of the nearest possible way to the target is ϑ, then the R_c selects the direction $d_c(\vartheta)$ as the next direction of the cluster of ant-bots, and broadcasts a direction message M_d, including the selected angle ϑ to its members for marking the target. All members, after receiving the M_d, set the same direction as the coordinator, by using an electronic compass by the $d_i(\vartheta)$. On the other hand, if its members recognize that the direction $d_c(\vartheta)$ has an obstacle, they determine which side of the R_c they are. Figure 6.11 shows that all R_i may distinguish themselves from the (x_i, y_i), (x_c, y_c) and the R_c's direction $d_c(\psi)$ into two situations: clockwise region and counterclockwise region. If the member R_i locates in the clockwise region of the R_c and meets the obstacles, it adjusts itself with the priority of the counterclockwise, as shown in Figure 6.10a. So does the member R_i locates in the clockwise region; the R_i locates in the counterclockwise region, and adjusts itself with the priority of the clockwise as shown in Figure 6.10b. With this adjustment, we can make sure that the R_i in the cluster does not disconnect with the R_c.

The direction of the cluster changes whenever the team meets an obstacle on the way. Since the CPPA is based on cooperative fashion, the members send the relative information to the coordinator when they meet obstacles. Therefore, when the R_i runs into an obstacle in a straight direction $d_i(\psi)$, it announces the information message M_i to the R_c for decision making. When the R_c receives the address of an obstacle, it updates its own map information, and determine whether they have to change a new direction or not. If the R_c decides to make a change for a new direction, it broadcasts this new information to its members with both the new direction $d_c(\psi + \Delta \gamma)$, and its current coordinate (x_c, y_c). Thus, the members move forward along the new direction announced by the R_c when they exceeded the (x_c, y_c). If the R_c does not broadcast a new direction, the R_i avoids the obstacles, and get closer to the R_c when its new direction $d_i(\psi + \Delta \gamma_i)$ has no obstacles. The algorithms of

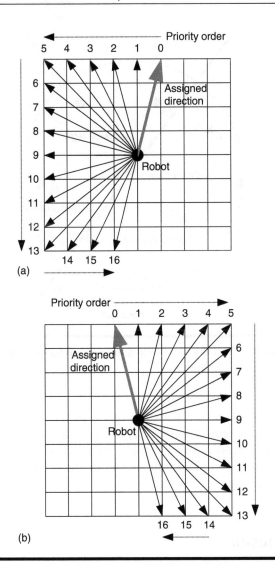

Figure 6.10 The two decision priorities of the obstacle avoiding in members located in counterclockwise and clockwise region based on the coordinator. (a) Counterclockwise region. (b) Clockwise region.

the coordinator and the members are shown in both Figures 6.12 and 6.13, respectively.

These processes proceed interactively until they reach the goal. Figure 6.14 shows an execution of the swarm-bot systems for the TGA game. This demonstration shows that these ant-bots autonomously coordinate with each other for a given mission. To alleviate the energy consumption, the direction to the target is only computed by the coordinator,

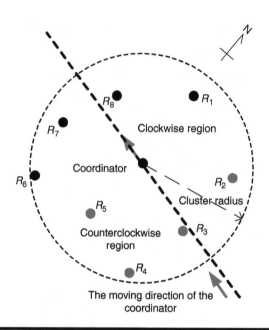

Figure 6.11 **The sketch of the clockwise and counterclockwise regions based on the direction of the coordinator.**

and the members follow the direction reach the goal. They only need to maintain its local information, and avoid the obstacles that meet on their way. The communication overhead is also reduced, since the members only communicate with the coordinator when they run into an obstacle in their straight direction.

6.7 Conclusion

This chapter presents the autonomous swarm-bot system, integrated with an IEEE 802.15.4[15] based wireless communication component and several sensors, and its implementation in a very compact size. This investigation brings innovative elements to collective and cooperative robotics, opening new research directions in swarm robotics, collective robotics, and distributed intelligence. Our innovation involves the integration of the characteristics of sensory devices, wireless communication devices, and mobility into a tiny ant-bot robot to create a fully distributed system. Particularly different is the mobility of each ant-bot module, and their high autonomy from the point of view of the sensors, and the computational power. Future work focuses on the research of the distributed intelligence algorithm, and energy efficiency.

THE COORDINATOR ALGORITHM

BEGIN
turn_left_counter ← ∅
turn_right_counter ← ∅
set Command_Type ← 11
$v = f((x_c, y_c), (x_t, y_t))$
broadcast $((x_c, x_c), v)$
while not reaching the Target or receive the M_m **do**
  if detecting an obstacle from sensor robots (x_o, y_o) **then**
    if OBSTACLE_DECISION(x_o, y_o) **then**
      if turn_right_counter flag is set **then**
        $v' = $ counterclockwise_choose_free_space (x_c, y_c, v)
      else
        if turn_left_counter flag is set **then**
          $v' = $ clockwise_choose_free_space (x_c, y_c, v)
        else
          $v' = $ choose_free_space (x_c, y_c)
          if v' is no more than v and turn_left_counter is not set **then**
            set the turn_left_counter flag
          else // v' is more than v and turn_right_counter is not set
            set the turn_right_counter flag
          set Command_Type ← 01
  else
    clear the turn_left_counter and turn_right_counter flags
    set Command_Type ← 00
  stack $((x_c, y_c), v, turn\_left\_counter, turn\_right\_counter) \leftarrow ((x_c, y_c)', v', turn\_left\_counter', turn\_right\_counter')$
  broadcast $((x_c, y_c), v)$
endwhile
END

Figure 6.12 The algorithm of the coordinator process.

THE MEMBER ALGORITHM

BEGIN

far_away ← ∅

while not reaching the Target **do**

  receive the broadcast information $((x_c, y_c), v)$

  if far_away flag is set and η is not bigger than h_u **then**

    $v = f((x_m, y_m), (x_c, y_c))$

  else

    if η is less than h_l

      set the far_away flag

      $v = f((x_m, y_m), (x_c, y_c))$

    else

      $v = v_c$

      clear the far_away flag

    constructing the minimap

    send the minimap back to its coordinator

    if OBSTACLE_DECISION(x_o, y_o) **then**

      $v' = f((x_m, y_m), (x_c, y_c))$

      if v' is no more than v **then**

        $v' =$ clockwise_choose_free_space (x_m, y_m, v)

      else // v' is more than v

        $v' =$ counterclockwise_choose_free_space (x_m, y_m, v)

endwhile

END

Figure 6.13   The algorithm of the member process.

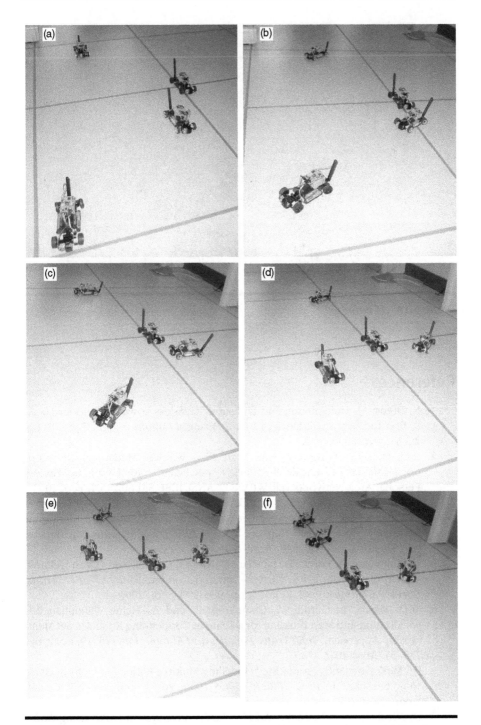

Figure 6.14 An arranging example of ant-bots for a 'L' word. (a) Initialization, (b) elected a coordinator, (c) arranging, (d) arranging, (e) arranging, (f) finished, (g) view from the sink node.

Figure 6.14 *(Continued)*

References

1. N. Bulusu, J. Heidemann, and D. Estrin, "GPS-less Lowcost Outdoor Localization for Very Small Devices," *IEEE Personal Commun.*, vol. 7, no. 5, pp. 28–34, Oct. 2000.
2. IEEE 802.15.4 Working Group, "Part 15.4: Wireless Medium Access Control (MAC) and Physical Layer (PHY) Specifications for Low Rate Wireless Personal Area Networks (LR-WPANs)," ANSI/IEEE Std 802.15.4, Feb. 2003.
3. P. Julian, A.G. Andreou, L. Riddle, S. Shamma, D.H. Goldberg, and G. Cauwenberghs, "A Comparative Study of Sound Localization Algorithms for Energy Aware Sensor Network Nodes," *IEEE Trans. Circuits and Systems*, vol. 51, no. 4, pp. 640–648, April 2004.
4. V. Jungnickel, A. Forck, T. Haustein, U. Kruger, V. Pohl, and C. von Helmolt, "Electronic Tracking for Wireless Infrared Communications," *IEEE Trans. Wireless Commun.*, vol. 2, no. 5, pp. 989–999, Sept. 2003.
5. M.D. Marco, A. Garulli, A. Giannitrapani, and A. Vicino, "Simultaneous Localization and Map Building for a Team of Cooperating Robots: a Set Membership Approach," *IEEE Trans. Robotics and Automation*, vol. 19, no. 2, pp. 238–24, April 2003.
6. P.J. McKerrow and S. min Zhu, "Modelling Multiple Reflection Paths in Ultrasonic Sensing," in *Proc. IEEE/RSJ IROS-1996*, vol. 1, pp. 284–291, Osaka, Japan, Nov. 1996.
7. D. Niculescu and B. Nath, "Ad Hoc Positioning System (APS) using AoA," in *Proc. IEEE INFOCOM 2003*, vol. 22, no. 1, pp. 1734–1743, San Francisco, CA, April 2003.

8. N.B. Priyantha, A. Chakraborty, and H. Padmanabhan, "The Cricket Location Support System," in *Proc. ACM MobiCom 2000*, pp. 32–43, Boston, MA, Aug. 2000.

9. A. Savvides, C.-C. Han, and M. Srivastava, "Dynamic Fine-Grained Localization in Ad-hoc Networks of Sensors," in *Proc. ACM MobiCom 2001*, pp. 166-179, Rome, Italy, July 2001.

10. S. Suksakulchai, S. Thongchai, D.M. Wilkes, and K. Kawamura, "Mobile Robot Localization Using an Electronic Compass for Corridor Environment," in *Proc. IEEE Int. Conf. Systems, Man and Cybernetics*, pp. 3354–3359, Nashville, TN, Oct. 2000.

Chapter 7

A Smart Blind Alarm Surveillance and Blind Guide Network System on Wireless Optical Communication

Ching-Chang Wong, Yih-Guang Jan, Rung-Hou Wu, Yang-Han Lee, Po-Jen Chuang, Shiann-Tsong Sheu, Huan-Chao Keh, Chih-Yung Chang, Kuei-Ping Shih, Jen-Shiun Chiang, Yung-Shan Chou, Hsien-Wei Tseng, Jheng-Yao Lin, and Cheng-Chang Ho

7.1 Introduction

This chapter discusses the critical technologies that are used in the development of optical wireless communication network, for the provision of safety and convenience to the vision-impaired population. The technologies of the hardware, firmware, decision algorithms, communication protocols, and application programs developed for sensor nodes in optical wireless communication network are applied for the smart blind-guidance network system. It follows the latest specifications, concepts, and trends in the design of sensor nodes to meet its requirements for applying in the next-generation wireless network. In addition, this technology is supplemented by the assistant equipment (cane) for the blind and the communication

191

interface protocol for the computer, which is specifically designed for the blind. The smart blind-guidance system is a combination of optics, microprocessor, voice acoustics, and wireless electronics. Moreover, it applies this kind of technology to study the grouping behavior of the blind-guidance robots and to develop a locating and addressing system. Digital signal processing (DSP) architecture is also being used to develop a fingerprint identification system, to process and to monitor identification in restricted areas. The combination of the fingerprint identification system and the current all-campus entry/exit monitoring systems will benefit the blind by using fingerprint identification to easily enter or exit the restricted areas. This system can also measure distance and environment parameters, collect data to furnish a complete pre-warning, guide, alarm, recovery and detection (GUARD) guide system for the blind, and provide a locating and addressing mechanism. It will give the vision-impaired group a higher level of service and protection when combined with the blind-guidance robot system.

Cane has for been long as the most commonly used aiding equipment for the blind, supporting them in their daily activities. The cane allows them to sense ground obstacles, unevenness of road surfaces, and potholes and steps along their walking paths. Although guide dogs also provide blind people guidance, they are becoming less preferable because of the long hours of training and the large financial burden involved. For instance, the Japanese Guide Dog Association (JGDA) conducted a census in 1991 and concluded that 20,000 out of 250,000 blind people can live independently with a guide dog. However there were only 7000 dogs available, which was obviously not enough to support the high numbers of the blind. Similarly, in 1999, a census conducted in Taiwan shows that 50,000 people needed the support of guide dogs. Unbelievably, Taiwan had only one guide dog! Clearly, the availability of guide dogs is a prodigious issue and it is about time an alternative was presented. This project discusses the advantage of "guide robots," which can be massively produced and at the same time solves the problems related to guide dogs.

Wireless optical communication and the traditional wireless network have different structures. The traditional wireless network is mainly used to send signals to hard-to-reach areas, therefore it is difficult to survive as a wire local network. Moreover, this system is most likely unable to be reused after being built. All these difficulties favor the production of wireless systems. Wireless optical communication network primarily uses sensors to detect the required information, and with the aid of network protocol, retrieves the information and saves it in the database.

Regarding the developing aspects of application, although the study of sensors is in accordance with the detect function, this project aims to study the hardware, firmware, communication protocols, and application programs for sensor nodes. By using the focusing and scattering properties of convex and concave lenses to differentiate the control of the safety coefficients, and by using the alarm surveillance system, we can instantly send blind-guidance robots to assist the people in need. This provides the vision-impaired persons with more safety, protection, and service. In addition, it is aided by the assistant equipment for the blind, which is embedded with personal basic information and the fingerprint identification system. This can verify the vision-impaired person's identity quite quickly. By combining this with the campus monitoring system, we can use the DSP architecture to develop a fingerprint identification system, which can process and monitor identity to ease the access into and out of restricted areas. This will allow us to fulfill the communication requirements of wireless optical communications.

7.2 The Manufacture of Wireless Optical Transceiver

The last step of communication network is arranging the connecting network with the optical, copper, and wireless networks, (Wi-Fi) being most commonly used. The Optic and copper line networks are not favored due to the high cost, time, and work required for operation. The installation of wireless network requires the use of bandwidths. Most bandwidths can only be utilized after registration and approval. Furthermore, the bandwidth and the rate of emission of signals of the wireless network are relatively low, which lowers its level of usage. Other than using optic, copper, or wireless networks, there is another option, which is using wireless optical communication for the access of the "last mile" network. The wireless optical communications, or free space optics (FSO), uses mainly laser diode or photodiode. Using air as a medium, FSO sends sound, image, or numerical data from the transmitter to the receiver. These signals can be sent as fast as 2.5 GBps. Since wireless optical communication sends signals from a particular point to another, it enables messages to be kept secret. Moreover, it costs 1/3 or 1/10 of using optic network and only requires 3 h to set up. Clearly, it is the best choice to make; it costs less, it is easier to maneuver, and it has a higher bandwidth than all the other described networks. Therefore, we gear our research toward wireless optical communication.

In this project, we want to create a guiding system for the blind using wireless optical communication to send signals. This system should contain numerous points of communications, and each point of communication needs to act as a wireless optical communication transceiver. Therefore, the transceiver should acquire the qualities of low power and low cost. In addition to the application of the blind guide system, we hope to expand the usage of the wireless optical communication transceiver on Wi-Fi network. Therefore, high performance of the transceiver is also the main focus of this project. We hope the transmission rate of the transceiver can reach 50 MBps.

We want the wireless optical transceiver to use the idea of wireless optical communication as the blind guide system. Through this system, blind people can use it to send the necessary information to the control center, which requires a wireless optical communication similar to Wi-Fi; it does not require high transceiver module specifications. Even though we do not need a high wireless optical communication, it plays a crucial part because wireless optical communication will be the critical technique for communication in the future. To pinpoint the exact location of the blind individual, we need to use a good transceiver and a good location mechanism; we prefer to use ultra wideband (UWB) to locate the target. Therefore, our focus is on the Wi-Fi wireless optical communication transceiver, which low cost and high performance. This transceiver can be used for blind guide system and can also be used for the future Wi-Fi wireless optical communication system.

We apply the study results of the front-end circuits of the analog receiver in wireless optical communication system to our design of the guide system transceiver. It includes the study of the general principle of the front-end circuits in the optical communication receiver, the optical detection device with silicon material, the design of voltage transimpedance amplifier (TIA) circuit by using current mirror, and the design of TIA by using single crystal inductance. All these studies lead to the fabrication of the front-end circuits with a low-power, high transmission rate, and a wideband analog receiver.

Figure 7.1 shows the conventional functional block diagram of an optical communication system. Input data in parallel form are converted into

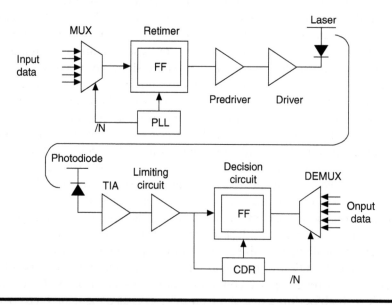

Figure 7.1 Function block diagram of optical communication system.

serial voltage signals after passing them through the multiplexer and modulator. Then they pass through the preamplifier and a driving amplifier to drive the laser diode to ignite the optical signal. This is the transmitter operation principle of an optical communication system. On the receiver end, the optical detector receives the optical signal, converts it into a small current signal, and further converts it into small voltage signals by passing it through the TIA. Afterwards, these voltage signals pass through a post amplifier and get converted into parallel data after passing through the demodulator. Figure 7.2 is the functional block diagram for a wireless optical communication system. Figure 7.3 is the front-end circuit of the analog receiver we designed and studied in this project.

We have prepared the specifications for the wireless optical transceiver and searched for optical detection devices based on these specifications. These devices include photodiode and TIA. TIA requires low power, but provides a high transmission rate and wideband.

In the selection of optical diode, we consider from the fabrication process, silicon PIN photodiode as our first choice. Based on the equivalent circuit of photodiode, as shown in Figure 7.4, it is necessary for us to select

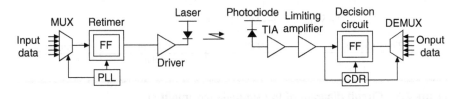

Figure 7.2　Function block diagram of wireless optical communication system.

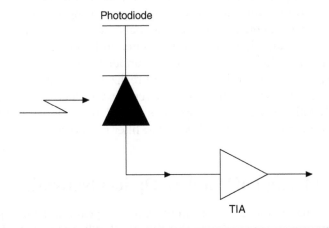

Figure 7.3　Function block diagram of the analog front-end circuit of wireless optical communication system.

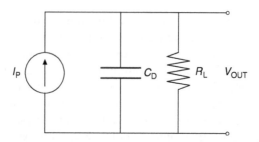

Figure 7.4 The equivalent circuit of PIN photodiode.

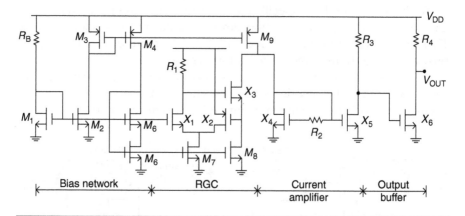

Figure 7.5 Circuit diagram of transimpedance amplifier.

the silicon PIN photodiode which will meet all IP and CD specifications. We are able to pick the right wavelength by referring to the graph of the response of wavelength of various materials' PIN photodiode. Figure 7.5 shows a low voltage, high transmission rate, and wideband TIA. Through the simulation of this circuit we gain an understanding of this transceiver system, which aids us in the design and realization of this kind of circuit. If we increase the inductance in the circuit it will increase the transmission bandwidth. However, if it operates at a high-frequency range, it will correspondingly increase the difficulty to filter out noise.

7.3 The Design of Wireless Optical Network

Technology has made great progress in recent years and blind-guidance equipments are no exceptions. But the blind still hopes that the blind-guidance equipments can clearly notify them of any obstructions in their walking path and any possibility of incurring danger. It should also be capable of indicating the correct path to the blind to save time. Furthermore, it should be able to lead the blind to their destinations without much difficulty. All these issues have impelled us to consider the main task of how to mass-manufacture practical but low-cost blind-guidance equipments, which will be more convenient to the blind. It will greatly remove the difficulties and inconveniences in the blind people's walking tasks if we can install a complete monitoring, locating, and addressing network system and give them multifunction assistant equipment such as the guidance cane. It will definitely be beneficial to the blind if we can invent a humanized blind-guidance machine to assist them whenever they need help.

Through the design of system communication protocol, the design of wireless optical network promotes the blind-guidance cane's reliability and safety. We want to build a system which provides the blind a safer, more reliable, and more convenient environment. The important problem regarding "direction" can be solved by using sensors and by using an entire series of positioning coordinates. This system will provide the blind people with information about their surroundings and guide them to their desired destinations through sound signals.

The sensor network of the whole GUARD system is designed in the project. We select the IEEE802.15.4-2003 (Zigbee) communication protocol as our main operation protocol for the sensor network to meet the requirements of network certification, network safety, error compatibility, system robustness, system expandability, cost, hardware, topology, working environment, and power consumption.[1–5]

Using the formulated transfer media access control protocol (MAC) as basic, we study the design of a high-efficient communication protocol, which will execute by means of embedded sensor nodes a method to provide sensor node communication. Wireless optical network design will achieve MAC by means of hardware type. MAC—integrated with upper-layer operation system, route protocol, and relative application algorithm—it is again integrated with lower-layer modulation and digital/analog transfer control and thus becomes the connecting bridge between these two layers. When considering sensor networks, we want to focus on one that functions easily, has small physical volume, and requires low power consumption. We are not looking to choose an operation processor unit that has exceptional functions. Rather, we are hoping to choose the best CPU for us to develop a set of specialized operation systems (OS) for wireless sensor networks (WSN). Moreover, we hope to develop a relative drive program and functional library, and to support the sensor function of sensor node under the OS architecture.

Currently, we have purchased the Zigbee development platform CC2420 DBK, as shown in Figure 7.6, from Chipcon Inc. to develop the communication protocol. This Zigbee platform carries Atmel's ATmega 128 microprocessor and Chipcon's CC2420 wireless transceiver. In this development

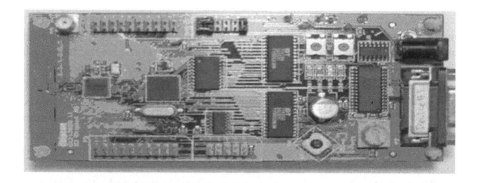

Figure 7.6   CC2420 DBK.

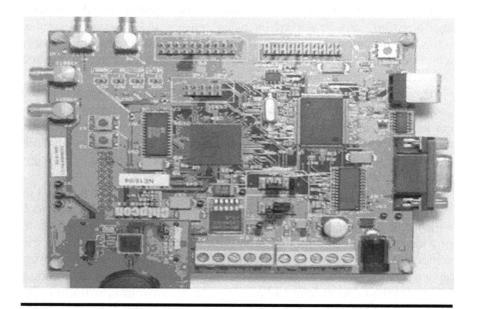

Figure 7.7   CC2420 DK.

platform, we can input the edited program codes to the platform to process the verification of the communication protocol, allowing us to standardize the wireless transmission between two transponders through the use of program codes. Also in the development of the programs, we adopt Chipcon's CC2420 DK, as shown in Figure 7.7, for error correction and for the verification of each process. In this platform we can extract signals in the Zigbee band and display the signals, through the GUI interface, as shown in Figure 7.8, at the PC port, in the IEEE 802.15.4 (Zigbee) standard packet format. Therefore, we can determine the correctness of the developed codes and decide if the communication protocol meets the IEEE standard.

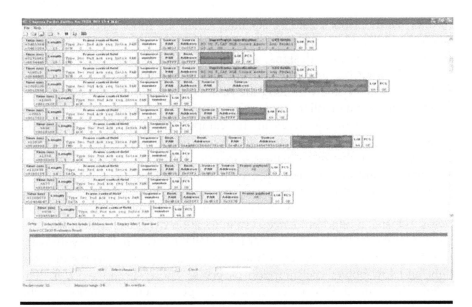

Figure 7.8 Graphic-user-interface (GUI) at PC port.

The design of wireless optical communication system has successfully completed the development of the IEEE 802.15.4 communication protocol, including the certification of the hardware. Moreover, in June 1994, Chipcon Inc. has decided to initiate the mass production of the second-generation hardware, which is a combination of microprocessor and transceiver (CC2430). This new product is compacted, therefore requiring less space and money. In different circumstances, the communication protocol proposed in this project will be flexible enough to make the necessary alternations.

7.4 Smart Wireless Optical Blind-Guidance Cane and Blind-Guidance Robot

In the designed and implemented guide system, the blind-guidance cane can provide voice-based location guidance information. In the future, it will combine the designed fingerprint identification algorithm and the campus entry/exit monitoring system to allow the blind to enter or exit the restricted area by using fingerprint identification. In the designed and implemented robot, it consists of an omni-directional moveable base mount mechanism and an all omnidirectional vision system. An omnidirectional movable base mount mechanism allows the robot to move more flexibly and the robot is more mechanized. An omnidirectional vision system allows the robot to extract information simultaneously from the surrounding environment to improve the robot's guiding accuracy when the robot guides the blind. Wireless optical transceiver enables more precise locating and addressing functions. In the future designs, we will not only integrate blind-guidance cane and robot, but also integrate the whole campus network system to realize the goal of safely guiding the blind to walk around the campus.

Installing a smart blind-guidance system[6–18] is based on the evaluation of safe and unsafe areas in the campus. We use this as a plan to consider the setup of this system, to design the blind-guidance cane guidance system, and to coordinate a wireless optical transceiver to propose safe and unsafe area design. The planning of setting up of this system, in combination with voice database, provides the vision-impaired persons the service of moving safety voice information. In the meantime, it also helps to install a blind-guidance system location database, which provides a smart wireless blind-guidance alarm surveillance system. Besides studying the design of blind-guidance robots, we also focus on concepts of radio light and low-power consumption as goals. These should work in harmony with camera, sensor setup, or other special-benefit millibot-specific applications.[19–23] For instance, we dispatch a team of robots to overcome obstacles from different environments and allow them to achieve the task of blind-guidance. Blind-guidance dogs have autonomous characteristics, meaning that the team will achieve the task even if one guide dog is unable to work properly. Moreover, they can efficiently and instantly collect information about the surroundings and can provide the best assistance to the blind. We want to integrate the formulated IEEE 802.15.4 with smart dust, optical wireless, and wireless optical sensor network system to achieve a smart blind-guidance alarm surveillance network system, which will provide the vision-impaired persons mobility services. We want to implement a guide system with blind-guidance cane, as shown in Figure 7.9, to provide location information through voice transmission and also to develop a fingerprint identification algorithm. Because the system needs to provide the specific location and address information, the wireless optical transceiver is also an important study subject.

The designed blind-guidance robot will move and guide the blind in a complicated environment by using the guidance of the vision and wireless optics. It can also use the real-time image transmission of the wireless network to execute the campus patrol duty when it is not in the blind-guidance situation. Because many unexpected obstacles exists in the campus, it becomes an important issue to know how to install the blind-guidance robots to set up the danger prewarning system for environment safety and how to implement the emergency-handling system to protect the safety of the vision-impaired. In this project, we set up the blind-guidance robots team to guide the blind through a complicated and dynamically changing environment, to arrive safely at his destination. The blind-guidance robots in the team will work in close cooperation with each other and organize

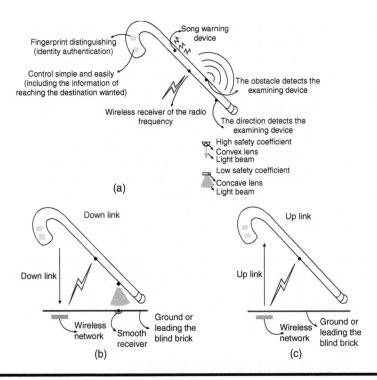

Figure 7.9   The design of a smart blind alarm surveillance and blind guide network system on wireless optical communication. (a) Smart cane. (b) Down link.

themselves to work as the most proper blind-guidance robots according to the variations in the actual environment and also by using the guidance of wireless opticals to lead the blind safely to his destination. When the robots are not on their guiding duty, they will share the duty of patrolling across the campus to improve the safety in the campus. The main focus of this project is on the design and fabrication of the robot, specifically the implementation of the main body of the robot and its vision system.

7.4.1   The Guide System with Blind-Guidance Cane

7.4.1.1   Multifunction Aid Equipment (Cane) for the Blind

The cane has been the most commonly used aid equipment for the blind to walk and move in their daily activities. Using a cane, the blind can sense ground obstacles, unevenness of road surface, and potholes and steps in his walking path. However, for the blind to possess such cane-handling techniques they need a lot of training and practice. For the blind to move safely and quickly, the cane needs to be designed to carry the capabilities of self-distance measurement and prewarning to give the blind an obstacle-free walking environment. Furthermore, the cane needs to be designed to obtain measures to clearly give the blind instructions on how to react under various situations. With this kind of cane designed, and the guide system implemented (by combining fingerprint identification system, personal identity verification system, and image identification processing), it will quickly recognize the blind person's identity. Consequently, it becomes an important task to design a perfect identity verification system, such as the fingerprint identification system, and the blind-aid equipment (cane) to reduce the blind person's training and practice time and also to make him safer in his movement. To give the blind a real-time service we need to coordinate the hardware and software apparatus such as to perfectly plan and program safety areas in the campus, to establish a database for the blind-guidance system (address, name of the location), to coordinate with the existing campus network to design a smart wireless blind-guidance GUARD system, to design and install optical network nodes and wireless network nodes to provide an unobstructed environment, to combine with the existing campus blind-aid equipment, and to combine the existing campus monitoring system.

7.4.1.2 Fingerprint Identification System

We design a fingerprint identification system based on the Nios/DSP architecture to process identity verification. It also combines the fingerprint identification system with to the existing campus entry/exit monitoring system to allow the blind to enter or exit the restricted area by using the fingerprint identification method.

7.4.2 Blind-Guidance Robot

7.4.2.1 The Fabrication of the Robot's Main Body

In the fabrication of the robot's main body we will directly use a single computer board in cooperation with the FPGA to design the control interface. With this kind of design it will effectively simplify the hardware circuits and reduce the robot size.

7.4.2.2 Vision System for the Robot

To equip the robot with a vision system to enable it to effectively adapt to environmental changes, the most important task is to complete the installation of the "eyes" for the robot. With the eyes installed in the robot we will be able to complete a dynamic and real-time image searching system, to identify the blind and his associated location and address in the campus environment. With this implementation, it will further allow the decision center to decide proper strategies, to correctly control the robot's movement. It will also allow us to study the robustness of the robot's vision system versus the light intensities.

7.4.2.3 Smart Cooperation Strategy for Multirobots

In this paragraph we will consider how to design a smart strategy to cooperate many robots. When the main hardware architecture system for the robot is finalized, we will allow the robot to carry out smart learning to enrich its knowledge. A distributed and autonomous architecture will be adopted for the development of the robot's artificial intelligence. It will solve the issues of robustness and adaptation that are generated from the use of the cooperation model. After a robot is fabricated we will use the characteristics similar to the immune algorithm to connect the behavior model of every robot to develop the most suitable cooperation model. In the entire process, we will let robots determine their cooperation model and let each robot determine its own behavior action. This is a challenging task in the study of artificial intelligence.

7.4.2.4 Smart Controller Units

The smart controller part of the conventional design method is to let the controller switch to different control condition or controller under different environments. It needs to consider and plan all possible occurring conditions thereby making the resulting robot control system very complicated and it will require the use of many different controllers to complete the control requirement. Consequently the design of smart controller requires fewer controllers to attain control capability under various environments. In this project we allow the robot to learn and to be trained under various environments to enforce the learning capabilities of the controller. After the smart controllers complete their training, they will be more capable of adapting themselves to environmental changes.

7.5 The Design of a Smart Guide System with Wireless Optical Blind-Guidance Cane and a Blind-Guidance Robot

7.5.1 The Design and Implementation of the Guide System with Blind-Guidance Cane

Figure 7.10 shows the block diagram of the design and implementation of a guide system with blind-guidance cane. This includes blind-guidance bricks, a blind-guidance cane, and a voice-broadcast main machine. The terminal of the blind-guidance cane continuously emits signals, and when

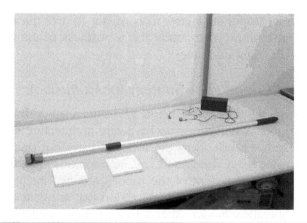

Figure 7.10 The guide system with blind-guidance cane.

Figure 7.11 A vision-impaired person and the guide system with blind-guidance cane.

the blind-guidance bricks receive this emitted signal they will reflect signals to turn on the voice-broadcast main machine, which will then broadcast the road information to guide the blind, as shown in Figure 7.11. Its detailed descriptions are as follows:

For the blind-guidance bricks we select a bandwidth of 125 Hz radio frequency idententification (RFID) and label it at every road as the identification device. Each RFID label has a set of codes to identify a node in Tamkang University. Because the code is 64 bits long we can therefore totally install 2^{64} nodes in the campus. We then embed the RFID label into each blind-guidance brick and seal the brick with waterproof bentonite. When the blind-guidance brick receives a signal triggered from the cane terminal the passive inducting coil in the RFID of the blind-guidance brick will generate an induced electromagnetic force and when this electromagnetic force passes through the voltage-stabilizing circuit, it will return the code in the label to the cane terminal for its detection and decision. The communication between the cane terminal and the blind-guidance brick is shown in Figure 7.12.

As Figure 7.12 depicts, when the cane's power is turned on, there will be two beeps and the red and green LED lights alternately twinkle. This makes sure the circuits are running in the correct direction. Afterwards, the microprocessor control unit (MCU)(89C2051) produces a carrier enable signal, a let read/write base station IC(U2270B) output, a 125 Hz carrier, and a repeatedly detect IC(U2270B) signal. When the read/write base station IC(U2270B) detects the alternate coding signal of transponder IC from the blind-guidance bricks, it will output this alternate coding signal from the second pin of the read/write base station IC(U2270B). The alternate coding signal of transponder IC is composed of 8-bit data, with series output in a rectangular-wave form. Each rectangular-wave high level signal has a different time interval. These signals are encoded by Manchester code. Therefore, as long as we know the time frame to maintain a high level

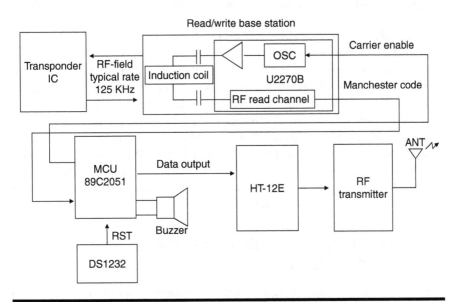

Figure 7.12   Hardware function block diagram of the cane terminal.

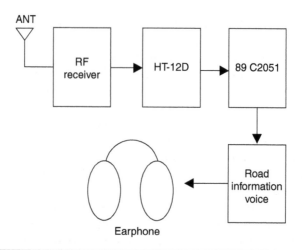

Figure 7.13   Hardware function block diagram of the main machine.

transmission, we will be able to know what data is received. We use the time function of MCU (89C2051) to achieve and accomplish this project.

When the received signals from the second pin of U2270B are at high level, MCU(89C2051) will record the TH0 and TL0 of TIMER 0 in a register. When eight sets of 8-bit data are completely received from the blind-guidance bricks, MCU(89C2051) will transfer this data by a binary code to an encoder (HT12E). By Transmission through wires to the main machine, the program will return to its initial position and the whole process repeats itself once again.

The blind person will carry a main machine in his waist, with the frame structure as shown in Figure 7.13. When the wireless module receives the information sent from the cane terminal, the MCU (89C2051) at the main machine will generate control signals according to the received code to select the correct road information, to control the voice-broadcast unit, and then to release this road information to the blind person through his earphone. We also integrate a set of fingerprint identification algorithm that will be perfected by hardware in the near future. We will then integrate this fingerprint identification system into the campus entry/exit monitoring system to help the blind person enter or exit the restricted area by using the fingerprint identification method.

7.5.2   The Design and Realization of the Blind-Guidance Robot

Figure 7.14 depicts the designed and implemented robot. In the design of the base module,[24–28] we design an omnidirectional, moveable, four-wheel mechanism so that the robot can move in any direction and

Figure 7.14 A blind-guidance robot.

can increase the flexibility of movement. In the design of a vision system for the robot,[29–31] we install an omnidirectional imaging system on the top of the robot head to allow the robot to extract the information surrounding him (information such as the identity of the vision-impaired person, blind-guidance bricks, and obstructions). The imaging system includes a dual curvature mirror and a camera. The dual curvature mirror is installed on the highest point of the robot and the camera is aligned with the dual curvature mirror to extract the image on the mirror. The omnidirectional imaging system has the merit and characteristics, which allow the camera, by utilizing the characteristic of the dual curvature mirror and the principle of the optical reflection, to simultaneously extract all the information surrounding the robot. Its horizontal vision range is 360° while its vertical vision range has the elevation angle ranging between −10° and 80°. Its extracted images are shown in Figure 7.15. It shows that relatively, in the omnidirectional imaging system, there is no dead vision angle, which ultimately improves the robot's guiding safety. The camera extracts images from all directions and through the image extraction box, it will convert the NTSC digital signal into USB2.0 signal. This signal is then transmitted to the main computer for further processing and analysis. In image identification, we utilize the principle of object edge detection in conjunction with the concepts of color discrimination and color separation to search and decide the location and direction of the blind-guidance brick. Moreover, in this project, we utilize the hardware sensing method to determine the object distance. This can improve the robot's emergency handling capability and stability when the robot is performing its guiding mission. As shown in Figures 7.16–7.18 we can identify the blind-guidance bricks from complex environments. By

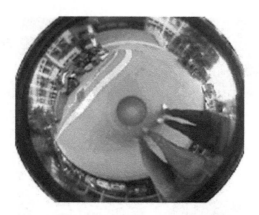

Figure 7.15   Omnidirectional images.

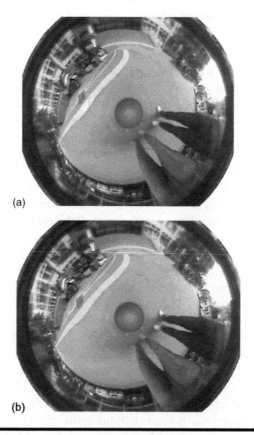

(a)

(b)

Figure 7.16   (a) The original omnidirectional image containing blind-guidance bricks. (b) The resulting identified images from original image in (a).

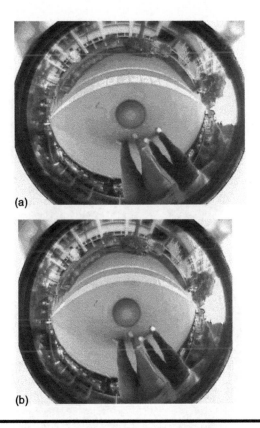

Figure 7.17 (a) The original omnidirectional image containing blind-guidance bricks. (b) The resulting identified images from original image in (a).

developing a proper guiding strategy, we can safely guide the blind to walk around the campus in the future.

In this project we have completed the design and implementation of a guide system with the blind-guidance cane. It not only provides a voicing location-guide system but also, in cooperation with the fingerprint identification system and the campus entry/exit monitoring system, allows the blind to enter or exit the restricted area by using fingerprint identification. The robot designed in this project is equipped with an omnidirectional movable base mount and an omnidirectional vision system. The omnidirectional moveable base mount allows the robot to move more flexibly and the robot is more mechanized. An omnidirectional vision system solves the disadvantage of being unable to extract all surrounding information in the general camera system. The designed robot will not suffer any possible vision dead angle, which improves the robot's guide safety when the robot is performing its guide mission. Because the light intensity will affect

Figure 7.18 **(a) The original omnidirectional image containing blind-guidance bricks. (b) The resulting identified images from original image in (a).**

the image detection probability we will emphasize this part of study to strengthen the image identification capability in the future. Moreover, we will work to integrate the guiding strategy of the blind-guidance cane, as shown in Figure 7.19, with other campus network systems to complete the plan of guiding the blind to walk safely around the campus.

7.6 Smart Wireless Optical Communication of Blind Alarm Surveillance System

The wireless optical communication on smart blind alarm surveillance and guide network system emphasizes the design and manufacture of the alarm and monitor control system. This alarm control system is programmed with

Figure 7.19   The integration of guide system with the blind-guidance cane and the guidance robot.

the position system and the protocol of devious route and broadcast; it can even act as the alarm and monitor for the blind walking on the campus. The project goal in the first year is to construct an ideal WSN, which has the multifunction characteristics of guide, alarm, recovery, and detection.

In the theoretical study, we proposed an installation technology of WSN. We analyzed the defaults and problems of the robot installation networks, and planned to use minimum numbers of sensor nodes to avoid obstacles and to reach full coverage. In addition, we also used artificial ways to install blind sensor network for future work. We analyzed the relationship between the sensor service and transmit range. We also analyzed how the sensor independently moves to achieve the complete cover service range to allow us to produce sensors, which can apply to the daily circumstances at Tamkang University campus. In the implementation study, we not only edited sensor and wireless communication module on present sensor node, but also processed wireless broadcast and point-to-point wireless transmission. Further, we targeted Tamkang University campus to implement this study on a blind-guidance system. We not only measured every building and working route on campus, but we also developed and edited a program of working route. All of these implementations allow the blind to move around the campus with the help and guidance of the GUI system.

As technology progresses, the blind population is provided with various safer and quicker services, without facing the obstacle of space. Using preceding experiences, we merge wireless and sensor technologies to install multifunction smart wireless blind alarm surveillance network system with guide, alarm, recovery, and detection. Through network installation we reinstall and renetwork the topology structure. This system will provide high-efficiency data transmission, lower power consumption, and sharing of bandwidth. By means of devious routes, route autorecovery, position track, and guide technologies, this system can safely guide the vision-impaired persons to reach their destination. It can also, according to the verification conditions of time, place, working speed, and the time of stay, provide alarm and surveillance services. Moreover, the vision-impaired persons can, by means of moving assistant equipment, set up the goal where they wants to go, e.g., class, dormitory, and restaurant. This system can dynamically install a safety walkway, provide automatic guidance to goal position, and support the application of both location- and context-aware services mapped to the Tamkang University campus at Tamsui. We installed a WSN for the vision-impaired. At the same time, since low electric consumption is a prerequisite, WSN can provide a long lifetime of services for vision-impaired persons. This design provides a longer lifetime for WSN and provides precise position information for vision-impaired persons. This will allow the vision-impaired persons walking on the campus to know their location at any time. Consequently, initialization of the WSN is needed to know the sensor node position information of WSN.

In summary, the initiation of an ideal WSN with guidance, alarm, recovery, and detection is the first of our study goals. We want to expand the functions of the smart wireless blind alarm surveillance and guide network system. We also want to design the protocols of the robot patrol network system, which can be used to study patrol network route, multicooperation patrol network, and also used against close loop. Finally, we want to design a perfectly smart wireless blind-guidance, alarm, and surveillance network system to create an obstacle-free life and learning space for the vision-impaired persons.

7.6.1 The Installation of Wireless Sensor Network for the Blind

We emphasize in this project the installation of a WSN for the blind, the default values setting of the network, the system realization, the system implementation, and the integration of the GUARD system. The installation of a WSN can be divided into two categories. One is using a manual method to randomly install the network while the other is using a robot-like method to progressively process the network installation. The ad hoc WSN system usually locates the network of sensor nodes to be installed through prior calculations. Then the sensor node processes independently the construction of network topology. Depending on its sensing mission, every WSN has different factors to consider in the construction of the network. In the previous design, every sensor node was designed to have its maximum coverage, and it was assumed that the sensor node possessed the moving capability so that the sensor node could be moved to undetectable areas to expand the whole WSN sensing range.

In addition to the WSN as discussed, a lot of researches emphasize on the subjects of no-rule, layered WSN, try to determine to install cluster in WSN, find out how to select the cluster head in the cluster, find out how to increase the lifetime of this kind WSN, and find out how to place the sinks. The layered WSN needs to have equilibrium in the number of members in the cluster (so that each cluster head will have almost the same lifetime) without considering how to select the cluster head. Consequently, the cluster head may consume too much power. Therefore, we also need to consider the selection of the cluster head so that it is possible to balance the power consumption in the WSN and to improve the lifetime of WSN.

To construct the best WSN with the best lifetime, we need to consider the installation method, which includes no-layered and layered WSN. We also hope the sensor node will have the maximum coverage in the WSN.

7.6.2 The Initialization of the Wireless Network for the Blind

In the stage of network management and maintenance, addressing and geographical information discovery are two important research topics in the setting of default parameter values of the network. In the addressing method, general ad hoc in the WSN usually takes an autonomous and distributed way to complete the network installation. Consequently, if each of the main machine in the network is identified by one unique identifier, it needs to be operated in the distributed way. Another way to address is to use the centralized operation. The network administrator takes control of the network management; he unifies the management and coordinates all members in the network. When conflict occurs he will notify the sender to improve the reliability of the information transmission.

The second research topic is the geographical information discovery. There have been many proposals regarding addressing techniques in the wireless detection network. The area-addressing system uses some reference points to aid in the location calculation. In other addressing system it uses ultrasonic detection method to measure the actual distance.

From the above considerations we know that when the complexity of network installation increases, the complexity of the installation or reinstallation of an ideal WSN system also increases. Therefore, it makes network management and maintenance more difficult. Although many research projects are under consideration, they do not have a WSN designed and installed for the blind. There still exists a lot of unresolved problems in the WSN installation. So we would like to conduct advanced researches and studies in the WSN system and hopefully, we can solve the theoretical and practical problems that may arise in the installation of WSN designed for the blind.

7.7 The Design and Implementation of a Smart Wireless Blind-Guidance Alarm Surveillance System

In the theoretical study we use the robot-like design method to install the sensor network and to develop its communication protocol to overcome the obstacles and to use the minimum number of sensor nodes to attain full coverage capability. In the implementation we extend and expand the functions of sink node and sensor node. We write a communication protocol for the sensing and transmitting of wireless signals between sensors and between the sensor and the sink node. We also aggressively develop the interfaces for the blind-guidance system to insure that this guide system will work closely with the blind sensor network.

7.7.1 The Installation of Wireless Sensor Network for the Blind

In the network, we set up one or many PCs or notebooks that are used as the sink nodes to connect with the modules of Mote MICA2.[32] Many Mote MICA2 modules form a sensor network. We also assemble Mote MICA2 modules to form a self-moving mobile unit to function as a mobile sensor and a mobile robot. The sensor board also possesses the six functions of sensing light intensity, sound, temperature, relative humidity, magnetic strength, and plane acceleration velocity. When the sensor board senses the environment information, it will transmit this information to the MICA2 through the analog to digital converter (ADC) for processing and calculating. The basic constituent elements in the sensor node are micro control unit (MCU), RF module, sensing component, power unit and I/O interface, as shown in Figure 7.20. The MCU is the data processing unit used to process the received or sensed data according to the requirement in the application layer. RF module is the wireless communication module used to make a point-to-point wireless communication. Sensor component is a sensor, and by using different kinds of sensors we can sense and get various environment information. I/O interface is the input/output port for the sensor node, and the power unit provides power to all constituent elements of the sensor node.

We use the self-moving mobile unit to replace the sensor node, which possesses sensing functions, to carry out the sensing operation by its partial movement or to search for the damaged sensor nodes via the patrolling

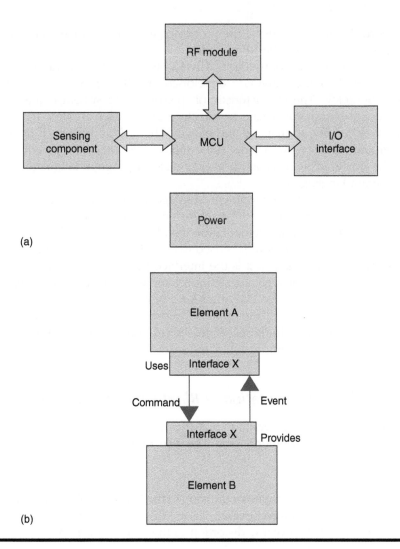

Figure 7.20 (a) Basic elements. (b) Interface architecture.

method. The self-moving mobile unit, through centralized computation and by considering the connectivity of the sensor network and its possible sensing range, uses the minimum power consumption to attain the tasks of full coverage and autonomous repairing. The components in the self-moving mobile unit consist of ultra sonic module, MICA2, MICA2 connector, motor control integrated circuit, motor, power unit, and location generator. It uses ultra sonic module to transmit ultrasonic waves to detect the existence of obstacles and measure their distances. It also conveys the information to the MICA2 for analysis and processing. The MICA2 is the data processing center to process information sent from all the modules and to make proper decisions. It can also transmit the processed information to all neighboring static sensors by using the RF module.

The software used in the Mote MICA2 module is the operating system implemented in the TinyOS[33] developed by the University of California at Berkeley. The characteristic of this operating system is that it has component-based small-scale architecture and it uses another scheduler, TinyOS scheduler, to manage and execute the task and event. It does not use virtual memory but adopts static memory arrangement to effectively use the power. The language it uses is nesC, a program language similar to program language C.

We use nesC as the development language. An application program written in nesC is an executable program formed from interconnections of one or more devices. The implementation also has two kinds of devices: one of them is the configuration. It is used to combine and define the interrelations between devices, i.e., it is the interface between the implementation of devices. It can, through the defined interfaces, make communications possible between devices. The other kind of device is the module. It generates a program code for the application program, i.e., the program code to define the function in the implementation of the interface. The module is the smallest device unit.

7.7.2 The Implementation of Blind-Guidance System

In the interface implementation for the blind-guidance system we use FLASH as the development tool. The use of FLASH has the following advantages:

- It uses the basis of vector drawing in the FLASH implementation, and therefore has more flexibility in the display presentation. Moreover, it will not display any distortion in the process of map enlargement or reduction.
- The use of FLASH has elegant effects in multimedia presentations and therefore it enables the acceleration of dynamic map development. It also generates a convenient function to enable the visualization of the information.
- FLASH has the integration capability to completely combine the functions of network and database. It can combine the users' certification systems and it also enables users, through network, to process system operations, instantly and without space limitations.

Therefore, as shown in Figure 7.21, when we install a WSN in Tamkang University, the main operations of the blind-guidance system will have the display as shown in Figure 7.22. It consists of two parts, namely, the

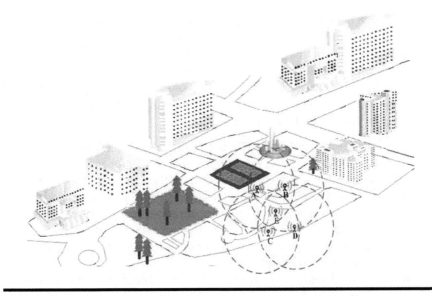

Figure 7.21   Sensor node installed in the campus of Tamkang University.

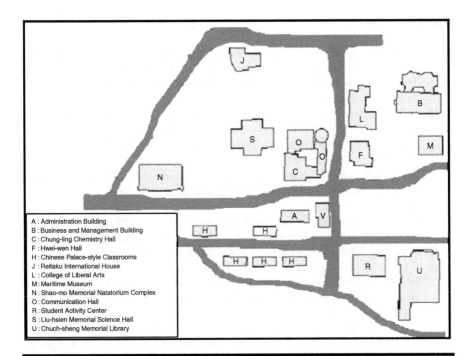

A : Administration Building
B : Business and Management Building
C : Chung-ling Chemistry Hall
F : Hwei-wen Hall
H : Chinese Palace-style Classrooms
J : Reitaku International House
L : College of Liberal Arts
M : Maritime Museum
N : Shao-mo Memorial Natatorium Complex
O : Communication Hall
R : Student Activity Center
S : Liu-hsien Memorial Science Hall
U : Chuch-sheng Memorial Library

Figure 7.22   Campus system interface of Tamkang University.

control-panel area and the map-display area. In the control-panel area, the contents displayed are controlled by the user. Depending on the user's requirements, he can change the contents in the display area. It will follow the user's control to display the information he requests for in the display area. The information includes the location of the blind person, the planned route, and the obstacle information. The following points are introductions of the main functions of the interfaces:

- Route display and plan: The user can, instantly, ask for the location of any blind person in the campus and search for his walking route. He can dynamically change or correct the guide path for the blind.
- Rotate/reduce/enlarge/move/restore: The user can display the map in various fashions such as using rotation, reduction, enlargement, and movement functions; he can dynamically adjust the appearance of the map. He can use the restoration function to restore the map into its default settings.
- Obstacle display and setting: When the blind person's walking is impeded due to construction, repairing and events held in the campus, the system will provide, display, and set the obstacle information for the blind person and help him to find an alternate walking route.

References

1. LAN/MAN Standards Committee of the IEEE Computer Society, *Standard for Part 15.4: Wireless Medium Access Control (MAC) and Physical Layer (PHY) Specifications for Low Rate Wireless Personal Area Networks (LR-WPANs) – D18*, February, 2003.
2. W. Ye, J. Heidemann, and D. Estrin, "An energy-efficient MAC protocol for wireless sensor networks," *INFOCOM 2002, Twenty-First Annual Joint Conference of the IEEE Computer and Communications Societies. Proceedings. IEEE*, vol. 3, pp. 1567–1576, 23–27 June, 2002.
3. G.J. Pottie and W.J. Kaiser, "Embedding the internet: Wireless integrated network sensors," *Communications of the ACM*, vol. 43, no. 5, pp. 51–58, May, 2000.
4. K. Sohrabi and G.J. Pottie, "Performance of a novel self-organization protocol for wireless ad hoc sensor networks," in *Proceedings of the IEEE 50th Vehicular Technology Conference*, pp. 1222–1226, 1999.
5. E.J. Duarte-Melo and M. Liu, "Analysis of energy consumption and lifetime of heterogeneous wireless sensor networks," *Global Telecommunications Conference, 2002. GLOBECOM '02. IEEE*, vol. 1, pp. 21–25, 17–21 November, 2002.
6. S. Shoval, I. Ulrich, and J. Borenstein, "NavBelt and the guide-cane [obstacle-avoidance systems for the blind and visually impaired]" *Robotics & Automation Magazine, IEEE*, vol. 10, no. 1, pp. 9–20, March, 2003.

7. A. Helal, S.E. Moore, and B. Ramachandran, "Drishti: An integrated navigation system for visually impaired and disabled" *Wearable Computers, 2001. Proceedings. Fifth International Symposium* 8–9, pp. 149–156, October, 2001.

8. I. Ulrich, and J. Borenstein, "The GuideCane-applying mobile robot technologies to assist the visually impaired" *Systems, Man and Cybernetics, Part A, IEEE Transactions* on, vol. 31, no. 2, pp. 131–136, March, 2001.

9. S. Ertan, C. Lee, A. Willets, H. Tan, and A. Pentland, "A wearable haptic navigation guidance system" *Wearable Computers, 1998. Digest of Papers. Second International Symposium,* pp. 164–165, October, 1998.

10. Y.-J. Kim, C.-H. Kim, and B.-K. Kim, "Design of auditory guidance system for the blind with signal transformation from stereo ultrasonic to binaural sound" *Department of Electrical Engineering & Computer Science Korea Advanced Institute of Science and Technology (KAIST)* 373-1 Kusong-dong, Yusong-gu, Taejon 305–701, Korea, 2001.

11. J.M. Benjamin, N.A. Ali, and A.F. Schepis, "A laser cane for the blind," *Proceedings of the San Diego Biomedical Symposium*, vol. 12, pp. 53–57, 1973.

12. K.B. Benson, *Audio Engineering Handbook*, McGraw-Hill Book Company, New York, 1986.

13. D. Bissitt, and A.D. Heyes, "An application of biofeedback in the rehabilitation of the blind," *Applied Ergonomics*, vol. 11, no. 1, pp. 31–33, 1980.

14. B.B. Blasch, R.G. Long, and N. Griffin-Shirley, "National evaluation of electronic travel aids for blind and visually impaired individuals: Implications for design," *RESNA 12th Annual Conference, New Orleans, Louisiana*, pp. 133–134, 1989.

15. J. Borenstein, and Y. Koren, "The vector field histogram – fast obstacle-avoidance for mobile robots," *IEEE Journal of Robotics and Automation*, vol. 7, no. 3, pp. 278–288, 1991.

16. J. Borenstein, and I. Ulrich, "The GuideCane – A computerized travel aid for the active guidance of blind pedestrians," *IEEE International Conference on Robotics and Automation*, Albuquerque, NM, pp. 1283–1288, April, 1997.

17. M.A. Ericson and R.L. Mckinley, "Auditory localization cue synthesis and human performance," *IEEE Proceedings of the National Aerospace and Electronics*, 1989.

18. R.G. Golledge, J.M. Loomis, R.L. Klatzky, A. Flury, and X.L. Yang, "Designing a personal guidance system to aid navigation without sight: Progress on GIS component," *International Journal of Geographical Information Systems*, vol. 5, no. 4, pp. 373–395, 1991.

19. L.E. Navarro-Serment, R. Grabowski, C.J.J. Paredis, and P.K. Khosla, "Millibots: The development of a framework and algorithms for a distributed heterogeneous robot team," in *IEEE Robotics and Automation Magazine*, vol. 9, no. 4, December, 2002.

20. H. Brown, J.M.V. Weghe, C.A. Bererton, and P.K. Khosla, IEEE/ASME, "Millibot trains for enhanced mobility," *Transactions on Mechatronics*, vol. 7, no. 4, pp. 452–461, December, 2002.

21. R. Grabowski, L.E. Navarro-Serment, C.J.J. Paredis, and P. Khosla, "Heterogeneous teams of modular robots for mapping and exploration," *Autonomous Robots – Special Issue on Heterogeneous Multirobot Systems*. vol. 8, no. 3, pp. 293–308, 2000.

22. C. Bererton, L.E. Navarro-Serment, R. Grabowski, C.J.J. Paredis, and P.K. Khosla, "Millibots: Small distributed robots for surveillance and mapping". *Government Microcircuit Applications Conference*, 20–23 March, 2000.

23. C.P. Diehl, M. Saptharishi, J.B. Hampshire II, and P.K. Khosla, "Collaborative surveillance using both fixed and mobile unattended ground sensor platforms," in *AeroSense '99*. SPIE. Orlando, Florida, 1999.

24. L. Husng, Y.S. Lim, D. Li, and C.E.L. Teoh, "Design and analysis of a four wheel omnidirectional mobile robot," *International Conference on Autonomous Robots and Agents*, 2004.

25. K.S. Byun, S.J. Kim, and J.B. Song, "Design of a four-wheeled omni-directional mobile robot with variable wheel arrangement mechanism," *International Conference on Robot & Automation*, 2002.

26. M. Wada and H.H. Asada, "Design and control of a variable footprint mechanism for holonomic omnidirectional vehicles and its application to wheelchairs," *IEEE Transactions on Robotics and Automation*, vol. 15, no. 6, 1999.

27. R. Holmberg and O. Khatib, "Development and control of a holonomic mobile robot for mobile manipulation tasks," *International Journal of Robotics Research*, vol. 19, no. 11, pp. 1066–1074, 2000.

28. H.B. Zhang, K. Yuan, and J.D. Liu, "A fast and robust vision system for autonomous mobile robots," *IEEE International Conference on Robotics, Intelligent Systems and Signal*, pp. 60–65, 2003.

29. J. Bruce, T. Balch, and M. Veloso, "Fast and inexpensive color image segmentation for interactive robots," *IEEE/RSJ International Conference on Intelligent Robots and Systems*, vol. 2, pp. 2061–2066, 2000.

30. T. Matsuoka, A. Motomura, and T. Hasegawa, "Real-time self-localization method in a dynamically changing environment," *IEEE/RSJ International Conference on Intelligent Robots and Systems*, vol. 2, pp. 27–31, 2003.

31. C.F. Marques and P.U. Lima, "Vision-based self-localization for soccer robots," *IEEE/RSJ International Conference on Intelligent Robots and Systems*, pp. 1193–1198, 2000.

32. J. Hill, P. BounadoMa, and D. Culler, "Active message communication for tiny network sensors," in *Proceedings of the Twentieth Annual Joint Conference of the IEEE Computer and Communications Societies (INFOCOM 2001)*, 2001.

33. TinyOS, http://webs.cs.berkeley.edu/tos/, Berkeley NEST.

WIRELESS LOCAL-AREA NETWORKS

Chapter 8

Opportunism in Wireless Networks: Principles and Techniques

Ahmad Khoshnevis and Ashutosh Sabharwal

8.1 Opportunism: Avenues and Basic Principles

In this chapter, we will study how the variation in the sources and channels can be exploited to improve the power and spectral efficiency of wireless networks. A typical multihop wireless network is displayed in Figure 8.1, with an inside look into a typical node. At any node, there can be multiple sources with multiple queues, either based on the data generated at the node (a user web-surfing) or generated at other nodes (forwarding traffic in a multihop network like ad hoc, sensor or mesh network). Most sources produce information in time-varying bursts. This is clearly evident in multimedia sources like voice, music and video, which when compressed yield a time-varying amount of information. Voice conversations have periods of silence and speech interspersed, clearly demarking periods of high information followed by low-information segments. The same time-varying burstiness is evident in a multitude of data sources ranging from web browsing to sensing natural scenery in sensor networks. In fact, the traffic in many networks is bursty at multiple timescales, evident from abundance of self-similar traffic models studied in wireline[6,39] and wireless networks.[21] Time-varying data rates of source imply a time-varying demand. Analogous to time-varying sources is the well-known property of

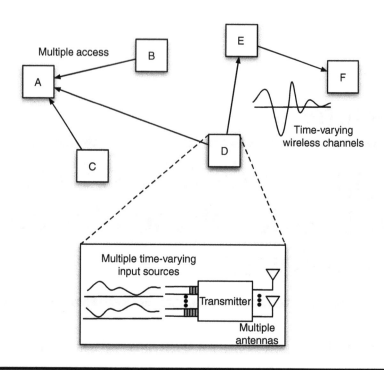

Figure 8.1   A typical multihop wireless network with multiple nodes. The figure also shows an expanded view of a typical node which could have multiple sources to be sent using a physical layer transmission over multiple antennas.

wireless channels—their quality—over time. The time-varying nature of the wireless channels can be attributed to the mobility of the users and their surrounding environments. Electromagnetic signals transmitted by a base station or by mobile users reach the intended receiver via several paths; the multiple paths are caused by reflections from man-made and natural objects. Since the length of the each path may be different, the resultant received signal shows a wide fluctuation in its power profile, thereby complicating the design of spectrally efficient systems. If the transmitter and/or receiver is equipped with multiple antennas, then between any two nodes, there are multiple wireless channels that may vary independently of each other depending on antenna configuration. This time-variability implies that channels have a data capacity which varies with time.

> *Thus, a typical wireless network consists of time-varying demand (sources) and time-varying capacity (channels).*

In this chapter, our goal will be to study techniques which exploit the inherent time-variability in networks to *improve* performance. Grouped together under the broad class called *opportunistic* techniques, the aim is to conserve resources under poor channel conditions (or high source demand) and judiciously use them in favorable conditions.

8.1.1 Degrees of Freedom

Performance of communication over wireless links can be characterized by the number of independent degrees of freedom along which a radio signal can be transmitted. The statistical degrees of freedom, available in spectral, temporal, and spatial domains describe the dimensions available to send a wireless signal.

Temporal and spectral dimensions. Consider any signal, $X(t)$, which is transmitted over a bandwidth of W Hz for time duration of T seconds. It is well known from Ref. [33] that any such signal has $\sim 2WT$ signaling dimensions. That is, only $2WT$ independent parameters are available at the transmitter to control the type of the signal that can be transmitted over the channel. Note that these dimensions are *not* statistical in form, i.e, they are available for any transmission as long as it is band- and time-limited. After transmission through the wireless channel, the signal undergoes distortions due to multipath propagation between the transmitter and the receiver. The number of dimensions that determine the communication system performance are the ones observable at the receiver. Thus, if some of the dimensions are lost during the transmission due to poor channel conditions, the effective number of dimensions available for communication is less than $2WT$. Since the channel variations are best characterized statistically because of random fluctuations of the channel gain and phases, the received degrees of freedom are also best characterized statistically.

Spatial dimensions. The $2WT$ formula gives the number of spectro-temporal dimensions for single antenna transmitter and receiver pairs. In wireless systems, spatial variations can also be exploited by the use of multiple transmit and receive antennas. For rich scattering environments, like those encountered in indoor channels or when the transmitters and receivers are located at low heights (potentially in dense deployments of sensor networks), each additional transmit–receive pair adds a new dimension. Thus, in a multiple-antenna system with n transmit and n receive antennas, we have $2nWT$ dimensions for signaling.[34,37]

Spatial degrees of freedom are also available in multiple-user systems. Owing to differences in topological environment, the propagation environment between any transmitter and receiver is different from that of any other transmitter–receiver pair. Consider node A in Figure 8.1 that is communicating with three nodes (B, C, and D). The three channels, A–B, A–C, and A–D vary independently of each other and thus have the same characteristics as a multiple-antenna system.

Source "degrees of freedom". In general, sources are not characterized in the same framework as the channels. However, there is certain "duality" between sources and channels, which allows a conceptual interpretation of many source opportunism-based techniques as duals of channel opportunism-based methods. For example, any arrival process $A(t)$ can be characterized by its temporal properties and spectrum, since it is simply a time-series like the channel $H(t)$ or the transmitted signal $X(t)$. Similarly, there can be multiple independent sources at any node (as shown in Figure 8.1), which are analogous to multiple independent channels as in multiple antenna links between two nodes.

However, there are some fundamental differences between sources and channels depending on the network usage. While the channel between any two nodes (as long as they are reasonably far away from each other) varies independently of the channel between any two other nodes, the sources at each node can be strongly correlated across nodes. For example in Figure 8.1, the data at node D travels to F via node E, which implies that there is a potential for correlation between the arrival processes at nodes D, E, and F. The only reason these arrival processes may not be perfectly correlated is that there may be random departures from each node due to channel time variations. Thus, the random nature of transmissions can "decorrelate" the arrival process across multiple nodes.[29]

8.1.2   Roadmap

> The general principle behind all opportunistic methods is to *adapt* transmitter actions based on the *current* conditions of the source and the channel.

For example, source burstiness has been conventionally exploited in communication networks by statistically multiplexing them over the same link using random access methods;[1] statistical multiplexing automatically provides more bandwidth to the user, who has high current data need. To understand the basic principles, we will study simple bare-bones formulations in depth, and then apply the lessons learnt to real networks. We will follow a rigorous approach to the derivation, inspired by the information–theoretic method. The outline for the rest of the chapter and the major milestones in it are as follows:

Source opportunism. Focusing on a single point-to-point link, we will first study (Section 8.2) how scheduling packets from a bursty source can be used to reduce the power consumption of any node. The opportunism can be attributed to allowing more delay in the system to avoid sending large bursts of packets in any single transmission slot.

Temporal–spatial opportunism over a single link. Our next focus will be on multiple antenna links and how temporal and spatial degrees of freedom can be exploited in an opportunistic manner (Section 8.3). The key concept is that if the transmitter can receive information (even if only a few quantized bits[17] about the channel condition, then it can adapt its physical layer parameters like modulation, code rate, and power to optimally use the channel conditions. Physical adaptation is commonly used in many current wireless networks (cellular networks like IS-95, 3G) and is proposed in upcoming standards (like IEEE 802.11n).

Temporal-spatial opportunism in ad hoc networks. The adaptive physical layer concepts are next applied to ad hoc networks, to allow opportunistic tuning of transmission rate (Section 8.4). Using the framework of IEEE 802.11 4-way handshake (request-to-send (RTS)-clear-to-send (CTS)-DATA-ACK), we discuss the Opportunistic Auto-Rate (OAR) MAC protocol that performs per packet rate adaptation to significantly improve the throughput compared to current systems.

Temporal-spectral-spatial opportunism in ad hoc networks. Finally, we show (Section 8.5) how all three dimensions can be fruitfully exploited to further increase the capacity of OAR protocol by a multichannel adaptation labeled Multichannel OAR (MOAR).

8.2 Source Opportunism

8.2.1 Problem Formulation

Consider a time-slotted system, each T_s seconds long, with an input buffer to queue arriving packets. In time slot n, there are a_n arriving packets, each of which is assumed to have the same number of information bits. The simplest source model assumes independent and identically distributed (i.i.d.) arrival from one time slot to another, according to the distribution $p(a_n)$. In more complex and realistic models, the arriving packets a_n can be correlated over time, such as signals generated by the measurements of a thermometer. Though correlated models are important for practical systems, the key issue which affects the performance of the system is whether the scheduler at the transmitter has knowledge of the distribution of the arrival process or not. Thus, our problem formulation and the subsequent developments in this section will explore how the scheduler structure differs with and without the knowledge of source statistics.

At every time slot, d_n packets are appropriately modulated as X_n and transmitted over an additive white Gaussian noise wireless channel, given by

$$Y_n = \sqrt{P_n}X_n + \epsilon_n \tag{8.1}$$

Thus, each transmission packet X_n is unit power and P_n is the power at which the packet is transmitted in slot n. Further the additive white

Gaussian noise component, ϵ_n is assumed to have a zero mean and a unit variance. With the above setup, the queue evolves according to the following equation,

$$x_{n+1} = \max(x_n + a_n - d_n, 0) \tag{8.2}$$

The transmission $\sqrt{P_n}X_n$ has to be designed such that the receiver can decode the packets *reliably*. We will use an information–theoretic notion of reliability, motivated by Ref. [27]. To invoke mutual information-based reliability, we require that the time slot duration, T_s, be long enough to encode d_n packets close to channel capacity. More precisely, the rate at which d_n packets can be reliably decoded when sent in a fixed length time slot T_s is given by

$$d_n = \log(1 + P_n) \tag{8.3}$$

where, for simplicity, we have assumed that the system bandwidth is 1 Hz. Equivalently, the power required to ensure reliable reception is given by

$$P_n = e^{d_n} - 1 \tag{8.4}$$

The relation between the number of scheduled packets d_n and the power required for their transmission shows that a linear increase in number of scheduled packets requires exponential increases in transmit power. The above convex relation between d_n and P_n will also be the reason why power control is important even *without* any channel variations.

Our objective is to minimize the average transmit power consumption of the transmitter in such a way that the packet delay does not exceed an average or a maximum delay bound,

$$P^*(D_0) = \min_{\Theta(D_0)} \lim_{n \to \infty} \mathbb{E}\{P_n\} \tag{8.5}$$

The set of schedulers $\Theta(D_0)$ consists of all those schedulers that (a) do not drop any packets and (b) ensure that the queueing delay is no more than D_0, which may be measured as an average or as the maximum packet delay. We will consider two important classes of schedulers. In the first class, the scheduler completely knows the packet arrival distribution and hence formulates its scheduling decisions based on source statistics. In contrast, the second class of schedulers works without any knowledge of the source statistics and hence aims to perform based only on current arrivals, without utilizing arrival distributions.

8.2.2 Fully Informed Schedulers

In this section, we will study the tradeoff between the average transmit power and the resulting average queuing delay of optimal schedulers.

Our approach will be to first (a) characterize the nature of schedulers that achieve the minimum in optimization criterion (8.5), (b) study the nature of optimal schedulers to derive suboptimal, computationally simple schedulers and lastly (c) derive an intuitive analytical approximation for the relation between average power and average queuing delay of optimal schedulers.

A scheduler is a mapping from the queue state (x_n) to the number of packet departures (d_n). Given the number of scheduled departures, the packet encoding and corresponding power P_n is decided by the Eq. (8.4). We will consider the set of randomized schedulers, $\mathcal{S} = \{\alpha | \alpha : x_n \mapsto d_n\}$, where a general scheduler mapping is defined by probabilities $\alpha_{j,i} = \text{Prob}(d_n = j | x_n = i)$. Since $\alpha_{j,i}$ are probabilities, $\sum_{j=0}^{i} \alpha_{j,i} = 1$. The set of valid schedulers is thus all those probabilities that satisfy the above properties. A scheduler is called a deterministic if $\alpha_{j,i} \in \{0, 1\}$.

The two key metrics of interest are average queuing delay and average power of the scheduler, which we will denote by $D_{avg}(\alpha)$ and $P_{avg}(\alpha)$. The work in Ref. [28] characterizes the subset of schedulers $\Theta \subset \mathcal{S}$, which does not lose any packet (no packet dropped at transmitter and enough power used for receiver to ensure reliable decoding). Furthermore, it is shown that the delay and power of any zero-outage randomized scheduler can be expressed as a convex combination of delay and power of deterministic zero-outage schedulers, as described by the following result:

Theorem 1 (Characterization of delay and power). *Consider a queue with finite buffer size L and an input process with no more than M packet arrivals in one slot. Define $\mathbb{W}$ as the set of deterministic schedulers. Then, for any randomized scheduler $\phi \in \Theta$, there exists $\eta_i' \in [0, 1]$, $i = 1, 2, \ldots, F' = |\mathbb{W} \cap \Theta|$, with $\sum_{i=1}^{F'} \eta_i' = 1$, such that*

$$D_{avg}(\phi) = \sum_{i=1}^{F'} \eta_i' D_{avg}(\gamma_i) \tag{8.6}$$

$$P_{avg}(\phi) = \sum_{i=1}^{F'} \eta_i' P_{avg}(\gamma_i) \tag{8.7}$$

where γ_i is a scheduler and $\gamma_i \in \mathbb{W} \cap \Theta$ for all i. Finally, the boundary of the achievable region in the delay-power plane is piecewise linear with the vertices achieved by deterministic schedulers.

Thus the average packet delay achieved by any zero outage randomized scheduler is given by a convex combination of the average packet delays achieved by all possible zero outage deterministic schedulers.

In addition, the same convex combination of the average powers of the zero outage deterministic schedulers gives the average power of zero outage randomized policy.

To compute the optimal scheduler (randomized) for a desired delay bound D_{avg}, a dynamic programming-based technique commonly known as value iteration algorithm (VIA)[31] can be employed. For large values of buffer sizes, L, the number of states in the VIA increase exponentially and hence computing the optimal scheduler is computationally intensive. For the cases where the arrival distribution is measured in real time and the optimal scheduler is adapted over time, the implementation of VIA can lead to an intractable design. There are two possible techniques to reduce the computational complexity of scheduler design.

A simple method for complexity reduction is to reduce the number of states in the VIA. The following state reduction is most useful for moderate to large delay scenarios. As delay increases beyond one time slot, the scheduler tends to take the same action in several consecutive queue states. Thus, to reduce the state diagram size, multiple states can be morphed into one state. For instance, for all states $x_n = j, \ldots, j + K$, the scheduler can be constrained to take the same action u_n, thereby reducing the state space. It is clear that the constrained system will be suboptimal, but the loss due to additional constraints can be minimized by appropriate choice of the number and lengths of constraint intervals.

The second suboptimal scheduler is labeled the *log-linear* scheduler, described as follows: For a queuing delay of one time slot the optimal scheduler flushes the buffer at all time slots, i.e., $d_n = x_n$ and the corresponding power required in each time slot is proportional to $e^{d_n}(x_n)$. As the delay increases, we observe that the optimal scheduler tends to choose $d_n(x_n)$ so that the power in each time slot is linearly proportional to x_n, thereby "equalizing" the power penalty in large buffer states x_n. For equalizing the power, the scheduler picks packets $d_n(x_n) \approx \log(x_n)$. Combined with the natural constraint that we cannot transmit more packets than available ($u_n \leq x_n$), we propose the following log-linear scheduler,

$$u_n = \min(x_n, \lfloor \log(\kappa x_n) \rfloor). \tag{8.8}$$

The parameter κ of the log-linear scheduler is chosen to meet the delay bound. For buffer states greater than $L - M$, the log-linear scheduler transmits at least $x_n - (L - M)$ packets to prevent buffer overflows. The log-linear scheduler greatly simplifies scheduler design since it requires only one parameter κ. The value of κ to achieve a certain average delay depends on the arrival distribution of the source. The following straightforward adaptive algorithm can be used to compute the value of κ dynamically to achieve any given average delay constraint based only on the knowledge of the mean arrival rate. In the adaptive scheduler, κ_n is a function of time and is updated in every time slot as given below:

$$\kappa_n = \kappa_{n-1} + \Delta_\kappa(\hat{D}_n - D_0) \tag{8.9}$$

where $\hat{D}_n$ is the sample average delay given by the ratio of the sample average buffer length to arrival rate. The performance of the adaptive scheduler is shown to numerically converge to the performance of a log-linear scheduler that has full knowledge of arrival statistics.[28] Figure 8.2 compares the performance of the log-linear and the adaptive log-linear scheduler with that of the optimal scheduler, and shows that the suboptimal methods perform very well.

In Ref. [28], an approximate relation between average power and average queuing delay is also derived for the optimal scheduler as

$$\log\left(1 + \frac{P_{\mathrm{av}}}{\sigma^2}\right) \approx \lambda + \frac{\sigma_a^2}{4\lambda D_{\mathrm{avg}}} \tag{8.10}$$

where P_{av} is the average transmit power and D_{avg} the average queuing delay. Further λ is the average arrival rate and σ_a^2 is the variance of the arrival process $\{a_n\}$. Thus, we can see that as the delay increases, the average transmit power decreases. Furthermore, if the traffic is more bursty, σ_a^2, the gains from larger delay are also larger.

Additional queuing delay allows the scheduler to smoothen the incoming traffic, thereby removing large swings in power

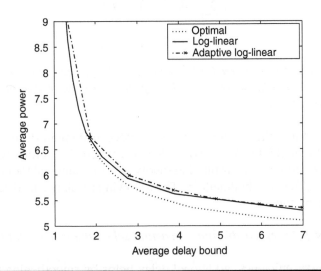

Figure 8.2 Comparison of log-linear and adaptive log-linear schedulers with corresponding optimal scheduler (buffer length $L = 50$, maximum arrival $M = 6$, uniform arrival distribution).

requirements needed if very bursty traffic is transmitted with low delays.

Thus, a scheduler acts as "filter" on the incoming packet arrivals, where the bandwidth of the filter is determined by the ratio of burstiness of the traffic to average delay. In the next section, we will further strengthen this filtering intuition and show that it is provably true that power-minimizing schedulers are always low-pass filters.

8.2.3 Uninformed Robust Schedulers

In this section, we consider the problem of minimizing average transmit power with a constraint on maximum delay, which any packet can tolerate. Let D_{max} denote the maximum delay constraint, then the following result can be proven.

Theorem 2 (Filter characterization[15]). *A scheduler which guarantees the maximum delay D_{max} for any arrived packet is a linear time variant filter of size D_{max} denoted by*

$$u_n = \beta_0^n x_n + \beta_1^n x_{n-1} + \cdots + \beta_{D_{max}-1}^n x_{t-D_{max}+1} \tag{8.11}$$

where the filter coefficients satisfy the following:

$$\sum_{i=1}^{D_{max}-1} \beta_i^{n+i} = 1, \quad \forall n \tag{8.12}$$

$$0 \le \beta_i^n \le 1, \quad \forall n, i \tag{8.13}$$

Furthermore, any time-variant filter of size D_{max} that satisfies the above constraints is a valid scheduler, which guarantees the maximum delay of D_{max} for any packet.

Theorem 2 turns the design of the guaranteed maximum delay scheduler into a problem of filter design with a linear structure. Therefore, we foresee the vast literature on linear-filtering theory as a fundamental tool in designing power-efficient schedulers.[9,25] The following result occurs directly from basic filtering theory:

Corollary 3 *Every feasible time-invariant scheduler is a low-pass filter.*

For power efficiency, additional delay helps by smoothening the input arrival process via queuing.[28] It is clear that by increasing the number of filter taps (equivalently increasing maximum possible scheduling delay), one can design better low-pass filters, leading to transmit power reduction.

The intuition behind the fact that the optimal average power scheduler would have smoother output sequence than the input sequence comes from the convexity of the objective function of the optimization problem in Eq. (8.5).

We first present upper and lower bounds on the performance of optimal schedulers, which will serve as benchmarks for the performance of the following scheduler designs:

Theorem 4 *The average power of the optimal schedulers can be bounded as*

$$\mathbb{E}^{l+1}\left[2^{\frac{a}{l+D_{\max}}}\right] \leq P_{\text{avg}} \leq \mathbb{E}^{D_{\max}}\left[2^{\frac{a}{D_{\max}}}\right] \qquad (8.14)$$

where the input arrival rates are i.i.d. random variables with distribution $f_A(a)$ and l is any positive number. The lower bound is tight when $D_{\max} = 1$ and the two bounds converge as $D_{\max} \to \infty$.

The filter design interpretation immediately allows us to design schedulers using filtering principles. The first such concept is that of linear time-invariant filters, which leads to the question of designing an optimal time-invariant scheduler (recall optimal schedulers are potentially time-varying filters, as shown in Theorem 2). It turns out that the optimal time-invariant scheduler is independent of input arrival distribution and is, in fact, a simple moving average filter.

Theorem 5 (Optimal time-invariant scheduler[15]). *The optimal time-invariant scheduler has the following form*

$$u_n = \frac{x_n + x_{n-1} + \cdots + x_{n-D_{\max}+1}}{D_{\max}} \qquad (8.15)$$

The performance of the optimal time-invariant scheduler in Theorem 5 is given by the upper bound in Theorem 4. In general, time-varying schedulers perform much better. The next result derives a time-varying scheduler, and adapts the coefficients β_i^n based on the current queue conditions to achieve improved performance.

Theorem 6 *At any given time slot n, let $(S_0^n, S_1^n, \ldots S_{D_{\max}-1}^n)$ denote the vector of the scheduled output service rates out of the queue from the previous time slot $n-1$, and let a_n be the new arrival which needs to be scheduled. The optimal output service rate for the current time n, d_n and the new vector of the scheduled service rate for time $n+1$, $(S_0^{n+1}, S_1^{n+1}, \ldots, S_{D_{\max}-1}^{n+1})$,* *are determined by the water-filling solution as*

$$d_n = (\mu - S_0^n)^+ \tag{8.16}$$

$$S_i^{n+1} = (\mu - S_{i+1}^n)^+ \forall 0 \le i \le D_{\max} - 2 \tag{8.17}$$

$$S_{D_{\max}-1}^{n+1} = 0 \tag{8.18}$$

$$d_n + \sum_{i=0}^{D_{\max}-1} = x_n \tag{8.19}$$

Therefore the coefficients of the optimal robust time-varying scheduler are given by

$$\beta_i^n = \frac{(\mu - S_{i+1}^n)}{x_n}, \quad \forall 0 \le i \le D_{\max} - 1 \tag{8.20}$$

The above discussion provides an extremely simple solution to the problem of robust scheduling for delay constrained inputs to a queue. There are some important observations that come out of the above solution. First, to find the optimal robust scheduler in Theorem 6, even though we did not use the filtering property of the scheduler, the optimal solution turns out to be exactly a linear time-variant filter of size $D_{\max}$, which is not surprising given the action of Theorem 2. Second, the optimal values of filter coefficient are exactly a function of the past $D_{\max} - 2$ values of the input arrivals and the queue backlog, which is expected, as we discussed earlier. Third, as we mentioned earlier, the optimal scheduler intuitively should try to make the output service rate as smooth as possible, in keeping with the low-pass property of the scheduler. Indeed, the water-filling solution is the best verification of this intuition. Fourth, the water-filling solution (shown in Figure 8.3) reveals that the optimal solution always tries to push the scheduled time of the packets as far as possible, nearer to the maximum value of the tolerable delay. Finally, Figure 8.4 shows the performance of the water-filling scheduler as compared to the upper and lower bounds.

8.3 Spatio-Temporal Opportunism over a Single Link

Much like source variations can be exploited to improve the performance of the wireless systems, channel time variations can also be productively used. Time variations occur in all degrees of freedom due to the physical mobility of users and their surroundings, resulting in channel conditions changing with time, frequency, and space. In this section, we will explore the fundamental ideas about exploiting channel variations for improved performance.

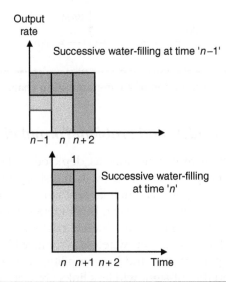

Figure 8.3 Water-filling solution for the time slot $n - 1$ and subsequent time slot n.

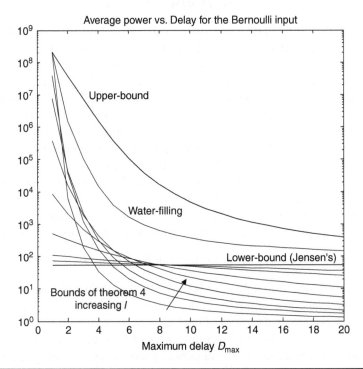

Figure 8.4 Performance of the optimal time-varying robust scheduler and the bounds on performance. Note that performance of the optimal time-invariant robust scheduler matches the upper bound.

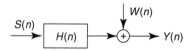

Figure 8.5   System representation of a communication channel.

8.3.1   *Channel Models and Problem Formulation*

In a traditional system representation, as depicted in Figure 8.5, a communication channel can be described by its transfer function $H(n)$. If the timescale of channel variation is much larger than the duration needed for transmission of a packet, then $H(n)$ can be modeled simply by a complex-valued multiplicative factor known as channel coefficient. Such a channel is referred to as a flat fading channel. Wireless systems with high data rate (in order of mega bits per second) and low mobility, such as indoor wireless LAN, wired line, and line-of-sight wireless links are examples of flat fading channel. The input–output relation of such a system can be expressed by

$$Y(n) = H(n)S(n) + W(n) \qquad (8.21)$$

where $H(n)$ is modeled by a random process. Distribution of $H(n)$ depends highly on the physical characteristics of the transmission path between the transmitter and the receiver and the objects surrounding the antennas at both ends. In an environment with a large number of scatterers (e.g., trees or furniture), $H(n)$ can be well approximated by a zero-mean complex Gaussian random process. Note that if we use an $N \times M$ matrix instead of a scalar for $H(n)$ in Eq. (8.21), and modify the input signal, $S(n)$, and additive noise, $W(n)$ to $M \times 1$ vectors then Eq. (8.21) represents a multiple antenna system with M transmit and N receive antennas.

The capacity, or equivalently the maximum achievable data rate with arbitrarily small reliability, is the time average of the maximum achievable data rate at each time. The maximum achievable data rate at each time instance, n, is given by the maximum mutual information expression[34]

$$I(S(n); Y(n)) = \log \det(I + H(n)Q(n)H^{\dagger}(n)) \qquad (8.22)$$

where $Q(n) = \mathbb{E}[S(n)S^{\dagger}(n)]$, and $\dagger$ is the complex conjugate transpose operation. Note that for a single antenna system where $S(n)$ is a scalar, $Q(n)$ is the total power of the input signal at time n. Similarly, for a multiple antenna system the diagonal elements of $Q(n)$ are the average power on each antenna. Therefore, the total power spent at time n can be expressed by the sum of the diagonal elements (power on each antenna) of $Q(n)$, or more precisely

$$P(n) = \text{tr}\{Q(n)\}$$

and the average power is given by simply a time average of the power dissipated at each time slot. That is

$$P_{av} = \lim_{T \to \infty} \frac{1}{T} \sum_{n=1}^{T} P(n)$$

$$\overset{(a)}{=} \mathbb{E}[P(H)] \tag{8.23}$$

where (a) is the consequence of the law of large numbers, and expectation in (a) is taken over by the distribution of H. In general the input signal can be decomposed into a power control factor and a fixed codebook, i.e.,

$$S(n) = P(n)x(n),$$

where $P(n)$ is the temporal adaptation scheme, which controls the power and phase of the input signal, and the input $x(n)$ is independent of channel condition with a unit average power $\mathbb{E}[xx^\dagger] = 1$. Note that in Eq. (8.23), $P(n)$ is replaced by $P(H)$, which indicates the power allocated for each realization of channel and does not concern any specific time slot, whereas $P(n)$ indicates the power allocated at time n. Hence, both $P(n)$ and $P(H)$ represent the same quantity from different views.

When the transmitter does not have any information about the channel, H, the best unbiased strategy is to divide the power equally among all antennas, and allocate the same amount of power at any given time instance. That is to say, for any given time slot n, the allocated power is P_{av} and the power allocated to each transmit antenna is P_{av}/M (M is the number of transmit antennas). But when channel state information is available at the transmitter, the allocated power can be adjusted such that it matches the channel condition. Therefore, at any time slot, n, the total power allocated will be a function of the channel realization at that time slot, $H(n)$, and the way the power is divided among transmit antennas is determined by the power control algorithm. This will be discussed in more detail in the next section.

The mutual information expression in Eq. (8.22) is the maximum achievable data rate at each time instance and because of $H(n)$ it is a time-varying quantity. To find the capacity of the communication link, it is enough to find the time average of expression (8.22), i.e.,

$$C(P_{av}) = \lim_{T \to \infty} \frac{1}{T} \sum_{n=1}^{T} I(S(n); Y(n))$$

$$= \lim_{T \to \infty} \frac{1}{T} \sum_{n=1}^{T} \log \det(I + H(n)Q(n)H^\dagger(n))$$

$$= \mathbb{E}[\log \det(I + HQH^\dagger)] \tag{8.24}$$

The capacity of a communication system is the solution of an optimization problem with a set of constraints. The solution as expected depends on the constraint set. What is described above is the capacity of a communication link with average power constraint. Interested readers can look at Ref. [5] for a complete treatment of the subject.

In designing a communication system two common schemes have been adopted by the system engineers. One scheme is based on achieving the highest possible net throughput, that is, to get as close as possible to the upper limit defined by Eq. (8.24). The performance metric used to evaluate different architectures and methods in this category is referred to as the *ergodic capacity* or average capacity given by expression (8.24).

The second scheme considers a fixed transmission rate, R. However, the capacity of the channel (expression (8.22) is time varying and it may not be able to support transmission rate R at all times. For the duration of time that the instantaneous capacity of the channel is smaller than the transmission rate, reliable communication is not possible, and transmission is said to be in outage. So the design issue is to pick the best rate R such that the net throughput of the system is maximized, or for any arbitrary rate R, find the adaptation scheme that minimizes the outage. We will not discuss this scheme; however, interested readers are referred to Refs. [3, 17]. Before delving into opportunistic resource allocation for maximizing throughput, we shall analyze the multiple antenna system in order to extract the important characteristics of channel matrix $H(n)$.

8.3.2 Multiple Antenna Representation

Till now, we have not discussed the differences of a multiple-antenna system with a single antenna. Consider the system model given in Eq. (8.21). Assume a multiple-antenna system in which $H(n)$ is an N by M matrix. Applying singular value decomposition to $H(n)$, we can replace $H(n)$ by its singular value decomposition (SVD) equivalent to get

$$Y(n) = U(n)D(n)V^{\dagger}(n)S(n) + W(n)$$

$$\tilde{Y}(n) = D(n)V^{\dagger}(n)S(n) + \tilde{W}(n) \tag{8.25}$$

where U and V are unitary matrices, and D is a diagonal $m \times m$ matrix with $m = \min(N,M)$. The second equality in Eq. (8.25) is obtained by multiplying both sides of the first equation with $U^{\dagger}$. This operation is possible when the receiver has a perfect knowledge of H. This is the case as there are channel estimate units at the receivers and training signals in the form of preambles in most of the current systems. The diagonal elements of D are denoted by $\lambda_i^{1/2}$, where λ_i's are the eigenvalues of $HH^{\dagger}$ when $N < M$ or

$H^\dagger H$ when $N \leq M$. Expanding Eq. (8.25) we get

$$\tilde{y}_1(n) = \lambda_1^{1/2}(n)\tilde{s}_1(n) + \tilde{w}_1(n)$$
$$\tilde{y}_1(n) = \lambda_1^{1/2}(n)\tilde{s}_1(n) + \tilde{w}_1(n)$$

$$\vdots$$

$$\tilde{y}_m(n) = \lambda_m^{1/2}(n)\tilde{s}_m(n) + \tilde{w}_m(n)$$

where $\tilde{s}_i(n)$ is the ith element of $V^\dagger(n)S(n)$. The multiple antenna system with M transmit and N receive antennas resemble m $(m = \min(M, N))$ independent single-antenna channels.

> These m *virtual* independent channels add new degrees of freedom (in addition to time and frequency) to the wireless system and are labeled as *spatial degrees of freedom*.

Note that the m channels vary independently of each other, thus increasing the probability that one or several of them are in good condition (compared to a single-antenna system), thereby creating more opportunities for transmitter adaptation. For example, the m channels can be used independently to increase the net throughput. In this case, if the net throughput of a single-antenna system is denoted by $C_1(P_{av})$, the net throughput of the multiple-antenna system can be at the most as high as $mC_1(P_{av})$. Alternately, the m channels can be used to add redundancy and improve the error performance of the system. In the next section, we explain in more detail the different ways in which the spatial degrees of freedom can be used and how their application may affect the net throughput.

8.3.3 Throughput Maximizing Resource Allocation

The problem of optimum power control for single- and multiple-antenna systems is posed rigorously and solved in Refs. [10, 13]. We state the problem and its solution briefly and explain the physical interpretation of the result. The optimum signaling that maximizes the throughput of a multiple-antenna communication system is the solution to the following optimization problem:

$$\min_{(P(n),X(n))} \quad I(S(n), Y(n)|H(n))$$

Subject to: $\quad \mathbb{E}[P(H)] \leq P_0$

The solution of the above optimization problem is known as matrix-water-filling given by

$$\frac{P_{i,i}(H)}{P_0/m} = \begin{cases} \frac{1}{\gamma_0} - \frac{1}{\bar{\gamma}\lambda_i} & \text{if } \lambda_i > \gamma_0 \\ 0, & \text{otherwise} \end{cases} \tag{8.26}$$

where $\bar{\gamma} = P_0/mN_0$, N_0 is the variance of the noise $W(n)$ and the γ_0 the solution to power constraint

$$\mathbb{E}[P_{i,i}(H)] = \frac{P_0}{m} \tag{8.27}$$

That is, the average power allocated to any given eigenvalue over the long run should be the same, and equal to $1/m$ of the total power P_0. For example, for a single-antenna system, the power allocation is depicted in Figure 8.6. The transmitter and receiver have perfect information about the channel state H. The transmitter adjust the transmission power and rate according to the power and rate expressions given in (8.26) and (8.22), respectively. The only practical issue of such system results from the variable transmission rate. It requires a large storage capacity at the transmitter to hold different codebooks for adjusting the rate and a complex receiver capable of decoding the codes of the variable rates.

> The channel-dependent power allocation across the m spatial degrees of freedom, $P_{i,i}(H)$ is an example of spatio-temporal opportunistic technique. The power allocation puts more power on those spatial channels which are in good condition and puts less on those in poor condition. Thus, the optimization is performed in both the dimensions of antennas and time.

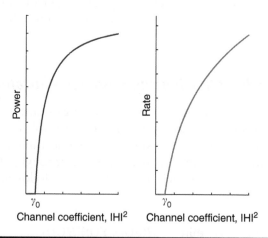

Figure 8.6 Power and rate adaptation based on channel condition for maximizing throughput.

The information at the transmitter is normally provided by the receiver through a feedback channel. To feedback the channel-state information perfectly requires an infinite amount of information to be fedback (a real number or matrix is represented with infinite number of bits). A more realistic scenario is to consider that only a few amount of bits are available to feedback information to the transmitter. In this case the receiver quantizes the channel state and sends back the quantized value using a finite number of bits. The transmitter then uses a power and a rate, which are quantized versions of the power and the rate as shown in Figure 8.6. A more complete treatment of the quantized power and rate allocation can be found in Ref. [22].

Spatial opportunism in some multiuser systems (multipoint-to-point or point-to-multipoint, like those in cellular systems) have also been studied[19,38] and implemented in some practical systems like Qualcomm HDR.[35] The strategy there is similar to the above methods with m independent basestation to mobile user links replacing the m virtual channels in multiple-antenna systems. The optimal method turns out to choose a single node, which has the highest channel gain, SNR_i to allow sending at the highest possible data rate, $\log(1 + SNR_i)$. The resulting use of multiple independent channels is also labeled *multiuser diversity* to emphasize that the multiple spatial channels are due to multiple nodes in the system. Mutliuser diversity systems are only proportionally fair, implying that their short-term fairness properties can be very poor, leading to large delays in the systems.

> Thus, performance improvement in channel condition-based adaptation inherently introduces more delay in the transmission, much like packet scheduling does when used to reduce transmission power.

8.4 Spatio-Temporal Opportunism in Ad Hoc Networks

In this section, we introduce a modification, OAR, to the existing IEEE 802.11x protocol. The key idea of OAR is to opportunistically exploit high-quality channels when they occur via transmission of multiple back-to-back packets. Consequently, nodes transmit more packets under high-quality channels than under low-quality channels. To explain the OAR mechanism, we shall briefly describe the underlying IEEE 802.11 protocol.

8.4.1 Review of IEEE 802.11

As described in Ref. [26], IEEE 802.11 media access is based on the RTS/CTS mechanism.[2] In particular, a transmitting node must first sense an idle channel for a time period of Distributed InterFrame Spacing (DIFS) after which it generates a random backoff timer chosen uniformly from the range $[0, w - 1]$, where w is referred to as the contention window.

At the first transmission attempt, w is set to CW_{min} (minimum contention window). After the backoff timer reaches 0, the node transmits a short RTS message. If successfully received, the receiving node responds with a CTS message. Any other node that hears either the RTS or CTS packet uses the data packet length information to update its Network Allocation Vector (NAV) containing the information of the period for which the channel will remain busy. Thus, all nodes including hidden node can defer transmission suitably to avoid collision. Finally, a binary exponential backoff scheme is used such that after each unsuccessful transmission, the value of w is doubled, up to the maximum value $2^m CW_{min}$, where m is the number of unsuccessful transmission attempts.

8.4.2 Multi-rate IEEE 802.11

The physical layers in IEEE 802.11a/b/g/n protocols are designed to have *multirate* transmission capabilities. So in good channel conditions, the transmitter can send data at a rate greater than the base rate (i.e., 2 Mbps in IEEE 802.11b, which is the protocol we will focus on in our discussion). Figure 8.7 shows a sample channel variation with time for a mobile speed of 2.5 m/s. The received power shows wide fluctuations such that the supported data rate varies between 5.5 and 2 Mbps with almost equal probability. In practice, depending on the existence and strength of a line-of-sight path and the distance from the transmitter, the rates that the channel

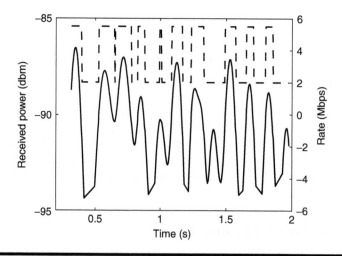

Figure 8.7   Illustration of channel condition variation.

can support may vary within the entire range of the lowest to highest possible data rate (the highest available rate in IEEE 802.11a is 54 and 11 Mbps for IEEE 802.11b). Auto Rate Fallback (ARF)[14] was the first commercial implementation that exploited multirate capability in IEEE 802.11. In ARF, senders use the error rates of previous transmissions to adaptively select future (attempted) transmission rates. That is, after a number of consecutive successful transmissions, the sender modifies its modulation scheme to increase transmission rate. Similarly, after consecutive losses it reduces the data rate. Consequently, if a mobile user has (for example) a perpetually high-quality channel, the user will eventually transmit at higher data rates. Receiver-Based Auto Rate (RBAR) was proposed in Ref. [12], in which the key idea is for receivers to control the transmission rate. To guarantee that all stations receive the messages (RTS, CTS, and ACK) error-free, all messages in IEEE 802.11 must be sent at the base rate. Using the received RTS message, the receiver determines the maximum possible transmission rate for a given acceptable bit error rate. The receiver sends back the calculated rate to the transmitter by adding it into a special field of the CTS message. Note that all other nodes overhearing this message are also informed of the modified transmission rate. This message is termed reservation-subheader (RSH) and is inserted preceding data transfer as illustrated in Figure 8.8. With the RSH message, overhearing nodes can modify their NAV values to the new potentially decreased transmission time. The explicit messaging in RBAR causes a quick adaptation to channel variations and extracts significant throughput gains as compared to ARF.

Although ARF and RBAR increase the data rate, at each access of channel they tend to send a single packet. Since the packet length is fixed, a channel

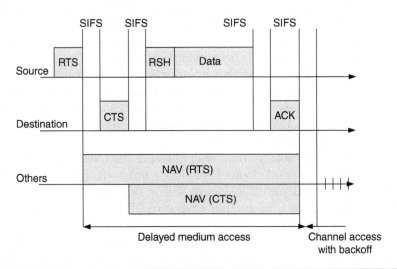

Figure 8.8 IEEE 802.11 with RBAR enhancement.

access with lower transmission rate keeps the channel for a longer time compared with higher data rate packet transmission. The key OAR mechanism is to allow a flow to keep the channel for multiple-packet transmission (instead of for a single packet) when channel quality is good and higher data rate is feasible. In OAR, the same amount of *time* is granted to a sender as if the sender is transmitting at the base rate. For example, if the base rate is 2 Mbps and the channel condition is measured such that transmission at 11 Mbps is feasible, the sender is granted a channel access time to send $\lfloor 11/2 \rfloor = 5$ packets.

Note that all three extensions of IEEE 802.11 protocol are examples of channel adaptation developed in Section 8.3.3. The difference is that optimum power allocation is not included and transmitted power is kept constant and only adaptive rate allocation is considered. The next section gives an insight on how OAR can be implemented and added to IEEE 802.11.[30]

8.4.3 OAR Protocol

In order to allow the sender to hold the channel for multiple packet transmission, the *fragmentation mechanism* of IEEE 802.11 can be exploited. The fragmentation mechanism—defined as part of the IEEE 802.11 standard and mandated for implementation—provides a simple and practical way for nodes to hold the channel for multiple packets when high data rates are measured. In particular, as in RBAR, the receiver indicates the available physical-layer rate via the RTS/CTS messages. If the data rate is above the base rate, the *more fragments* flag in the *frame control* field of the MAC header is set by the sender until ⌊transmission_rate/base_rate⌋ packets are transmitted. The duration field carried by the data and the subsequent ACK is also updated to indicate that the medium will be busy until the end of the next data packet. Thus, as described above, each data packet and ACK serve as a virtual RTS/CTS so that no additional RTS/CTS frames need to be generated after the initial RTS/CTS handshake. The sender must also set the *fragment number* subfield in the *sequence control* field of the MAC header to 0. This prevents the receiver from treating the data packet as a part of an actual fragmented packet.

To illustrate the protocol time line, consider the following example in which node 1 has a good channel (11 Mbps) and node 2 has a poor channel (5.5 Mbps). As depicted in Figure 8.9, in one case, node 1 uses the OAR modification and, in another case uses RBAR. Using OAR, node 1 retains channel access and sends four more packets without any channel contention. Whereas, RBAR goes into contention immediately after the transmission of node 1 is completed. The backoff counter for node 2 was frozen while node 1 was transmitting, and node 1 picks a new backoff counter after finishing its transmission. Thus, with high probability, node 2

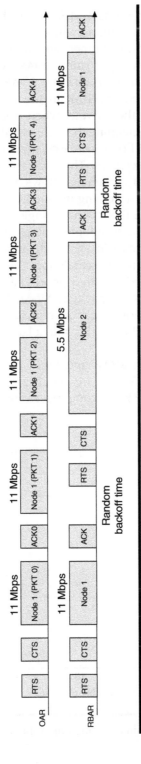

Figure 8.9 Illustration of OAR and RBAR time-lines for a two-node system. Node 1 is in a better channel state than node 2 (11 versus 5.5 Mbps).

gets access to the channel and sends its data at 5.5 Mbps. Thus, RBAR loses in throughput owing to extra contention after every packet, in addition to not capitalizing fully when good channels are encountered.

Figure 8.10 shows the throughput gain of OAR over RBAR for different numbers of flows. Observe that in all cases, OAR results in significant throughput gains as compared to RBAR in the range of 42–56%. OAR extracts this gain by holding the channel when it is good to the longest extent possible, subject to maintaining the base-rate time shares. Moreover, observe that as the number of flows increases, the throughput gain of OAR as compared to RBAR also increases. This increase is due to two factors: First, with a higher number of contending flows, the fraction of time that a flow with a good channel is accessing the medium is higher, resulting in increasing gains. That is, since flows send a single packet at the base rate, but up to five consecutive packets at higher rates, additional users provide more opportunities for a flow to be in a good-channel state, thereby extracting further gains from opportunistic scheduling. Second, OAR reduces the contention overhead since nodes with a good channel exploit the channel for transmission of additional packets. Since IEEE 802.11 as well as RBAR have increasing contention times for an increasing number of nodes, OAR extracts increased throughput gains with a larger number of nodes by decreasing contention time.

Finally, note that for both RBAR and OAR, the throughput gains as compared to base-rate IEEE 802.11 are significant. For example, for 10 flows, RBAR obtains a gain of 230% throughput as compared to IEEE 802.11,

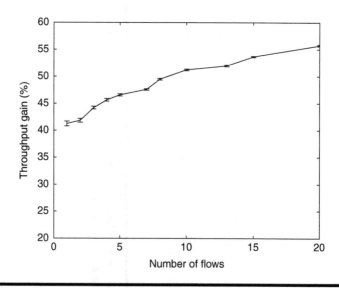

Figure 8.10   OAR throughput gain as a function of the number of flows.

and OAR obtains an additional throughput gain of 51% as compared to RBAR, or 398% above IEEE 802.11.

8.5 Spatiotemporal-Spectral Opportunism in Ad Hoc Networks

In many wireless systems, the available spectrum is more than what is used, thus the system has multiple channels (frequency bands) available for transmission. If the channel condition of all the frequency bands are known at the transmitter, then sender can choose the channel with best quality in order to maximize its throughput. However, knowledge of channel qualities is not free and there is a cost (requires time and power) associated with channel measurement.

The key idea of MOAR is to allow nodes to find the band with the best channel quality. The search should be done in an optimal way such that the net throughput is maximized while the measurement overhead is kept as low as possible.

8.5.1 MOAR Protocol

In this section, we describe how MOAR employs a band-skipping technique within the IEEE 802.11 framework.* All nodes initially reside on a single common frequency band, known as the *home band*. DATA transmission is preceded by the sender transmitting an RTS packet to the receiver on the home band. On reception of the RTS frame, the receiver makes the decision to skip by comparing the measured SNR to a channel skip threshold.† If the measured SNR is low, the sender and receiver skip to a new channel in search of a better quality channel, whereas if the measured SNR is high, data are transferred on the current frequency channel as in the OAR protocol, in which nodes transfer multiple back-to-back packets in proportion to their channel quality.

On making the decision to skip, the receiver selects a band to skip to and piggy-backs this channel on the CTS packet. After transmitting the CTS frame, the receiver immediately skips to the new frequency channel and waits for another RTS from the receiver for a time equal to the CTS timeout value as mandated by the IEEE 802.11 standard. Since we assume that in a realistic setting channel conditions on other frequency channels

---

* Although our discussion of MOAR is within the context of the RTS/CTS mechanism within the DCF mode of IEEE 802.11, the concepts are equally applicable to other RTS/CTS-based protocols such as SRMA[36] and FAMA.[8]

† A reasonably accurate estimate of the received SNR can be made from physical-layer analysis of the PHY layer preamble to each packet.

are unknown, the band to which the receiver decides to skip is selected *randomly* from among the available frequency bands. Yet, if information regarding channel conditions or interference on some other band is known (e.g, in a wireless LAN scenario the Access Point (AP) may have information regarding interference on other bands), the receiver can take that into account to make a better decision about which band to skip to. However, for the purpose of this discussion we do not require the existence of such information.

If, after skipping to a new band, the receiver does not receive another RTS from the sender within a CTS timeout period, the receiver node switches back to the home band and starts contending for channel access as mandated by the IEEE 802.11 standard.

Once the sender receives confirmation of the choice of band to skip to from the receiver (via a CTS frame), it immediately skips to that band. Note that the time elapsed for switching channels is $\sim 1\,\mu s$[7] and has negligible overhead. After skipping to the selected channel, the transmitter and receiver renegotiate the data rate via another RTS/CTS exchange, which also serves the dual purpose of measuring the channel. In case, the channel quality on the new band is measured to be below the skip threshold, the sender–receiver pair can choose to skip again in search of a better quality channel.

Since RTS/CTS exchange prior to any channel skip is done at the base rate on the home band, all nodes within radio range of the receiver and the transmitter can also decode these packets. However, some nodes (including nodes within radio range of the sender but outside the radio range of the receiver) may be unable to hear the CTS packet and are unable to detect whether a decision to skip bands was made or not. Moreover, even though nodes within radio range of the receiver can correctly decode a CTS packet and infer that a decision to skip has been made, they are unable to set a correct defer time since it is not known a *priori* how many times the sender–receiver pair may skip in search of a better quality channel. This can lead to problems similar to the hidden terminal problem.[2]

To solve the problem mentioned above, all MOAR nodes upon reception of an RTS/CTS packet defer (via NAV) for a fixed amount of time corresponding to a maximum time, D_{skip}, necessary for the transmitter and receiver to skip (multiple times, if required) to a better quality channel and finish the DATA/ACK transmission. D_{skip} is given by

$$D_{skip} = K\,T_{overhead} + T_D \tag{8.28}$$

where K is the maximum number of allowed channel skips, $T_{overhead}$ the time taken for RTS/CTS exchange at base rate, including all the defer timers (EIFS, SIFS, DIFS, etc.) as mandated by the IEEE 802.11 standard. The time period T_D represents the time to send data packet at the base rate.

We refer to D_{skip} as a *temporary reservation*, to denote the fact that the reservation is not an actual reservation but represents a maximal amount of reservation time. A temporary reservation serves to inform the neighboring nodes that a reservation has been requested but the duration of the reservation is not known. Any node that receives the temporary reservation is required to treat it the same as an actual reservation with regard to later transmission requests; that is if a node overhears a temporary reservation it must update its NAV so that any later requests it receives that would conflict with the temporary reservation must be denied. Thus the temporary reservation serves as a placeholder until either a new reservation is received or is canceled. If the sender–receiver pair decide not to skip bands then they can proceed with the DATA/ACK exchange on the home band as dictated by OAR in which case other nodes can replace the temporary reservation with the exact reservation, as carried in the DATA/ACK packets.

Once the transmitter and the receiver conclude the DATA/ACK transmission by skipping to one or more bands, they return to the home band. The ACK for the final packet (recall in OAR, nodes can send multiple back-to-back packets) is sent in the home band at base rate by the sender, and the sender rebroadcasts the same ACK. The two-way ACK in the home-channel is necessary to ensure that all the nodes within the range of either the sender and/or the receiver can correctly infer the end of channel skipping and cancel the temporary reservation timer. In case, a node is unable to hear either the updated reservation or the DATA/ACK transmission signalling the end of the temporary reservation, it would be able to contend for the channel again after the temporary reservation has expired.

8.5.2 Performance Study of MOAR

Here we consider random topologies representative of a wireless LAN and consider a scenario where the link distance is uniformly distributed in a circular area with diameter 250 m. We fix the Ricean fading parameter to $K = 4$ and also set the size of the estimation window to 60 packets, as discussed in the previous section. Figure 8.11 shows the average throughput gain of MOAR over OAR (computed on per-flow basis and then averaged over flows) as well as the 95% confidence interval values of the percentage gain for each number of flows. The curve labeled "Look-ahead" represents the genie-aided protocol in which channel state information for all the 11 bands is known *a priori*, and thus flows can at the maximum skip once to the band with a known higher rate than the present band. This serves as an upper bound to the gain that MOAR can extract over OAR. We also implement the optimal skipping rule and plot the throughput gains of MOAR with optimal skipping over OAR. The opportunistic gain that MOAR can extract is dependent upon the distance between the sender and receiver of a flow. For a given random topology, some of the flows are located in

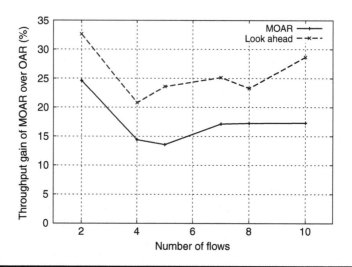

Figure 8.11 Throughput gain of MOAR for random fully connected topologies.

a region where the opportunistic gain obtained by skipping bands is not significant. These nodes, besides contributing little to the net overall gain that MOAR can obtain, actually reduce the opportunistic gain for better located nodes. The reason for this can be attributed to the random nature of the MAC. Whenever nodes with lower opportunistic gain access the medium, nodes that are better located to exploit the opportunistic gain through band-skipping defer medium access. Thus, the net opportunistic gain that can be obtained by exploiting channel diversity is reduced. However, on average MOAR still outperforms OAR by 14–24%. Also note that the gain of MOAR with optimal skipping is very close to the maximum gain achievable if the channel conditions on all the 11 bands are known *a priori*. Thus, in realistic systems where channel state information on other channels may be unavailable, the optimal skipping rule can still enable MOAR to capture most of the performance gains available via opportunistic skipping.

8.6 Conclusions

In this chapter, we discussed the general ideas of adaptation to exploit the inherent time-variability of sources and channels in wireless networks. The significant gain in all cases, results from making use of opportunistic algorithms as default candidates for any new design of high-performance network elements.

It is necessary to say that the adaptive algorithms are not limited to what is mentioned in this chapter. For example in Refs. [18, 38] an adaptive beamforming scheme is investigated. Beamforming algorithms match the phase of the input signal to the phase of the channel, such that the received signals (in a multiple-antenna system) are combined constructively. Opportunism in scheduling can be exploited in many ways. In this chapter, we talked about source scheduling of a single queue. Opportunistic scheduler can be extended to a multiuser scheduler,[20,23] in addition to best-user scheduling, adding new spatial degrees of freedom even in single-antenna systems, often referred to as *multiuser diversity*.[11,40]

In a multiple-antenna system another way of exploiting the spatial degrees of freedom is by antenna selection in which the transmitter only uses one or a subset of one of a set possible transmit antennas.[24,32] On higher networks layers, opportunistic routing can be done in which the best route or a subset of routes among all possible routes is chosen.[4,16] We believe that there are numerous other ways that one can use the information about source and channel, and adapt the transmission scheme in a beneficial way.

References

1. D.P. Bertsekas, and R. Gallager. *Data Networks*. Prentice-Hall, New York, 1992.
2. V. Bharghavan, S. Demers, S. Shenker, and L. Zhang. MACAW: A media access protocol for wireless LANs. In *Proceedings of ACM SIGCOMM '94*, London, UK, 1994.
3. E. Biglieri, G. Caire, and G. Taricco. Limiting performance of block-fading channels with multiple antennas. *IEEE Transactions on Information Theory*, 47(4):1273–1289, May 2001.
4. C. Cetinkaya, and E. Knightly. Opportunistic traffic scheduling over multiple network paths. *IEEE INFOCOM'04*, 3:1928–1937, 2004.
5. T.M. Cover, and J.A. Thomas. *Elements of Information Theory*. Wiley, New York, 1991.
6. M.E. Crovella, and A. Bestavros. Self-similarity in world wide web traffic: Evidence and possible causes. *IEEE/ACM Transacations on Networking*, 5(6):835–846, Dec. 1997.
7. R. Garces, and J.J. Garcia-Luna-Aceves. Collision avoidance and resolution multiple access for multichannel wireless networks. In *Proceedings of IEEE INFOCOM 2000*, pp. 595–602, Tel Aviv, Israel, 2000.
8. J.J. Garcia-Luna-Aceves, and C. Fullmer. Floor acquisition multiple access (FAMA) in single-channel wireless networks. *ACM Mobile Networks and Application (MONET), Special issue on Ad Hoc Networks*, 4:157–174, 1999.
9. J.C. Geromel, M.C. De Oliveira, and J. Bernussou. Robust filtering of discrete-time linear systems with parameter dependent lyapunov functions. *SIAM Journal on Control and Optimization*, 41(3):700–711, 2002.
10. A.J. Goldsmith, and P.P. Varaiya. Capacity of fading channels with channel side information. *IEEE Transactions on Information Theory*, 43(6):1986–92, Nov. 1997.

11. M. Grossglauser, and D.N.C. Tse. Mobility increases the capacity of adhoc wireless networks. *IEEE Transactions on Networking*, 10(4):477–486, Aug. 2002.

12. G. Holland, N. Vaidya, and P. Bahl. A rate-adaptive MAC protocol for multihop wireless networks. In *Proceedings of ACM MOBICOM'01*, Rome, Italy, 2001.

13. S.K. Jayaweera, and H.V. Poor. Capacity of multiple-antenna system with both receiver and transmitter channel state information. *IEEE Transactions on Information Theory*, 49(10):2697–2709, Oct. 2003.

14. A. Kamerman, and L. Monteban. WaveLAN II: A high-performance wireless LAN for the unlicensed band. *Bell Labs Technical Journal*, 2(3):118–133, Summer 1997.

15. M.A. Khojastepour, and A. Sabharwal. Delay-constrained scheduling: Power efficiency, filter design and bounds. *IEEE INFOCOM*, 2004.

16. A. Khoshnevis, and A. Sabharwal. Network channel estimation in cooperative wireless networks. *Canadian Workshop on Information Theory*, May 2003.

17. A. Khoshnevis, and A. Sabharwal. On the asymptoti performance of multiple antenna channels with fast channel feedback. *IEEE Transactions on Information Theory*, 2005 (submitted).

18. I.-M. Kim, S.-C. Hong, S.S. Ghassemzadeh, and V. Tarokh. Opportunistic beamforming based on multiple weighting vectors. *IEEE Transaction on Wireless Communications*, 4(6):2683–2687, Nov. 2005.

19. R. Knopp, and P.A. Humblet. Information capacity and power control in single-cell multiuser communications. In *Proc. ICC*, 331–335, June 1995.

20. C. Li, and X. Wang. Adaptive opportunistic fair scheduling over multiuser spatial channels. *IEEE Transaction on Cp, imocatopms*, 53(10):1708–1717, Oct. 2005.

21. Q. Liang. Ad hoc wireless network traffic: Self-similarty and forecasting. *IEEE Communications Letters*, 6(7):297–299, July 2002.

22. L. Lin, R.D. Yates, and P. Spasojevic. Adaptive transmission with discrete code rates and power levels. *IEEE Transactions on Communications*, 51(12):2115–2125, Dec. 2003.

23. X. Liu, E.K.P. Chong, and N.B. shroff. Opportunistic transmission scheduling with resource-sharing constraints in wireless networks. *IEEE Journal on Selected Areas in Communicaitons*, 19(10):2053–2064, Oct. 2001.

24. Y. Liu and E. Knightly. Opportunistic fair scheduling over multiple wireless channels. *IEEE INFOCOM'03*, 2:1106–1115, 30 March–3 April 2003.

25. S.O.R. Moheimani, A.V. Savkin, and I.R. Petersen. Robust filtering, prediction, smoothing and observability of uncertain systems. *IEEE Transactions on Circuits and Systems*, 45(4):446–457, 1998.

26. B. O'Hara, and A. Petrick. *IEEE 802.11 Handbook, A Designer's Companion*. Washington, DC, IEEE Press, 1999.

27. L.H. Ozarow, S. Shamai, and A.D. Wyner. Information theoretic considerations for cellular mobile radio. *IEEE Transactions on Information Theory*, 43(2):359–378, May 1994.

28. D. Rajan, A. Sabharwal, and B. Aazhang. Delay-bounded packet scheduling of bursty traffic over wireless channels. *IEEE Transactions on Information Theory*, 50(1):125–144, January 2004.

29. A. Sabharwal, A. Khoshnevis, and E.W. Knightly. Opportunistic spectral usage: Bounds and a multi-band csma/ca protocol. *IEEE/ACM Transactions on Networking*, 2005 (Submitted).

30. B. Sadeghi, V. Kanodia, A. Sabharwal, and E. Knightly. OAR: An opportunistic autorate media access protocol for ad hoc networks. *To appear in ACM/Baltzer Mobile Networks and Applications (MONET): Special Issue on Selected Papers from MobiCom 2002*, 2003.

31. L.I. Sennott. *Stochastic Dynamic Programming and the Control of Queueing Systems*. Wiley, New York, 1999.

32. N. Sharma and L.H. Ozarow. A study of opportunism for multiple-antenna systems. *IEEE Transaction on Information Theory*, 51(5):1804–1814, May 2005.

33. D. Slepian. On bandwidth. *Proceedings of IEEE*, 64(3):292–300, March 1976.

34. I.E. Telatar. Capacity of multi-antenna Gaussian channels. Technical report, AT&T Bell Labs, 1995. (Appeared in *European Transactions on Telecommunications*, 10(6):585–595, 1999.)

35. TIA/EIA IS-856. CDMA 2000: high rate packet data air interface specification. *Std*, November 2000.

36. F.A. Tobagi and L. Kleinrock. Packet switching in radio channels: Part III—polling and (dynamic) split-channel reservation multiple access. *IEEE Transactions on Communications*, 24(8):832–845, 1976.

37. D. Tse and P. Viswanath. *Fundamentals of Wireless Communications*. New York, NY, Cambridge University Press, 2005.

38. P. Viswanath, D.N.C. Tse, and R. Laroia. Opportunistic beamforming using dumb antennas. *IEEE Transaction on Information Theory*, 48(6):1277–1294, June 2002.

39. W. Willinger, M.S. Taqqu, R. Sherman, and D.V. Wilson. Self-similarity through high-variability: Statistical analysis of ethernet LAN traffic at the source level. *IEEE/ACM Transacations on Networking*, 5(1):71–86, Feb. 1997.

40. D. Wu and R. Negi. Utilizing multiuser diversity for efficient support of quality of service over a fading channel. *IEEE Transaction on Vehicular Technology*, 55(3):1298–1206, May 2005.

Chapter 9

Localization Techniques for Wireless Local Area Networks

Shih-Lin Wu and Chao-Hsin Chang

9.1 Introduction

The rapid progress of wireless communication and the availability of many lightweight, small-size, and portable computing devices have made a great impact on community. These technologies have made the dream of *communication anytime and anywhere* possible. People using a mobile device can surf the web as well as talk with their friends while they are moving all over the world. To provide more value-added applications, such as geographical navigation for tourists, advertising messages for local potential customers, 911 emergency service for subscribers, etc., we need to acquire the location information of a mobile user (MU) with a mobile device.

In recent years, many localization techniques have been proposed and, based on measured media, can be generally divided into two classes: *dedicated* and *nondedicated* systems. The active badge system[1] is an early and famous dedicated localization technique. It uses infrared (IR) as the measured medium. In the system, a badge equipped by a user emits an IR signal with its unique identifier. Sensors installed at known locations within an area pick up the unique identifier and relay this information to a server to determine the user position. Although the techniques using the dedicated infrastructure for localization purpose can provide a higher degree of accuracy, they usually need cost-expensive, energy-consuming, and scale-limited equipments.

Localization techniques using nondedicated systems can be devised in a variety of ways depending on estimated accuracy, cost, and ease of implementation. Many traditional localization techniques are proposed for wide-area cellular systems, such as angle of arrival (AoA), time of arrival (ToA), and time difference of arrival (TDoA). The well-known global positioning system (GPS) is one of the most successful systems but can be used only in outdoor environments. Because accurate time synchronization is needed for ToA/TDoA, special hardware for AoA, and because the estimated accuracy of a location is seriously affected by the factor line of sight (LoS) of radio frequency (RF) signal for all of these techniques, the traditional techniques may not be suitable and feasible, and may be inaccurate for indoor environments. Recently, many research works have focused their attention on the received signal strength (RSS) of the (RF) signals provided by wide-deployed wireless infrastructures, such as IEEE 802.11 WLANs[2-6] and radio frequency identification (RFID).[7] We will introduce each of these techniques in the next section.

9.2   Nondedicated Localization Techniques

In this section, we mainly survey nondedicated localization techniques using RF as a measurement medium. To ease the understanding of the techniques, we first review the fundamental principles that are employed in traditional localization techniques. And then we give an introduction of recent researches and discuss their limitations.

9.2.1   Fundamental Principles

There are two fundamental principles used in traditional localization techniques: the *triangulation* and the *trilateration*. Below, we will review the two fundamental principles and give some samples for each principle.

9.2.1.1   Triangulation

The triangulation method employs antenna arrays and tries to estimate the direction of arrival of the signal from a to-be-located radio source. At least two such estimates are required from two antenna arrays located at two different locations. The position of the signal source can be estimated at the intersection of the lines of directions from the two antenna arrays.

AoA, also called direction of arrival (DoA), is a localization approach belonging to the triangulation methods. A receiver (e.g., a base station) uses the angle of the received signals from a mobile device to limit the location of a mobile device along a line. Thus, at least two base stations are required to locate the mobile device by taking the intersection of two lines of directions from the two antenna arrays. The base stations have to be equipped with an antenna array or a directional antenna for calculations of the angle of the received signals. Figure 9.1(a) illustrates a mobile device that is located by the base stations BS_1 and BS_2. The angles a_1 and a_2 can be estimated by an antenna array or a directional antenna.

9.2.1.2 Trilateration

The trilateration method employs a distance between an MU and a base station. With the distance as the radius and the base station as the center,

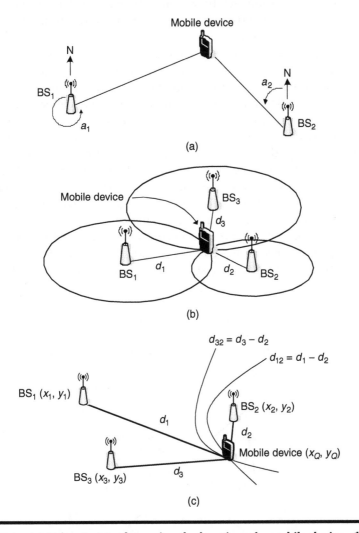

(a)

(b)

(c)

Figure 9.1   (a) Using AoA to determine the location of a mobile device. (b) Locating a mobile user by the intersection of three circles with radii d_1, d_2, and d_3. (c) Using a pair of time differences to locate a mobile user.

a circle is formed. As shown in Figure 9.1(b), trilateration locates an MU through three base stations. The three distances form three circular equations, and the location of the MU is determined by the intersection point of the three circles. The distance can be estimated by the RF propagation time from sender to receiver such as ToA, the difference between the arrival times of the signal from the MU at two base stations such as TDoA, and the RSS due to attenuation such as path loss model.[8] In the following subsection, we will review the estimated distance methods.

ToA approach uses the propagation time of the signals received from a mobile device to estimate the distance between a base station and a mobile device. Figure 9.1(b) illustrates how the three base stations use ToA parameter to determine the distances d_1, d_2, and d_3 between the mobile device and the three base stations BS_1, BS_2, and BS_3, respectively. The three distances form three circular equations, and the location of the mobile device is determined by the intersection node of the three circles.

TDoA approach is similar to ToA and uses the characteristic of a hyperbola to estimate the location of a mobile device. The characteristic of a hyperbola is the difference between d_1 and d_2, and is a constant in Figure 9.2. d_1 is the distance between any node in hyperbola and the focal point BS_1. d_2 is the distance between any node in hyperbola and the focal point BS_2. TDoA locates a mobile device by a pair of time differences. A time difference is a difference between the signal propagation time measured at two base stations. Figure 9.1(c) illustrates how the mobile device is located by TDoA. The time difference $t_{12} = t_1 - t_2$, where t_1 and t_2 are the propagation times measured at BS_1 and BS_2, respectively. d_{12} is the difference between the distance d_1 and d_2 and is represented as follows:

$$d_{12} = ct_{12} = c(t_1 - t_2) = ct_1 - ct_2 = d_1 - d_2$$
$$= \sqrt{(x_1 - x_Q)^2 + (y_1 - y_Q)^2} - \sqrt{(x_2 - x_Q)^2 + (y_2 - y_Q)^2}$$

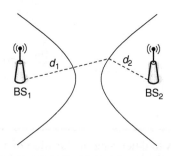

Figure 9.2   The difference between d_1 and d_2 is a constant.

where c is the velocity of light. Thus (x_Q, y_Q) can be calculated by simultaneous equations d_{12} and d_{32}:

$$\begin{cases} d_{12} = \sqrt{(x_1 - x_Q)^2 + (y_1 - y_Q)^2} - \sqrt{(x_2 - x_Q)^2 + (y_2 - y_Q)^2} \\ d_{32} = \sqrt{(x_3 - x_Q)^2 + (y_3 - y_Q)^2} - \sqrt{(x_2 - x_Q)^2 + (y_2 - y_Q)^2} \end{cases}$$

The path loss model is another distance estimated method. One can use the RSS of the RF signal measured by at least three base stations. The method is similar to ToA that ranges the distance between an MU and a base station and then uses the distance to form a circular equation. Note that the path loss model transforms the RSS into distance rather than the signal propagation time. The path loss model (also called signal attenuation model or radio propagation model) uses a mathematical model to represent the relation between the transmitted and received power level and is represented as follows:

$$P_r = P_t \left(\frac{\lambda}{4\pi d} \right)^n g_t g_r$$

where P_t and P_r are the power levels at which the packet is transmitted and received on the sender and receiver, respectively, d is the distance between the sender and receiver, λ the wavelength of RF signal, g_t and g_r are the antenna gain of the sender and receiver, respectively, and n is a path loss coefficient and ranges from 2 to 6 depending on the physical environment. Typically the value of n is 2 in free space. There is another path loss model[9] as follows:

$$S = S_0 - 10\alpha \log \frac{d}{d_0}$$

where S is the received power at the receiver, d the distance between the sender and receiver, S_0 the received power at a reference distance d_0, and α a path loss coefficient. In fact, the above two models are not different but use the distinct unit on the power level. The former uses watt while the latter uses dBm. The unit dBm represents a measured power level in decibels relative to 1 milliwatt.

Three distances can form three circles and their equations so a unique intersection of the circles can be determined theoretically. In practice, the distance estimation may be inaccurate since the RF signals suffer from multipath and noise interferences, especially in an indoor environment. Thus, an MU could be situated in an area, like the gray area in Figure 9.3. The centroid of the gray area is estimated to be the location of the MU.

9.2.2 Radio Map

The main idea of this kind of approach is to build a signal strength model and estimate the location of an MU via finding the best match from the

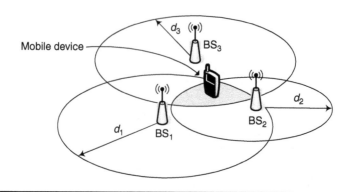

Figure 9.3 A mobile user may be situated in an area constructed by three circular equations.

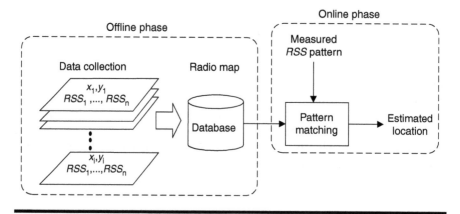

Figure 9.4 Radio map approaches consist of an offline phase and an online phase.

signal strength model. Figure 9.4 illustrates the approach that usually comprises two stages: an offline phase and an online phase. The offline phase is also called the training phase, and its objective is to construct a radio map. The online phase, also called the location determination phase, estimates the location of an MU based on the radio map. We describe how to build a radio map based on the collected signal strengths and predict a location of an MU in the next paragraph.

During the offline phase, the first step is data collection. We collect the RF signal strengths that are broadcasted by the IEEE 802.11 access points at different predefined locations. The information of signal strengths is usually recorded as the format (ss_i, x, y, θ), where ss_i is the RSS from AP_i, $i = 1, \ldots, m$, (x, y, θ) is the location coordinate and orientation of an MU. Note that if the RSS value received from an access point is too weak to be measured by the MU, we assign the ss_i with a minimum value. Therefore, we obtain the relationship between the locations and RSS. During the online phase, an MU measures the RSS from the access point and estimates its location based on the radio map. The localization algorithm uses the measured RSS to find a best matching location from the radio map and predicts if the MU is in the location.

In the rest of this subsection, we introduce several localization algorithms that are based on the concept of radio map.

9.2.2.1 Nearest Neighbors

The nearest neighbor in signal space (NNSS) technique[2] is one of the method that estimates the location of an MU during the online phase. The concept of NNSS is to calculate the distance (in signal space) between the RSS measured by an MU and each RSS recorded in the radio map, and then estimate the location which has the minimal distance, named *the smallest Euclidean distance*. The distance is calculated according to Euclidean distance, as follows:

$$E = \sqrt{\sum_{i=1}^{m}(SS_{\text{MU}_i} - SS_i)^2}$$

where SS_i is the recorded RSS in the radio map and SS_{MU_i} the RSS measured by an MU.

In addition to using only one minimal distance (i.e., single nearest neighbor), there is a method that picks multiple nearest neighbors in signal space, called NNSS-AVG[2,3]. The method is to choose k nearest neighbors (i.e., k training locations in the radio map), and then obtain an estimated location via averaging the coordinates of the k neighbors. We usually choose a small value of k, not large k because the locations far removed from the true location would decrease the accuracy of estimation. Figure 9.5 illustrates how an estimated location is calculated by averaging the coordinates of the $k = 3$ nearest neighbors.

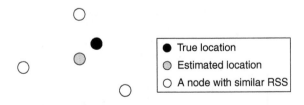

Figure 9.5 **An example using multiple nearest neighbors to estimate the location of an MU.**

9.2.2.2 Smallest Polygon

The idea of smallest polygon[10] is similar to NNSS-AVG. The difference between the NNSS-AVG and the smallest polygon is that the NNSS-AVG focuses on the signal domain, but the smallest polygon focuses on the space domain. After estimating some candidate locations via certain methods (like the smallest Euclidean distance), the NNSS-AVG makes various polygons from the candidate locations and then selects a smallest polygon and obtains the location of the MU from the centroid of the smallest polygon. Figure 9.6 assumes that the candidate locations A, B, C, and D are selected by the similar RSS (i.e., smaller Euclidean distance), and the candidate locations form the variant triangles, $\triangle ABC$, $\triangle ABD$, $\triangle BCD$, and $\triangle ACD$. The smallest polygon method selects the triangle $\triangle BCD$ with the smallest area and estimates the centroid of $\triangle BCD$ to be the location of the MU.

In addition, the approximate point-in-triangulation (APIT)[11] is a localization algorithm applied in wireless sensor network. Figure 9.7 illustrates how the APIT algorithm locates a node through various triangles that are formed by the anchors. These anchors are a small part of sensing nodes

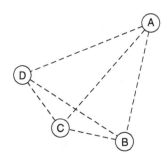

Figure 9.6 The candidate locations A, B, C, and D form variant triangles. $\triangle BCD$ is the smallest area and the centroid of $\triangle BCD$ is the estimated location.

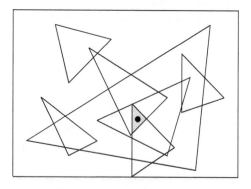

Figure 9.7 An overview of APIT algorithm.

in a network. They usually equip powerful equipments such as GPS or high-powered transmitters. The APIT algorithm locates a node according to the following steps: (1) Beacon exchange, (2) APIT testing, (3) APIT aggregation, and (4) center of gravity (COG) calculation.

Initially, each anchor broadcasts beacons that include the location information where the anchor is. A node who wants to know its location could receive many beacons from the audible anchors. Then a node builds various triangles from a set of the N audible anchors. Figure 9.8 illustrates the APIT testing aims to determine whether a node is inside or outside the triangular regions. The area of the maximum overlap is the most probable area where a node is present during APIT aggregation. Figure 9.9 illustrates the APIT aggregates of the outcome of the APIT testing through a grid SCAN approach. The SCAN approach uses a grid array to represent the maximum area where a node likely resides. Finally, the centroid of the area is treated as the location of an MU.

9.2.2.3 Probabilistic Techniques

Deterministic techniques[2,3] store the scalar values of a measured RSS from each access point for each predefined location in the radio map, i.e.,

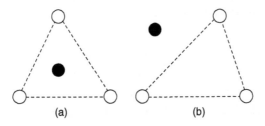

(a)   (b)

Figure 9.8 **(a) A node is inside a triangle. (b) A node is outside a triangle.**

0	0	0	0	0	0	0	0	0	0	0	0	0
0	0	-1	-1	0	0	0	0	0	0	1	0	0
0	0	-1	-1	0	0	1	1	0	1	1	0	0
0	-1	-1	-1	0	0	1	1	2	1	1	0	0
0	-1	-1	-1	-1	0	1	2	2	2	2	0	0
0	-1	-1	-1	-1	0	2	2	2	2	2	1	0
0	-1	-1	-1	-1	0	2	2	2	2	2	1	0
0	0	0	0	0	0	1	1	2	2	1	0	0
0	0	0	0	0	0	0	0	0	0	0	0	0

Figure 9.9 **Using SCAN approach to execute APIT aggregation.**

deterministic techniques average the RSS samples from an access point at a location and record the average as a measured value of RSS for this access point at a location. But the radio maps of the probabilistic techniques[4,5,12–14] store the RSS distributions from the access points. Figure 9.10 shows that the values of RSS received from an access point at a fixed location are unfixed values,[6,14] that vary with time and environment factors. Therefore, probabilistic techniques gather the statistics of RSS samples and analyze the percentages of the probable RSS. This kind of method usually assumes that the received values of RSS from an access point at a location look like a Gaussian probability distribution. They compute the mean and variance according to the percentages of the received RSS during the offline phase. On the other hand, the radio maps of probabilistic techniques store the means and variances of RSS samples from the access points at the predefined locations. Probabilistic techniques then employ the measured RSS to calculate the probabilities of the predefined locations according to the radio map and choose the most probable location to be the estimated location of an MU.

Some savants think that it would be better to treat directly each RSS sample from an access point as a significant information rather than to average the RSS samples. The sensor model[4] that differs from the deterministic techniques is an example of such localization methods that adopt this idea. The sensor model uses Bayes' rule to compute the state probability π'_i of an MU at s_i as follows:

$$\pi'_i = \frac{\pi_i \Pr(o_j \mid s_i)}{\sum_{i=1}^{n} \pi_i \Pr(o_j \mid s_i)}$$

where $s_i = (x_i, y_i, \theta_i)$ is a state (i.e., a predefined location) of the state space $S = \{s_i \mid i = 1, \ldots, n\}$, o_j an observation of observation space

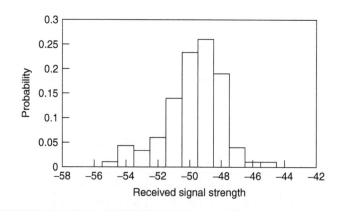

Figure 9.10 The RSS values received from an access point at a fixed location vary with time.

$O = \{o_j | j = 1, \ldots, m\}$, $\Pr(o_j | s_i)$ a conditional probability of getting an observation o_j at state s_i, state vector $\pi = \{\pi_i \mid i = 1, \ldots, n\}$ a probability distribution over various states, and state probability $\pi_i = \Pr(s_i = s^*)$ where s^* is a predicted location of an MU, not the true location. The sensor model assumes that the distribution π is uniform because it assumes that an MU loses his/her way, so every location is equally likely. An observation o contains k RSS samples and is represented as a vector

$$o = \langle k, f_1, \ldots, f_N, (b_1, \lambda_1), \ldots, (b_k, \lambda_k) \rangle$$

where N is the total number of the access points represented in the observation, f_q the number of times for the qth access point, $b_l, l = 1, \ldots, k$ represents the access point in the lth RSS sample, and λ_l represents the signal strength in the lth RSS sample. Note that each access point could appear in the observation four times at most.

During the offline phase, the sensor model obtains an observation at each state. For each access point the sensor model builds two histograms for that state. The first is the distribution of the number of times over the sampled observations. The second is a distribution of the observed signal strengths. Therefore, the sensor model stores the two distributions in the radio map. During the online phase, the sensor model calculates $\Pr(o | s_i)$ based on the two distributions.

$$\Pr(o | s_i) = \left(\prod_{q=1}^{N} \Pr(f_q | s_i) \right) \left(\prod_{l=1}^{k} \Pr(\lambda_l | b_l, s_i) \right)$$

where $\Pr(f_q | s_i)$ is the conditional probability of the number of times for the qth access point at state s_i and $\Pr(\lambda_l | b_l, s_i)$ the conditional probability that the signal strength of access point b_l is λ_l at state s_i.

Finally, the sensor model computes the probability distribution over the states, π', based on Bayes' rule and selects the most probable state to be the location of an MU.

Most indoor localization methods partition the environment into a grid that is 1–2 m apart. Such grid-based methods attempt to estimate which grid is the location of an MU, and they need higher accuracy. However, for variant requirements, some indoor localization methods do not desire an accurate coordinate but just need to know the region (cell) where an MU stays. The region maybe a room, an office, or a hallway segment in a building. Such kinds of localization methods are topology-based methods. The Gaussian fit sensor model[5] is a topology-based localization scheme. The model partitions the environment into several regions and the precision of the method can only locate an MU within a space of a room, an office or a hallway segment. The algorithm of Gaussian fit sensor model collects the RSS samples from the access points for each region rather than each grid, and it fits the RSS distributions for each access point to the Gaussian distributions.

During the offline phase, the Gaussian fit sensor model collects the RSS samples from the access points at region r_i and computes the mean $\mu_{i,j}$ and standard deviation $\sigma_{i,j}$ from the RSS samples. It builds a Gaussian distribution for each access point at region r_i by the $\mu_{i,j}$ and $\sigma_{i,j}$. The Gaussian distribution $G_{i,j}(v)$ is represented as follows:

$$G_{i,j}(v) = \int_{v-\frac{1}{2}}^{v+\frac{1}{2}} \frac{e^{(\mu_{i,j}-x)/2\sigma_{i,j}^2}}{\sigma_{i,j}\sqrt{2\pi}}\, dx$$

where v is the value of signal strength. While it is good as the sensor model, the Gaussian fit sensor model uses Bayes' rule to compute the probability distribution over the regions, $\pi' = \{\pi_i' \mid i=1,\ldots,n\}$, given region r_i and access point b_j as follows:

$$\pi_i' = \frac{\pi_i \Pr(o_j \mid r_i)}{\sum_{i=1}^{n} \pi_i \Pr(o_j \mid r_i)}$$

where

$$\Pr(o_j \mid r_i) = \Pr((b_j, v) \mid r_i)$$
$$= \frac{G_{i,j}(v) + \beta}{N_{i,j}}$$

β is a small constant which represents the probability of observing an artifact. $N_{i,j}$ is used to normalize $\Pr((b_j, v) \mid r_i)$ such that $\sum_{v=0}^{255} \Pr((b_j, v) \mid r_i) = 1$. Finally, the Gaussian fit sensor model chooses the most probable region from the probability distribution π' and estimates whether an MU is in that region.

9.2.3 Reference Point (RP)

In this section, we describe a localization method based on an idea of reference. The concept of reference is to locate an MU according to the real-time environment. Some literatures[4–6,14] mention a key point at which the signal strengths vary over different time periods at a fixed location, as shown in Figure 9.11. The radio maps are built based on the past RSS patterns during the offline phase, but not real time. Therefore, the radio maps are not completely suitable for localization during online phase.

Based on the reasons above, reference tags are used on location-based services. This location system[7] includes some RFID readers and many active

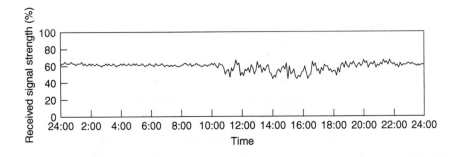

Figure 9.11 **The RSS values received from an access point over 24 h at a fixed location.**

RFID tags. This system differs from signal strength approach, because the current RFID systems do not support the value of signal strengths. Instead, it adapts eight power levels to indicate different detection ranges. Therefore, we use the Euclidean distance to represent the location relationship E_j between a tracking tag and a reference tag r_j, $j \in (1, m)$, as follows:

$$E_j = \sqrt{\sum_{i=1}^{n} (\theta_i - S_i)^2}$$

where S_i is the signal strength of the tracking tag observed on reader i, $i \in (1, n)$, and θ_i the signal strength of the reference tag observed on reader i.

Then, we select k-nearest reference tags (x_j, y_j) from $(E_1, E_2, \ldots, E_m)$ to estimate the coordinate (x, y) of the tracking tag, as follows:

$$(x, y) = \sum_{j=1}^{k} \omega_j (x_j, y_j)$$

where ω_j is the weighting factor according to the jth reference tag:

$$\omega_j = \frac{1/E_j^2}{\sum_{j=1}^{k} 1/E_j^2}$$

9.2.4 Hybrid

Here we introduce a hybrid method that combines the radio maps and the reference points. This localization method[6] adapts the temporal radio maps to locate an MU using regression analysis and reference points. Figure 9.12 simply illustrates how to reconstruct the adaptive temporal radio maps. We will describe two approaches: a linear approach based on multiple

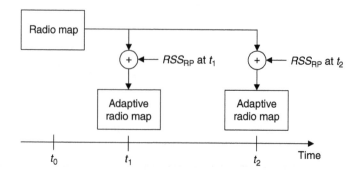

Figure 9.12   Reconstructing an adaptive temporal radio map by the radio map built during the offline phase and the RSS of the reference points.

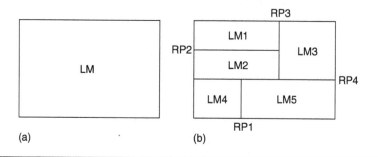

(a)   (b)

Figure 9.13   For an access point, multiple regression treats the entire space as a single linear model, but several linear models by a model tree. (a) Multiple regression. (b) Model tree.

regression, and a nonlinear approach based on model tree. The difference between a multiple regression, and a model tree is the processing in state space. A multiple regression treats the whole state space as a single linear model. But a model tree partitions the whole state space into several regions, each region is a linear model. Figure 9.13 shows the difference between the two approaches.

Initially, we place some reference points in the environment. Therefore, we know the number of reference points, but the physical locations of reference points are not necessary. As is the radio map method, the multiple regression method is composed of two phases: the offline and the online phase.

During the offline phase, we collect the RSS patterns in the predefined locations, and the reference points do the same work simultaneously. After doing the RSS patterns collection, we have a radio map that consists of the RSS collected by an MU and the reference points from each access point in the predefined locations.

9.2.4.1 Multiple Regression

In this method, we learn the relationship of the RSS patterns between the predefined locations and the reference points using the regression analysis, as follows:

$$RSS_{MUj} = \alpha_{0j} + \alpha_{1j}RSS_{RP1j} + \alpha_{2j}RSS_{RP2j} + \cdots + \alpha_{nj}RSS_{RPnj} + \varepsilon_j$$

where RSS_{MUj} is the RSS collected by the MU from the jth BS and RSS_{RPij}, $i = 1,\ldots,n$, the RSS collected by the ith RP from the jth access point. The α_{ij}, $i = 1,\ldots,n$, is the regression coefficient, which represents the independent contributions of each reference point to the RSS_{MUj}. When all the α_i, $i = 1,\ldots,n$, are equal to 0, α_0 is called the intercept. ε_j is the random error, which is usually assumed to be normally distributed with zero mean and variance σ^2.

Then, we obtain the regression coefficients $\alpha_j = (\alpha_{0j}, \alpha_{1j},\ldots,\alpha_{nj})^T$ for each $AP_j, j = 1,\ldots,m$, by executing the *least square estimation*. Therefore, we have a set of regression coefficients, $\alpha_1,\ldots,\alpha_m$ in each predefined location.

During the online phase, we rebuild the adaptive temporal radio maps to locate an MU. The adaptive temporal radio maps consist of the RSS measured by the reference points and the computed regression coefficients, i.e., the adaptive temporal radio maps are a set of estimated RSS patterns for each predefined location. We compute the Euclidean distance E_k between the RSS measured by an MU and the estimated RSS for each predefined location k. Finally, we decide a location l_k with a minimal E_k as the most probable position of an MU.

$$E_k = \sqrt{\sum_{j=1}^{m}(RSS_{MUj} - RSS_{kj})^2} \qquad \text{for each location } k$$

where RSS_{kj} is the estimated RSS pattern that is introduced by $\alpha_{kj} = (\alpha_{0kj}, \alpha_{1kj},\ldots,\alpha_{nkj})^T$ and the RSS measured by the reference points. RSS_{MUj} is the RSS measured by an MU.

9.2.4.2 Model Tree

In this method, we create a model tree for each access point to learn the relationship of RSS patterns between the predefined locations and the reference points. Each model tree looks like the tree structure in Figure 9.14. Tree construction works as follows: we place all of the RSS samples in the root node and try to break up the RSS samples using every possible binary split on every RP. We select the splitting point that partitions the samples into two parts in such a way that it minimizes the expected variances (V_{exp})

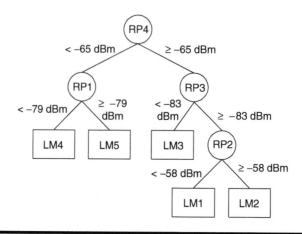

Figure 9.14 A structure of a model tree.

for each part. The $V_{\exp}$ is computed as follows:

$$V_{\exp} = \frac{1}{N_L + N_R}(N_L\hat{\sigma}_L^2 + N_R\hat{\sigma}_R^2)$$

where N_L and N_R are the numbers of RSS samples belonging to the left child node and the right child node, respectively. $\hat{\sigma}_L^2$ and $\hat{\sigma}_R^2$ are the variances of the predicted values at the two children nodes, respectively.

$$\hat{\sigma}_L^2 = \frac{1}{N_L}\sum_{n\in L}(y_n - \hat{\mu}_L)^2, \qquad \hat{\sigma}_R^2 = \frac{1}{N_R}\sum_{n\in R}(y_n - \hat{\mu}_R)^2$$

where y_n is the value of each RSS sample, and $\hat{\mu}_L$ and $\hat{\mu}_R$ are the means of the values at the left and the right child node, respectively.

Each of the new branches executes the splitting process. The process goes on until each node reaches a user-specified minimum node size and becomes a leaf node.

9.2.5 Path Loss

The physical environment in a building differs from the free space because of the multipath phenomena such as the reflection, diffraction, and scattering of the RF signals. Therefore, the path-loss model in the real world is not as simple as the model in free space. The interferences introduced by the obstacles (like walls) and the noise must be considered. The wall attenuation factor (WAF) model[2] proposed in RADAR is represented as follows:

$$P_d = P_{d_0} - 10\alpha\log\frac{d}{d_0} - \begin{cases} nW\,\text{WAF}, & nW < C \\ C\,\text{WAF}, & nW > C \end{cases}$$

where P_d is the received power of the receiver at distance d from the sender, P_{d_0} the received power at a reference distance d_0, α a path loss coefficient, WAF the wall attenuation factor, nW the number of walls between the sender and the receiver, and C the maximum number of walls. The values of α and WAF depend mainly on the physical environment such as the building layout and construction material. The WAF can be determined by the following experiment: (1) We measure the RSS at the receiver when the sender and the receiver have line of sight. (2) We measure the RSS at the receiver given the various number of walls between the sender and the receiver. (3) We determine the WAF by averaging the differences between these values of RSS.

In addition to taking the walls between the sender and receiver into account, the path-loss model[15] considers the interference of multipath fading and noise, and is represented as follows:

$$(S+N) = S = S_0 - 10\alpha \log \frac{d}{d_0} + X_\sigma = S + X_\sigma$$

where $(S+N)$ is the RSS received at the receiver, X_σ a Gaussian random variable with zero mean, and standard deviation σ (i.e., $N(0,\sigma)$). This model is built in the physical environment that is in the line of sight between the nodes. Figure 9.15 illustrates this environment. There are n locations including three access points and $n-3$ known locations in the environment. During the offline phase, the model collects the RSS $(S+N)_{ji}$, where $(S+N)_{ji}$, $i=1,\ldots,n$, is measured at the n known locations from the $j=1,\ldots,3$ access points. Given a fixed and known transmitted power,

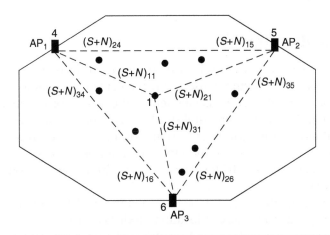

Figure 9.15   Collecting the RSS from the 3 access points at the 6 known locations.

a path-loss coefficient, and an ideal propagation, the ideal RSS S_i is theoretically calculated using the known distance d_i between the access point and the location i. Thus the estimated signal strength $\hat{N}_i$, of the noise for a d_i is calculated as follows:

$$\hat{N}_i = (S + N)_i - S_i$$

Feature function (FF) is the distribution of the ratio of noise power to received signal power at any given distance. The path loss model uses the concept of FF to map a d_i to the ratio of $\hat{N}_i$ to $(S + N)_i$ as follows:

$$\delta(d_i) = \frac{\hat{N}_i}{(S + N)_i}$$

and then the model gets a set of $\delta(d_i)$.

During the online phase, an MU who is at an unknown location k measures the RSS from the three access points. The path-loss model calculates the distance d_j by the RSS $(S + N)_j$ and gets a $\delta(d_j)$ through the set of $\delta(d_i)$. Therefore, $\tilde{S}_j$, the RSS used to estimate the true distance $\tilde{d}_j$ is calculated as follows:

$$\tilde{S}_j = (S + N)_j - N_j$$

where

$$N_j = (S + N)_j * \delta(d_j)$$

Finally, the path-loss model uses trilateration to determine the location of an MU by the three $\tilde{d}_j$ between the three access points and the location j. Once a location j is calculated, the same information updates the set of $\delta(d_i)$.

9.3 Location Tracking

The previous section discusses how to locate an MU using various localization methods. These methods use the RSS received at the time to estimate the location of an MU. They usually ignore the past information about the user locations. Even the probabilistic localizations[4,5] use π to represent the probability distribution over various locations where an MU may be present. Such π is usually assumed to be of a uniform distribution because the methods assume an MU loses his/her way. Thus the locations in π are equally likely.

The advantage of using the past information to determine the location of an MU is greater accuracy of location prediction. Thus we can guess the newest location of an MU based on the past movement (trajectory) and eliminate the erroneous guesses via physical constraints e.g., it is impossible that an MU randomly moves (walking or running inside a building)

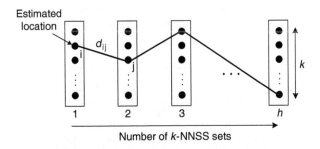

Figure 9.16 Viterbi-like algorithm overview.

between two locations which are very distant from each other in two continuous units of time. In general, the present location S_t of an MU is close to the previous location S_{t-1}. Erroneous estimations may be alleviated by such a concept.

The rest of this section introduces some algorithms with regard to location tracking like the Viterbi-like algorithm and Markov-based algorithm.

9.3.1 Viterbi-Like Location Tracking

Viterbi-like algorithm is based on NNSS algorithm. The algorithm selects k locations with similar RSS (a k-NNSS set) in each time instant and gathers a set of h k-NNSS sets during h unit times. The history vector of the h k-NNSS sets is shown in Figure 9.16. There are edges between the consecutive NNSS sets and each edge between two locations at two consecutive NNSS sets represents the Euclidean distance between two endpoints of the edge. The history vector always maintains h k-NNSS sets, i.e., the newest k-NNSS set is updated to the history vector, and the oldest one is removed from the history vector. Viterbi-like algorithm computes the shortest path between the oldest and the newest sets and estimates that this shortest path is the most likely trajectory of an MU. Once the shortest path is computed, the location at the start of this path is estimated to be the location of an MU.

9.3.2 Markov-Based Location Tracking

Markov-based algorithm is proposed in Ref. [5] that treats each region (it may be a room, an office, or a segment of a hallway) as a location. The algorithm adopts the concept of Markov chain and assigns a probability value for a user to stay behind or move to one of the adjacent locations. Figure 9.17(a) illustrates that the physical environment is divided into several regions, and they are mapped into the finite state machine in Figure 9.17(b). The probability matrix A represents the Markov chain, i.e.,

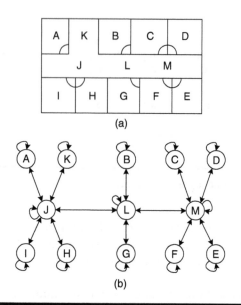

Figure 9.17 Markov-based algorithm overview. (a) The physical environment. (b) The corresponding Markov chain.

the finite state machine. Each state represents a region, and each edge A_{ij} represents the probability of a user moving from state i to state j. The location of an MU is determined from the state distribution π_i^t (see the details in Section 9.2.2.3). The state distribution π_i^t is computed by observations $o_1, \ldots, o_k$, as follows:

$$\pi_i^t = \frac{\prod_{j=1}^{k} P(o_j \mid s_i))\pi_i^{t-1^+}}{\eta}$$

where

$$\pi_i^{t-1^+} = A\pi_i^{t-1}$$

In the implementation of Ref. [5], each A_{ii} is assigned a fixed probability, and each A_{ij} (j is one of the adjacent region for i, $j \neq i$) is equally assigned the remaining probability.

9.4 Conclusion

Using the existent and widespread IEEE 802.11 WLANs as the infrastructure of a localization method possesses the following benefits. Many devices and buildings already have wireless capability with IEEE 802.11 adapters, so the system does not need any hardware cost for localization and can save a lot of battery power. Moreover, if we take RSS as a major parameter to estimate the position of an MU, the localization will be easy to design because most of the adapters provide the measurement of RSS. On the contrary, other localization systems need extra cost to install dedicated hardware. For example, the active badges system[1] uses the IR sensors, and the Cricket system[16,17] uses the ultrasound emitters and receivers.

Performance of a localization system can be evaluated with several factors such as accuracy, cost, time, and energy consumption. The most important factor of a localization system is accuracy of the estimated position. The more the accuracy of the estimated position the greater is the precision of the predicted distance based on the measured RSS. In a word, we need a more sophisticated radio-propagated model for wireless communication. But the radio-propagated model is affected by many factors such as mutipath fading, moving object, wall, and temperature. On the other hand, the reference point method, by comparing the received RSS of an MU with base stations on line, may provide more accurate positions without dealing with the cumbersome problems of the radio-propagated model.

References

1. R. Want, A. Hopper, V. Falco, and J. Gibbons. The active badge location system. *ACM Transactions on Information Systems*, Vol. 10, No. 1, pp. 91–102, 1992.
2. P. Bahl, and V.N. Padmanabhan. RADAR: An in-building RF-based user location and tracking system. *IEEE INFOCOM*, Vol. 2, pp. 775–784, March 2000.
3. P. Bahl, V.N. Padmanabhan, and A. Balachandran. Enhancements to the RADAR user location and tracking system. Technical Report MSR-TR-00-12, Microsoft Research, February 2000.
4. A.M. Ladd, K. Bekris, A. Rudys, G. Marceau, L.E. Kavraki, and D.S. Wallach. Robotics-based location sensing using wireless ethernet. *ACM MOBICOM*, pp. 227–238, September 2002.
5. A. Haeberlen, E. Flannery, A.M. Ladd, A. Rudys, D.S. Wallach, and L.E. Kavraki. Practical robust localization over large-scale 802.11 wireless networks. *ACM MOBICOM*, pp. 70–84, September 2004.
6. J. Yin, Q. Yang, and L. Ni. Adaptive temporal radio maps for indoor location estimation. *IEEE PerCom*, pp. 85–94, March 2005.
7. L.M. Ni, Y. Liu, Y.C. Lau, and A.P. Patil. LANDMARC: Indoor location sensing using active RFID. *IEEE PerCom*, pp. 407–415, March 2003.
8. E.K. Wesel. *Wireless Multimedia Communications: Networking Video, Voice, and Data*. Reading, MA: Addison-Wesley, 1998.
9. G.L. Stuber. *Principle of Mobile Communication*. Netherlands: Kluwer Academic Publishers, 1996.

10. D. Pandya, R. Jain, and E. Lupu. Indoor location estimation using multiple wireless technologies. *IEEE PIMRC*, Vol. 3, pp. 2208–2212, September 2003.

11. T. He, C. Huang, B.M. Blum, J.A. Stankovic, and T. Abdelzaher. Range-free localization schemes for large scale sensor networks. *ACM MOBICOM*, pp. 81–95, September 2003.

12. M. Youssef, and A. Agrawala. On the optimality of WLAN location determination systems. Technical Report UMIACS-TR 2003-29 and CS-TR 4459, University of Maryland, College Park, March 2003.

13. M. Youssef, and A. Agrawala. Handling samples correlation in the horus system. *IEEE INFOCOM*, Vol. 2, pp. 1023–1031, March 2004.

14. M. Youssef, A. Agrawala, and A.U. Shankar. WLAN location determination via clustering and probability distributions. *IEEE PerCom*, pp. 143–150, March 2003.

15. R. Singh, M. Gandetto, M. Guainazzo, D. Angiati, and C.S. Ragazzoni. A novel positioning system for static location estimation employing WLAN in indoor environment. *IEEE PIMRC*, Vol. 3, pp. 1762–1766, September 2004.

16. N.B. Priyantha, A. Chakraborty, and H. Balakrishman. The cricket location-support system. *ACM MOBICOM*, pp. 32–43, August 2000.

17. N.B. Priyantha, A. Miu, H. Balakrishman, and S. Teller. The cricket compass for context-aware mobile applications. *ACM MOBICOM*, pp. 1–14, July 2001.

Chapter 10

Channel Assignment in Wireless Local Area Networks

Alan A. Bertossi and M. Cristina Pinotti

10.1   Introduction

The tremendous growth of wireless networks requires an efficient use of the scarce radio spectrum allocated to wireless communications. However, the main difficulty against an efficient use of radio spectrum is given by *interferences*, caused by unconstrained simultaneous transmissions, which result in damaged communications that need to be retransmitted leading to a higher cost of the service. Interferences can be eliminated (or at least reduced) by means of suitable *channel assignment* techniques. Indeed, co-channel interferences caused by frequency reuse is one of the most critical factors on the overall system capacity in the wireless networks. The purpose of channel assignment algorithms is to make use of radio propagation loss characteristics to increase the radio spectrum reuse efficiency and thus to reduce the overall cost of the service.

The channel assignment algorithms partition the given radio spectrum into a set of disjoint channels that can be used simultaneously by the stations while maintaining acceptable radio signals. By taking advantage of physical characteristics of the radio environment, the same channel can be reused by two stations at the same time without interferences (*co-channel stations*), provided that the two stations are spaced sufficiently apart.

The minimum distance at which co-channels can be reused with no interferences is called *co-channel reuse distance*.

The interference phenomena may be so strong that even different channels used at near stations may interfere if the channels are too close. Since perfect filters are not available, interference between close frequencies is a serious problem, which can be handled either by adding guard frequencies between adjacent channels or by imposing channel separation. In this latter approach, followed in the present chapter, channels assigned to near stations must be separated by a gap on the radio spectrum—counted in a certain number of channels—which is inversely proportional to the distance between the two stations. In other words, the channels $f(u)$ and $f(v)$ assigned to the stations u and v at distance i must verify $|f(u) - f(v)| \geq \delta_i$ when a minimum *channel separation* δ_i is required between stations at distance i. The purpose of channel assignment algorithms is to assign channels to transmitters in such a way that the co-channel reuse distance and channel separation constraints are verified and the difference between the highest and lowest channels assigned is kept as small as possible.

Formally, the channel assignment problem can be modeled as an appropriate coloring problem on an undirected graph $G = (V, E)$ representing the wireless network topology, whose vertices in V correspond to stations, and edges in E correspond to pairs of stations whose transmission areas intersect. Specifically, given a vector $(\delta_1, \delta_2, \ldots, \delta_t)$ of nonincreasing positive integers, and an undirected graph $G = (V, E)$, an $L(\delta_1, \delta_2, \ldots, \delta_t)$ coloring of G is a function f from the vertex set V to the set of nonnegative integers $\{0, \ldots, \lambda\}$ such that $|f(u) - f(v)| \geq \delta_i$ whenever $d(u, v) = i$, $1 \leq i \leq t$, where $d(u, v)$ is the distance (i.e., the minimum number of edges) between the vertices u and v. An *optimal* $L(\delta_1, \delta_2, \ldots, \delta_t)$ coloring for G is one minimizing λ over all such colorings. Note that the co-channel reuse distance is actually $t + 1$ but one need not explicitly state the co-channel reuse constraint because it is implied by the channel separation constraints. Note also that, since the set of colors includes 0, the overall number of colors involved by an optimal coloring f is in fact $\lambda + 1$ (although, due to the channel separation constraint, some colors in $\{1, \ldots, \lambda - 1\}$ might not be actually assigned to any vertex). Thus, the channel assignment problem consists of finding an optimal $L(\delta_1, \delta_2, \ldots, \delta_t)$ coloring for G.

When the separation vector $(\delta_1, \delta_2, \ldots, \delta_t)$ is equal to $(1, 1, \ldots, 1)$, the channel assignment problem has been widely studied in the past.[1–5] In particular, the intractability of optimal $L(1, \ldots, 1)$ coloring, for any positive integer t, has been proved by McCormick.[4] Optimal $L(1, \ldots, 1)$ colorings, for any positive integer t, have been proposed for rings, bidimensional grids, and honeycomb grids,[2,6] and for trees and interval graphs.[7] Moreover, when the separation vector is equal to $(\delta_1, 1, \ldots, 1)$, optimal $L(\delta_1, 1, \ldots, 1)$ colorings have been proposed for rings, bidimensional grids, and cellular grids.[8,9] Optimal solutions have been proposed for the $L(\delta_1, \delta_2)$ coloring problem on rings[10] and on bidimensional and cellular grids.[11] This latter paper provided also an optimal $L(2, 1, 1)$ coloring for bidimensional grids. The $L(2, 1, 1)$ coloring problem has been also optimally solved for cellular grids, honeycomb grids, and rings.[8] The $L(2, 1)$ coloring has been investigated.[12–15] Bodlaender et al.[12] proved that the $L(2, 1)$ coloring problem is *NP*-hard for planar, bipartite, and chordal graphs, and presented approximate solutions for outerplanar, permutation and split graphs. Moreover, $L(2,1)$ colorings for unit interval graphs and trees have been found, respectively, by Sakai[15] and by Chang and Kuo.[13] A recent annotated bibliography on the $L(\delta_1, \delta_2)$ coloring problem for several special classes of graphs can be found in the literature.[16]

As a related case, when $(\delta_1, \delta_2) = (0, 1)$, the $L(0, 1)$ coloring problem models that of avoiding only the so-called *hidden interferences*, due to stations which are outside the hearing range of each other and transmit to the same receiving station. Optimal $L(0, 1)$ colorings have been provided for bidimensional grids,[17] whereas the intractability of optimal $L(0, 1)$ coloring has been proved,[18] where also optimal solutions for rings and complete binary trees were given.[18] As another related case, observe that the classical minimum vertex coloring problem on undirected graphs arises when $t = 1$ and $\delta_1 = 1$. Thus, the minimum vertex coloring problem consists in finding an optimal $L(1)$ coloring.

For arbitrary graphs and general separation vectors, the $L(\delta_1, \delta_2, \ldots, \delta_t)$ coloring problem is usually addressed by means of heuristic approaches, like genetic algorithms, taboo search, saturation degree, simulated annealing, and ants heuristics, just to name a few.[19] However, approximation algorithms have been proposed for trees and interval graphs.[20]

This chapter deals with the channel assignment problem for general separation vectors on some relevant classes of graphs—rings, grids, trees, and interval graphs—which occur in modeling realistic wireless network topologies. Indeed, rings are perhaps the most used topologies for local area networks, grids represent tessellations of the plane with regular polygons (like squares or triangles), trees model hierarchical topologies, and interval graphs model wireless networks serving narrow surfaces, like highways or valleys confined by natural barriers (e.g., mountains or lakes). It is still unknown whether finding optimal $L(\delta_1, \ldots, \delta_t)$ colorings on such classes of graphs is polynomially time-solvable or not. While the $L(\delta_1, \delta_2)$ coloring problem can be solved in polynomial time grids[11] and rings,[10] some authors conjecture that it is *NP*-hard for trees and unit interval graphs when $\delta_2 > 1$.[16]

In the rest of this chapter, several algorithms are reviewed.[8,11,14,20,21] First, some preliminary results useful to derive upper and lower bounds on the largest color needed are summarized in Section 10.2, where the notions of t-simplicial and strongly simplicial vertices are also recalled. Then, Section 10.3 presents optimal $O(n)$ time algorithms for solving the $L(2, 1)$,

$L(2, 1, 1)$, and $L(\delta_1, 1, \ldots, 1)$ coloring problems of rings with n vertices. Section 10.4 revises optimal algorithms for the $L(\delta_1, \delta_2)$ coloring problem on both bidimensional and cellular grids, and for the $L(\delta_1, 1, \ldots, 1)$ coloring problem on bidimensional grids. All the algorithms are based on arithmetic progressions and take $O(rc)$ time to color the entire grid of r rows and c columns. Sections 10.5 and 10.6 present two polynomial time algorithms to approximate the $L(\delta_1, \ldots, \delta_t)$ coloring problem on interval graphs and trees, respectively. The algorithms give the same constant approximation, which depends on t and $\delta_1, \ldots, \delta_t$, but they run in $O(n(t + \delta_1))$ and $O(nt^2\delta_1)$ time, respectively, where n is the number of vertices. A better approximation for the $L(\delta_1, \delta_2)$ coloring of unit interval graphs is also given in Section 10.5.2. Finally, conclusions are offered in Section 10.7.

10.2 Preliminaries

Throughout this chapter, it is assumed that G is a connected undirected graph with at least two vertices and that the separations verify $\delta_1 \geq \delta_2 \geq \cdots \geq \delta_t$. When $\delta_1 = \delta_2 = \cdots = \delta_t = 1$, the $L(1, \ldots, 1)$ coloring problem reduces to the classical vertex coloring problem on the tth *power* G^t of the graph G. The vertex set of G^t is the same as the vertex set of G, while the edge uv belongs to the edge set of G^t if and only if the distance $d(u, v)$ between the vertices u and v in G is at most t. Now, colors must be assigned to the vertices of G^t so that every pair of vertices connected by an edge are assigned different colors and the minimum number of colors is used. Hence, the role of *maximum cliques* in G^t is apparent for deriving lower bounds on the minimum number of channels for the $L(1, \ldots, 1)$ coloring problem on G. A *clique* K for G^t is a subset of vertices of G^t such that for each pair of vertices in K there is an edge. Clearly, a clique of size k in the power graph G^t implies that at least k different colors are needed to color G^t. In other words, the size of the largest clique in G^t is a lower bound for the $L(1, \ldots, 1)$ coloring problem on G. Of course, a lower bound for the $L(1, \ldots, 1)$ coloring problem is also a lower bound for the $L(\delta_1, 1, \ldots, 1)$ coloring problem, with $\delta_1 \geq 1$.

A simple lower bound for the $L(2, 1)$ coloring problem, which can be applied to the vertex with maximum degree of any graph, can be derived as follows. Consider the *star* graph S_ρ which consists of a *center* vertex v with degree ρ, and ρ *ray* vertices of degree 1.

Lemma 10.1[14]   *Let the center v of S_ρ be already colored. Then, the largest color λ required for an $L(2, 1)$ coloring of S_ρ is at least*

$$\lambda = \begin{cases} \rho + 1 & \text{if } f(v) = 0 \text{ or } f(v) = \rho + 1 \\ \rho + 2 & \text{if } 0 < f(v) < \rho + 1 \end{cases}$$

For any value of $t \leq |V|$, let $\lambda^*_{G,t}$ denote the minimum value of λ over all the $L(1, \ldots, 1)$ colorings $f : V \to \{0, \ldots, \lambda\}$ of $G = (V, E)$. Note that: (i) $\lambda^*_{G,1} \geq 1$ since G is assumed to be connected and to have at least 2 vertices; (ii) $\lambda^*_{G,t} = \lambda^*_{G',1}$; and (iii) $\lambda^*_{G,t} + 1$ is at least as large as the size ω_{G^t} of the largest clique of the power graph G^t.

Lemma 10.2[20] *Any $L(\delta_1, \delta_2, \ldots, \delta_t)$ coloring requires at least $\max_{1 \leq j \leq t} \{\delta_j \lambda^*_{G,j}\}$ as the largest color.*

Proof Since $\delta_1 \geq \delta_2 \geq \cdots \geq \delta_t$, any $L(\delta_1, \delta_2, \ldots, \delta_j)$ coloring, for any value of $j \leq t$, requires at least as many colors as any $L(\delta_j, \delta_j, \ldots, \delta_j)$ coloring, which, in its turn, requires at least $\delta_j \lambda^*_{G,j}$ as the largest color. □

Given $G = (V, E)$, let S be a subset of V, and let $G[S]$ denote the subgraph of G induced by S, i.e. $G[S] = (S, \{uv \in E : u, v \in S\})$. A vertex x of G is called *t-simplicial* when, for every pair of vertices u and v such that $d(x, u) \leq t$ and $d(x, v) \leq t$, it holds also that $d(u, v) \leq t$. A vertex x is called *strongly simplicial* when x is t-simplicial for any value of t.

Lemma 10.3[21] *Given $G = (V, E)$ and an integer t, let v be a t-simplicial vertex of G. Consider $G' = G[V - \{v\}]$ and let f' be an optimal $L(1, \ldots, 1)$ coloring of G' using $\lambda^*_{G',t}$ as the largest color. Define an $L(1, \ldots, 1)$ coloring f of V extending f' to v so that*

$$f(x) = \begin{cases} \min\{i : i \neq f'(u) \text{ for each } u \in G' \text{ with } d(u, v) \leq t\} & \text{if } x = v, \\ f'(x) & \text{if } x \in V - \{v\} \end{cases}$$

Then f is an optimal $L(1, \ldots, 1)$ coloring for G.

Note that verifying whether a vertex is t-simplicial or not can be done in polynomial time.[7] Therefore, Lemma 10.3 implies the existence of an algorithm that optimally solves in polynomial time the $L(1, \ldots, 1)$ coloring problem using exactly ω_{G^t} colors for any class of graphs closed under taking induced subgraphs and with the property that every graph of that class has a t-simplicial vertex. The next lemma shows that there is always an $L(\delta_1, \delta_2, \ldots, \delta_t)$ coloring for such a class of graphs where the largest used color is bounded from above by a function of the clique sizes ω_{G^j} and of the separations δ_j, with $1 \leq j \leq t$.

Lemma 10.4[20] *Given $G = (V, E)$ and t, let v be a t-simplicial vertex of G, and consider $G' = G[V - \{v\}]$. Then, there is an $L(\delta_1, \delta_2, \ldots, \delta_t)$ coloring such that $f(v) = c$, where $c \in \{0, 1, \ldots, \lambda^*_{G,t} + 2(\delta_t - 1)\lambda^*_{G,t} + \sum_{j=1}^{t-1} 2(\delta_j - \delta_{j+1})\lambda^*_{G,j}\}$.*

Proof   Let $N_j(v) = \{u \in V : d(u, v) \leq j\}$. Since v is t-simplicial, any two vertices in $N_j(v)$ are at distance at most j, for every $1 \leq j \leq t$. Hence $N_t(v)$ is a clique of G^t, and thus $|N_t(v)| \leq \omega_{G^t}$. Therefore at most $|N_t(v)| - 1 \leq \lambda^*_{G,t}$ colors have been used for $N_t(v) - \{v\}$. Each of them forbids $2(\delta_t - 1)$ colors due to the δ_t-separation constraint, and overall $2(\delta_t - 1)\lambda^*_{G,t}$ colors are forbidden. Moreover, for any $1 \leq j \leq t - 1$, v is j-simplicial, and hence $N_j(v)$ is a clique of G^j, and $|N_j(v)| - 1 \leq \lambda^*_{G,j}$. Since each color assigned to a vertex of $N_j(v) - \{v\}$ forbids $2(\delta_j - \delta_{j+1})$ colors, $2(\delta_j - \delta_{j+1})\lambda^*_{G,j}$ colors are forbidden due to the δ_j-separation constraint. Before coloring v, the total number of used and forbidden colors is $\lambda^*_{G,t} + 2(\delta_t - 1)\lambda^*_{G,t} + \sum_{j=1}^{t-1} 2(\delta_j - \delta_{j+1})\lambda^*_{G,j}$. Therefore, there is at least an available color c in $\{0, 1, \ldots, \lambda^*_{G,t} + 2(\delta_t - 1)\lambda^*_{G,t} + \sum_{j=1}^{t-1} 2(\delta_j - \delta_{j+1})\lambda^*_{G,j}\}$ that can be assigned to v.   □

The results of Lemmas 10.1 and 10.2 will be used in Sections 10.3 and 10.4 to obtain lower bounds on the number of colors needed by any $L(2, 1)$ coloring of rings and $L(\delta_1, 1, \ldots, 1)$ coloring of grids. Moreover, the properties stated in Lemmas 10.2–10.4 will be exploited in Sections 10.5 and 10.6 to derive lower and upper bounds for the $L(\delta_1, \delta_2, \ldots, \delta_t)$ coloring problem on interval graphs and trees.

10.3   Rings

A *ring* R of size $n \geq 3$ is a sequence of n vertices, indexed consecutively from 0 to $n - 1$, such that vertex i is connected to both vertices $(i - 1)$ mod n and $(i + 1)$ mod n.

In this section, three algorithms are presented to optimally solve the $L(2,1)$-, $L(2,1,1)$- and $L(\delta_1, 1, \ldots, 1)$ coloring problems on rings. Each algorithm colors a single vertex in $O(1)$ time and thus colors the entire ring in $O(n)$ time.

By Lemma 10.1, the largest color used by any $L(2,1)$ coloring of a ring is at least 4, as one can easily check by observing that in a ring $\rho = 2$ and there is a vertex v that must be colored $0 < f(v) < 3$. An optimal solution has been provided by Griggs and Yeh,[14] who color each vertex i as follows:

$$f(i) = \begin{cases} 0 & \text{if } i \equiv 0 \mod 3 \\ 2 & \text{if } i \equiv 1 \mod 3 \\ 4 & \text{if } i \equiv 2 \mod 3 \end{cases}$$

However the above coloring is redefined on the ring tail depending on whether $n \equiv 1$ mod 3 or $n \equiv 2$ mod 3. In the first case, $f(n - 4), \ldots, f(n - 1)$

become

$$f(i) = \begin{cases} 0 & \text{if } i = n - 4 \\ 3 & \text{if } i = n - 3 \\ 1 & \text{if } i = n - 2 \\ 4 & \text{if } i = n - 1 \end{cases}$$

In the second case, $f(n-2)$ and $f(n-1)$ are modified as

$$f(i) = \begin{cases} 1 & \text{if } i = n - 2 \\ 3 & \text{if } i = n - 1 \end{cases}$$

For the $L(2,1,1)$ coloring problem on rings, the same lower bound previously discussed for the $L(2,1)$ coloring holds and an optimal coloring can be derived as follows. Let $n \geq 12$ and $\theta = 4 \left(\lfloor \frac{n}{4} \rfloor - (n \bmod 4) \right)$. Then assign to each vertex i the color:

$$f(i) = \begin{cases} 0 & \text{if } i \equiv 0 \bmod 4 \text{ and } i < \theta, \text{ or } (i - \theta) \equiv 0 \bmod 5 \text{ and } i \geq \theta \\ 1 & \text{if } i \equiv 2 \bmod 4 \text{ and } i < \theta, \text{ or } (i - \theta) \equiv 3 \bmod 5 \text{ and } i \geq \theta \\ 2 & \text{if } (i - \theta) \equiv 1 \bmod 5 \text{ and } i \geq \theta \\ 3 & \text{if } i \equiv 3 \bmod 4 \text{ and } i < \theta, \text{ or } (i - \theta) \equiv 4 \bmod 5 \text{ and } i \geq \theta \\ 4 & \text{if } i \equiv 1 \bmod 4 \text{ and } i < \theta, \text{ or } (i - \theta) \equiv 2 \bmod 5 \text{ and } i \geq \theta \end{cases}$$

The above algorithm colors the first θ vertices repeating $\lfloor \frac{n}{4} \rfloor - (n \bmod 4)$ times the color sequence 0,4,1,3 of length 4, while the remaining vertices are colored repeating $n \bmod 4$ times the sequence 0,2,4,1,3 of length 5. It is worth noting that, for $n < 12$, an optimal coloring requires a larger number of colors.[8]

10.3.1 Optimal $L(\delta_1,1,\ldots,1)$ coloring of Rings

In this subsection, an optimal $L(\delta_1, 1, \ldots, 1)$ coloring of rings is discussed. Assuming a sufficiently large ring, the following lower bound holds:

Lemma 10.5[2] *Given $n \geq t + 2$, $t \geq 2$, and $\delta_1 \geq 1$, the largest color used by any $L(\delta_1, 1, \ldots, 1)$ coloring is at least* $t + \left\lceil \frac{n \bmod (t+1)}{\lfloor \frac{n}{t+1} \rfloor} \right\rceil$.

Proof Observe that any $L(\delta_1, \ldots, 1)$ coloring needs at least as many colors as any $L(1, \ldots, 1)$ coloring. One notes that in an $L(1, \ldots, 1)$ coloring each color may appear at most $\tau = \lfloor \frac{n}{t+1} \rfloor$ times, and thus at least $\lceil \frac{n}{\tau} \rceil$ colors are needed. Since $n = \lfloor \frac{n}{t+1} \rfloor (t+1) + (n \bmod (t+1))$, it follows that at least $\lceil \frac{n}{\tau} \rceil = (t+1) + \lceil \frac{n \bmod (t+1)}{\lfloor \frac{n}{t+1} \rfloor} \rceil$ colors are required. Therefore, the largest color is at least $t + \lceil \frac{n \bmod (t+1)}{\lfloor \frac{n}{t+1} \rfloor} \rceil$. □

The optimal $L(\delta_1, 1, \ldots, 1)$ coloring algorithm for rings acts as follows. Let $\zeta = t + \lceil \frac{n \bmod (t+1)}{\lfloor \frac{n}{t+1} \rfloor} \rceil$. If $n \equiv 0 \bmod (t+1)$ then $\zeta = t$ results, and one assigns to each vertex i the color:

$$
f(i) = \begin{cases}
i \lfloor \frac{\zeta}{2} \rfloor \bmod (t+1) & \text{if (t is even) or (t is odd and $\lfloor \frac{\zeta}{2} \rfloor$} \\
& \text{is odd)} \\
i \lfloor \frac{\zeta}{2} \rfloor \bmod (t+1) & \text{if t is odd and $\lfloor \frac{\zeta}{2} \rfloor$ is even and} \\
& i \equiv b \bmod (t+1), \text{ for } 0 \le b \le \lfloor \frac{t}{2} \rfloor \\
(i \lfloor \frac{\zeta}{2} \rfloor + 1) \bmod (t+1) & \text{if t is odd and $\lfloor \frac{\zeta}{2} \rfloor$ is even and} \\
& i \equiv b \bmod (t+1), \text{ for } \lceil \frac{t}{2} \rceil \le b \le t
\end{cases}
$$

If $n \not\equiv 0 \bmod (t+1)$, let $\theta = \lfloor \frac{n}{\lambda} \rfloor - n \bmod \zeta$. If $i \ge \theta\zeta$, then

$$
f(i) = \begin{cases}
(i - \theta\zeta) \lfloor \frac{\zeta}{2} \rfloor \bmod (\zeta+1) & \text{if (ζ is even) or} \\
& \text{(ζ is odd and $\lfloor \frac{\zeta}{2} \rfloor$ is odd)} \\
(i - \theta\zeta) \lfloor \frac{\zeta}{2} \rfloor \bmod (\zeta+1) & \text{if ζ is odd and $\lfloor \frac{\zeta}{2} \rfloor$ is even and} \\
& i \equiv b \bmod (\zeta+1), \text{ for } 0 \le b \\
& \le \frac{\zeta+1}{2} - 1 \\
\left((i - \theta\zeta) \lfloor \frac{\zeta}{2} \rfloor + 1\right) \bmod (\zeta+1) & \text{if ζ is odd and $\lfloor \frac{\zeta}{2} \rfloor$ is even and} \\
& i \equiv b \bmod (\zeta+1), \text{ for } \frac{\zeta+1}{2} \\
& \le b \le \zeta
\end{cases}
$$

otherwise (that is, when $i < \theta\zeta$)

$$
f(i) = \begin{cases}
0 & \text{if } i \equiv 0 \bmod \zeta \\
(i+1) \lfloor \frac{\zeta}{2} \rfloor \bmod (\zeta+1) & \text{if (ζ is even) or} \\
& \text{(ζ is odd and $\lfloor \frac{\zeta}{2} \rfloor$ is odd) and} \\
& i \not\equiv 0 \bmod \zeta \\
(i+1) \lfloor \frac{\zeta}{2} \rfloor \bmod (\zeta+1) & \text{if ζ is odd and $\lfloor \frac{\zeta}{2} \rfloor$ is even and} \\
& i \equiv b \bmod \zeta, \text{ for } 1 \le b \le \frac{\zeta+1}{2} - 1 \\
\left((i+1) \lfloor \frac{\zeta}{2} \rfloor + 1\right) \bmod (\zeta+1) & \text{if ζ is odd and $\lfloor \frac{\zeta}{2} \rfloor$ is even and} \\
& i \equiv b \bmod \zeta, \text{ for } \frac{\zeta+1}{2} \le b \le \zeta
\end{cases}
$$

The above $L(\delta_1, 1, \ldots, 1)$ coloring algorithm assumes $t \geq 2$ and when $n \geq t + 2$ works for any value of $\delta_1 \leq \lfloor \frac{\varsigma}{2} \rfloor$. Note that when $t = 2$ or when $t = 3$ and $n \equiv 0 \mod 4$, such an algorithm solves the $L(1,1)$- and $L(1,1,1)$-coloring problems, but not the $L(2,1)$- and $L(2,1,1)$ colorings because in such a case $2 \leq \varsigma \leq 3$ and thus $\delta_1 = 1$. However, such cases are optimally solved by the previous algorithms.

The correctness and optimality of the algorithm is based on the following lemma:

Lemma 10.6[8]   *Given $n \geq t + 2$, and $t \geq 2$, let $\varsigma = \sigma - 1 + \left\lceil \frac{n \bmod \sigma}{\lfloor \frac{n}{\sigma} \rfloor} \right\rceil$.*

- *If (ς is even) or (ς is odd and $\lfloor \frac{\varsigma}{2} \rfloor$ is odd), then $i \lfloor \frac{\varsigma}{2} \rfloor \bmod (\varsigma + 1)$ assumes all the values in the range $[0, \varsigma]$ while i varies within the range $[0, \varsigma]$.*
- *If (ς is odd and $\lfloor \frac{\varsigma}{2} \rfloor$ is even), then $i \lfloor \frac{\varsigma}{2} \rfloor \bmod (\varsigma + 1)$ assumes all the even values in the range $[0, \varsigma - 1]$ while i varies within the range $[0, \frac{\varsigma + 1}{2} - 1]$.*

It is worth noting that the three algorithms seen in this section solve the $L(2, 1, \ldots, 1)$ coloring problem for any $t \geq 2$. In general, given any $\delta_1 \geq 1$, the $L(\delta_1, 1, \ldots, 1)$ coloring problem can be optimally solved for every $t \geq 2\delta_1$, independently on the size n of the ring and using as few colors as the $L(1, 1, \ldots, 1)$ coloring problem.[8] Nevertheless, the $L(\delta_1, 1, \ldots, 1)$ coloring problem may be also solved for values of t smaller than $2\delta_1$ for suitable values of $n < (2\delta_1 + 1)^2$.

10.4   Grids

A *bidimensional grid B* of size $r \times c$ has r rows and c columns, indexed respectively from 0 to $r - 1$ (from top to bottom) and from 0 to $c - 1$ (from left to right), with $r \geq 2$ and $c \geq 2$. A generic vertex u of B will be denoted by $u = (i, j)$, where i is its row index and j is its column index. All internal vertices, i.e., those not on the borders, have degree 4. In particular, an internal vertex $u = (i, j)$ is adjacent to the vertices $(i - 1, j), (i, j + 1), (i + 1, j)$, and $(i, j - 1)$.

A *cellular grid C* of size $r \times c$, with $r \geq 2$ and $c \geq 2$, is obtained from a bidimensional grid B of the same size augmenting the set of edges with left-to-right *diagonal* connections. Specifically, each vertex $u = (i, j)$ of C is also connected to the vertices $v = (i - 1, j - 1)$ and $z = (i + 1, j + 1)$. Hence, each vertex has degree 6, except for the vertices on the borders.

10.4.1 Optimal L(δ_1,δ_2) coloring of Grids

Optimal solutions for the $L(\delta_1, \delta_2)$ coloring problems on both bidimensional and cellular grids have been provided by Van Den Heuvel et al.[11] They have shown that optimal $L(\delta_1, \delta_2)$ colorings can be obtained by *arithmetic progression*, namely, determining three nonnegative integers a, b, and m such that the color assigned to any vertex $u = (i, j)$ is calculated as $f(u) = f(i, j) = (ai + bj) \bmod m$, where $m - 1$ is the largest used color.

Precisely, such parameters for bidimensional grids are $a = \delta_1$, $b = \delta_1 + \delta_2$, and $m = 2\delta_1 + 3\delta_2$, and thus

$$f(i, j) = (\delta_1 i + (\delta_1 + \delta_2)j) \bmod (2\delta_1 + 3\delta_2)$$

In contrast, for cellular grids, an optimal $L(\delta_1, \delta_2)$ colorings can be obtained by a different arithmetic progression

$$f(i, j) = \begin{cases} ((2\delta_1 + \delta_2)i - \delta_1 j) \bmod (4\delta_1 + 3\delta_2) & \text{if } \delta_1 \le \frac{3}{2}\delta_2 \\ (5\delta_2 i - 2\delta_2 j) \bmod 9\delta_2 & \text{if } \frac{3}{2}\delta_2 \le \delta_1 \le 2\delta_2 \\ ((2\delta_1 + \delta_2)i - \delta_1 j) \bmod (3\delta_1 + 3\delta_2) & \text{if } \delta_1 \ge 2\delta_2 \end{cases}$$

Clearly, $O(1)$ time is spent to color a single vertex, and thus the overall time complexity of the algorithm is $O(n)$, where $n = rc$.

10.4.2 Optimal L (δ_1,1,...,1) coloring of Bidimensional Grids

In this subsection, the $L(\delta_1, 1, \ldots, 1)$ coloring problem is dealt with. Although this problem can be optimally solved for both bidimensional grids[8] and cellular grids,[9] only the algorithm for bidimensional grids is shown here since it is considerably simpler to be described than its counterpart for cellular grids.

A lower bound is given by the following lemma.

Lemma 10.7[8] *If $\delta_1 \le \left\lceil \frac{(t+1)^2}{2} \right\rceil$, any $L(\delta_1, 1, \ldots, 1)$ coloring of a bidimensional grid B of size $r \times c$, with $r \ge t + 1$ and $c \ge t + 1$, requires at least $\left\lceil \frac{(t+1)^2}{2} \right\rceil - 1$ as the largest color.*

Indeed, consider a generic vertex $x = (i, j)$ of B, and its opposite vertex at distance t on the same column, i.e., $y = (i - t, j)$. All the vertices of B at distance t or less from both x and y are mutually at distance t or less. Therefore, in the power graph B^t, they form a clique of size $\left\lceil \frac{(t+1)^2}{2} \right\rceil$, and thus $\lambda_{B^t,1}^* \geq \left\lceil \frac{(t+1)^2}{2} \right\rceil - 1$. By Lemma 10.2, any $L(\delta_1, 1, \ldots, 1)$ coloring on B requires at least $\max\{\delta_1 \lambda_{B,1}^*, \lambda_{B^t,1}^*\} = \max\left\{\delta_1, \left\lceil \frac{(t+1)^2}{2} \right\rceil - 1\right\} = \left\lceil \frac{(t+1)^2}{2} \right\rceil - 1$ as the largest color.

The following optimal $L(\delta_1, 1, \ldots, 1)$ coloring is based on arithmetic progression and works for $t \geq 2$ and $\delta_1 \leq \lfloor \frac{t}{2} \rfloor$.

If t is even, then assign to each vertex $u = (i, j)$ the color:

$$f(i, j) = \left(\left(\frac{t}{2} + 1 \right) i + \frac{t}{2} j \right) \bmod \left\lceil \frac{(t+1)^2}{2} \right\rceil$$

If t is odd, let $i' = i \bmod (t+1)$ and $j' = j \bmod (t+1)$.
If t is odd and $\lfloor \frac{t}{2} \rfloor$ is even, then

$$f(i, j) = \begin{cases} \lfloor \frac{t}{2} \rfloor (\lceil \frac{t}{2} \rceil i' + j') \bmod \frac{(t+1)^2}{2} & \text{if } 0 \leq i' \leq \lfloor \frac{t}{2} \rfloor \text{ and } 0 \leq j' \leq \lfloor \frac{t}{2} \rfloor \\ & \text{or } \lceil \frac{t}{2} \rceil \leq i' \leq t \text{ and } \lceil \frac{t}{2} \rceil \leq j' \leq t \\ (\lfloor \frac{t}{2} \rfloor (\lceil \frac{t}{2} \rceil i' + j') + 1) & \text{if } 0 \leq i' \leq \lfloor \frac{t}{2} \rfloor \text{ and } \lceil \frac{t}{2} \rceil \leq j' \leq t \\ \quad \bmod \frac{(t+1)^2}{2} & \text{or } \lceil \frac{t}{2} \rceil \leq i' \leq t \text{ and } 0 \leq j' \leq \lfloor \frac{t}{2} \rfloor \end{cases}$$

If both t and $\lfloor \frac{t}{2} \rfloor$ are odd, then

$$f(i, j) = \begin{cases} \lfloor \frac{t}{2} \rfloor (\lceil \frac{t}{2} \rceil i' + j') \bmod \frac{(t+1)^2}{2} & \text{if } 0 \leq i' \leq \lfloor \frac{t}{2} \rfloor \text{ and } 0 \leq j' \leq \lfloor \frac{t}{2} \rfloor \\ & \text{or } \lceil \frac{t}{2} \rceil \leq i' \leq t \text{ and } \lceil \frac{t}{2} \rceil \leq j' \leq t \\ \lfloor \frac{t}{2} \rfloor \left(\frac{(t+1)^2}{4} - 1 + \lceil \frac{t}{2} \rceil i' + j' \right) & \text{if } 0 \leq i' \leq \lfloor \frac{t}{2} \rfloor \text{ and } \lceil \frac{t}{2} \rceil \leq j' \leq t \\ \quad \bmod \frac{(t+1)^2}{2} & \text{or } \lceil \frac{t}{2} \rceil \leq i' \leq t \text{ and } 0 \leq j' \leq \lfloor \frac{t}{2} \rfloor \end{cases}$$

The correctness and optimality of the above coloring algorithm has been proved.[8] It is worth noting that arithmetic progression is followed on the whole grid when σ is odd. In contrast, when σ is even, the coloring covers the bidimensional grid B with a tessellation of basic tiles T of size $\sigma \times \sigma$, each consisting of four subtiles of size $\frac{\sigma}{2} \times \frac{\sigma}{2}$, and each following arithmetic progression.

Given any $\delta_1 \geq 1$, the above algorithm optimally solves the $L(\delta_1, 1, \ldots, 1)$ coloring problem for every $t \geq 2\delta_1$. As a consequence, given $\delta_1 = 2$, the algorithm solves the $L(2, 1, \ldots, 1)$ coloring problem for every $t \geq 4$. Hence, when $\delta_1 = 2$, the only values of t not covered by the algorithm are 2 and 3. However, in such cases, the $L(2, 1)$- and $L(2, 1, 1)$ coloring problems had been solved by Van Den Heuvel et al.[11] Therefore, the $L(2, 1, \ldots, 1)$ coloring problem can be optimally solved for any value of t.

10.5   Interval Graphs

A graph $G = (V, E)$ is termed an *interval graph* if it has an *interval representation*, namely, if each vertex of V can be represented by an interval of the real line such that there is an edge $uv \in E$ if and only if the intervals corresponding to u and v intersect. More formally, let the graph $G = (V, E)$ have n vertices. Two integers l_v and r_v, with $l_v < r_v$, (the *interval endpoints*) are associated to every vertex v of G, and there is an edge $uv \in E$ if and only if $l_u < l_v < r_u$ or $l_u < r_v < r_u$. Without loss of generality, one can assume that all the $2n$ interval endpoints are distinct and are indexed from 1 to $2n$.

Several alternative characterizations of interval graphs have been proposed so far in the literature.[20] Polynomial-time algorithms that recognize interval graphs and compute their interval representations are known.[22,23] Polynomial-time algorithms are also known for the classical vertex coloring problem on interval graphs.[24] Since it is known that a power of an interval graph is also an interval graph, the $L(1, \ldots, 1)$ coloring problem of an interval graph G can be solved in polynomial time[7] by coloring the interval graph G^t.

The following lemma shows how to locate the strongly simplicial vertex of an interval graph, which will be used to find an approximate $L(\delta_1, \delta_2, \ldots, \delta_t)$ coloring.

Lemma 10.8[21]   *In an interval graph, the vertex with maximum left endpoint is strongly simplicial.*

Lemma 10.8 suggests one scan the vertices of an interval graph by increasing left endpoints since, in this way, the t-simplicial vertex v of the induced subgraph $G[\{1, \ldots, v\}]$ is processed at each step, for $1 \le v \le n$.

10.5.1   Approximate $L(\delta_1, \ldots, \delta_t)$ coloring of Interval Graphs

Consider the interval representation of G, and assume that the intervals (vertices) are indexed by increasing left endpoints, namely $l_1 < l_2 < \cdots < l_n$. For each endpoint k, an interval v is called *open* if $l_v \le k < r_v$ and *deepest* if it is open and its right endpoint is maximum.

The algorithm to be presented maintains a family of $t + 1$ sets of colors, called *palettes*, denoted by $P_0, P_1, \ldots, P_t$. The palette P_0 is initialized to the set of colors $\{0, 1, \ldots, U\}$, where $U = \lambda^*_{G,t} + 2(\delta_t - 1)\lambda^*_{G,t} + \sum_{j=1}^{t-1} 2(\delta_j - \delta_{j+1})\lambda^*_{G,j}$. As shown by Lemma 10.4, such a color set is sufficiently large to obtain a legal $L(\delta_1, \ldots, \delta_t)$ coloring for G. The palette P_0 contains the readily usable colors. A color can leave P_0 because it is assigned to an interval. In such a case, the color is inserted into P_t and it will go downward through all the previous palettes before being reusable.

Algorithm Interval-Coloring $(G = (V, E), t, \delta_1, \ldots, \delta_t)$;

$U := \lambda^*_{G,t} + 2(\delta_t - 1)\lambda^*_{G,t} + \sum_{j=1}^{t-1} 2(\delta_j - \delta_{j+1})\lambda^*_{G,j}$;
$L_v := \emptyset$ for every vertex $v = 1, \ldots, n$;
$P_0 := \{0, 1, \ldots, U\}$ and $P_j := \emptyset$ for $j = 1, \ldots, t$;
TABOO$[\gamma] := 0$ for $\gamma = 0, \ldots, U$;
MAX $:= 0$;
$\delta_{t+1} := 0$;
for $k := 1$ to $2n$ do
  if $k = l_v$ for some v, then
    extract a color c from P_0;
    $f(v) := c$;
    for each color $\max\{0, c - \delta_1 + 1\} \leq \gamma \leq \min\{c + \delta_1 - 1, U\}$ do
      if $\gamma \in P_0$ then extract γ from P_0;
      TABOO$[\gamma] :=$ TABOO$[\gamma]+1$;
    insert color c into both L_v and P_t;
    if $r_v >$ MAX then set MAX $:= r_v$ and DEEP $:= v$;
  otherwise, if $k = r_v$ for some v, then
    for each color c in L_v do
      let j be such that $c \in P_j$;
      extract c from P_j;
      for each color $\max\{0, c - \delta_{t-j+1} + 1\} \leq \gamma \leq \max\{0, c - \delta_{t-j+2}\}$
      or $\min\{c + \delta_{t-j+2}, U\} \leq \gamma \leq \min\{0, c + \delta_{t-j+1} - 1, U\}$ do
        if $\gamma \neq c$ then TABOO$[\gamma] :=$ TABOO$[\gamma]-1$;
        if TABOO$[\gamma] = 0$ then insert γ into P_0;
      if $j > 1$ then
        insert c into both P_{j-1} and L_{DEEP}
      else
        TABOO$[c] :=$ TABOO$[c]-1$;
        if TABOO$[c]=0$ then insert c into P_0;

Figure 10.1 **The approximate algorithm for $L(\delta_1, \ldots, \delta_t)$ coloring of interval graphs.**

Precisely, the palette P_t includes the colors used for the currently open intervals, while the generic palette P_i, with $1 \leq i \leq t - 1$, contains the colors that could be reused as soon as all the next i consecutive deepest intervals will be ended. A color can leave P_0 without being assigned to any interval just because another used color forbids it. A counter is used to keep track of how many used colors currently forbid it.

Figure 10.1 illustrates the algorithm for $L(\delta_1, \ldots, \delta_t)$ coloring of interval graphs, called *Interval-Coloring*. The algorithm scans the $2n$ interval endpoints from left to right. Whenever a new interval v begins, that is a left endpoint l_v is encountered, v is colored by a color c extracted from the palette P_0 and, if needed, the deepest interval is updated. The used color c is put both in the palette P_t and in the set L_v of colors which depend on vertex v. Moreover, all the colors γ with $|\gamma - c| < \delta_1$ are forbidden by c, and thus their counters are incremented. Whenever an interval v ends, that is a right endpoint r_v is encountered, every color c belonging to L_v is deleted from its current palette, say P_j. Since the δ_{t-j+1}-separation constraint due to c does not hold anymore, all the colors γ with $\delta_{t-j+2} \leq |\gamma - c| < \delta_{t-j+1}$

are no more forbidden by c, and their counters are decremented. A color γ becomes available whenever its counter reaches 0, and in such a case it is reinserted in P_0. Whenever j is larger than 1, the color c, previously extracted from P_j, is moved to palette P_{j-1}, and inserted in the set L_{DEEP} of the colors which depend on the current deepest interval DEEP. If j is equal to 1, the color c becomes reusable, and thus it is inserted into P_0 provided that its counter becomes 0.

Lemma 10.9[20] *Consider the Interval-Coloring algorithm at the beginning of iteration k with $k = l_v$ for some interval v to be colored. Consider any color $c \in P_j$, and let w be the rightmost interval colored by c. Then v is at distance $t - j + 1$ from w, that is $d(w, v) = t - j + 1$.*

Lemma 10.10[20] *Consider the Interval-Coloring algorithm at the beginning of iteration k, and let c be any color.*

- ■ *If $c \in P_0$, then c is readily usable, and it does not forbid any other color.*
- ■ *If $c \in P_j$ with $j > 0$, then c cannot be used, and it forbids all the colors γ such that $c - \delta_{t-j+1} + 1 \leq \gamma \leq c + \delta_{t-j+1} - 1$.*
- ■ *If $c \notin P_0 \cup \cdots \cup P_t$, then it is forbidden, but it does not forbid any other color.*

In practice, the above two lemmas guarantee that the Interval-Coloring algorithm finds a legal $L(\delta_1, \ldots, \delta_t)$ coloring, that is one verifying all the separation constraints. Instead, the next theorem provides a bound on the ratio U/L between the largest used color $U = \lambda^*_{G,t} + 2(\delta_t - 1)\lambda^*_{G,t} + \sum_{j=1}^{t-1} 2(\delta_j - \delta_{j+1})\lambda^*_{G,j}$, given by Lemma 10.4, and the lower bound $L = \max_{1 \leq j \leq t} \{\delta_j \lambda^*_{G,j}\}$, provided by Lemma 10.2.

Theorem 10.1[20] *Let $\delta_m \lambda^*_{G,m} = \max_{1 \leq j \leq t} \{\delta_j \lambda^*_{G,j}\}$, and recall that $\delta_{t+1} = 0$. The Interval-Coloring algorithm gives an α-approximation with $\alpha = \min \{2t, \frac{2\delta_{m+1} - 1}{\delta_t} + \frac{2(\delta_1 - \delta_{m+1})}{\delta_m}\}$.*

It is worth noting that the Interval-Coloring algorithm provides a 4-approximation for the $L(\delta_1, \delta_2)$ coloring problem, as one can easily check by setting $t = 2$ in the formula of α given by Theorem 10.1. However, a better 3-approximation has been found,[21] even for arbitrary t, when $\delta_1 \geq 1$ and $\delta_2 = \cdots = \delta_t = 1$. As regard to the time complexity, one can prove that the Interval-Coloring algorithm runs in $O(n(t + \delta_1))$ time. Such a result is based on the fact that U can be computed in $O(nt)$ time[7] and that, between two consecutive assignments of the same color c to two intervals, $O(t)$ time is spent for at most $t + 1$ insertions and extractions of c through the palettes and $O(\delta_1)$ time is paid for updating the TABOO counters. All the details can be found in the literature.[20]

10.5.2 Approximate $L(\delta_1, \delta_2)$ coloring of Unit Interval Graphs

Consider now the $L(\delta_1, \delta_2)$ coloring problem on the class of *unit interval graphs*. This is a subclass of the interval graphs for which all the intervals have the same length, or equivalently, for which no interval is properly contained within another. Recalling that vertices are assumed to be indexed by increasing left endpoints, the main property of unit interval graphs is that whenever $v < u$ and $vu \in E$, then the vertex set $\{v, v+1, \ldots, u-1, u\}$ forms a clique and $u \le v + \lambda_{G,1}^*$ (as a consequence, the maximum vertex w at distance 2 from v verifies $w \le v + 2\lambda_{G,1}^*$). Assume that the unit interval graph to be colored is not a path since otherwise the optimal $L(\delta_1, \delta_2)$ coloring algorithm for paths can be applied.[11]

A linear time algorithm has been given,[20] which colors each vertex v in $O(1)$ time as follows. If $\delta_1 > 2\delta_2$, then assign to vertex v the color

$$
f(v) = \begin{cases} \delta_1(\lambda_{G,1}^* - p) & \text{if } 0 \le p \le \lambda_{G,1}^* \\ \delta_1(\lambda_{G,1}^* - p) + \delta_2 & \text{if } \lambda_{G,1}^* + 1 \le p \le 2\lambda_{G,1}^* + 1 \end{cases}
$$

where $p = (v-1) \bmod (2\lambda_{G,1}^* + 2)$. Otherwise, namely when $\delta_1 \le 2\delta_2$, then color v as

$$
f(v) = (2\delta_2(v-1)) \bmod (2\delta_2\lambda_{G,1}^* + 3\delta_2).
$$

The algorithm colors the vertices in a cyclic way. When $\delta_1 > 2\delta_2$, the vertices are colored by repeating the following sequence of length $2\lambda_{G,1}^* + 2$:

$$
0, \delta_1, 2\delta_1, \ldots, \lambda_{G,1}^*\delta_1, \delta_2, \delta_1 + \delta_2, \delta_1 + 2\delta_2, \ldots, \lambda_{G,1}^*\delta_1 + \delta_2.
$$

Instead, when $\delta_1 \le 2\delta_2$, the vertices are colored by repeating the sequence of length $2\lambda_{G,1}^* + 3$:

$$
0, 2\delta_2, 4\delta_2, \ldots, 2(\lambda_{G,1}^* + 1)\delta_2, \delta_2, 3\delta_2, 5\delta_2, \ldots, 2\lambda_{G,1}^*\delta_2 + \delta_2.
$$

The algorithm runs in $O(n)$ time and provides a 3-approximation.[21] In particular, it uses either at most δ_2 additional colors with respect to the optimum, when $\delta_1 > 2\delta_2$, or at most $2\delta_2$ additional colors when $\delta_1 \le 2\delta_2$. When $\delta_1 = 2$ and $\delta_2 = 1$, the algorithm uses at most 2 extra colors with respect to the optimum; as found by Sakai.[15]

10.6 Trees

An undirected graph $T = (V, E)$ is a *free tree* when it is connected, and it has exactly $|V| - 1$ edges. Given a vertex v of a free tree T, $Adj(v)$ denotes the set of vertices adjacent to v. Given also an integer t, $N_t(v)$ denotes the set of vertices at distance at most t from v. Clearly, $Adj(v) = N_1(v) - \{v\}$. A *rooted tree* is a free tree in which a vertex r is identified as a *root*, and all the other vertices are ordered by levels, where the level $\ell(v)$ of a vertex v is equal to the distance $d(r, v)$. Thus, all the vertices adjacent to v are partitioned into its *father*, denoted by *father*(v), which is at level $\ell(v) - 1$, and into its children, which are at level $\ell(v) + 1$. The *height* h of a tree T is the maximum level of its vertices. For each vertex v of T, let $anc_i(v)$ denote the ancestor of v at distance i from v (which clearly is at level $\ell(v) - i$). Of course, $anc_1(v) = father(v)$, and $anc_0(v) = v$. Moreover, $lca(u, v)$ denotes the *lowest common ancestor* of u and v, that is the vertex with maximum level among all the common ancestors of both u and v. Finally, given a vertex v of T, T_v denotes the induced subtree rooted at v consisting of all the vertices having v as an ancestor.

To derive an approximate $L(\delta_1, \ldots, \delta_t)$ coloring of a rooted tree, the following lemma is useful since it shows how to locate a strongly simplicial vertex:

Lemma 10.11[21] *In a rooted tree of height h, any vertex at level h is strongly simplicial.*

Lemma 10.11 suggests visiting the tree in breadth-first-search order, namely scanning the vertices by increasing levels. Hereafter, it is assumed that the vertices are numbered according to the breadth-first-search order, obtained by starting the visit from the root. Precisely, the vertices are numbered level by level, and those at the same level from left to right. In this way, when a vertex v is considered, v is a t-simplicial vertex of the subtree $T[\{1, 2, \ldots, v\}]$ induced by the first v vertices of T, for $1 \leq v \leq n$.

10.6.1 Approximate $L(\delta_1, \ldots, \delta_t)$ coloring of Trees

Consider a double implementation of T, where T is viewed both as a free tree and as the rooted tree T_1. Specifically, $Adj(v)$, $father(v)$, and $\ell(v)$ are maintained for each vertex v. As before, the palette P_0 of readily usable colors is maintained, which is initialized to the set $\{0, 1, \ldots, U\}$, where U is the upper bound given by Lemma 10.4. Again, a counter TABOO$[c]$ keeps track of how many used colors forbid color c.

The *Tree-Coloring* algorithm, illustrated in Figure 10.2, uses three procedures: *Ancestor*, *Up-Neighborhood-BFS*, and *Clear-BFS*, depicted in Figures 10.3, 10.4, and 10.5, respectively. Tree-Coloring first performs a

---

Algorithm Tree-Coloring $(T = (V, E), t, \delta_1, \ldots, \delta_t)$;

$U := \lambda^*_{G,t} + 2(\delta_t - 1)\lambda^*_{G,t} + \sum_{j=1}^{t-1} 2(\delta_j - \delta_{j+1})\lambda^*_{G,j}$;
$P_0 := \{0, 1, \ldots, U\}$;
TABOO$[\gamma] := 0$ for $\gamma = 0, \ldots, U$;
$u := 1$; $\delta_0 := 0$;
for $v := 1$ to n do
   $x := $ Ancestor(u, v);
   if *new* then Clear-BFS(u);
   Up-Neighborhood-BFS(v, x);
   extract a color c from P_0;
   $f(v) := c$;
   TABOO$[c] := $ TABOO$[c] + 1$;
   $u := v$;

Figure 10.2   The approximate Tree-Coloring algorithm for $L(\delta_1, \ldots, \delta_t)$ coloring of trees.

Function Ancestor (u, v);

$w := u$; $y := u$; $up := \min\{t, \ell(u)\}$;
if $\ell(v) = \ell(u) + 1$
   then $i := 1$; $x := father(v)$
   else $i := 0$; $x := v$;
while $x \neq w$ and $i < up$ do
   $y := w$; $w := father(y)$;
   $x := father(x)$; $i := i + 1$;
if $x \neq w$ or $\ell(v) = \ell(u) + 1$
   then *new* := true
   else *new* := false;
return(x)

Figure 10.3   The Ancestor procedure to find the vertex x with maximum level between $lca(u,v)$ and $anc_t(v)$. If $x = lca(u,v)$ then y is the child of x on the path between u and x.

---

preprocessing to compute the upper bound U, which depends on λ_{Tj}, for each i, $1 \leq j \leq t$.

The algorithm scans the n vertices according to the BFS numbering. At each iteration, v represents the vertex to be colored next, while u is the last colored vertex. Hence, $u = v - 1$, for $2 \leq v \leq n$. To color v, one needs to determine the set of colors already used and forbidden in the neighborhood $N'_t(v) = N_t(v) \cap T[\{1, \ldots, v\}]$. Such a set of colors depends only on the distance $d(z, v)$ for each vertex $z \in N'_t(v)$, and it is computed incrementally with respect to the neighborhood $N'_t(u)$ of the last colored vertex u.

The behavior of the Tree-Coloring algorithm depends on whether $N'_t(v)$ and $N'_t(u)$ do intersect or not, and on whether u and v are at the same level or not. When $\ell(v) = \ell(u) + 1$ or when $\ell(v) = \ell(u)$ and $N'_t(v) \cap N'_t(u) = \emptyset$,

---

Procedure Up-Neighborhood-BFS (v, x);

$dist(v) := 0; M := \emptyset;$
Enqueue(Q, v);
while $Q \neq \emptyset$ do
   $w := $ Dequeue(Q);
   if $w = x$ and not *new* then $S := \{y\}$ else $S := Adj(w)$;
(1) for each $z \in S$ do
     if $\ell(x) \leq \ell(z) \leq \ell(v)$ and $z < v$ and z is not marked then
       if $w = v$ or $0 < dist(w) < t$ then
(2)   for each $f(z) - \delta_{dist(z)} + 1 \leq \gamma \leq f(z) + \delta_{dist(z)} - 1$ do
         if $\gamma \neq f(z)$ then
           TABOO$[\gamma] := $ TABOO$[\gamma] - 1$;
           if TABOO$[\gamma] = 0$ then insert γ into P_0;
       $dist(z) := dist(w) + 1$; mark z;
(3)   for each $f(z) - \delta_{dist(z)} + 1 \leq \gamma \leq f(z) + \delta_{dist(z)} - 1$ do
         if $\gamma \neq f(z)$ then
           if TABOO$[\gamma] = 0$ then extract γ from P_0;
           TABOO$[\gamma] := $ TABOO$[\gamma] + 1$;
       if not *new* and $w \neq v$ and $(dist(w) = 0$ or $dist(w) = t)$ and $dist(z) > 0$ then
(4)   for each $f(z) - \delta_{dist(z)} + 1 \leq \gamma \leq f(z) + \delta_{dist(z)} - 1$ do
         TABOO$[\gamma] := $ TABOO$[\gamma] - 1$;
         if TABOO$[\gamma] = 0$ then insert γ into P_0;
       $dist(z) := 0$; mark z;
     Enqueue(Q, z);
   insert w into M;
  for each $z \in M$ do unmark z;

Figure 10.4   The modified BFS procedure to compute distances and forbid colors for vertices in $N'_t(v)$ and clear distances and colors for vertices in $N'_t(u) - N'_t(v)$.

Procedure Clear-BFS (u);

Enqueue(Q, u);
while $Q \neq \emptyset$ do
   $w := $ Dequeue(Q);
   for each $z \in Adj(w)$ do
     if $dist(z) > 0$ then
       for each $f(z) - \delta_{dist(z)} + 1 \leq \gamma \leq f(z) + \delta_{dist(z)} - 1$ do
         TABOO$[\gamma] := $ TABOO$[\gamma] - 1$;
         if TABOO$[\gamma] = 0$ then insert γ into P_0;
       $dist(z) := 0$;
       Enqueue (Q, z);

Figure 10.5   The procedure to clear distances and colors for vertices in $N'_t(u)$ when $\ell(v) = \ell(u) + 1$ or $N'_t(v) \cap N'_t(u) = \emptyset$.

the variable *new* is set true by the Ancestor function, the counters of all the used and forbidden colors in the old neighborhood $N'_t(u)$ are decremented by the Clear-BFS procedure, while the distances and the forbidden colors in $N'_t(v)$ are computed from scratch by the Up-Neighborhood-BFS procedure. Indeed, although the neighborhoods $N_t'(v)$ and $N_t'(u)$ may intersect when $\ell(v) = \ell(u) + 1$, each node z belonging to the intersection has $d(z, v) \neq d(z, u)$, and thus its distance and its forbidden colors have to be recomputed. In particular, the empty intersection between $N_t'(v)$ and $N_t'(u)$ is recognized by the Ancestor function when $x = anc_t(v)$ is deeper than $lca(u, v)$, the least common ancestor between u and v. When $\ell(v) = \ell(u)$ and $N_t'(v) \cap N_t'(u) \neq \emptyset$, $x = lca(u, v)$ is deeper than $anc_t(v)$, and x belongs to the shortest path $sp(u, v)$ between u and v. Consider the vertices y and y' which are the children of x on the paths $sp(u, x)$ and $sp(v, x)$, respectively. In this case, the Ancestor function sets *new* to false and returns $x = lca(u, v)$ along with the vertex y. Indeed, y and y' are the roots of the subtrees $T_y \cap N_t'(v)$ and $T_{y'} \cap N_t'(u)$ containing each vertex $z \in N_t'(v) \cap N_t'(u)$ such that $d(z, u) \neq d(z, v)$. The Up-Neighborhood-BFS procedure updates only the colors already used and forbidden by vertices in $T_y \cap N_t'(v)$ and in $T_{y'} \cap N_t'(u)$, leaving unchanged those colors used and forbidden by vertices in $(N_t'(v) \cap N_t'(u)) - (T_y \cup T_{y'})$ because for each vertex z in such a subset $d(z, v) = d(z, u)$ holds. Moreover, the procedure introduces new forbidden colors due to the vertices in $N_t'(v) - N_t'(u)$, and finally frees the colors no longer used or forbidden by vertices in $N_t'(u) - N_t'(v)$.

Precisely, the Up-Neighborhood-BFS procedure is invoked with the aim of computing the distances from v of each vertex z in $N_t'(v)$, and accordingly changing the TABOO counters. This is done in Loop (1) during a Breadth-First-Search starting from vertex v in which the label $dist(z)$ is set to $d(z, v)$, and each separation constraint is updated by first decrementing in Loop (2) the counters of the colors γ with $0 < |f(z) - \gamma| < \delta_{d(z,u)}$, and then incrementing in Loop (3) those of the colors with $0 < |f(z) - \gamma| < \delta_{d(z,v)}$. Summarizing, when *new* is true, the procedure computes from scratch all the distances and forbidden colors for all vertices in $N_t'(v)$. In contrast, when *new* is false, distances and forbidden colors are computed from scratch only for vertices in $N_t'(v) - N_t'(u)$, they are updated only for a subset of vertices in $N_t'(v) \cap N_t'(u)$, and they are cleared in Loop (4) for those in $N_t'(u) - N_t'(v)$.

Lemma 10.12[20] *Let the Tree-Coloring algorithm be at iteration v just before coloring vertex v, and consider any vertex z in $T[\{1, \ldots, v\}]$. Then*

$$dist(z) = \begin{cases} d(z, v) & \text{if } z \in N_t'(v) \\ 0 & \text{otherwise} \end{cases}$$

Lemma 10.13[20] *Let the Tree-Coloring algorithm be at iteration v just before coloring vertex v, and consider any color c.*

■ If c is assigned to a vertex $z \in N_t'(v)$, then $c \notin P_0$ and it forbids only the colors γ such that $c - \delta_{d(z,v)} + 1 \leq \gamma \leq c + \delta_{d(z,v)} - 1$.

■ If c is not assigned to any vertex in $N_t'(v)$ and $c \notin P_0$, then c is forbidden by at least a color assigned to a vertex in $N_t'(v)$, but c does not forbid any color.

■ If $c \in P_0$, then c is readily usable and it does not forbid any color.

In practice, the above lemma guarantees that a legal $L(\delta_1, \delta_2, \ldots, \delta_t)$-coloring is found, namely, a color c is assigned to a vertex v only when it satisfies all the separation constraints due to colors assigned to vertices at distance at most t from v. In particular, any vertex already colored c is at distance greater than t from v. As regard to the approximation ratio, the Tree-Coloring algorithm provides exactly the same α-approximation as the Interval-Coloring algorithm (see Theorem 10.1). In particular, the same 4- and 3-approximations hold for $L(\delta_1, \delta_2)$- and $L(\delta_1, 1, \ldots, 1)$-colorings, respectively. In addition, when $t = 2$ and T is a binary tree then a 10/3-approximation is achieved for the $L(\delta_1, \delta_2)$ coloring problem.[20]

The Tree-Coloring algorithm runs in $O(nt^2 \delta_1)$ time. This result is based on the fact that computing U takes $O(nt^2)$ time,[7] while all the updates of $dist(x)$ for a given vertex x require $O(t^2 \delta_1)$ time. Indeed, while coloring any vertex v at a given level ℓ, $dist(x)$ can assume $O(t)$ different values, one for each possible $lca(x, v) = anc_k(x)$ with $0 \leq k \leq \lfloor \frac{t}{2} \rfloor$. Moreover, all the vertices at level ℓ for which $dist(x)$ has a given value are split into at most two sequences of consecutive vertices and thus $O(\delta_1)$ time is paid only at the beginning of each subsequence. Since x can be involved in coloring vertices in at most levels $\ell(x), \ell(x) + 1, \ldots, \ell(x) + t$ and there are n vertices, the overall time taken by the algorithm is $O(nt^2 \delta_1)$.

10.7 Conclusion

This chapter has considered the channel assignment problem for various separation vectors and several particular network topologies—rings, grids, trees, and interval graphs. Specifically, $O(n)$ time algorithms have been presented for optimally solving the $L(2, 1)$-, $L(2, 1, 1)$-, and $L(\delta_1, 1, \ldots, 1)$ coloring problems on rings. Moreover, optimal solutions for the $L(\delta_1, \delta_2)$ coloring problem on both bidimensional and cellular grids, and for the $L(\delta_1, 1, \ldots, 1)$ coloring problem on bidimensional grids have been described, which are based on arithmetic progressions and take $O(rc)$ time to color the entire grid of r rows and c columns. Then, based on the notion of strongly-simplicial vertices, $O(n(t + \delta_1))$ and $O(nt^2 \delta_1)$ time algorithms have been proposed to find α-approximate $L(\delta_1, \ldots, \delta_t)$ colorings on trees and interval graphs, respectively, where α is a constant depending on t and $\delta_1, \ldots, \delta_t$. In particular, when $t = 2$, such algorithms provide 4-approximate $L(\delta_1, \delta_2)$ colorings, while they yield 3-approximate $L(\delta_1, 1, \ldots, 1)$ colorings when δ_1 is the only separation greater than 1. For $t = 2$ and binary trees, a 10/3-approximation is achieved. Moreover, an $O(n)$ time 3-approximate algorithm giving an $L(\delta_1, \delta_2)$ coloring of unit interval graphs has also been presented.

The main results reviewed in this chapter are summarized in Table 10.1. Several questions remain open. For instance, at the best of our knowledge, no results have been presented so far on the $L(\delta_1, \ldots, \delta_t)$ coloring problem for rings and grids. In addition, it is still unknown whether finding optimal $L(\delta_1, \delta_2)$ colorings of unit interval graphs or trees is NP-hard or not. Finally, as a matter of further research, one could devise better approximate algorithms for finding $L(\delta_1, \ldots, \delta_t)$ colorings of interval graphs and trees.

Table 10.1 Main Results on the Channel Assignment Problem Reviewed in the Present Chapter, where $\zeta = t + \left\lceil \frac{n \bmod (t+1)}{\lfloor \frac{n}{t+1} \rfloor} \right\rceil$ and

$$\alpha = \min\left\{ 2t, \frac{2\delta_{m+1} - 1}{\delta_t} + \frac{2(\delta_1 - \delta_{m+1})}{\delta_m} \right\}$$

Networks	$L(\delta_1, \delta_2)$		$L(\delta_1, 1, \ldots, 1)$		$L(\delta_1, \delta_2, \ldots, \delta_t)$	
Rings	Optimum	[10]	Optimum if $\delta_1 \leq \lfloor \frac{\zeta}{2} \rfloor$	[8]	Open	
Bidimensional grids	Optimum	[11]	Optimum if $\delta_1 \leq \lfloor \frac{t}{2} \rfloor$	[8]	Open	
Cellular grids	Optimum	[11]	Optimum	[9]	Open	
Trees	4-approximation	[20]	3-approximation	[20]	α-approximation	[20]
Interval graphs	4-approximation	[20]	3-approximation	[20]	α-approximation	[20]

References

1. R. Battiti, A.A. Bertossi, and M.A. Bonuccelli, "Assigning Codes in Wireless Networks: Bounds and Scaling Properties", *Wireless Networks*, Vol. 5, 1999, pp. 195–209.
2. A.A. Bertossi and M.C. Pinotti, "Mappings for Conflict-Free Access of Paths in Bidimensional Arrays, Circular Lists, and Complete Trees", *Journal of Parallel and Distributed Computing*, Vol. 62, 2002, pp. 1314–1333.
3. I. Chlamtac and S.S. Pinter, "Distributed Nodes Organizations Algorithm for Channel Access in a Multihop Dynamic Radio Network", *IEEE Transactions on Computers*, Vol. 36, 1987, pp. 728–737.
4. S.T. McCormick, "Optimal Approximation of Sparse Hessians and its Equivalence to a Graph Coloring Problem", *Mathematical Programming*, Vol. 26, 1983, pp. 153–171.

5. A. Sen, T. Roxborough, and S. Medidi, "Upper and Lower Bounds of a Class of Channel Assignment Problems in Cellular Networks", *Proceedings of IEEE INFOCOM'98*, 1998.

6. A.A. Bertossi, M.C. Pinotti, R. Rizzi, and A.M. Shende "Channel Assignment for Interference Avoidance in Honeycomb Wireless Networks", *Journal of Parallel and Distributed Computing*, Vol. 64, No. 12, 2004, pp. 1329–1344.

7. G. Agnarsson, R. Greenlaw, and M.M. Halldorson, "On Powers of Chordal Graphs and Their Colorings", *Congressus Numerantium*, Vol. 144, 2000, pp. 41–65.

8. A.A. Bertossi, M.C. Pinotti and R.B. Tan, "Channel Assignment with Separation for Interference Avoidance in Wireless Networks", *IEEE Transactions on Parallel and Distributed Systems*, Vol. 14, 2003, pp. 222–235.

9. M.V.S. Shashanka, A. Pati, and A.M. Shende, "A Characterisation of Optimal Channel Assignments for Cellular and Square Grid Wireless Networks", *Mobile Networks and Applications*, Vol. 10, 2005, pp. 89–98.

10. J.P. Georges and D.W. Mauro, "Generalized Vertex Labelling with a Condition at Distance Two", *Congressus Numerantium*, Vol. 109, 1995, pp. 141–159.

11. J. Van den Heuvel, R.A. Leese, and M.A. Shepherd, "Graph Labelling and Radio Channel Assignment", *Journal of Graph Theory*, Vol. 29, 1998, pp. 263–283.

12. H.L. Bodlaender, T. Kloks, R.B. Tan, and J. van Leeuwen, "Approximations for λ-Coloring of Graphs", *The Computer Journal*, Vol. 47, 2004, pp. 193–204.

13. G.J. Chang and D. Kuo, "The $L(2, 1)$-Labeling Problem on Graphs", *SIAM Journal on Discrete Mathematics*, Vol. 9, 1996, pp. 309–316.

14. J.R. Griggs and R.K. Yeh, "Labelling Graphs with a Condition at Distance 2", *SIAM Journal on Discrete Mathematics*, Vol. 5, 1992, pp. 586–595.

15. D. Sakai, "Labeling Chordal Graphs: Distance Two Condition", *SIAM Journal on Discrete Mathematics*, Vol. 7, 1994, pp. 133–140.

16. T. Calamoneri, "The $L(h, k)$-Labelling Problem: An Annotated Bibliography", *The Computer Journal*, Vol. 49, 1996, pp. 585–608.

17. T. Makansi, "Transmitted Oriented Code Assignment for Multihop Packet Radio", *IEEE Transactions on Communications*, Vol. 35, 1987, pp. 1379–1382.

18. A.A. Bertossi and M.A. Bonuccelli, "Code Assignment for Hidden Terminal Interference Avoidance in Multihop Packet Radio Networks", *IEEE/ACM Transactions on Networking*, Vol. 3, 1995, pp. 441–449.

19. K.I. Aardal, S.P.M. van Hoesel, A.M.C.A. Koster, C. Mannino, and A. Sassano, "Models and Solution Techniques for Frequency Assignment Problems", *4OR*, Vol. 1, 2003, pp. 261–317.

20. A.A. Bertossi and M.C. Pinotti, "Approximate $L(\delta_1, \delta_2, \ldots, \delta_t)$ coloring of Trees and Interval Graphs", Networks, to appear.

21. A.A. Bertossi, M.C. Pinotti, and R. Rizzi, "Channel Assignment with Separation on Trees and Interval Graphs", *3rd International Workshop on Wireless, Mobile and Ad Hoc Networks (WMAN)*, IEEE IPDPS, Nice, April 2003.

22. K.S. Booth and G.S. Lueker, "Linear Algorithms to Recognize Interval Graphs and Test for the Consecutive Ones Property", *Seventh Annual ACM Symposium on Theory of Computing*, Albuquerque, New Mexico, 1975, pp. 255–265.

23. D.G. Corneil, S. Olariu, and L. Stewart, "The Ultimate Interval Graph Recognition Algorithm?", *Proceedings of the Ninth Annual ACM-SIAM Symposium on Discrete Algorithms*, San Francisco, 1998, pp. 175–180.

24. A. Brandstädt, V.B. Le, and J.P. Spinrad, *Graph Classes: A Survey*, SIAM, Philadelphia, PA, 1999.

Chapter 11

MultiChannel MAC Protocols for Mobile Ad Hoc Networks

Shih-Lin Wu and Jhen-Yu Yang

11.1   Introduction

A *mobile ad hoc network* (MANET) consists of a set of mobile hosts operating without the aid of established infrastructure and of centralized administration (e.g., base stations or access points). Communication is done through wireless links among mobile hosts using by their antennas. Owing to concerns such as radio power limitation and channel utilization, a mobile host may not be able to communicate directly with other hosts in a *single-hop* fashion. In this case, a *multihop* scenario occurs, where the packets sent by the source host must be relayed by several intermediate hosts before reaching the destination host. The applications of the MANET are getting more and more important, especially in emergencies, and in military, entertainment, and outdoor business environments, where instant fixed infrastructure or centralized administration is too difficult or too expensive to install. Issues related to MANET have been studied intensively.[1–7]

A *medium access control* (MAC) protocol is used to resolve potential contention and collision when using the communication medium. The common channel is assumed to be shared by mobile hosts in some MAC protocols, which we do not discuss. Such protocols are *single-channel* MAC protocols. Some single-channel MAC protocols such as IEEE 802.11[8,9] standard still have some problems. The most serious problem with such protocols is that the network performance will degrade quickly as the network load increases due to the rapidly raised contentions and collisions of transmitted packets. Some other problems also limit these typical single-channel protocols, for instance, hidden- and exposed-terminal problems.

To overcome the problems in single-channel MAC protocols, utilizing *multiple channels* is a favorite solution. With the advance of technology, empowering a mobile host to access multiple channels is already feasible. There are some advantages in using multiple channels. The first advantage is that while the maximum throughput of a single-channel MAC protocol is limited by the bandwidth of the channel, the throughput may be increased immediately if hosts are allowed to utilize multiple channels. For example, IEEE 802.11b standard provides three nonoverlapping frequency channels. Moreover, IEEE 802.11a standard provides 12 nonoverlapping frequency channels that can be used concurrently without interference. If only one channel is used, the total throughput will be limited and the remaining channels wasted. Second, as shown in Refs. [10,11], use of multiple channels will facilitate less normalized propagation delay per channel than the single-channel counterparts, where the normalized propagation delay is defined as the ratio of the propagation time to the packet transmission time. Therefore, this reduces the probability of collisions. Third, since using a single channel to support quality of service (QoS) is difficult, it is easier to do so by using multiple channels such as QoS routing protocols.[12]

We categorize a mobile host based on its capability to access multiple channels as follows: *Single transceiver*, a mobile host that can only access one channel at a time, and *Multiple-transceivers*, a mobile host that can access multiple channels simultaneously. Note that this is not necessarily equivalent to the single-channel model, because the transceiver is still capable of switching from one channel to another.

A *multichannel MAC protocol* (MCMAC), which can operate on multiple channels, typically needs to address two tasks: *channel assignment* and *medium access*. The former is to decide which channels can be used by which hosts, and the latter is to resolve the potential contention or collision problem when communicating in a particular channel.

11.2 Design Issues of Multichannel Protocols

Using multiple transceivers in multichannel wireless ad hoc networks is a straightforward design and may simplify the difficulties of designing one, but it brings hardware size and power-consumption problems. A well-designed MCMAC protocol should try to reduce the number of transceivers equipped in mobile hosts to the minimum and to increase the aggregation throughput. It is a difficult task to diminish the transceiver number but indeed necessary for a multichannel MANET due to the battery capacity of mobile hosts.

Designing a suitable channel assignment algorithm for a multichannel MANET is an important task, one which should consider many aspects of the problem. To prevent the data package collisions, two neighboring nodes should be assigned to different channels if their transmission time is overlapping. Some of the classical multichannel MAC Protocols implement the channel assignment by maintaining a *free channel list* (FCL) in each host. A FCL could helps hosts in collecting and updating the channel information in local areas. Based on this information, the sender and receiver can pick a suitable channel for their communication. Another type of MAC protocols proposes hopping techniques in which hosts are not necessary for maintaining and updating the neighboring channel information.

To use FCL in a MCMAC, it could be divided into two categories. One category is negotiating and assigning channel on-demand, which allows hosts to immediately choose a suitable channel for data transmission while they have data packages to send. The other lets the hosts negotiate and assign channels during a periodical negotiating time. The former implies to a completely distributed negotiation by which hosts can derive a channel for data transmission in time but typically require one or more fixed control channels for negotiation purposes, while the latter restricts the hosts. Negotiating and channel-assigning powers during a specific period. Hosts may delay transmission, buffer the data package, and wait for the incoming negotiation period if they do not catch up with the previous one. (To design a multichannel MAC protocol exploiting hopping techniques, one has to take the synchronization of nodes and the extra channel switching time into consideration while repeatedly hopping between channels.

11.3 Multichannel Protocols

Multichannel protocols can be classified based on their general principles of operation into three types: *dedicated control channel, split phase,* and *channel hopping*. We will describe the spirit of each type and illustrate them by some MCMAC below.

11.3.1 Dedicated Control Channel

The protocols DCA,[13] DCA-PC,[14] GRID,[15] GRID-B,[16] Bi-MCMAC,[17] DPC,[18] and AMNP,[19] belong to the category of dedicated control-channel protocols. According to their hardware architecture, we further categorize them into two types: single transceiver and multiple transceivers. In the multiple transceivers category, for example—the two transceivers category—DCA, DCA-PC, GRID, GRID-B, Bi-MCMAC, AMNP, one of them is tuned to a dedicated control channel to sense the control packet exchanges and to transmit control packets. The other is tuned to choose among data channels for data transmission. The sensing operation will be performed by the control channel transceiver even if data transmission is ongoing. By overhearing the exchange of control packages, containing the intended data, between neighboring nodes, nodes can update their own FCL. Data transmission is performed after the sender and receiver exchange the control packages. The receiver randomly determines a channel which is free in both FCLs and tunes the data transceiver to the selected channel for transmission. It is worth paying attention to the loss-of-information problem in some single-transceiver protocols (Bi-MCMAC, AMNP,[20]): nodes cannot sense the wireless medium to update their FCL during their data transmissions. It will cause error in the FCL, and will result in data package collisions continuously. We consider this problem as an open issue in Section 11.5, and particularly illustrate it in Bi-MCMAC.

11.3.1.1 On-Demand Channel Assignment

DCA is briefly shown in Figure 11.1. Node B sends a packet to node C while host A and host D are node B's and node C's neighbors respectively. In the first step, node B sends a control package request to sent (RTS) including its FCL. We assume that the contents of FCL_B is {1,2,4,7}, and it is correctly received by node C. Let us suppose that the FCL of node C is {2,4,5,7,11}; now, it will randomly select a free data channel which is also in FCL after receiving the RTS packet. The selected data channel (D_{select}) is for data transmission by node B and C later, and it must be from {2,4,7} in this example. D_{select} is included in the clear to send (CTS) packet transmitted from node C to B in the second step, and it will be received by the neighboring nodes around node C (e.g., node D) to prevent the D_{select} from being used during node B's and node C's data transmissions. The reservation (RES) packet and the DATA package are transmitting at the same time in the third step, and they will not collide with each other because they are transmitting in different channels. The RES packet is transmitted in the control channel to notify the neighboring nodes to update their FCLs, and the data package is transmitted in D_{select}. Finally, node C sends an acknowledgment (ACK) package to finish the transmission.

A more detailed illustration is shown in Figure 11.2. In this example, we assume that there are three independent channels (channels 0 to 2). Channel 0 is a control channel used for exchanging control packets, while channel 1 and channel 2 are data channels for data transmissions/receptions. At first, node B initiates a new transmission by exchanging RTSCTS-RES packets with A. Node C overhears the RTS packet exchange from B and then sets a *network allocation vector (NAV)* for channel control in its FCL. A channel cannot be used if its corresponding NAV is not zero. As time goes by, the corresponding NAVs of the channel will reduce to zero and nodes can resume their tasks. We can also find out if the NAV set for channel 1 is performed by C after it overhears the RES packet

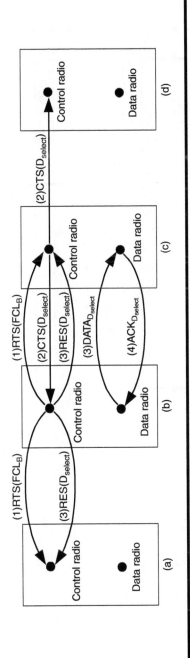

Figure 11.1   The operation of DCA.

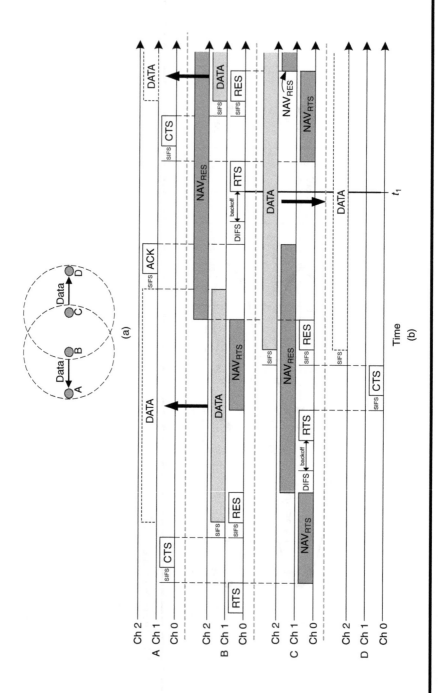

Figure 11.2 An example of DCA.

exchange from B. By observing the new transmission initiated by B at t_1, the vector will enable collision-free communications by C and D's. Because two transceivers are being used, the NAV to channel 1 can be updated successfully by B even if it had been busy transmitting in channel 1.

11.3.1.2 DCA-PC

The authors of DCA-PC propose a multichannel protocol with power control to increase channel reuse in wireless ad hoc networks. The general idea of DCA-PC is to dynamically adjust the radio power of transmitting data packages. Each node keeps a table called POWER[...], which contains the suitable radio power levels used for the data transmissions between different neighboring nodes. By monitoring the communications around itself in the control channel, the POWER table can be effortlessly generated. The protocol keeps the spirit of its purpose such that the appointed sending/receiving data of nodes cannot interrupt the ongoing data transmissions of these nodes or have a chance to be interrupted by the sending/receiving data of other nodes. Figure 11.3(a) and Figure 11.3(b) demonstrate the results of nodes with typical IEEE 802.11 carrier sense multiple access with collision avoidance (CSMA/CA) MAC and DCA-PC, respectively. In Figure 11.3(a), node A is transmitting data to B after the successful exchange of RTS/CTS control packages. During the transmission, other nodes inside the transmission range of them could not initiate transmissions by sending RTS or CTS packages. Accordingly, nodes C, D, E, and F should wait till node A receives the ACK package from node B. However, nodes could concurrently transmit in Figure 11.3(b) because of the benefits of DCA-PC MAC. Nodes A, B and C, D could transmit data packages concurrently in the same data channel (e.g., channel 1) by controlling the data sending power of data transceiver. In this situation, the transmission from node E to node F will interfere with the transmissions of nodes A and B even if node E

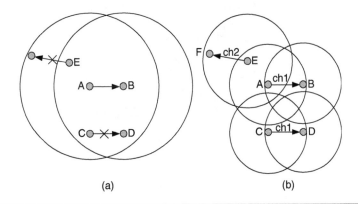

(a)   (b)

Figure 11.3 The operation of DCA-PC.

decreases the data-transmitting power. Nevertheless, it is still possible for node E to send data to F concurrently by using a different data channel (e.g., channel 2).

According to the channel assignment algorithm of DCA-PC, nodes will pick the first data channel which will not interrupt the data transmissions of ongoing nodes or have a chance to be interrupted by the sending/receiving data of other nodes. Channels that conform to the above conditions will be the candidates and the first among of them will be selected for the data transmission. A possible way to increase the throughput is for the neighboring nodes that are transmitting data to be taken into consideration. Assume that two nodes (say X and Y) are parts of a sender and receiver pair, and that they are in the process of selecting a channel as the data channel from several candidate channels. If there is an ongoing transmission by two nodes (say C and D), and if at least one of the nodes X and Y is in the control packages' transmission range but outside the data packages' transmission range of C and D, then the channel used by C and D is the more suitable one for X and Y compared to the original one alloted by the channel-selection algorithm to increase the channel reuse if available.

For example, the related locations of nodes and the channel utilization are shown in Figure 11.4. Following DCA-PC, the candidate channels of nodes X and Y are {1,2,3} and channel 1 must be selected as their data channel. In our viewpoint, channels 2 and 3 have higher priority than channel 1 because these two channels are used near by them but are still available. More importantly, channel 3 will be the best choice in this case if we count the selected times on them. Finally, nodes X and Y select

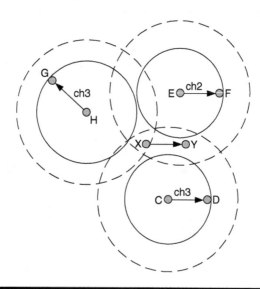

Figure 11.4 Improving the channel selection manner of DCA-PC.

channel 3 as their data channel and increase the channel reuse successfully. (DCA-PC succeeds in merging the power control mechanism into the typical multi-channel MAC, but some drawbacks are that the hardware of the dynamically adjusting transceiver should be supported and that the heedless channel selection algorithm) We have proffered a superior channel selection method above to improve on the original DCA-PC.

11.3.1.3 GRID and GRID-B

GRID and GRID-B are two multichannel protocols for ad hoc networks based on location information. GRID-B is a sequel to GRID. Each mobile host is equipped with a *global positioning system* (GPS) to obtain its location information. Figures 11.5(a) and (b) are illustrations of GRID and we will give a brief explanation below. We can see that the geographic area is divided into several logical grids in Figure 11.5(a). By definition, the assumed data channel number is nine, so the logical grids are assigned their corresponding numbers from 1 to 9. If a host has data packages for another node, it will use the assigned grid number as its data channel. Because the predefined channel number in each grid forms a well-regulated channel reuse pattern, it can increase the channel reuse and the throughput quite a lot.

The detailed protocol of GRID is as follows: For a mobile host A to communicate with host B, A will first send an RTS to B. This RTS will also carry the channel number that A intends to use in its subsequent transmissions. The intended channel number, we have explained above, depends on its location. Then B will match this request with its FCL which contains the state of each channel. If the state of the intended channel is free, B will reply a CTS to A. All this will happen on the control channel by using the

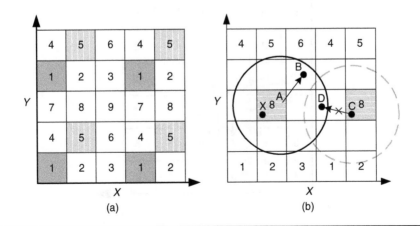

(a)   (b)

Figure 11.5   Illustration of GRID.

control transceiver. Similar to the IEEE 802.11, the purpose of the RTS/CTS dialogue is to warn the neighbors of A and B not to interfere in their subsequent transmissions, except for the host which is still allowed to use the channels different from those indicated in the RTS and CTS packets. Finally, the transmission of a data packet will occur on the data channel by using the data transceiver. The control transceiver is still active and can collect the channel usage information and dynamically renew its channel usage list.

We give a simple example of GRID in Figure 11.5(b). Node A first initiates a transmission to B by sending an RTS packet which contains the intended data channel 8. Owing to all the channels being free in the FCL of B, it immediately replies a CTS packet to inform the neighboring nodes that data channel 8 will be used for a while. After receiving the CTS packet, A could start its data transmission in channel 8 with B. Now, assume that another node C intends to communicate with D and then sends an RTS packet to D. Obviously, node D cannot allow this communication because channel 8 is not free in its channel usage list due to the information in the RTS and CTS packets sent by nodes A and B. Hence, it will not reply a CTS to node C and this communication will be aborted. If node X has a data packet to send during the communications by A and B, it will also not be allowed to do so because channel 8 is already in use by A and B. However, it is of a regrettable—that the senders cannot initiate a transmission if there is another sender which is already in transmission in the same grid. The equipped GPS device is also an extra expense and it will consume additional energy.

GRID-B allows the nodes following some predefined rules to borrow a channel from another grid. Figure 11.6 shows a scenario where a node

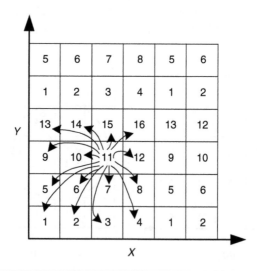

Figure 11.6   Illustration of GRID-B.

wants to communicate with an another node with $n = 16$ channels. We illustrate one of the borrowing rules, named *GRID–B$_{ss}$*. That is, let i be the channel assigned to a grid. The host in this grid will try to borrow channels in the order $i+1, i+2, \ldots, n, 1, 2, \ldots, i-1$. Assuming that there is a sender in grid 11 in tending to send a data package to another. It will follow this channel borrowing rule: {11,12,13,14,15,16,1,2,3,4,5,6,7,8,9,10} and try to get a usable channel for communication. For example, if the assigned channel (channel 11) has been used by other nodes, it could first try to borrow channel 12 for transmission. If that one has already been used, it will try channel 13, channel 14, and so on till it acquires an employable channel. The advantage of GRID-B is that nodes can still communicate with other nodes even if the originally assigned channel is occupied by neighboring nodes in the same grid or has been borrowed by foreign nodes, but the channel pattern has some probability to be disarranged in high traffic load environments. In addition, the extra cost of equipping a GPS is still a vexing disadvantage in GRID-B.

11.3.1.4 Bi-MCMAC

Bidirectional multichannel MAC protocol is one which attempts to improve *transmission control protocol (TCP)* performance on multihop wireless ad hoc networks. Trying to equip only one transceiver, it faces more challenges such as losing channel information, which may cause more data packages to collide when compared with protocols using multiple transceivers. Inevitably, this problem exists in Bi-MCMAC and the authors try to decrease its impact by using some simple channel assignment schemes.

Figure 11.7 demonstrates the Bi-MCMAC protocol. We assume that there are one control and only two data channels for conciseness of explanation. Channel 0 and channel 1, 2 are the control channel and, the data channels for sending/receiving control, and data packages, respectively. We also omit the random backoff mechanism from the figure for simplicity. As Figure 11.7(a) shows, TCP data flows take place from source A to destination D continuously. D will reply *TCP acknowledgment (TCP-ACK)* packets back to the source A after receiving the corresponding data packages intact. We know that the data and TCP-ACK flows, traveling in opposite directions, will contend the wireless medium and collide with each other, hence, to resolve the unfavorable phenomenon above, Bi-MCMAC allows receivers to immediately reply some packets (such as TCP-ACK packets) after receiving the data package from the sender.

Just as in the DCA, each node constructs an FCL and maintains it by overhearing the control packets sent by neighboring nodes. The CTS and *channel reservation notification* (CRN) packets that indicate the intended data channel are broadcasted to the neighboring nodes. Other nodes overhearing these packets remove the indicated data channel from their FCL,

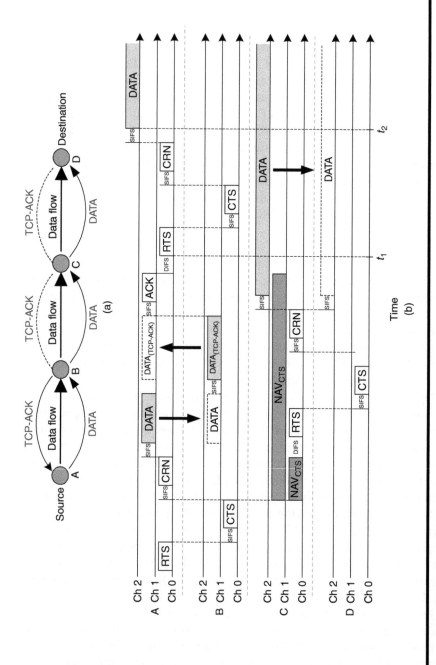

Figure 11.7   Illustration of Bi-MCMAC.

and set the NAV timer for the reserved data channel. In Figure 11.7(b), node A first sends an RTS including its FCL to B. Node B will compare the FCL of A with its own to obtain a usable data channel on both sides. In this example, both data channel 1 and 2 are usable and B chooses channel 1 as their data channel. After successfully choosing the data channel, B replies a CTS packet which embeds the choice. Node C overhears the CTS packet setting the NAV timer in the control channel and the data channel 1. This choice is then announced in a CRN packet broadcasted by A to inform neighboring nodes for updating their FCLs. After that, the control phase ends and the data exchange phase starts immediately. Both A and B then switch to data channel 1 to transmit/receive data packages. Since the receiver B also has a data package for the sender A (in this case, it is a TCP-ACK), B will send it back to A immediately after receiving the data package from A. Finally, A sends an ACK packet to complete the transmission. We can also see that after A and B finished their control phase and switched to channel 1, C and D could initiate a new transmission by the same process. Since channel 1 was not free in C's FCL anymore, C and D could only choose data channel 2 for transmission.

Now, we illustrate the potential problem of Bi-MCMAC as follows. We do not show the first one in the picture because of the limited size of the figure. Hence we expound it through theory. If node B intends to communicate with C at time t_1, it will initiate a control handshake to C by sending an RTS packet. Obviously, C is busy communicating with D but this information is unknown to B. Therefore, there will be no CTS reply to B and the RTS will cause neighboring nodes to set their control channel NAVs. More unfortunately, it will temporarily block the control channel during each of attempt by B to send an RTS if C sends a long-enough data package. It is because node B was busy communicating with A and has lost the control packets exchanged between C and D previously. We illustrate another potential problem of Bi-MCMAC below—losing channel information—which is more severe as it causes a collision on the data channel. A new transmission initiated by node A is at time t₁. Node B responds to A with a CTS packet including the randomly chosen data channel—channel 2. Node A then finishes the control phase by broadcasting the CRN packet which embeds the selected data channel. When A starts its data transmission to B at time t₂, the transmitted data package collides with that of C in data channel 2. The reason is—during the data phase, nodes cannot overhear the control packets transmitted in the control channel (channel 0).

11.3.2 Split Phase

In the split phase approach such as MMAC[21] and MAP,[22] time is divided into two phases—appointed phase and data transmission phase. Typically, devices equip a single transceiver. During the appointed phase, nodes exchange control packages in a common channel and contract a specific data channel for transmission from several data channels. After that, nodes follow their selections to switch to the selected data channel for sending or receiving the data package. The common channel becomes a data channel so that it is also a candidate and can be chosen for data transmissions in the data transmission phase. Sometimes, the environment is assumed to be a one-hop wireless network (e.g., MAP) so that nodes can hear each other's signals in the common channel during the appointed phase. This oversimplified assumption constrains the employment of the nodes to a narrow area since—nodes form a one-hop wireless network. Another drawback is the bandwidth waste except for the common channel during the appointed phase. The idling of channels during the appointed phase is one of the main reasons for the performance to go down. Actually, the synchronization costs are usually neglected in this approach and this makes the protocol loose.

11.3.2.1 MMAC

MMAC also requires only one transceiver per host, and solves the multichannel loss of information problem that we have described in Bi-MCMAC. The largest difference between MMAC and typically dedicated control channel protocols such as DCA and Bi-MCMAC is that MMAC needs all hosts to synchronize the time and to reserve the intended data channel during a regular duration called *announcement traffic indication message (ATIM) window*. Generally speaking, time is divided into several continuous beacon intervals which are formed by the ATIM window and the remaining periods. To take notice of the ATIM window in this paper, we will have nothing to do with power-saving-related works. The ATIM window is for hosts to negotiate the data channel, which the hosts will switch to after the ATIM, and within which they contend to transmit/receive data packages. In each beacon interval, the remaining time period except the ATIM is available for hosts to compete to transmit/receive in their data channels selected during ATIM window.

Like several other multichannel protocols, MMAC need hosts to construct an FCL-like list called preferable channel list (PCL) for the requirement of negotiating and reserving the intended data channel. In MMAC, the PCL adds the concept of priority for channels to balance the channel load and to reduce the contention costs when hosts contend the wireless medium. Every channel in the PCL in one of these three states:

- High preference (HIGH): This channel has already been selected by the host for use in the current beacon interval.
- Medium preference (MID): This channel has not yet been taken by other hosts for use in the transmission range of the host.

■ Low preference (LOW): This channel is already taken by at least one of the host's immediate neighbors.

Here we briefly illustrate the operations of hosts while negotiating the data transmitted channel during ATIM windows. If a source node (say A) wants to send a data package to a destination (say B), it will notify B by sending an ATIM packet including its PCL. Upon receiving the ATIM packet, B selects one channel based on its own and on A's PCLs. After that, B will reply an ATIM-ACK packet embedding the selected channel. The selection rules will be described later. On receiving the ATIM-ACK, A will comprehend which channel it should switch to at the end of ATIM window to compete for the medium to communicate with B and then it will broadcast an ATIM-RES packet to notify its neighboring nodes.

In this paragraph, we illustrate the state transitions of channels in PCL and the channel selection rules of MMAC. When the node is powered up, all the channels in the PCL are reset to MID state at the start of each beacon interval. A channel will be changed to HIGH state in the PCLs of both the source and the destination after they have finished their negotiation and agree upon that one for data transmission. If a node obtains the selected channel by overhearing ATIM-ACK or ATIM-RES packets, the state transition of this channel can have three possibilities: First, the channel which was previously in HIGH state stays in the high state, second, the channel which was previously in the MID state changes to LOW and the associated counter is set to 1, Last and third, the channel which was already in the LOW state, would have its associated counter incremented by 1.

■ If there is a HIGH state channel in the receiver's PCL, this channel is selected.
■ Or else, if there is a HIGH state channel in the sender's PCL, this channel is selected.
■ Or else, if there is a channel which is in the MID state at both the sender and the receiver nodes, it is selected. If there are more than one, one of them is randomly selected.
■ Or else, if there is a channel which is in the MID state at only one end, it is selected. If there are more than one, one of them is randomly selected.
■ If all of the channels are in the LOW state, sum up the sender's and the receiver's associated counters and select the one with the least load.

Figure 11.8 is a simple example of MMAC, assuming that nodes B and C have data packages to send to A and D separately. During ATIM window, all nodes should negotiate in a common channel (e.g., channel 1 in this example). At first, B sends an ATIM packet including its own PCL in which

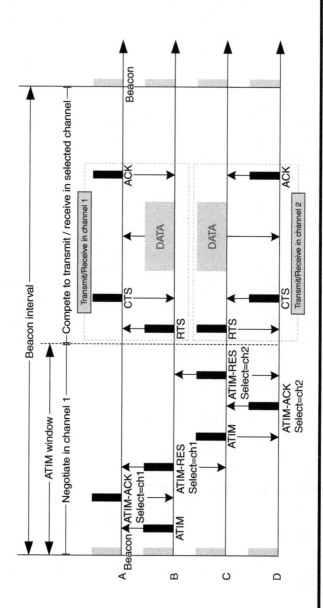

Figure 11.8 Illustration of MMAC.

channels are all in MID state. Since all channels in A's PCL are in MID state too, A follows the channel selecting rules by arbitrarily picking a channel—channel 1—after receiving the ATIM packet and replies an ATIM-ACK. B then repeats this selection by broadcasting an ATIM-RES packet in which channel 1 is embedded. This information will be overheard by node C and the state of channel 1 in its PCL is set as LOW. Therefore, a channel besides channel 1 is randomly selected during the negotiation phase of C and D because their states are all in MID. C and D finally choose channel 2 as their intended channel in this example. After ATIM window ends, nodes switch to their intended channels and contend the medium to transmit/receive data packages following the *four-way handshake (RTS-CTS-DATA-AKE)* steps defined in IEEE 802.11 standards.

11.3.3 Channel Hopping

In the channel hopping approach such as CHMA[23] and SSCH,[24] all available channels are regarded as data channels. In both the MCMAC protocols, devices are arranged as a transceiver and then hop among all channels. All hosts hop following the same hopping sequences in CHMA, while the hopping sequences can either be the same or be different in SSCH. Both approaches have the advantages of a single transceiver and greater availability of channels for data transmissions. However, it is regrettable that the channel hopping approach usually needs global time synchronization to keep its protocols working correctly.

11.3.3.1 CHMA

CHMA is a multichannel protocol that allows hosts to equip just one transceiver by following a common channel-hopping sequence. At any given time, all hosts that are not sending or receiving data must listen on the common channel hop. If a sender has a data package for another node, it sends an RTS control packet over the current channel hop. Notice that the dwell time of each channel hop needs to be only long enough to enable a control packet such as the RTS to be received by a destination node. After that, all the hosts hop to the next channel hop and the intended receiver sends a CTS packet to the source node over the same common channel hop if the RTS is received successfully. The sender and the receiver will then proceed to exchange data over the same channel hop whereas all the other hosts hop immediately to the next channel hop. There are two cases in which the nodes should backoff and not initiate a new transmission before the backoff timer reduces to zero: (1) A host (say A) sends an RTS to another one (say B) which is busy transmitting data to another node. In this case, B does not receive the RTS and does not reply a CTS packet back to A in the next channel hop. Hence, A will backoff due to the silence of the

	Time slot					
	t_0	t_1	t_2	t_3	t_4	
Channel 0	RTS(A→B)				RTS(E→F)	
Channel 1		CTS(B→A)		DATA(A→B)		CTS(E→F)
Channel 2			RTS(C→D)			
Channel 3				CTS(D→C)	DATA(C→D)	

Figure 11.9 Illustration of CHMA.

wireless medium. (2) Two hosts send RTS packets on the same channel hop to their destinations but these are in the same neighborhood and a collision occurs. In this case, both the senders should rejoin the channel hop and try to send an RTS at a later periods after the backoff timer reduces to zero.

In Figure 11.9, we illustrate the CHMA. Assuming the number of available channels is four and that all nodes follow the channel hopping sequence $0 \rightarrow 1 \rightarrow 2 \rightarrow 3 \rightarrow 0....$ We can see that A initiates a transmission by sending an RTS packet to B at t_0 in channel 0. All nodes then hop to channel 1 and B replies a CTS packet back to A at t_1. After the successful exchange of control packets, A and B stay in channel 1 to transmit/receive data packages and will rejoin the channel hop when the transmissions are over. During the transmissions of A and B, other nodes (C, D, E, and F) can simultaneously initiate other transmissions and transmit data packages in other channels.

11.3.3.2 SSCH

SSCH is a link-layer protocol that increases the capacity of wireless networks by utilizing frequency diversity. Nodes using SSCH hop across channels in such a manner that those desiring to communicate overlap, while disjointed communications mostly do not overlap, and hence do not interfere with each other. SSCH exploits a simple mathematical property to achieve this efficacy and will be illustrated later.

In SSCH, a slot is the time spent on a single channel and the authors choose the slot duration to be of 10 min in order to amortize the overhead of channel switching. It is equivalent to 35 maximum length packet transmissions at 54 Mbps bandwidth in IEEE 802.11a. Each node maintains a channel-hopping list generated by two parameters (channel, seed), denoted as (x_i, a_i). The node then increments each of the channels in its schedule using the seed and repeats the process continuously,

$$x_i \leftarrow (x_i + a_i) \bmod n$$

where n is a prime, the same as the channel number. x_i will form a periodical cycle among $[0, n-1]$ such as $(1,2,0,1,2,0,...)$ while $n = 3$.

Figure 11.10(a) is an instance of nodes incrementing their channels. At first, each node randomly chooses a (channel, seed) pair and gets hopping according to the selected channel and seed. Nodes learn each others' schedules by periodically broadcasting their channel schedule (channel, seed) pairs and offset them within this cycle.

There are four possibilities with channel-seed pairing between two nodes: (1) having the same channels and seeds, (2) having both different channels and seeds different, (3) having the same channels but different seeds, and (4) having the same seeds but different channels. Nodes will never overlay during a cycle only in the fourth case, but will overlay at least once in the other cases. To solve the fourth case, the authors add a parity slot at the end of the cycles and the channel assignment is given by the seed number of the node $(x_{parity} = a_i)$ so that nodes can overlay even if they belong to the fourth case. Nodes can have more than one schedule and iterate among these schedules. For example, assume that (x_1, a_1) and (x_2, a_2) are two (x_i, a_i) pairs of a node. It will generate

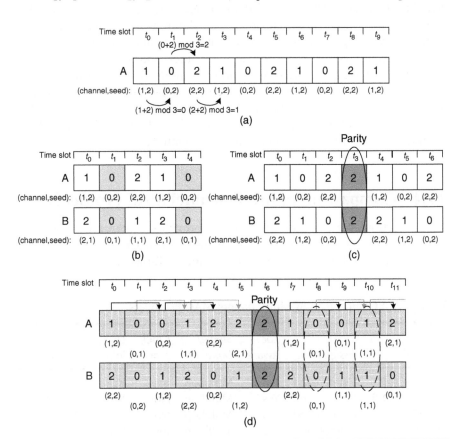

(a)

(b)

(c)

(d)

Figure 11.10   Illustration of SSCH.

two channel hopping lists denoted by $(x_{11}, a_1),(x_{12}, a_1),(x_{13}, a_1), \ldots$ and $(x_{21}, a_2),(x_{22}, a_2),(x_{23}, a_2), \ldots$, respectively, and the node will iterate channel-switching operations through $(x_{11}, a_1) \rightarrow (x_{21}, a_2) \rightarrow (x_{12}, a_1) \rightarrow (x_{22}, a_2) \rightarrow (x_{13}, a_1) \rightarrow (x_{23}, a_2)$. In the multiple schedule case, the parity slot is the one which the node has iterated through every channel on each of its schedules.

In Figure 11.10(b), we can see that the case in which nodes A and B overlap in channel 0 is (2). Figure 11.10(c) demonstrates the fourth case in which two nodes that have the same seed but different initial channels will never meet in periodical cycles but will meet in the parity slot. Recall that nodes can learn each others' schedules by periodically broadcasting their channel schedule per slot. The property that nodes will overlap at least once in a cycle (parity slot is included) ensures that works. At a high level, a node would change its schedule for another where it has straightforwardnes traffic. In brief, if node A has traffic to send to B and knows B's hopping schedule, it will probably be able to do so quickly by changing its own schedule to match B's. SSCH also introduces the partial synchronization technique for a node to synchronize with different nodes simultaneously. A node can change just one of its schedules to match another's and leave the other schedules to synchronize with other nodes. This technique is illustrated in Figure 11.10(d). Assuming that node B has data for node A, it changes its second schedule to match A's schedule (0,1) at slot t_8 and leaves the first schedule for the need to synchronize with other nodes. After synchronizing with node A, node B would follow the four-way handshake (RTS-CTS-DATAAKE) steps to communicate with A.

11.4   Comparison of Multichannel MAC Protocols

In this section, we compare each of the previously mentioned protocols according to some major factors in multichannel MANETs. We explain these ones below and the results are demonstrated in Table 11.1.

Table 11.1   Comparison of Multichannel MAC Protocols

Protocol	Assignment	Loc. aware	Single/ Multihop	Clock sync.	Losing info.	Transceivers
DCA, DCA-PC	On-demand	No	Multihop	No	No	2
GRID, GRID-B	On-demand	Yes	Multihop	No	No	2
Bi-MCMAC	On-demand	No	Multihop	No	Yes	1
MMAC	On-demand	No	Multihop	Yes	No	1
CHMA, SSCH	Dynamic	No	Multihop	Yes	–	1

- *Assignment*: On-demand means hosts will be assigned a data channel only when they need it for transmitting data packages. Dynamic means hosts will dynamically be assigned a channel for transmission even if they have no package to send. CHMA and SSCH belong to the dynamic category as the hosts in these two approaches can hop among all the data channels even if they have no package to send.
- *Location awareness (loc. aware)*: Do hosts need to know their own location information?
- *Single hop or multihop (single/multihop)*: A single hop or multihop environment: which is the protocol designed for?
- *Clock synchronization (clock sync.)*: Does it need hosts to be clock synchronized?
- *Losing channel information (losing info.)*: Is loss of channel information inherent in the protocol?
- *Transceivers*: What is the transceiver number of each host?

11.5 Open Issues

We list some open issues related to MCMAC below. They may be a problems appearing in existing MCMAC protocols, or an aspects which are worth thinking over. Many other issues are left for observing and solving in the future.

11.5.1 Losing Channel Information

As we described in Bi-MCMAC, losing channel information will result in two problems: sending an RTS packet to a node which is busy communicating with another node in a data channel, and collisions of data packages. The reasons for these problems is that hosts are equipped with a single transceiver. Hence hosts cannot sense and collect the channel usage data negotiated in the common channels near them while transmitting or receiving data packages in data channels. A possible solution is that nodes which have finished one communication wait for a maximal data transmission period. After the waiting, nodes can proceed to that original tasks to initiate a new transmission. This will cause an increase of package delay and throughput descent both of which need to be solved.

11.5.2 Disordered Channel Reuse

In personal communication systems, a geographic area is divided into several logical triangles which form a hexagonal region. In a hexagonal region, each triangle can be assigned to different channels to increase the channel reuse and this is usually done by a base station. In a multichannel MANET, the channel reuse is usually disordered due to there being no base station to coordinate the channel assignment. It is feasible for each host to be equipped with a GPS (illustrated in GRID and GRID-B) to acquire their corresponding geographic information which can assist them in choosing a suitable data channel for communications so as to increase the throughput. How to maintain the channel reuse pattern properly without being equipped with any additional device is a good issue which deserves to be mentioned.

11.5.3   Scheduling in MultiChannel MANET

In Section 11.3.2, we have described the main principle of split phase and illustrated the MAC operation in MMAC. Hosts exchange control packets in a common channel during the negotiation phase and switch to their appointed channels to compete with the wireless medium for communication after the negotiation phase. The competition and the RTS/CTS packets exchange in the data exchange phase are redundant if we compare MMAC with some multichannel MAC protocols such as DCA and Bi-MCMAC. To accomplish a transmission, MMAC needs six types of control package exchanges (ATIM, ATIM-ACK, ATIM-RES, RTS, CTS, and ACK) while DCA and Bi-MCMAC need only four. If the data channels and the starting/ending times of nodes, which have data packages to be sent, can be arranged during the negotiation phase, nodes will no longer compete in the data exchange phase and the redundant RTS/CTS packet can be economized. A well-arranged schedule can greatly increase the throughput, but it is arduous to be accomplished because the multihop environment is complex.

11.6   Conclusions

In this chapter, some fundamental MCMAC protocols for MANETs have been introduced and discussed. Comparison between MCMAC protocols on six major factors is also demonstrated in Section 11.4. We have referred several existing problems and issues related to MCMAC; all of them are worth being resolved or thought over. A well-designed MCMAC should have at least three properties: It should be simple, power-efficient, and have high performance.

References

1. E.-S. Jung and N.H. Vaidya. An Efficient MAC Protocol for Wireless LANs. *IEEE INFOCOM*, Vol. 3, pp. 1756–1764, 2002.

2. C.R. Lin and M. Gerla. MACA/PR: An Asynchronous Multimedia Multihop Wireless Network. *IEEE INFOCOM*, Vol. 1, pp. 118–125, 1997.

3. A. Nasipuri, J. Zhuang, and S.R. Das. A Multichannel CSMA MAC protocol for Multihop wireless Networks. *IEEE WCNC*, Vol. 3, pp. 1402–1406, 1999.

4. Y.-C. Tseng, C.-S Hsu, and T.-Y Hsieh. Power-saving protocol for IEEE 802.11-Based Multi-Hop Ad Hoc Networks. *IEEE INFOCOM*, Vol. 1, pp. 23–27, 2002.

5. J.-P. Sheu, C.-H. Liu, S.-L. Wu, and Y.-C. Tseng. A Priority MAC Protocol to Support Real-Time Multimedia Traffic in Ad Hoc Networks. *ACM Wireless Networks (Selected paper from EW 2002)*, Vol. 10, pp. 61–69, 2004.

6. S.-L. Wu, P.-C. Tseng, and Z.-T. Chou. Distributed Power Management Protocols for Multi-Hop Mobile Ad Hoc Networks. *Computer Networks*, Vol. 47, No. 1, pp. 63–85, 2005.

7. S.-Y. Ni, Y.-C. Tseng, Y.-S. Chen, and J.-P. Sheu. The Broadcast Storm Problem in a Mobile Ad hoc Network. *ACM Wireless Networks*, Vol. 8, pp. 158–167, 2002.

8. IEEE Std 802.11b-1999. Wireless LAN Medium Access Control (MAC) And Physical Layer (PHY) Specifications: High-speed Physical Layer Extension In The 2.4 GHz Band. *IEEE*, 1999.

9. IEEE Std 802.11a-1999. Wireless LAN Medium Access Control (MAC) and Physical Layer (PHY) Specifications: High-speed Physical Layer In The 5 GHz Band. *IEEE*, 1999.

10. M. A. Marsan and D. Roffinella. Multichannel local area networks protocols. *IEEE Journal on Selected Areas in Communications*, Vol. 1, pp. 885–897, (1983).

11. T. Makansi. Transmitter-oriented code assignment for multihop radio networks. *IEEE Transactions on Communications*, Vol. 35, No. 12, pp. 1397–1382, 1987.

12. C.R. Lin and J.-S. Liu. QoS Routing in Ad Hoc Wireless Networks. *IEEE Journal on Selected Areas in Communications*, Vol. 17, No. 8, pp. 1426–1438, 1999.

13. S.-L. Wu, C.-Y Lin, Y.-C. Tseng, and J.-P. Sheu. A new multi-channel MAC protocol with on-demand channel assignment for multi-hop mobile ad hoc networks. *International Symposium on Parallel Architectures, Algorithms and Networks (I-SPAN)*, pp. 232–237, 2000.

14. S.-L. Wu, Y.-C. Tseng, C.-Y. Lin, and J.-P. Sheu. A multi-channel MAC protocol with power control for multi-hop mobile ad hoc networks. *The Computer Journal*, Vol. 45, No. 1, pp. 101–110, 2002.

15. Y.-C. Tseng, S.-L. Wu, C.-M. Chao, and J.-P. Sheu. An efficient MAC protocol for multi-channel mobile ad hoc networks based on location information. *International Journal of Communication Systems*, Vol. 19, pp. 877–896, 2005.

16. Y.-C. Tseng, C.-M. Chao, S.-L. Wu, and J.-P. Sheu. Dynamic Channel Allocation with Location Awareness for Multi-hop Mobile Ad Hoc Networks. *Computer Communications*, Vol. 25, No. 7, pp. 676–688, 2002

17. T. Kuang and C. Williamson. A bidirectional multi-channel MAC protocol for improving TCP performance on multihop wireless ad hoc networks. *ACM MSWiM*, pp. 301–310, 2004.

18. W.-C. Hung, K.L. Eddie Law, and A. Leon-Garcia. A dynamic multi-channel MAC for ad-hoc LAN. *21st Biennial Symposium on Communications*, pp. 31–35, 2002.

19. J. Chen and Y.-D. Chen. AMNP: Ad hoc multichannel negotiation protocol for multihop mobile wireless networks. *IEEE ICC*, Vol. 6, pp. 3607–3612, 2004.

20. A. Baiocchi, A. Todini, and A. Valletta. Why a multichannel protocol can boost IEEE 802.11 performance. *ACM MSWiM*, pp. 143–148, 2004.

21. J. So and N. Vaidya. Multi-channel MAC for ad hoc networks: Handling multi-channel hidden terminals using a single transceiver. *ACM MobiHoc*, pp. 222–233, 2004

22. J. Chen, S.-T. Sheu, and C.-A Yang. A new multichannel access protocol for IEEE 802.11 ad hoc wireless LANs. *IEEE PIMRC*, Vol. 3, pp. 2291–2296, 2003.

23. A. Tzamaloukas and J.J. Garcia-Luna-Aceves. Channel-hopping multiple access. *IEEE ICC*, Vol. 1, pp. 415–419, 2000.

24. P. Bahl, R. Chandra, and J. Dunagan. SSCH: Slotted seeded channel hopping for capacity improvement in IEEE 802.11 ad-hoc wireless networks. *ACM MobiCom*, pp. 216–230, 2004.

Chapter 12

Enhancing Quality of Service for Wireless Ad Hoc Networks

Hao Zhu and Guohong Cao

A set of enhancements to the IEEE 802.11 standard, viz., the IEEE 802.11e have been proposed to meet the increasing demand for quality of service (QoS) in wireless ad hoc networks. The standard provides the means for service differentiation by using multiple traffic categories at each node, where each traffic category has its own individual parameters such as priority, inter-frame space, and contention window size. After each successful transmission, the contention window size will be decreased based on a static equation, which may result in poor channel utilization, and decreased system throughput. In this chapter, we propose a new protocol, called enhanced distributed coordination function with dual-measurement (EDCF-DM), to address this issue. EDCF-DM is based on the idea of reducing the number of idle slots by dynamically varying the contention window size according to the current traffic state of the traffic categories at each node. Meanwhile, it carefully adapts the contention window size based on the network condition of the system to avoid incurring extra collisions. Simulation results demonstrate that EDCF-DM provides a good service differentiation and outperforms the standard 802.11e in terms of channel utilization, throughput, and packet delay.

12.1 Introduction

With the tremendous growth of applications available over wireless networks, it is envisioned that wireless access will be considered as another hop of the communication path. Since many applications have (QoS) requirements such as delay and throughput, it is imperative that the wireless part of the communication should be able to support QoS similar to wired networks.[1] To achieve such goals, the medium access control (MAC) protocol should provide an efficient mechanism to share the limited spectrum among all mobile nodes, together with simplicity of operation, high system throughput, and good service differentiation for flows with different priorities.

The IEEE 802.11[2] for wireless local area networks (WLANs) is one of the most widely deployed wireless techniques.[3,4] It allows people to implement a wireless network in one of two possible configurations: the *infrastructure* mode or the *ad hoc* mode. Under the infrastructure mode, all nodes reside in a particular region where all communication must go through the access point. If the connection between a node and the access point is lost, the node cannot transmit any packets. Under the ad hoc mode, all nodes can form an ad hoc network spontaneously without any centralized control. Even if a node loses direct connections with some nodes, it is still possible for the node to communicate with others through multihop connections. This feature allows ad hoc networks to be flexibly deployed in scenarios such as battlefields, emergencies, etc., where no preestablished infrastructure exists.

The IEEE 802.11 standard includes a set of protocols that are responsible for medium access control. The basic access mechanism for ad hoc networks is the *distributed coordination function* (DCF), which uses the carrier sense multiple access with collision avoidance (CSMA/CA). However, it has been found[5] that, in legacy DCF, the packet delay exponentially grows as the amount of traffic contending the channel increases, which means the legacy DCF does not support adequate degrees of service differentiation[1,6,7] for real-time applications such as home networking, video-on-demand, and real-time voice-over-IP. To address this issue, many medium access protocols have been proposed to enhance QoS provisions under DCF.[1,6,7] The priority of each flow is controlled by different backoff times of CSMA/CA. As a result, high-priority flows can have more chances to grab the medium than low-priority flows, and get better QoS. Following the same principle, the IEEE task force group has proposed a set of QoS enhancements to IEEE 802.11.[8,9] Among them, the enhanced DCF (EDCF) protocol, which is an extension to DCF and is completely distributed, adds many new and necessary features to the current IEEE 802.11 standard. EDCF provides a flexible and distributed solution to service differentiation by introducing the concept of prioritized traffic categories. By assigning different interframe spaces and contention window sizes to different priorities, the high-priority flow is granted faster access to the medium than the low-priority flow.

Similar to the DCF, the EDCF is a contention-based MAC protocol. Thus, it is important to carefully control the backoff time of each flow to achieve good system throughput and channel utilization. As shown in Ref. [10, 11], aggressively decreasing the backoff time may increase the number of collisions a great deal and significantly decrease the system throughput. On the other hand, changing the backoff time too conservatively could cause low channel utilization. In this chapter, we study how to improve the performance of EDCF in terms of channel utilization and service differentiation in ad hoc wireless networks. We design a new protocol, called EDCF with dual-measurement (EDCF-DM), to achieve this goal. EDCF-DM can significantly increase the channel utilization by dynamically modifying the contention window size according to the collision rate and the current state of each traffic category. When the collision rate is high, the contention window size of each flow will be changed slowly to avoid further collisions. When the higher-priority queue is empty, the contention window size of the low-priority flow will be decreased faster than that when the high-priority flow has packets to send, so that the wasted, idle time slots can be reduced. Meanwhile, the service differentiation between traffic categories is maintained by monitoring the higher-priority traffic and retaining the arbitration interframe space (AIFS)[8] of each traffic category. We evaluate the proposed protocol through extensive simulations. The simulation results show that, compared to the standard EDCF and the adaptive EDCF (AEDCF)[11], EDCF-DM achieves a marked improvement in throughput, and channel utilization, with a good service differentiation.

The remainder of the chapter is organized as follows: Section 12.2 provides an overview of the 802.11 DCF as well as the 802.11 EDCF, and the issues of 802.11 EDCF. Section 12.3 gives a detailed description of the EDCF-DM protocol. The simulation results are given in Section 12.4. Section 12.5 concludes the chapter.

12.2 Background

12.2.1 The 802.11 DCF Protocol

The standard DCF protocol is described in Ref. [2]. The protocol requires that each node should sense the medium before sending the packets. This is called carrier sense multiple access. The DCF also supplements a collision avoidance mechanism to reduce the probability of collisions caused by multiple nodes detecting the channel as free at the same time. In particular, after a transmitting node senses an idle channel for a time period of a *distributed interframe space* (DIFS), it backs off for a time period that is chosen uniformly from the range of 0 to its contention window size (CW). Only after the channel remains idle for the backoff period, the node is allowed to initiate the transmission. Otherwise, the backoff timer is frozen when the channel is sensed busy, and reactivated when the channel becomes idle again for more than a DIFS. After each successful data transmission, the window size is set to CW_{min}, which denotes the prespecified minimum contention window. To solve the hidden terminal problem,[12] and to improve the system performance when the packet size is large,[13] DCF defines an additional four-way handshaking technique used optionally for a data packet transmission. As shown in Figure 12.1, after the backoff time expires, the node sends a request-to-send (RTS) packet to the receiver. If the receiver successfully receives the RTS, it replies a clear-to-send (CTS) packet after a time period of *short interframe space* (SIFS). When the sender receives the CTS, it transmits the impending data packet. For the purpose of reliability, the receiver needs to reply an acknowledgment (ACK) after it receives the packet correctly. Any other node overhearing either the RTS or the CTS extracts the information contained in the packet and updates its *network allocation vector* (NAV), which indicates the time period reserved for data transmissions. Then, the node defers its transmission until its NAV expires. For each transmission failure, which may be caused by collisions or channel errors, a binary exponential backoff is applied to double the backoff window, and the window size is bounded by the maximum contention window (denoted by CW_{max}).

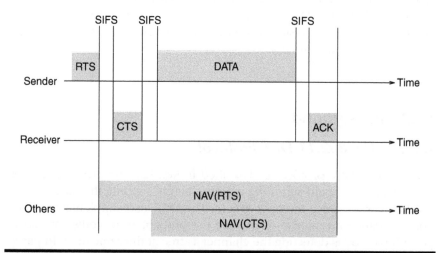

Figure 12.1   The illustration of 802.11 DCF four-way handshaking.

12.2.2 The Enhanced DCF of IEEE 802.11e

Since the original 802.11 DCF does not support any priority mechanisms, and all nodes have the same priority to access the shared channel in a contention manner, the QoS requirements (e.g., throughput and delay) of real-time applications may not be satisfied when the number of competing nodes is large.[5] (To support real-time applications, it is desirable to give higher priority of medium access to real-time traffic than best-effort traffic. Barry et al.[6] proposed priority schemes by differentiating the initial and the maximum backoff window sizes. Aad and Castelluccia[1] proposed a service-differentiation scheme by assigning SIFS to high-priority traffic. Following these principles, IEEE 802.11e Task Group proposed the EDCF protocols[10] for supporting real-time applications in WLANs, and wireless ad hoc networks.

The EDCF is a contention-based channel access protocol. The proposed scheme provides capability for up to eight types of traffic classes or categories (TC). It assigns smaller CW sizes for higher-priority classes and larger sizes to low-priority classes, giving high-priority classes a small lower bound of CW sizes. IEEE 802.11e also proposes the use of different IFS according to the priority of each TC. As shown in Figure 12.2, instead of DIFS used in DCF, an AIFS would be used for each TC. The TC with the smallest AIFS will have the highest priority. Each AIFS is equal to the DIFS time plus some (possibly zero) time slots. A big difference between DCF and EDCF is that when the medium is detected as idle for a period of AIFS, the backoff counter is reduced by one at the beginning of the last slot interval of the AIFS period, while in the legacy DCF, the backoff counter is reduced by one at the beginning of the first slot interval after the DIFS period.[9]

As shown in Figure 12.3, each TC within every node behaves like a virtual node and independently contends for access to the medium.

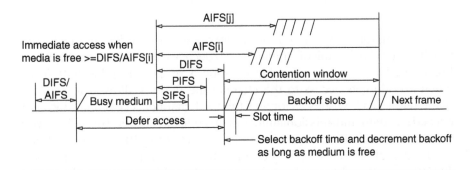

Figure 12.2 The relationships of IFS.

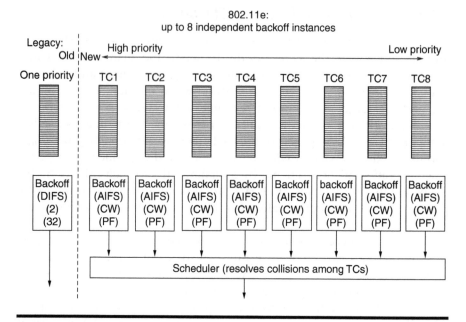

Figure 12.3 The organization of traffic category in IEEE 802.11e.

The backoff process is also carried out individually after detecting if the medium is idle for a time equal to its AIFS. Thus each TC_i is individually parameterized as: $AIFS_i$, $CW_{i,\min}$, $CW_{i,\max}$, and PF_i, where PF_i is the persistence factor of TC_i. Conflicts between virtual nodes within each node are resolved by granting access to the higher-priority transmission. With EDCF, after each successful transmission of TC_i, the corresponding CW_i will be set to $CW_{i,\min}$. Once a transmission fails, CW_i will be calculated as follows:

$$CW_i = \min\{CW_{i,\max}, CW_i \times PF_i\} \qquad (12.1)$$

After waiting for $AIFS_i$, each backoff timer is set to a random number from $[1, CW_i + 1]$ with the unit of a time slot. To decrease delay, jitter, and achieve higher medium utilization, packet bursting is proposed in the draft. Thus when a node gets access to the channel the node is allowed to send as many frames as it wishes as long as the total channel access time does not exceed a certain transmission limit, and no collision occurs. This is achieved via a feature called the transmission opportunity (TXOP), which is defined as an interval of time when a node has the right to initiate transmissions. TXOPs are granted through contention in the ad hoc mode and through hybrid coordination function (HCF) polling in the infrastructure mode.

12.2.3 Issues of 802.11 EDCF

For contention-based medium access mechanisms (i.e., DCF of IEEE 802.11), there are two major factors affecting the system throughput: transmission failure due to collisions and the wasted idle slots due to backoffs at each contention period. These problems exist in EDCF of IEEE 802.11e since it is contention-based. However, these two problems are inherently conflicting, which means reducing the backoff time could increase the number of collisions and vice versa. As a result, it is desirable to carefully control the backoff time at each contention period to achieve a good throughput.

Since the backoff time is directly related to the contention window size, we aim at designing a good scheme to manage contention window under EDCF. Generally, the management of contention window needs to be improved from two aspects: First, as stated in Ref. [11], the window size should be adjusted according to the network condition. Specifically, when the system is heavily loaded, the value of CW_i should be adjusted more slowly rather than setting to $CW_{i,min}$ upon a successful transmission. Second, one new feature of EDCF is that different TCs are assigned different values of parameters and then have different priorities. The performance obtained is not optimal since the parameters do not adapt to the traffic state of each TC. For example, suppose there are two TCs: TC_i and TC_j, where TC_i has higher priority than TC_j. When TC_i does not have data to send, CW_j can be decreased faster after each successful transmission. As a result, the number of wasted idle slots during backoff can be reduced, which increases system throughput and channel utilization. Based on these observations, it is important to design a new scheme that adapts the contention window size to both the traffic state of each TC at each node and the network condition of the system.

12.3 The Proposed EDCF-DM Protocol

To take into account the network condition and the traffic state for contention window adaptation, we design a dual-measurement scheme to get the related online information. The architecture of EDCF-DM is shown in Figure 12.4. In this section, we describe the scheme and give the algorithm used by the EDCF-DM protocol.

12.3.1 Network Condition Measurement

It is well known that, with contention-based MAC protocol, the number of contentions could significantly increase as the system is heavily loaded.

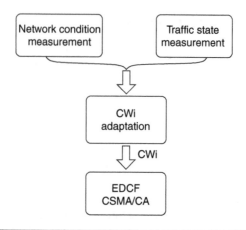

Figure 12.4 The architecture of the EDCF-DM scheme.

Similar to Ref. [11], we use the number of recent collisions as an indicator of the network condition. Specifically, the time domain is divided into continuous measurement windows (MWs) with the specified window size. When the jth measurement window, denoted by MW_j expires, the node summarizes the network condition indicator $\alpha(j)$ during MW_j by calculating

$$\alpha(j) = \frac{n_j(\text{collisions})}{n_j(\text{packets successfully sent})} \tag{12.2}$$

where n_j(event) is the number of the events occurring during MW_j. Since $\alpha(j)$ itself cannot precisely represent the long-term network condition, we also apply the exponentially weighted average method to smooth the measurement result of each measurement window. In particular, the average network condition indicator after MW_j, denoted by $\alpha_{\text{avg}}(j)$ is calculated by

$$\alpha_{\text{avg}}(j) = \phi \times \alpha_{\text{avg}}(j-1) + (1-\phi) \times \alpha(j) \tag{12.3}$$

where $0 < \phi < 1$. In this chapter, we set ϕ to be 0.8. At a time, the network condition is reflected by the value of α_{avg}. A small α_{avg} means the network condition is good.

12.3.2 Traffic State Measurement

As stated in Section 12.2.3, when the traffic density of each high-priority TC is low, reducing the contention window size of the low-priority TC could improve the system throughput and channel utilization. Similar to the principle of network condition measurement, each node maintains a traffic state indicator for each TC_i, denoted by β_i. During the jth measurement window, $\beta_i(j)$ is equal to the number of transmitted or received packets of each high-priority TC. As one optimization, TC_i does not need to compute the number of transmitted packets of all the higher TCs. Instead, suppose TC_i has lower priority than TC_j if and only if $i > j$, $\beta_i(j)$ can be computed by

$$\beta_i(j) = n_j(\text{packets of } TC_{i-1}) + \sum_k \beta_k(j), k < i - 1 \qquad (12.4)$$

Since the traffic state of each TC may change quite fast (e.g., tens of milliseconds), unlike the computation of α_{avg}, we use the most recent value of β_i as the measurement result to capture the chance to adjust the contention window in time.

12.3.3 The EDCF-DM Protocol

With the measurements of network condition and traffic state, we propose a new dual-measurement based EDCF protocol, called EDCF-DM. The key idea of EDCF is that: based on the measurement results, the extent of changing CW_i is adapted. When the workload of the system is high, CW_i is changed slowly to avoid further collisions. For TC_i, if other high-priority TCs have low traffic density, CW_i can be decreased faster to reduce the number of wasted idle slots. The speed is controlled by the control factor of TC_i, denoted by σ_i, which is dynamically assigned different values according to the measurement results. In particular, when there is no collision during the last measurement window and no higher TCs have packets to transmit or receive, σ_i is calculated by

$$\sigma_i = \min\{(1 + (i \times 2)) \times \alpha_{\text{avg}}, \sigma_{\min}\} \qquad (12.5)$$

where $\sigma_{\min}$ is the specified system parameter. Otherwise, σ_i is calculated by

$$\sigma_i = \min\{(1 + (i \times 2)) \times \alpha_{\text{avg}}, \sigma_{\max}\} \qquad (12.6)$$

where $\sigma_{\max}$ is the specified system parameter. After a successful transmission, according to σ_i, CW_i is changed as follows:

$$CW_i = \max\{CW_{i,\min}, CW_i \times \sigma_i\} \qquad (12.7)$$

When a transmission fails due to collision, similar to standard EDCF, CW_i is updated according to Eq. (12.1). The formal description of EDCF-DM is shown in Figure 12.5.

<u>Initialization:</u>
Set the parameters of each TC_i:
$AIFS_i, CW_{i,min}, CW_{i,max}, PF_i$;

When the jth measurement window expires:
get $n_j(packet)$ and $n_j(collision)$;
calculate $\alpha(j)$ according to Eq (12.2);
update α_{avg} according to Eq (12.3);
for (each TC_i)
   calculate $\beta_i(j)$ according to Eq (12.4);
   if $(n_j(collision) == 0$ && $\beta_i(j) == 0)$
     update σ_i according to Eq (12.5);
   else
     update σ_i according to Eq (12.6);

After a successful transmission in TC_i:
adjust CW_i according to Eq (12.7);

After a collision in TC_i:
adjust CW_i according to Eq (12.1);

Figure 12.5   The EDCF-DM protocol.

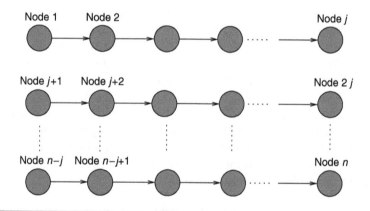

Figure 12.6   The simulation topology.

12.4   Performance Evaluation

12.4.1   Simulation Setup

We evaluate the performance of EDCF-DM through simulations by using ns-2.[14] The simulation duration is 500 s, the channel capacity is set to be 36 Mbps, the slot time is 9 μs, and the SIFS is equal to 16 μs. The simulation topology is shown in Figure 12.6. In particular, there are n nodes in the area, and each node is assumed to stay within the transmission range of other nodes. Each node generates *three* different flows representing three uniquely prioritized traffic categories viz. high, medium, and low. We use

Table 12.1   The Control Overhead of *r*PCF

Parameters	High	Medium	Low
CW_{min}	5	15	31
CW_{max}	200	500	1023
AIFS(μs)	34	43	52
PF	2	4	5
Packet size (bytes)	160	1280	1000
Packet interval (ms)	20	10	n/a
Data rate (kbps)	64	1024	n/a

two constant bit rate (CBR) sources for the high- and medium-priority flows, and a best effort source for low-priority traffic. The values of the parameters used for each traffic category are listed in Table 12.1. As can be seen, the data rate for each node is at least 1088 kbps, which is a sum of the CBR traffic categories at each node. Following our simulation setup, as the number of nodes increase, the number of flows keeps increasing by a factor of 3. As the number of nodes increases from 5 to 50, the workload of CBR flows changes approximately from 15 to 150% of the channel capacity.

We first determine the proper value of each parameter used in EDCF-DM. Then, we compare EDCF-DM with the standard EDCF,[8] and the Adaptive EDCF (AEDCF).[11] The main difference between EDCF-DM and EDCF is that EDCF-DM adapts CW_i according to the network condition. The main difference between EDCF-DM and AEDCF is that EDCF-DM adapts CW_i with the consideration of the traffic condition. In addition to throughput and channel utilization, we use packet delay as one of the performance metrics. In the following subsections, we show the simulation results of throughput, packet delay, and channel utilization, respectively.

There are three important parameters of the EDCF-DM protocol: the measurement window size, σ_{max}, and σ_{min}. With the results of Ref. [11], a good value of σ_{max} is 0.8. With extensive simulations, we choose the measurement window size and σ_{min} to be 1000 × time slot, and 0.6, respectively.

12.4.2   Throughput Comparisons

In this subsection, we show the throughput of these three different schemes. For the purpose of clearance, we give the results of each traffic category separately. The simulation results are shown in Figure 12.7. As shown in Figure 12.7(a), there is only a small change in throughput for the high-priority traffic. This is because we avoid any modification

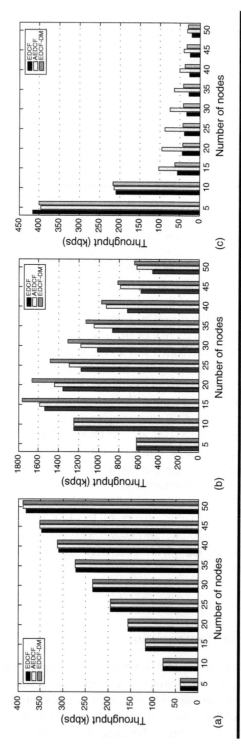

Figure 12.7 The throughput comparison of EDCF, AEDCF, and EDCF-DM. (a) High priority traffic. (b) Medium priority traffic. (c) Low priority traffic.

of parameters for the high-priority class, which acts as a reference to all the other lower-priority classes. Any visible improvements on the highest-priority class can be attributed to the adaptation mechanism according to the network condition, which decreases the probability of collision when the workload is high. For example, when the number of nodes is 50, both AEDCF and EDCF-DM achieve better throughput than EDCF.

From Figure 12.7(b), we can see that, for the medium-priority traffic, the throughput of EDCF-DM is better than that of EDCF and AEDCF when the number of nodes is greater than or equal to 15. Especially, when the number of nodes varies between 15 and 35, which means that the workload is medium, the throughput improvement of EDCF-DM is significant. Under EDCF-DM, when the workload of the high-priority traffic is low, the medium-priority flows can exploit more chances of medium access than AEDCF since the CW_i can be decreased faster. As a result, the throughput of medium-priority traffic is better. One the other hand, as the workload of high-priority traffic increases, the opportunity of fast decreasing CW_i reduces, and then the throughput gain of EDCF-DM becomes less.

The throughput of the low-priority traffic is shown in Figure 12.7(c). Surprisingly, AEDCF achieves the best throughput for the low-priority traffic. Compared with EDCF, when the workload of the system is not too low (i.e., the number of nodes is greater than 5), the number of collisions is reduced under AEDCF by adapting CW_i according to the network condition. The reason for less throughput under EDCF-DM is that under EDCF-DM, with the consideration of the traffic state of higher-priority traffic, CW_i of the low priority traffic is changed more slowly than that under AEDCF. However, this indicates that EDCF-DM can provide better service differentiation than AEDCF. For example, when the number of nodes is 20, the throughput difference between the medium-priority traffic and the low-priority traffic under EDCF-DM is much larger than that under AEDCF.

In terms of system throughput, by simply adding the throughput of each traffic category together, it is easy to see that EDCF-DM has higher system throughput than EDCF and AEDCF.

12.4.3 *Packet Delay Comparisons*

In this subsection, we compare the average packet delay under these three schemes. Since the low priority traffic is the best effort traffic, each packet delay is dominated by the queuing delay. Thus, we do not show the packet delay of the low-priority traffic. As shown in Figure 12.8(a), under most workloads, AEDCF achieves the best packet delay for high-priority traffic, and the delays under EDCF and EDCF-DM are almost the same. The reason why AEDCF outperforms EDCF has been explained in the previous subsection and is still valid here. For EDCF-DM, the longer packet delay is mainly

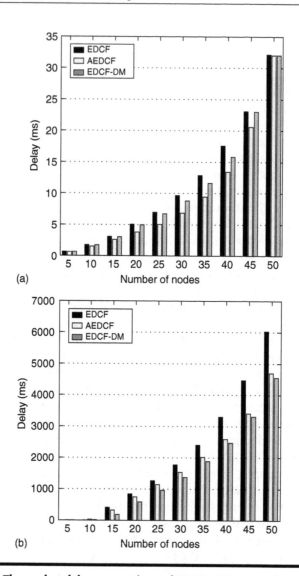

Figure 12.8 The packet delay comparison of EDCF, AEDCF, and EDCF-DM. (a) High-priority traffic. (b) Medium-priority traffic.

due to the aggressively changing CW_i for medium-priority traffic. Medium-priority flows under EDCF-DM have more chances to transmit data than under AEDCF. As a result, the high-priority packet will experience longer delay under EDCF-DM.

The packet delay of medium-priority traffic is shown in Figure 12.8(b). We can see that EDCF-DM has the best performance. Since both EDCF-DM and AEDCF control CW_i according to the network condition, as the workload increases, the packet delays of medium-priority traffic under EDCF-DM and AEDCF are shorter than that under EDCF. According to the traffic state of high-priority traffic, EDCF-DM adjusts CW_i of medium-priority traffic more aggressively than AEDCF. Consequently, under EDCF-DM, a medium-priority flow could grab the channel faster, which results in shorter packet delay.

12.4.4 Channel Utilization Comparisons

In addition to throughput and packet delay comparisons, we further evaluate channel utilization of these three schemes. As shown in Figure 12.9(a), under most workloads, EDCF-DM, and AEDCF have much better channel utilization than EDCF. This is because both EDCF-DM and AEDCF adjust

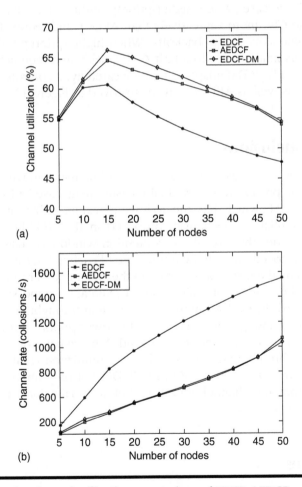

Figure 12.9 The channel utilization comparison of EDCF, AEDCF, and EDCF-DM. (a) Channel utilization. (b) Collision rate.

CW_i after a successful transmission according to the network condition. However, EDCF just blindly resets CW_i to $CW_{i,min}$ after a successful transmission. As a result, when the workload is not very low, the numbers of collisions under EDCF is much greater than that under EDCF-DM and AEDCF. To prove this claim, we show the collision rate of each scheme in Figure 12.9(b). The collision rate is calculated as the average number of collisions per second. From the figure, we can see that the number of collisions is significantly reduced under EDCF-DM and AEDCF.

From Figure 12.9(a), we can also see that EDCF-DM achieves better channel utilization than AEDCF. As explained before, EDCF-DM adapts CW_i more aggressively than AEDCF when high-priority traffics are low. As the workload of high-priority traffics increases, which is correspondent to the number of nodes, the differences in channel utilization between EDCF-DM and AEDCF decrease. When higher-priority traffic is not high, EDCF-DM can decrease CW_i more aggressively than AEDCF and reduce the number of wasted time slots due to backoffs. When higher-priority traffic is high, EDCF-DM has little chance to adjust CW_i quickly and then acts similar to AEDCF. Since EDCF-DM adapts CW_i quite carefully, it controls the number of collisions as well as AEDCF, which is proved in Figure 12.9(b).

12.5 Conclusions

In this chapter, we focused on QoS and real-time issues in wireless ad hoc networks. We proposed a new dual-measurement based EDCF scheme, called EDCF-DM, which has the following features: first, it retains the service differentiation mechanism proposed in IEEE 802.11e; second, it dynamically varies the size of the contention window by using an adaptive approach based on both the traffic state of each traffic category at each active node and the network condition of the system, (as a result, EDCF-DM can reduce the number of collisions and maintain a good service differentiation); third, EDCF-DM improves the system throughput and channel utilization by granting better access to the lower-priority flows when the higher-priority flows have no data to send. We evaluate the performance of EDCF-DM through extensive simulations. Simulation results show that, compared with EDCF and AEDCF, EDCF-DM not only improves the system throughput and channel utilization, but also maintains a good service differentiation.

References

1. I. Ada and C. Castelluccia, "Differentiation Mechanisms for IEEE 802.11," *IEEE INFOCOM'01*, 2001.

2. IEEE, "Wireless LAN Medium Access Control (MAC) and Physical Layer (PHY) Spec," *IEEE 802.11 Standard*, 1999.
3. Jim Geier, "Wireless LANs, 2nd edition," *Sams Publishing*, 2002.
4. J. Walrand and P. Varaiya, "High-Performance Communication Networks, 2nd edition," Morgan Kaufmann Publishers, 2000.
5. Y. Xiao, "An Analysis for Differentiated Services in IEEE 802.11 and IEEE 802.11e Wireless LANs," *IEEE ICDCS'04*, 2004.
6. M. Barry, A. T. Campbell and A. Veres, "Distributed Control Algorithms for Service Differentiation in Wireless Packet Networks," *IEEE INFOCOM'01*, 2001.
7. X. Yang and N.H. Vaidya, "Priority Scheduling in Wireless Ad Hoc Networks," *ACM MobiHoc'02*, June 2003.
8. IEEE Work Group, "Draft Supplement to Standard for Telecommunications and Information Exchange between Systems-LAN/MAN Specific Requirements — Part 11: Wireless LAN Medium Access Control (MAC) and Physical Layer (PHY) Specifications: Medium Access Control (MAC) Enhancements for Quality of Serice (QoS)," *IEEE 802.11e Draft 3.1*, May 2002.
9. S. Mangold, S. Choi, P. May, O. Klein, G. Hiertz and L. Stibor, "IEEE 802.11e Wireless LAN for Quality of Service," *Proceedings of European Wireless*, February 2002.
10. Y. Kwon, Y. Fang and H. Latchman, "A Novel MAC Protocol with Fast Collision Resolution for Wireless LANs," *IEEE INFOCOM'03*, 2003.
11. L. Romdhani, Q. Ni and T. Turletti, "Adaptive EDCF: Enhanced Service Differentiation for IEEE 802.11 Wireless Ad Hoc Networks," *IEEE WCNC'03*, 2003.
12. V. Bharghavan, A.J. Demers, S. Shenker and L. Zhang, "A media access protocol for wireless lans," *The Proceedings of ACM SIGCOMM.*, pp. 212–225, 1994.
13. G. Bianchi, "Performance analysis of the ieee 802.11 distributed coordination function," *IEEE Journal on Selected Areas in Communications*, vol. 18, no. 3, pp. 535–547, 2000.
14. VINT group, "UCB/LBNL/VINT Network Simulator — ns (Version 2)," http://mash.cs.berkeley.edu/ns.

Chapter 13

QoS Routing Protocols for Mobile Ad Hoc Networks

Chih-Shun Hsu, Jang-Ping Sheu, and
Shen-Chien Tung

13.1 Introduction

One wireless network architecture that has attracted much attention recently is the *mobile ad hoc network (MANET)*, which consists of mobile hosts only. Since there is no base station in the MANET, each mobile host must act as a router to forward other hosts' packets. MANET can be deployed quickly, and thus has applications in such as battlefields or disaster areas. The design of efficient routing protocols is a critical issue for all types of networks. Compared with the traditional wired network, the MANET has no fixed topology. Therefore, the source-initiated on-demand routing protocol,[20] which establishes the route between the source and destination only when the source demands that, becomes the most popular routing protocol in the MANET. However, these routing protocols use the best-effort approach to transmit message and cannot guarantee the quality of the transmission.

As the bandwidth of the wireless channel increased, variety of services can be provided in the wireless network, such as the video conference, network phone, and online video games. Some of these services require the guarantee of a certain bandwidth or a bounded delay, otherwise, the quality of these services will be unacceptable. Therefore, quality of service (QoS) becomes an important issue in the MANET and the QoS routing is the most important issue. The QoS routing protocols for the MANET have been studied extensively.[3–6,8,10–14,16–18,21–30] Since the MANET has no fixed topology, flooding is regarded as the most reliable way to discover the QoS route but is expensive. Therefore, many works try to reduce the cost of establishing the QoS routes or focus on establishing stable (or secure) QoS routes.[3,6,8,11,19,21,24,25] These works are efficient but cannot provide QoS-guaranteed transmissions.

In the MANET, since the radio signals within two hops may interfere with each other, channel reservation is one of the best approach to avoid interference. It is shown in Ref.[2] that the best technique to guarantee QoS is attained with appropriate resource reservation. Therefore, several works provide QoS-guaranteed transmissions by reserving resources.[4,5,10,14,16–18,22,23,30] In Ref. [5], the hop-by-hop bandwidth reservation is done by exchanging Request To Send(RTS)/Clear to Send(CTS) packets, which may block the channel for a long period of time and cause the fairness problem. The QoS routing protocol proposed in Ref. [23] is designed for a WLAN and thus cannot work properly in a multihop MANET. The reservation-based QoS routing protocols for the Time Division Multiple Access(TDMA) based MANETs are proposed in Refs. [4, 10, 14, 16–18, 22, 30]. These protocols can avoid contentions and collisions and thus are more efficient than the other QoS protocols, but they require the MANET to be synchronized. The major drawback of these protocols is that they do not provide a clear guide to reserve proper time slot and thus lower down the route-establishment probability.

To improve previous works, we propose a novel QoS routing protocol for TDMA-based MANETs. Our goal is to discover a stable route, which satisfies the bandwidth requirement. In this chapter, we propose an algorithm to reserve the proper time slot and thus raises the route-establishment probability. Besides, we propose an algorithm to guide the destination host to choose the route that is most likely to satisfy the QoS requirement.

The proposed QoS routing protocol consists of two phases: the route-discovery and bandwidth-reservation phases. In the route-discovery phase, we propose a novel algorithm to calculate the available bandwidth of the possible QoS routes, so that the destination host can choose the route that is most likely to satisfy the QoS requirement. After the QoS route has been chosen, the bandwidth-reservation phase follows. In the bandwidth-reservation phase, we propose a novel algorithm to calculate the weight of each available time slot. The hosts in the QoS route will reserve the time slot with the lowest weight, which will block the fewest number of free time slots of neighboring hosts, and thus raises the route-establishment probability. Simulation results show that the proposed protocol can achieve much higher route-establishment probability and lower packet loss rate than those of the Forwarding Algorithm (*FA*) proposed in Ref. [30] with a reasonable cost.

The rest of this chapter is organized as follows. Reviews of the QoS routing protocols are given in Section 13.2. In Section 13.3, we present our QoS-routing protocol. Simulation results are presented in Section 13.4. Section 13.5 concludes this chapter.

13.2 Reviews of the QoS Routing Protocols

13.2.1 QoS Routing Protocols for MANETs

The QoS routing protocols for the MANETs have been studied extensively.[3–6,8,11–14,16–18,21–30] A ticket-based QoS routing protocol is proposed in Ref. [3]. This protocol uses a limited number of tickets as probes to search several qualified paths and thus reduces the cost of establishing the QoS routes. To further improve the ticket-based protocol, the authors in Ref. [12] use the location and QoS information to guide the tickets and thus reduce the overhead. Based on dynamic source routing (DSR) protocol, the authors in Ref. [29] adopt fuzzy logic to select the appropriate QoS route from multiple paths. This protocol considers not only the bandwidth and end-to-end delay of routing, but also the cost of the path. It can tolerate a high degree of information imprecision by adopting the fuzzy logic module. A core-extraction distributed routing algorithm (CEDAR) for QoS routing is proposed in Ref.[24]. This protocol maintains a connected dominating set, named as the "core." The bandwidth of links and the route request packets are transmitted only along the "core" and thus reduces the overhead to discover QoS routes. A predictive location-based QoS routing protocol based on a low-cost location-resource update protocol is proposed in Ref. [21]. This protocol uses the predicted locations to predict future routes before the route is actually broken and thus reduce the cost to reestablish route. In REf. [8], the authors use a simple spatial probabilistic locality model to enhance the performance of traditional on-demand routing protocols. Based on the probabilistic model, we can evaluate the stability of the routing path and choose the routing path that is highly probable to avoid traffic bottlenecks. A trigger-based routing protocol for supporting real-time applications in MANETs is proposed in Ref. [6]. It uses local nodal mobility and location information to maintain a route for each session by invoking reroute routine before any link fails, and thus reduces the nodal database size and control overheads. An adaptive QoS routing scheme based on the prediction of the local performance is proposed in Ref. [25]. The QoS performance information is built based on each node's mobility and the QoS routing is adaptive to the changes of the mobile network. A secure form of QoS-guided route discovery in on-demand routing for MANETs is proposed in Ref. [11]. It uses symmetric cryptographic to reduce computation overhead and control the source to forward route request packets to reliable nodes only and thus avoid denial-of-service attacks. In Ref. [27], an admission control protocol, named as contention-aware admission control protocol (CACP), is proposed to support QoS in MANETs. CACP provides admission control for flows in a single-channel MANET based on the knowledge of a node's local resources and the effect of admitting the new flow on neighboring nodes. It is shown that by controlling bandwidth allocation, delay and jitter can also be controlled. In Ref. [28], the authors discuss the key design issues for QoS routing and review some previous works by addressing the issues of route selection and QoS constraints. A novel on-demand delay-constrained QoS routing protocol is then presented. Finally, the authors discuss some possible future directions for providing efficient QoS routing support in MANETs. All of the above works either try to reduce the cost of establishing the QoS routes or focus on establishing stable (or secure) QoS routes. None of the above works tries to reserve the required resources and thus cannot provide QoS-guaranteed transmissions. In Ref. [26], the authors address several MAC and QoS issues for delay-sensitive real-time traffic and enhance distributed medium access techniques for real-time transmissions in IEEE 802.11 single-hop MANETs. A priority scheme is proposed to differentiate the delay-sensitive real-time traffic from the best-effort traffic. In the proposed priority scheme, instead of retransmission, a smaller backoff window size is adopted for the real-time traffic. A simple scheme, which sets the initial window size and the retry limit according to the QoS requirement, is proposed to provide an upper bound on frame-dropping probability for real-time-traffic. An admission control scheme is also proposed to guarantee throughput and delay. However, these schemes are design for single-hop MANETs.

13.2.2 *Reservation-based QoS Routing Protocols*

Several works provide QoS-guaranteed transmissions by reserving resources.[4,5,10,14,16–18,22,23,30] A QoS routing protocol based on the CSMA/CA and reservation mechanisms is proposed in Ref. [5]. In this protocol, each host exchanges the QoS routing table and reservation table periodically, which is costly. In the beginning, the source host chooses the QoS route according to its QoS routing table and reservation table, and then the hosts in the chosen route make the hop-by-hop bandwidth reservation by exchanging the RTS/CTS packets. In Ref. [23], the authors propose a distributed bandwidth reservation protocol to support both nonreal-time(data) and real-time (voice) traffics over ad hoc WLAN. The voice packets always have the higher priority than ordinary data packets to access the channel.

The host, which has real-time traffics, may reserve the channel and transmits its voice packets in a round-robin manner. This protocol can only work properly in a WLAN. A QoS routing protocol uses signal strength to choose the long-lived feasible path proposed in Ref. [13]. This protocol reserves the resources by exchanging the *QoS_Reserve* and *QoS_Ack* packets. Several on-demand reservation-based QoS routing protocols for TDMA-based MANET are proposed in Refs. [4, 10, 14, 16–18, 22, 30]. These QoS routing protocols are more efficient than the other routing protocols because they can avoid a number of contentions and collisions. A multicluster packet radio network architecture is presented in Ref. [10]. The cluster heads serve as a local coordinator to schedule channel and enhance the spatial reuse of time slots and codes. The proposed network relies on both time division and code division access schemes to support multimedia traffic and thus increases hardware cost. Besides, it needs to maintain the cluster structure. An on-demand QoS routing protocol that contains bandwidth calculation and slot reservation for multihop MANETs is proposed in Ref. [18]. An improvement of the work in Ref. [18] is proposed in Ref. [17], which can accept more requests in the MANET. Both of the above works allow multiple sessions to share a common TDMA slot via CDMA. The authors in Refs. [1–9], use CDMA-over-TDMA as the MAC protocol, which may avoid interference, and thus simplify the selection and reservation of time slot, but this will increase the hardware cost. Besides, it will slow down the actual data transmission rate. In a CDMA-based network, we use a code word to represent a bit. The higher the host density, the more orthogonal codes are needed and thus the length of the code word becomes longer and the actual data transmission rate becomes lower. A TDMA-based QoS routing protocol that reserves route by addressing both the hidden- and exposed-terminal problems is proposed in Ref. [16]. This protocol uses the two-hop information to calculate the route bandwidth more accurately, and thus the precious wireless bandwidth can be better utilized. However, this protocol does not provide a clear guide to reserve the proper time slots. An on-demand QoS routing protocol, named as SBR (signal-to-interference (SIR) and bandwidth routing) is proposed in Ref. [14]. The SBR protocol reserves bandwidth by allocating time slots and ensures SIR by assigning adequate powers at the intermediate nodes between a source and a destination. The power-assignment method used in SBR supports finding routes that satisfy the SIR requirements and reduces the level of interference. The SBR protocol can establish multiple paths for a single connection to reduce the probability of call denials. A QoS protocol that grants high priority to important nodes' communications in TDMA-based MANET is proposed in Ref. [4]. This protocol prereserved bandwidth at intermediate nodes in a quadrangle-shaped area formed between important nodes. It uses each node's geographic location information to minimize potential scheduling conflicts for transmissions. A distributed end-to-end bandwidth allocation algorithm for

TDMA-based MANET is proposed in Ref. [22]. This protocol eliminates both the hidden terminal and the shortcut collision problems by introducing the link contention graph. The basic idea of this protocol is to iteratively decompose the contention graph into local clique vertex by vertex from the source to the destination. Then contentions are resolved in each local clique. A forwarding algorithm(*FA*) is proposed in Ref. [30] to calculate the effective bandwidth of a route. When the route request packet is transmitted, the effective bandwidth is also calculated according to previous two links' free time slots. After receiving several route request packets, the destination will choose a route whose effective bandwidth meets the QoS requirement and then transmit the route reply packet along the chosen route. On receiving the route reply packet, the host will randomly reserve effective time slots for QoS routing. There are two drawbacks of this protocol. First, the effective bandwidth calculated by *FA* is not accurate enough because it only considers the interference caused by previous two links in the route. Second, this protocol does not provide a clear guide to reserve the proper time slots.

13.3   Our QoS Routing Protocol

13.3.1   System Model

We intend to design an on-demand bandwidth reservation QoS routing protocol for TDMA-based multihop MANETs. Our goal is to provide a total solution for bandwidth-guaranteed service. The MANET is modeled as a graph $G = (V, E)$, where V is the set of all the mobile hosts and E the set of all the links in the MANET. If two mobile hosts, say x and y, are in the communication range of each other then $(x, y) \in E$. All the links in the MANET are bidirectional. The set of neighboring hosts of host x is defined as $NB(x) = \{y \in V : (x, y) \in E\}$. Since the MANET has no fixed topology, it is almost impossible to guarantee QoS in a high mobile environment.[2] Therefore, our QoS routing protocol is designed for a MANET with low mobility. We assume that each host maintains the information of its neighbors, including the availability of time slots. The information is gathered by broadcasting the hello packet periodically. The MANET is synchronized according to the protocols proposed in Refs. [7, 15]. The TDMA-based channel is divided into several frames with fixed length. Each frame consists of two subframes: the control frame and the data frame.[10] In the control frame, each host has a dedicated control time slot and thus can avoid contention and collision. The host can reliably transmit the control packets in its dedicated control time slot, such as the hello packet, the route request packet, the route reply packet, and the route error packet. As for the data frame, it consists of a fixed number of data time slots. The set of the data time slots is denoted as $S = \{s1, s2, \ldots, sk\}$, where k is the number of data time slots in a data frame. When a host wants to transmit (or receive) data packets, it may use its control time slot to reserve the desired data time slots, use the reserved time slot to transmit (or receive) data packets, and no acknowledgment is needed. Since the data packet is much longer than the control packet, the length of the data time slot is much longer than that of the control time slot.

13.3.2 *Motivation*

Calculating the effective bandwidth of the feasible routes accurately and providing a clear guide to reserve the proper time slots are the most important issues in the route discovery and the bandwidth-reservation phases, respectively. With the accurate effective bandwidth of the feasible routes, the destination is able to choose the route that is most likely to satisfy the QoS requirement. Reserving the proper time slots may block fewer number of neighboring hosts' free time slots and thus raise the route-establishment probability.

In Figure 13.1(a), host A wants to send data packets to host C. Since the shaded time slots are not available, host A can reserve time slots 1, 2, 3, and 4 for transmitting messages to host B, and host B can reserve time slots 3, 4, 5, and 8 for transmitting messages to host C. The effective bandwidth for links $<A, B>$ and $<B, C>$ is 4. However, time slots 3 and 4 are shared by links $< A, B >$ and $<B, C>$. Time slots 3 and 4 cannot be reserved for both links $<A, B>$ and $<B, C>$ simultaneously, so the actual effective bandwidth for path $<A, B, C>$ is 3 as shown in Figure 13.1(b). Figure 13.2 shows the effects of reserving different time slots. As shown in Figure 13.2(a), the available time slot set of links $<A, B>$ and $<C, D>$ are $\{s2, s3, s4\}$ and $\{s2\}$, respectively. If time slot $s2$ is reserved for link $<A, B>$, there will be no available time slot for link $<C, D>$ as shown in Figure 13.2(b). However, if time slot $s3$ or $s4$ is reserved for link $<A, B>$, time slot $s2$ can be reserved for link $<C, D>$ as shown in Figure 13.2(c).

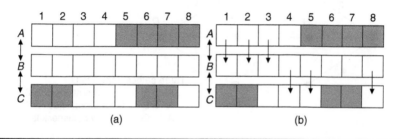

Figure 13.1 Free time slots of links $<A, B>$ and $<B, C>$.

Our Novel QoS Routing Protocol

routing protocol consists of two phases: the route-discovery
the bandwidth-reservation phase. In the route-discovery phase,
the AODV or DSR-like on-demand routing protocol to discover
le route and the min–max approach to choose the route that
kely to satisfy the QoS requirement. Figure 13.3 is a simplified
example to choose the QoS route. The minimum bandwidth
$<A,E,F,D>$, $<A,B,C,D>$, and $<A,G,H,D>$ is 2, 1, and 3,
vely. The minimum bandwidth of path $<A,G,H,D>$ is greater
e other two paths, so we will choose path $<A,G,H,D>$ as the

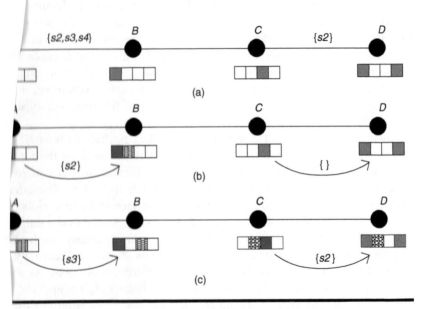

e 13.2   Reserve the proper time slot for links $<A,B>$ and $<C,D>$.

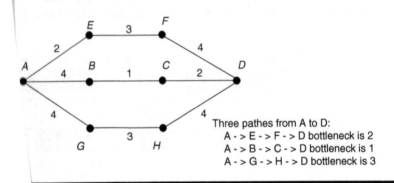

Three pathes from A to D:
A -> E -> F -> D bottleneck is 2
A -> B -> C -> D bottleneck is 1
A -> G -> H -> D bottleneck is 3

.3   A simplified min–max example to choose the QoS route.

QoS route. In the bandwidth-reservation phase, the hosts in the chosen path calculate the weight of each available time slot according to how many neighboring hosts' free time slots are blocked by this time slot. The time slot with the lowest weight will be reserved and thus leaves more free time slots for other requests. In the following subsections, we will show the data structures and the basic rules those are used in our protocol and then we will describe the route discovery and bandwidth-reservation phases of our protocols. Finally, we will describe the maintenance of the QoS route.

13.3.4 Data Structures

For any host x in the MANET, host x needs to maintain two $n \times k$ tables as follows:

- $TS_x[1, \ldots, n, 0, \ldots, k]$: a table that records the availability of the transmitting time slots. The *id*s of the neighboring hosts are recorded in column 0 of the table. If the pth neighbors of host x cannot use the qth time slot for transmission then $TS_x[p, q] = 0$, otherwise, $TS_x[p, q] = 1$.
- $RS_x[1, \ldots, n, 0, \ldots, k]$: a table that records the availability of the receiving time slots. The *id*s of the neighboring hosts are recorded in column 0 of the table. If the pth neighbors of host x cannot use the qth time slot for reception then $RS_x[p, q] = 0$, otherwise, $RS_x[p, q] = 1$.

where n is the total number of hosts within host x's communication range (including host x) and k is the total number of data time slots in a data frame.

The contents of tables *TS* and *RS* of hosts A, B, C, and D are shown in Figure 13.4. For the ease of maintaining tables *TS* and *RS*, we can combine the two tables with the neighbor list. Each record of the neighboring host consists of three fields: the *id* of the neighboring host, the transmitting time slot availability, and the receiving time slot availability of the neighboring host. To save space, we can use a bitmap to represent the time slot availability. If the ith bit equals to 1, it indicates that the ith time slot is available; otherwise, it is not available.

The data structures of the route request (*RREQ*) and route reply (*RREP*) packets are shown as follows:

- *RREQ* packet: contains seven tuples: $< src, dest, b, seq\_no, LS\_Table, BN, TTL >$, where *src* and *dest* are the *id*s of the source and destination, respectively, b is the required bandwidth, *seq_no* the sequence number of the QoS session, the *LS_Table* records the time slot availability of the previous five links which the *RREQ* packets have

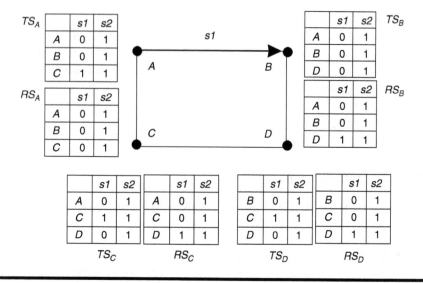

Figure 13.4   Tables *TS* and *RS* of hosts *A, B, C,* and *D*.

passed through, *BN* is the minimum effective bandwidth of all the links in the path, and the lifetime of the packet is recorded in *TTL*.
■ *RREP* packets contains eight tuples: $< src, dest, b, seq\_no, previous\_hop, TS_x, RS_x, tra\_slots >$, where *previous_hop* is the *MAC* address of the host in the previous hop of the QoS route, *tra_slots* the set of the time slots that the host reserves for transmission.

To reduce the size of the *RREQ* packet, we can use a bitmap to represent the *LS* table. If there are k data time slots in a data frame, the maximum size of the *LS* table is only $5k$ bits.

13.3.5   Fundamental Rules

In the TDMA-based MANET, the mobile host can reserve a free data time slot to transmit or receive data packets and no acknowledgment is required. Therefore, the neighboring hosts may reserve the same time slot for transmission or reception. A data time slot can be reserved for transmission or reception if it satisfies the following conditions:

C1. The time slot is not used by the sender and receiver.
C2. None of the sender's neighboring host has reserved this time slot for reception.
C3. None of the receiver's neighboring host has reserved this time slot for transmission.

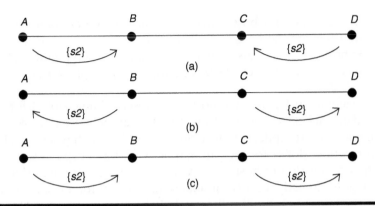

Figure 13.5 (a) Hosts *A* and *D* can use time slot {s2} simultaneously; (b) the exposure terminal problem; (c) the hidden terminal problem.

Condition C1 must be satisfied because a time slot cannot be used for transmission and reception simultaneously; condition C2 must be satisfied; otherwise, the data packets sent by the sender will collide with the data packets sent to the sender's neighbors; condition C3 must be satisfied; otherwise, the data packets sent to the receiver will collide with the data packets sent by the receiver's neighbors. For example, in Figure 13.5(a), hosts *A* and *D* can reserve the same time slot for sending messages to hosts *B* and *C*, respectively, and no collision occurs. Figure 13.5(b) is an example of the exposure-terminal problem. Although hosts *B* and *C* are neighbors, they can reserve the same time slot for sending messages to hosts *A* and *D*, respectively. Figure 13.5(c) is an example of the hidden-terminal problem. If hosts *A* and *C* reserve the same time slot to transmit messages, collision occurs in host *B*. Therefore, hosts *A* and *C* cannot reserve the same time slot to transmit messages.

A time slot cannot be used for transmission if the sender has already reserved this time slot for transmission or reception, or any of the sender's neighboring host has reserved this time slot for reception. Similarly, a time slot cannot be used for reception if the receiver has already reserved this time slot for transmission or reception, or any of the receiver's neighboring host has reserved this time slot for transmission. For any two neighboring hosts *x* and *y*, assume that hosts *x* and *y*'s time slot availability are recorded in the *i*th and *j*th rows of host *x*'s tables (TS_x and RS_x), respectively. Host *x* updates its *TS* and *RS* tables according to the following rules:

R1. If host *x* is going to reserve time slot *q* for transmission, host *x* sets $TS_x[i, q]$ and all RS_x in column *q* as 0.

R2. If host *x* is going to reserve time slot *q* for reception, host *x* sets all TS_x in column *q* and $RS_x[i, q]$ as 0.

R3. If host y notifies host x that it is going to reserve time slot q for transmission, host x sets $RS_x[i, q]$, $TS_x[j, q]$, and $RS_x[j, q]$ as 0.

R4. If host y notifies host x that it is going to reserve time slot q for reception, host x sets $TS_x[i, q]$, $TS_x[j, q]$, and $RS_x[j, q]$ as 0.

For example, in Figure 13.4, host A has reserved time slot $s1$ for transmitting data packets to host B. Hosts A and B can only use time slot $s2$ for transmission and reception because time slot $s1$ has been reserved. Host C cannot use time slot $s1$ for reception; otherwise, the reception will collide with the transmission of host A. However, host C can use time slot $s1$ for transmission. Similarly, host D cannot use time slot $s1$ for transmission; otherwise, the transmission will collide with the reception of host B. However, host D can use time slot $s1$ for reception.

13.3.6 Route-Discovery Phase

The goal of the route-discovery phase is to discover a QoS route that is most likely to satisfy the QoS requirement. Assume that there are k data time slots in a data frame, the required bandwidth is b, host x's receiving time slot availability is recorded in the ith row of table RS_x (denoted as $RS_x(i)$) and host y's transmitting time slot availability is recorded in the jth row of table TS_x (denoted as $TS_x(j)$), where $y \in NB(x)$. The time slot availability (denoted as SL_{y-x}, which contains k bits) of link $< y, x >$ can be calculated as $SL_{y-x} = RS_x(i) \cap TS_x(j)$. If the mth bit of SL_{y-x} equals to 1, it indicates that the mth time slot is available for host y to transmit data packets and host x to receive data packets, and thus can be used for transmission in link $< y, x >$. In the beginning of the route-discovery phase, the source host s broadcasts the *RREQ* packet. On receiving the *RREQ* packets from host y, host x works as follows:

> **If** (host x is not the destination and has not broadcasted its own *RREQ* packet) or (host x is the destination and has not transmitted its own *RREP* packet) **then**
>> **If** it is the first time that host x received the *RREQ* packet **then** host x sets $BN(s, x) = 0$, where $BN(s, x)$ is the minimum effective bandwidth of all the links in the path $< s, x >$.
>> **Endif**
>> Calculate the number of free time slots in link $< y, x >$ (denoted as $S_f(y, x)$) by counting the number of 1s in SL_{y-x}, where $SL_{y-x} = RS_x(i) \cap TS_x(j)$.
>> **If** $(S_f(y, x) < b)$ **then** discard the *RREQ* packet.
>> **Else**
>>> **If** host y's *LS* table contains 5 rows **then** delete the first row from host y's *LS* table

> **Endif**
> *Temp = Calculate_Bottleneck(x, y)*
> **If** *Temp ≠ invalid* and (*Temp > BN(s, x)*) **then** $BN(s, x) =$
> *Temp, previous_hop = y*, and *x*'s *LS* table = *y*'s *LS* table
> ∪ SL_{y-x}
> **Endif**

Endif
If $(BN(s, x) > 0)$ and (host *x*'s dedicated control time slot
arrives) **then**

> **If** host *x* is the destination **then** host *x* transmits its *RREP*
> packet with empty *tra_slots* to *previous_hop* and starts the
> bandwidth-reservation phase
> **Else** host *x* broadcasts its own *RREQ* packet.
> **Endif**

Endif
Else discard the *RREQ* packet.
Endif

The major idea of the above algorithm is that, host *x* will choose the neighboring host *y*, which has the greatest value of *BN(s, x)*, as its previous hop and combines host *y*'s *LS* table with its own *LS* table before host *x* broadcasts its own *RREQ* packet. The destination will wait for a certain period of time and then chooses the route that is most likely to satisfy the QoS requirement. When the *RREP* packet is transmitted along the *previous_hop*, a QoS route can be reserved. The key issue of the above algorithm is the calculation of the effective bandwidth of the bottleneck for a given path. In a path, the links within two hops cannot use the same time slots. Therefore, when calculating the effective bandwidth of a link, we need to gather the time slot availability of previous and next two hops. However, if the path length is less than 5, we will only consider the time slot availability of all the links in the path. Assume that the *LS_table* contains *r* rows, where $0 \le r \le 5$. The details of the *Calculate_Bottleneck* algorithm is shown as follows:

> **Algorithm:** *Calculate_Bottleneck(x, y)*
> *x*'s *LS_Table = y*'s *LS* table $\cup SL_{y-x}$
> **If** $(r < 3)$ **then**
>
> > **If** (*x* is the destination) **then**
> > Calculate the minimum effective bandwidth (denoted as *BW*) of
> > the last *r* links.
> > Return the minimum *BW* of the last *r* links.
> > **Else** return invalid.
> > **Endif**

Else

> **If** x is the destination **then** calculate the minimum effective bandwidth (denoted as BW) of the last three links.
> **Else** calculate the effective bandwidth of the link whose time slot availability is recorded in the $(r - 2)$th row of x's *LS_Table*.
> **Endif**
> Return min$(BN(s, y), BW)$

Endif

The key issue of the above algorithm is the calculation of the effective bandwidth of a given link. Since the time slot availability of the previous r links is recorded in the *LS* table, the effective bandwidth of a given link can be derived from the *LS* table. The number of 1s in a row is the number of free time slots in the link. The number of 1s in a column within previous and next two rows is the number of hosts those share the same time slot. Assume that the time slot availability of link $<y, x>$ is recorded in the tth row of x's *LS_Table*, where $1 \le t \le r$. The algorithm to calculate the effective bandwidth of the link $<y, x>$ is shown as follows:

Algorithm: *Calculate_Bandwidth(t)*

Initial: $n = 0$ and $BW = 0$, where n is the number of hosts those share the same time slot, BW the effective bandwidth of the link, and r the number of rows in x's *LS_Table*.

For $i = 1$ to k **do**

> **If** the ith bit of the tth row in x's *LS_Table* is 1 **then**
>
> > **If** $(r \le 4$ and $t \le 3)$ **then** $u = 1$
> > **Else** $u = t - 2$
> > **Endif**
> > **For** $j = u$ to r **do if** the ith bit of the jth row in x's *LS_Table* is 1 **then** $n = n + 1$
> > **Endfor**
>
> **Endif**
> $BW = BW + \frac{1}{n}$

Endfor

Return BW.

Note that, even when the path length is less than 5, we can still apply the above algorithm to calculate the effective bandwidth of a certain link. Figure 13.6 shows the contents of the *LS* table and how we calculate the bottleneck of a path. Host A is the starting point of the path and host G is the destination. The set below the link contains the free time slots of the link, which can be derived from tables *TS* and *RS*. When the *RREQ*

$S = \{s1, s2, s3, s4\}$

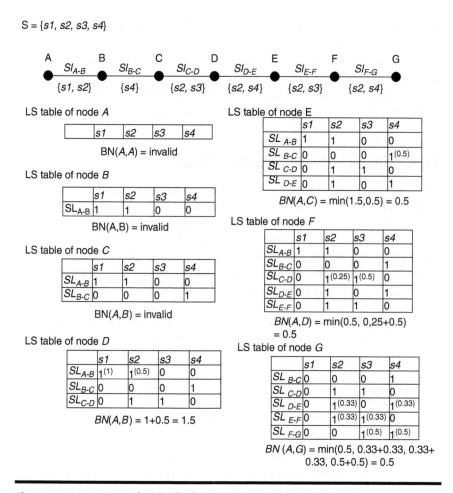

Figure 13.6 An example of calculating the bottleneck of a path.

packet has been transmitted from host A to hosts B and C, since hosts A, B, and C have no enough information to calculate the effective bandwidth of any link, the value of BN is set as invalid. However, when the *RREQ* packet has been transmitted to host D, host D has gathered the time slot availability of links $<A, B>$, $<B, C>$, and $<C, D>$ and thus has enough information to calculate the effective bandwidth of link $<A, B>$. In link $<A, B>$, time slots $s1$ and $s2$ can be used for transmission. Time slot $s1$ is not shared by other links in the path, so its effective bandwidth is 1; time slot $s2$ is shared by link $<C, D>$, so its effective bandwidth is 0.5, and the total effective bandwidth of link $<A, B>$ (denoted as $BW(A, B)$) is 1.5. The bottleneck of path $<A, B>$ (denoted as $BN(A, B)$) is 1.5. Similarly, when host E has received the *RREQ* packet, it has enough information to calculate the effective bandwidth of link $<B, C>$. In link $<B, C>$, time slot $s4$ can be used for transmission. Time slot $s4$ is shared by link $<D, E>$, so its effective bandwidth is 0.5. We have $BW(B, C) = 0.5$ and $BN(A, C) = \min(BW(A, B), BW(B, C)) = \min(1.5, 0.5) = 0.5$. We can calculate $BN(A, D)$, $BN(A, E)$, $BN(A, F)$, and $BN(A, G)$. $BN(A, G)$ is the effective bandwidth of the bottleneck in the path. The destination can choose the QoS route according to the BN value of different routes.

13.3.7 Bandwidth-Reservation Phase

After the destination discovers the QoS route and transmits the *RREP* packet, the bandwidth-reservation phase starts. The goal of the bandwidth-reservation phase is to reserve the proper time slots for each link in the QoS route so that more free time slots are left for other requests and thus raises the route-establishment probability. Assume that b is the required bandwidth. On receiving the *RREP* packet from host z, where host z's *previous_hop* is y and host y's *previous_hop* is x, host y works as follows:

> Update the TS_y and RS_y tables according to the TS_z, RS_z, and *tra_slots* of the *RREP* packet sent by host z.
> Update SL_{y-z} according to the new TS_y and RS_y.
> Calculate the number of free time slots in link $<y, z>$ (denoted as $S_f(y, z)$).
> **If** $(S_f(y, z) < b)$ **then** transmit the *fail_to_reserve* packet to host x.
> **Else**
>> Let $TS_{yz} = TS_y \cup TS_z$ and $RS_{yz} = RS_y \cup RS_z$.
>> Calculate the weights of all the free time slots in link (y, z).
>> Select b time slots with the least total weights for reservation.
>> Update the TS_y and RS_y tables according to the reservation.
>> Add the *ids* of the reserved time slots to host y's *tra_slots*.
>> **If** host y is not the source **then** transmit the *RREP* packet to host x.
>> **Else** the bandwidth reservation phase is over and start to transmit data packets in the reserved time slots.
>> **Endif**
> **Endif**

When host z overhearing the *RREP* packet sent by host y, host z will reserve the time slots in *tra_slots* for reception and will notify its neighboring hosts about the reservation in its dedicated control time slot. The other neighboring hosts, overhearing the *RREP* packets, will update their *TS* and *RS* tables according to the TS_z, RS_z, and *tra_slots* of the *RREP* packet. If host x receives the *fail_to_reserve* packet, it will forward the packet to its *previous_hop* until it reaches the source. If the source receives the *fail_to_reserve* packet or does not receive the *RREP* packet for a certain period of time, it will consider that the route establishment has failed. The key issue of the above algorithm is to evaluate the weight of a given free time slot in a given link. Assume that we want to evaluate the weight of the ith time slot in link $< y, z >$, the details of the algorithm is shown as follows:

> **Algorithm:** *Calculate_Weight*(y, z, i)
>
> Count the total number of the ith bit (denoted as *Weight*) in tables TS_{yz} and RS_{yz} that needs to be updated when the ith time slot is going to be reserved by host y for transmission and host z for reception.
>
> Return *Weight*.

The weight of the time slot is the number of the neighboring hosts' free time slots blocked by the reservation of this time slot. Therefore, our goal is to choose a time slot that has already been used or blocked by most of the neighboring hosts and thus leaves more free time slots for other requests. For example, in Figure 13.4, if host C wants to establish a QoS route to host D, host C broadcasts an *RREQ* packet and host D may receive the *RREQ* packets from hosts C and B. The *BN* of the path $<C, D>$ is 2 and the *BN* of the path $< C, A, B, D >$ is 0.33. Therefore, host D will choose host C as its *previous_hop* and send its *RREP* packet to host C. On receiving the *RREP* packet from host D, host C combines the *TS* and *RS* tables of hosts C and D as shown in Figure 13.7(a) and then calculates the weights of time slot $s1$ and $s2$ as shown in Figure 13.7(b). The weight of time slot $s1$ is 2 because reserve time slot $s1$ will only block host C's transmitting time slot and host D's receiving time slot. The weight of time slot $s2$ is 6 because reserve time slot $s2$ will block hosts B, C, and D's transmitting time slots and hosts A, C, and D's receiving time slots. Therefore, hosts C and D will reserve time slot $s1$ for transmission and reception, respectively. After reserved time slots $s1$, time slot $s2$ is free for all the links. However, if time slot $s2$ is reserved for hosts C and D, there is no free time slot for links $< A, C >$, $< C, A >$, $< D, B >$, $< B, D >$, $< B, A >$, and $< D, C >$.

13.3.8 Route Maintenance

Since the MANET has no fixed topology, the QoS route may become broken. Only the receiver is able to realize that the link between the sender and the receiver is broken. If the receiver does not receive the data packet in the reserved time slot, it considers the link is broken and uses its dedicated control time slot to broadcast a route error (*RERR*) packet with a finite

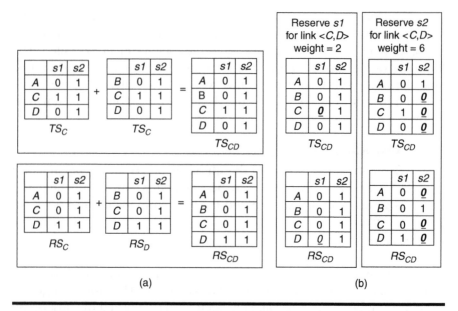

Figure 13.7 **(a) Combines *TS* and *RS* tables of hosts *C* and *D* and get *TS_CD* and *RS_CD*. (b) Calculate the weights for reserving s1 and s2.**

TTL, which is set according to the mobility of the mobile host. Any host in the QoS route will forward the *RERR* packet to the source. On receiving the *RERR* packet, the source can reestablish a new QoS route or try to repair the original QoS route according to the existing on-demand routing protocols. When a host in the QoS route has received the last data packet, it will release the reserved time slot by sending a *bandwidth_released* packet in its dedicated control time slot. On receiving the *bandwidth_released* packet, each host will update its *TS* and *RS* tables.

13.4 Simulation Results

To evaluate the performance of our QoS routing protocol, we have developed a simulator using C. We compare the performance of our protocol with that of the *FA* protocol,[30], because *FA* protocol is the latest and best one. Besides, it has the same system model as our protocol. Assume that the area size of the MANET is 1000 m × 1000 m, the transmission radius is 250 m, the transmission rate is 11 M bits/s, the packet size is 1 kbytes. there are 40 data time slots in a data frame, and the length of a data time slot is 5 ms. In each session, the sources and the destination are randomly selected. Each simulation run lasts for 120 s and each result is obtained from the average of 1000 simulation runs.

Five parameters are tunable in our simulations:

- Number of packets in a session: each source will generate 5000–20,000 packets for transmission.
- Traffic load: during the simulation, 15–30 sessions are generated.
- Mobility: host mobility follows the random way-point model. The pause time is set to 30 s. When moving, a host will move at a speed between 5 and 10 m/s.
- Bandwidth requirement: the bandwidth requirement is 1–4 slots/session.
- Number of hosts: the number of mobile hosts in the MANET is 25–50.

Four performance metrics are used in the simulations:

- Route establishment probability: the total number of successfully established routes over the total number of requests.
- Packet loss rate: $(TP - RP)/TP$, where TP is the total number of packets transmitted by the source and RP is the total number of packets received by the destination.
- Route delay: the time from the source transmits the $RREQ$ packet to the time that 200 data packets have been transmitted by the source.
- Control overhead: The cost to transmit a data byte, which is defined as the total amount of control messages over the total amount of data packets.

For simplicity, the FA protocol requesting 1, 2, and 4 time slots are denoted as FA_QoS1, FA_QoS2, and FA_QoS4, respectively. Our protocol requesting 1, 2, and 4 time slots are denoted as QA_QoS1, QA_QoS2, and QA_QoS4, respectively. We name our protocol as the quantitative approach (QA) protocol, because in our protocol, the route selection and the time slot reservation are all based on the quantitative calculation of the effective bandwidth and time slot's weight.

13.4.1 Impact of Bandwidth Requirement

To observe the impact of bandwidth requirement, we vary the bandwidth requirement between 1–4 slots/session. As Figure 13.8 shows, higher bandwidth requirement do decrease the route establishment probability because more time slots are reserved and thus fewer free time slots are left. When the bandwidth requirement is raised to 4 slots/session, the bandwidth is not enough to satisfy the bandwidth requirement and thus the route establishment probability drops sharply. Our protocol can achieve much

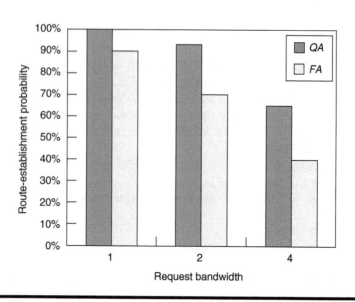

Figure 13.8 Route establishment probability vs. bandwidth requirement (50 hosts, 15 sessions, 5000 packets/session, speed = 5 m/s).

higher route establishment probability than the *FA* protocol does, because our protocol can choose the QoS route that is most likely to satisfy the QoS requirement and reserve the time slot that blocks least neighboring host's free time slots. When the bandwidth requirement does not exceed 2 slots/session, the route establishment probability of our protocol is higher than 90%.

Figure 13.9 shows the impact of the bandwidth requirement to the packet loss rate. As the bandwidth requirement increases the packet loss rate increases. Higher bandwidth requirement may saturate the network and thus causes higher packet loss rate. The packet loss rate of our protocol is much lower than that of the *FA* protocol because our QoS route is more stable and our protocol can achieve higher route establishment probability.

Table 13.1 and Table 13.2 show the impact of the bandwidth requirement to the route delay and control overhead, respectively. As the bandwidth requirement increases, the route delay decreases. With a higher bandwidth, 200 packets can be transmitted in a shorter period of time and thus reduces the route delay. As for the control overhead, the bandwidth requirement has slight impact on it. The route delay of our protocol is slightly higher than that of the *FA* protocol. The control overhead of our protocol is quite close to that of the *FA* protocol, which indicates that our protocol can achieve high performance with low cost.

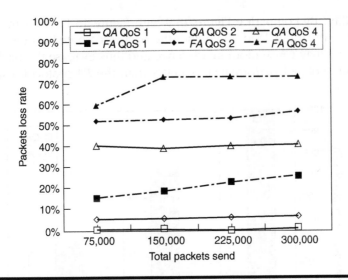

Figure 13.9   Packet loss rate vs. bandwidth requirement (50 hosts, 15 sessions, speed = 5 m/s).

Table 13.1   Route Delay vs. Bandwidth Requirement (50 Hosts, 15 Sessions, 5000 Packets/Session, Speed = 5 m/s)

Protocol	QoS1	QoS2	QoS4
QA	1.031	0.501	0.284
FA	1.029	0.495	0.281

Table 13.2   Control Overhead vs. Bandwidth Requirement (50 Hosts, 15 Sessions, 5000 Packets/Session, Speed = 5 m/s)

Protocol	QoS1	QoS2	QoS4
QA	0.007	0.008	0.007
FA	0.009	0.007	0.007

13.4.2   Impact of Traffic Load

Figure 13.10 and Figure 13.11 show the route establishment probability and packet loss rate in the high traffic load environment. By comparing Figure 13.8 with Figure 13.10, we can see that as the number of sessions increases, the route establishment probability decreases. Higher traffic load will occupy more time slots and thus decreases the route establishment probability. By comparing Figure 13.9 with Figure 13.11, we can see that the higher the traffic load, the higher the packet loss rate becomes. Higher traffic load may cause more interference and thus increase the packet loss rate. Our protocol performance is better than the *FA* protocol, especially

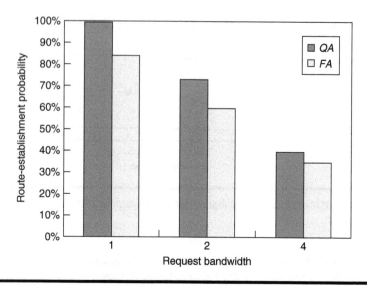

Figure 13.10   Route-establishment probability vs. traffic load (50 hosts, 30 sessions, 5000 packets/session, speed = 5 m/s).

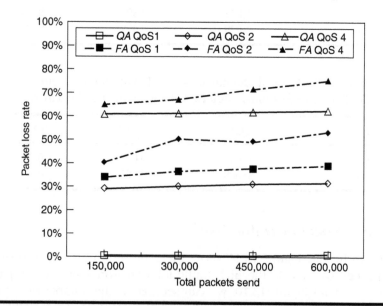

Figure 13.11   Packet loss rate vs. traffic load (50 hosts, 30 sessions, speed = 5 m/s).

when the traffic load is high, because we provide a clear guide to reserve the proper time slots and thus leave more free time slots for other requests.

13.4.3 Impact of Host Density

Figure 13.12 and Figure 13.13 show the route establishment probability and packet loss rate in the low host density environment. By compare Figure 13.8 with Figure 13.12 and Figure 13.9 with Figure 13.13, we can see that as the number of hosts increases, there will be more optional route. This raises the route establishment probability and decreases the packet loss rate. Our protocol performs better than the *FA* protocol even when the host density is low. Our bandwidth-reservation strategy will reserve the time slot that causes least interference to the neighboring hosts. Therefore, we can keep more routes free and thus perform much better than the *FA* protocol. The *FA* protocol calculates the effective bandwidth only according to previous two links' free time slots, and randomly reserves effective time slots. Therefore, packets may be collided by hidden terminals. The higher the host density, the more packets may collide by hidden terminals.

13.4.4 Impact of Mobility

Figure 13.8 and Figure 13.14 show the route establishment probability in the low- and high-mobility environments, respectively. As the

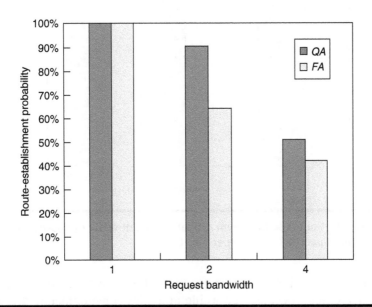

Figure 13.12 **Route-establishment probability vs. host density (25 hosts, 15 sessions, 5000 packets/session, speed = 5 m/s).**

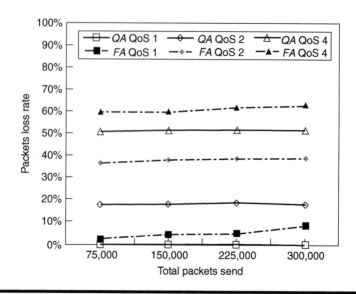

Figure 13.13   Packet loss rate vs. host density (25 hosts, 15 sessions, speed = 5 m/s).

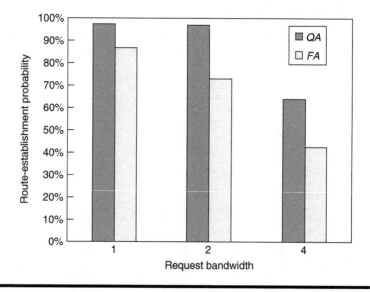

Figure 13.14   Route-establishment probability vs. mobility (50 hosts, 15 sessions, 5000 packets/session, speed = 10 m/s).

mobility becomes higher the route establishment probability of our protocol decreases, because in a high-mobility environment, the route is more likely to break before the route reply packet is sent to the source.

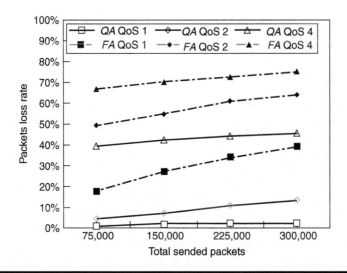

Figure 13.15   Packet loss rate vs. mobility (50 hosts, 15 sessions, speed = 10 m/s).

The packet loss rates in the low- and high-mobility environments are shown in Figure 13.9 and Figure 13.15, respectively. The higher the mobility, the higher the packet loss rate of our protocol becomes, because the link of a QoS route in a high-mobility environment is more likely to break and thus raises the packet loss rate. Our protocol performs much better than the *FA* protocol in terms of route establishment probability and packet loss rate irrespective of whether the mobility is low or high.

13.5   Conclusions

In this chapter, we have reviewed the QoS routing protocols for MANETs and proposed a novel QoS routing protocol for TDMA-based multihop mobile ad hoc networks. We have pointed out two important issues, the effective bandwidth calculation problem and the bandwidth-reservation problem. As far as we know, the two problems have not been addressed seriously in the literature. Our QoS routing protocol contains two phases: the route-discovery phase and the bandwidth-reservation phase. In the route-discovery phase, we adopt the AODV or DSR-like on-demand routing protocol to discover the feasible route and the min–max approach to choose the route that is most likely to satisfy the QoS requirement. In the bandwidth-reservation phase, we proposed an algorithm to evaluate the weight of each available time slot according to how many neighboring hosts' free time slots are blocked by this time slot. The time slot with the lowest weight will be reserved and thus leaves more free time slots for other requests. Simulation results show that the proposed protocol can achieve high route establishment probability and low packet loss rate with a reasonable cost.

References

1. A.A. Bertossi and M.A. Bonuccelli. Code Assignment for hidden terminal interference avoidance in multihop radio networks. *IEEE/ACM Transactions on Networks*, 3:441–449, Aug. 1995.
2. S. Chakrabarti and A. Mishra. QoS issues in ad hoc wireless networks. *IEEE Communications Magazine*, 142–148, Feb. 2001.
3. S. Chen and K. Nahrstedt. Distributed quality-of-service in ad-hoc networks. *IEEE Journal on Select Areas in Communications*, 17:1488–1505, Aug. 1999.
4. X. Chen, W. Liu, Y.-G. Fang, and M.C. Yuang. Supporting QoS with location aware prereservation in mobile ad hoc networks. *Proceedings of IEEE International Conference on Communications*, 6:20–24, Jun. 2004.
5. P.-H. Chuang, H.-K. Wu, and M.-K. Liao. Dynamic QoS allocation for multimedia ad hoc wireless networks. *Proceedings of International Conference on Computer Communications and Networks*, pp. 480–485, Oct. 1999.
6. S. De, S.K. Das, H.-Y. Wu, and C.-M. Qiao. Trigger-based distributed QoS routing in mobile ad hoc networks. *ACM SIGMOBILE Mobile Computing and Communications Review*, 6:22–35, 2002.
7. A. Ebner, H. Rohling, M. Lott, and R. Halfmann. Decentralized slot synchronization in highly dynamic ad hoc networks. *Proceedings of the 5th International Symposium on Wireless Personal Multimedia Communications*, 2:494–498, Oct. 2002.
8. E.S. Elmallah, H.S. Hassanein, and H.M. AboElFotoh. Supporting QoS routing in mobile ad hoc networks using probabilistic locality and load balancing. *Proceedings of IEEE Global Telecommunications Conference*, 5:2901–2906, Nov. 2001.
9. J.J. Garcia-Luna-Aceves and J. Raju. Distributed assignment of codes for multihop packet-radio networks. *Proceedings of IEEE Military Communications Conference*, pp. 450–454, Nov. 1997.
10. M. Gerla and J.T.-C. Tsai. Multicluster, mobile, multimedia radio network. *Wireless Networks*, 1:255–265, 1995.
11. Y.-C. Hu and D.B. Johnson. Securing quality-of-service route discovery in on-demand routing for ad hoc networks. *Proceedings of the 2nd ACM workshop on Security of ad hoc and sensor networks*, 2004.
12. C. Huang, F. Dai, and J. Wu. On-demand location-aided QoS routing in ad hoc networks. *Proceedings of International Conference on Parallel Processing*, 1:502–509, Aug. 2004.
13. Y.-K. Hwang and P. Varshney. An adaptive QoS routing protocol with dispersity for ad-hoc networks. *Proceedings of the 36th Annual Hawaii International Conference on System Sciences*, pp. 302–311, Jan. 2003.
14. D.-W. Kim, C.-H. Min, and S.-H. Kim. On-demand SIR and bandwidth-guaranteed routing with transmit power assignment in ad hoc mobile networks. *IEEE Transactions on Vehicular Technology*, 53:1215–1223, Jul. 2004.
15. J.E. Kleider and S. Gifford. Synchronization for broadband OFDM mobile ad hoc networking: simulation and implementation. *Proceedings of IEEE*

International Conference on Acoustics, Speech, and Signal Processing, 4:3756–3759, May 2002.

16. W.H. Liao, Y.C. Tseng, and K.P. Shih. A TDMA-base bandwidth reservation protocol for QoS routing in a wireless mobile ad hoc network. *Proceedings of IEEE International Conference on Communications*, 5:3186–3190, 2002.

17. C.R. Lin. On-demand QoS routing in multihop mobile networks. *Proceedings of the Joint Conference of the IEEE Computer and Communications Societies (INFOCOM)*, 3:1735–1744, 2001.

18. C.R. Lin and J.-S. Liu. QoS routing in ad hoc wireless networks. *IEEE Journal on Selected Areas in Communications*, 17:1426–1438, Aug. 1999.

19. G.V.S. Raju, G. Hernandez, and Q. Zou. Quality of service routing in ad hoc networks. *Proceedings of IEEE Wireless Communications and Networking Conference*, 1:263–265, Sep. 2000.

20. E. M. Royer and C.-K. Toh. A review of current routing protocols for ad hoc mobile wireless networks. *IEEE Personal Communications*, 6:46–55, Apr. 1999.

21. S.H. Shah and K. Nahrstedt. Predictive location-based QoS routing in mobile ad hoc networks. *Proceedings of IEEE International Conference on Communications*, 2:1022–1027, 2002.

22. W.-J. Shao, V.O.K. Li, and K.-S. Chan. A distributed bandwidth reservation algorithm for QoS routing in TDMA-based mobile ad hoc networks. *Workshop on High Performance Switching and Routing*, pp. 317–321, May 2005.

23. S.-T. Sheu, T.-F. Sheu, C.-C. Wu, and J.-Y. Luo. Design and implementation of a reservation-based MAC protocol for voice/data over IEEE 802.11 ad-hoc wireless networks. *Proceedings of IEEE International Conference on Communications*, 6:1935–1939, Jun. 2001.

24. R. Sivakumar, P. Sinha, and V. Bharghavan. CEDAR: a core-extraction distributed ad hoc routing algorithm. *IEEE Journal on Selected Areas in Communications*, 17:1454–1465, Aug. 1999.

25. H.X. Sun and H.D. Hughes. Adaptive QoS routing based on prediction of local performance in ad hoc networks. *Proceedings of IEEE Wireless Communications and Networking Conference*, 2:1191–1195, Mar. 2002.

26. Y. Xiao and Y. Pan. Differentiation, QoS guarantee, and optimization for real-time traffic over one-hop ad hoc networks. *IEEE Transactions on Parallel and Distributed Systems*, 16:538–549, Jun. 2005.

27. Y.-L. Yang and R. Kravets. Contention-aware admission control for ad hoc networks. *IEEE Transactions on Mobile Computing*, 4:363–377, Jul.–Aug. 2005.

28. B.-X. Zhang and H.T. Mouftah. QoS routing for wireless ad hoc networks: problems, algorithms, and protocols. *IEEE Communications Magazine*, 43:110–117, Oct. 2005.

29. X. Zhang, S. Cheng, M.-Y. Feng, and W. Ding. Fuzzy logic QoS dynamic source routing for mobile ad hoc networks. *Proceedings of International Conference on Computer and Information Technology*, pp. 652–657, Sep. 2004.

30. C. Zhu and M.S. Corson. QoS routing for mobile ad hoc networks. *Proceedings of the Joint Conference of the IEEE Computer and Communications Societies (INFOCOM)*, 2:958–967, Jun. 2002.

Chapter 14

Energy Conservation Protocols for Wireless Ad Hoc Networks

Shu-Min Chen, Sheng-Po Kuo, and Yu-Chee Tseng

14.1 Introduction

Energy conservation is a critical issue in mobile ad hoc networks (MANETs) since mobile nodes usually operate on batteries. A lot of research works have been dedicated to this issue. In this chapter, we organize solutions on this topic in three parts: *power management, power control*, and *topology control protocols*. Power management is to decide when a mobile node can go to sleep to save energy. Power control is to decide a mobile node's transmission power to reduce interference and to save energy. The main purpose of topology control is to adjust a network's connectivity.

Section 14.2 describes the basic power-saving mechanism of the IEEE 802.11 standard. To solve the clock asynchronism problem, *asynchronous* and *semiasynchronous* protocols are then presented. Section 14.3 reviews some *power control* protocols in the medium access control (MAC) layer. Section 14.4 presents two types of topology control protocols to achieve power saving: *topology control by power management* and *topology control by power control*. Finally, Section 14.5 draws some conclusions.

14.2 Power Management

Generally speaking, power management in ad hoc networks means that a station can decide when to switch its radio transceiver off to save energy.

However, turning its transceiver off means that other stations cannot send their packets to the station, if any. Therefore, it should be done with caution to avoid possible delays. The power management issue is challenging especially in a multihop ad hoc network because hosts may sometimes lose synchronization.

In this section, the relationship of synchronization and power management will be investigated. First, we will describe the power-saving mechanism in the IEEE 802.11 standard. IEEE 802.11 actually relies on clock synchronization to achieve power management. For a multihop mobile ad hoc network, clock synchronization is sometimes difficult to achieve. To solve this problem, two advanced, *asynchronous* and *semiasynchronous*, techniques will be presented.

14.2.1 The Power-Saving Mechanism of IEEE 802.11 Ad Hoc Networks

In the IEEE 802.11 ad hoc mode, stations cooperate to support the power-saving mechanism since there is no infrastructure. Active stations will buffer packets for those stations in the sleep mode and try to notify them for data transmission. Sleeping stations will also wake up periodically to listen to the possible notification messages. The notification messages are called *ad hoc traffic indication messages* (*ATIM*). Stations can exchange ATIMs only during the ATIM window, a time window with a fixed length. During ATIM windows, all stations should be active to listen to possible ATIMs. Stations which have buffered packets for other stations compete to send ATIMs to notify the target stations not to enter the sleep mode. Those stations that receive the ATIMs will reply an acknowledgment (ACK) packet to the sender and keep active in the rest of the beacon interval. An example is shown in Figure 14.1. In the first beacon interval, both stations *A* and *B* can go in to the sleep mode since they do not receive any ATIM frame. In

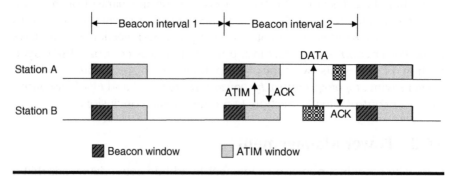

Figure 14.1   Power management in IEEE 802.11 ad hoc networks.

the second beacon interval, however, station A has to stay active after the ATIM window since it received an ATIM frame from station B. Then A and B can exchange DATA and ACK frames after the ATIM window.

Clock synchronization is essential for power management in an IEEE 802.11 ad hoc network because each station determines the beginning and the end of an ATIM window by maintaining a timer. If stations cannot synchronize their clocks, the power-saving mechanism may not work properly. The IEEE 802.11 standard specifies a distributed *timing synchronization function (TSF)* to fulfill clock synchronization in an *independent basic service set (IBSS)*. IBSS is a fully connected ad hoc network in which all stations are within the transmission range of each other. Each station maintains a TSF timer with modulus 2^{64} counting in increments of microseconds (μs). The value of the TSF timer is the summation of a variable *offset* and the station's clock. Clock synchronization is achieved by periodically exchanging timing information through *beacon frames*. A beacon frame contains a timestamp declaring when the beacon was sent. After a station receives a beacon and finds that its own TSF timer is slower than the timestamp specified in the beacon, it will add the timing difference to its offset. This implies that clocks only move forward, but never backward.[1]

At the beginning of each beacon interval, all stations contend with each other to send a beacon frame. First of all, they will wait a random delay. If a beacon arrives before its random delay timer expires, the station will give up its beacon transmission during this beacon interval. Otherwise, when its random delay timer expires, the station transmits a beacon with a timestamp which is equal to the value of the TSF timer at the time the first bit of the timestamp was transmitted to the physical layer plus the transmitting station's delay through its physical layer from the MAC-PHY interface to its interface with the wireless medium. Upon receiving a beacon, the station will adjust the timestamp by adding an amount equal to the delay through the physical layer, and then set its TSF timer to the adjusted timestamp if the value of the adjusted timestamp is larger than its TSF timer. Therefore, if faster stations fail in contending to transmit their beacon frames for some time, they may go out of synchronization.

Analysis and simulation in Ref. [1] shows that the synchronization problem gets more serious especially when the size of the IBSS becomes larger. This means that the IEEE 802.11 TSF does not scale very well. In addition, because stations can only set their timers forward, the station with the fastest clock may suffer from asynchronism with a high probability if it fails to transmit beacons for too many beacon intervals. To alleviate these asynchronism problems, an *adaptive timing synchronization procedure (ATSP)* is proposed Ref. [1]. The main idea of ATSP is to give faster stations a higher priority to send beacons. Based on this idea, each station i is assigned a parameter $I(i)$ to determine how often it should participate

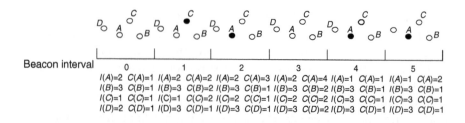

Figure 14.2 An example of the ATSP algorithm. The black node (if any) in each beacon interval is the node which wins the beacon transmission in that interval.

in beacon contention. Station i contends for beacon transmission every $I(i)$ beacon intervals. Let I_{max} be the maximum value of $I(i)$ and $C(i)$ be the counter in station i that counts the number of beacon intervals. Initially, $I(i)$ is randomly generated between 1 and I_{max} and $C(i) = 1$. In each beacon interval, station i participates in beacon contention iff $C(i)$ mod $I(i) = 0$. When station i receives a beacon with a larger timer than its own one, its timer will be set to the received timer value. Then, its priority will be decreased through increasing $I(i)$ by 1 if $I(i)$ is less than I_{max}, and its counter $C(i)$ will be reset to 0. On the other hand, if station i does not receive any larger timer than its own for I_{max} consecutive beacon intervals, it decreases $I(i)$ by 1 if $I(i)$ is larger than 1 and sets $C(i)$ to 0. At the end of each beacon interval, each station increases its $C(i)$ by 1.

Consider the example in Figure 14.2, where $I_{max} = 3$. The order of clock speed is $A > B > C > D$. Initially, the value of the timer is $A > B > C > D$, and $I(A) = 2, I(B) = 3$, $I(C) = 1$, and $I(D) = 2$. At the beacon interval 1, since $C(C)$ mod $I(C) = 0$, only C participates in beacon contention, and thereby sends its beacon. This causes D to increase its $I(D)$ by 1 and set $C(D)$ to 0. At the end of this beacon interval, every station increases its own counter $C(i)$ by 1. At the beacon interval 2, both A and C participate in beacon contention. In this example, A sends its beacon first, resulting in $I(C) + 1$, and $C(B) = C(C) = C(D) = 0$. Note that $I(B)$ and $I(D)$ cannot be increased since they are equal to I_{max}. At the beacon interval 3, according to the rules of ATSP, no station contends to send their beacons. At the beacon interval 4, note that since $I(A)$ has remained an unchanged value for $3(= I_{max})$ continuous beacon intervals, A will decrease $I(A)$ by 1, and sets $C(A)$ to 0. Finally, the fastest station A will participate in the beacon contention in every beacon interval.

With ATSP, the fastest station will have a very high probability of successfully sending its beacons, thereby synchronizing all other stations. Compared to the IEEE 802.11 TSF, ATSP provides a simple but effective solution to improve the clock synchronization mechanism.

14.2.2 Asynchronous Power-Saving Protocols

The clock asynchronism problem in IEEE 802.11 TSF may worsen in a multihop MANET since packet delays may vary due to unpredictable mobility and radio interference. Although ATSP has been proposed in Ref. [1] to alleviate the clock asynchronism problem, it is designed for single-hop ad hoc networks. Power management protocols that can function well in a MANET without the support of clock synchronization are needed. We call such protocols *asynchronous* power-saving protocols. This means that stations do not need to synchronize their timers with each other. However, stations should be able to discover their neighbors without problems even if they go to sleep sometimes. Note that without precise clock information, a station may not be aware of its neighbors because they may potentially miss each other's awake period. Incomplete and incorrect neighbor information may be harmful to most current routing protocols. For example, the route discovery procedure may incorrectly report the non-existence of routes to certain destination stations even when routes actually exist. This problem was first reported in Ref. [2].

Three asynchronous protocols are proposed, namely *dominating-awake-interval, periodically-fully-awake-interval,* and *quorum-based* protocols in Ref. [2]. Before explaining these protocols, we first point out several guidelines used in the design. First, multiple beacons are allowed in an ATIM window. To prevent the inaccurate-neighbor problem, a station should insist more on sending beacons. Specifically, a station should not inhibit its beacon in an ATIM window even if it has heard others' beacons. This will allow others to be aware of its existence. Second, when a station hears another station's beacon, it should be able to derive that station's wake-up pattern based on their time difference. This will allow the former to send buffered packets to the latter in the future. Note that such prediction is not equal to clock synchronization since stations do not try to adjust their clocks. Last, like the IEEE 802.11 standard, time is divided into fixed-length beacon intervals. In each beacon interval, there are three windows, which are called *active window, beacon window,* and *multihop traffic indication message (MTIM) window.* During the active window, the station should turn on its receiver to listen to any packet and take proper actions as usual. The beacon window is for the station to send its beacon, while the MTIM window is for other stations to send their MTIM frames to the station. MTIM frames serve the similar purpose as ATIM frames in IEEE 802.11 (we use MTIM to emphasize that it is for multihop MANETs). Excluding these three windows, a station with no packet to send or receive may go to the sleep mode. The structure of beacon interval may vary for different protocols. The following notations are used in our discussion: *BI*, length of a beacon interval; *AW*, length of an active window; *BW*, length of a beacon window; and *MW*, length of an MTIM window.

Since these asynchronous power-saving protocols do not count on clock synchronization, the wake-up patterns of any two stations must overlap with each other at a certain point of time, no matter how their clocks drift away. These protocols differ mainly on their wake-up patterns. The basic idea of the *dominating-awake-interval* protocol is to order a station to stay awake long enough to ensure that neighboring stations can hear from each other and, if desired, deliver buffered packets. The term "dominating-awake" implies that a station should stay awake for at least half of a *BI* in each beacon interval. This guarantees that a station's beacon window always overlaps with any neighboring station's active window, and vice versa. The protocol is formally derived as follows. To satisfy the "dominating-awake" property, the inequality $AI \geq BI/2 + BW$ must be satisfied. This guarantees that any two stations' active windows always have some overlapping. However, this does not guarantee that a station's beacon window overlaps with any neighboring station's active window. Let us consider Figure 14.3, where beacon windows always appear at the beginning of beacon intervals. In this case, station B can hear station A's beacons, but A always misses B's beacons. To solve this problem, beacon intervals are alternatively labeled as *odd* and *even* intervals. Odd and even intervals have different structures. Each odd beacon interval starts with an active window. The active window is led by a beacon window followed by an MTIM window. Each even beacon interval also starts with an active window, but the active window is terminated by an MTIM window followed by a beacon window (see the illusion in Figure 14.4). With this design, a station is able to receive its neighbors' beacon frames in every two beacon intervals, if there is no collision in receiving the latter's beacons.

It is clear that the dominating-awake-interval protocol is not very energy-efficient since stations have to keep active at least half of the time. To reduce on-duty time, in the *periodically-fully-awake-interval* protocol, two types of beacon intervals are designed: *low-power* intervals and *fully awake* intervals. Each low-power interval starts with an active window, which

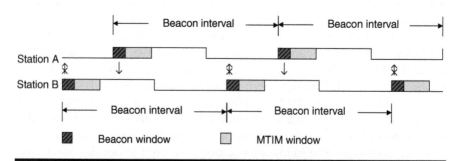

Figure 14.3   An example of the dominating-awake-interval protocol in which beacon windows are arranged at the beginning of beacon intervals.

contains a beacon window followed by an MTIM window, such that $AW = BW + MW$. In the rest of the time, the station can go to the sleep mode. Each fully awake interval also starts with a beacon window followed by an MTIM window. However, the station must remain awake in the rest of the time, i.e., $AW = BI$. A fully awake interval arrives periodically every N beacon intervals, and the rest of the beacon intervals are low-power intervals. Figure 14.5 shows an example with $N = 4$ intervals. Intuitively, the low-power intervals are for a station to send out its beacons to inform others of its existence. The fully awake intervals are for a station to discover who are in its neighborhood. It is not hard to see that a station with a fully awake interval always has an overlap with any station's beacon windows, no matter how their clocks drift away. Compared to the dominating-awake-interval protocol, which requires a station to stay awake more than half of the time, this protocol can save more energy as long as $N > 2$. However, the response time to get aware of a newly appearing station could be as long as N beacon intervals.

In the previous two protocols, a station has to contend to send a beacon in each beacon interval. To alleviate the contention, the quorum-based protocol is proposed. A *quorum* is a set of entities from which one

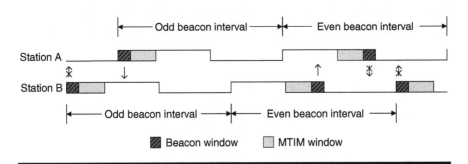

Figure 14.4 An example of the dominating-awake-interval protocol in which beacon windows are alternated between odd and even beacon intervals.

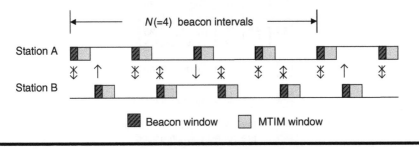

Figure 14.5 An example of the periodically-fully-awake-interval protocol with $N = 4$.

has to obtain permission to perform some critical action. Typically, two quorum sets always have nonempty intersection so as to guarantee the atomicity of a transaction. This is a classical problem in the design of distributed operating systems.[3,4] Here, the concept of quorum is adopted to design stations' wake-up patterns so as to guarantee that a station's beacons can always be heard by other stations' active windows. The quorum structure used in this protocol is as follows: Beacon intervals are divided into groups such that each continuous m^2 beacon intervals are called a *group*, where m is a global parameter. In each group, the m^2 intervals are arranged as a two-dimensional $m \times m$ array in a row-major manner. On the $m \times m$ array, a station can arbitrarily pick one column and one row of entries as its quorum members. These $2m - 1$ intervals are called *quorum intervals*. The remaining $m^2 - 2m + 1$ intervals are called *nonquorum intervals*. An example of $m = 6$ is in Figure 14.6. In this example, station A picks the second column and the fourth row as its quorum intervals, while station B picks the fourth column and the sixth row as its quorum intervals. When perfectly synchronized, intervals 21 and 31 are their intersections.

The structures of quorum and nonquorum intervals are formally defined as follows: Each quorum interval starts with a beacon window followed by an MTIM window. After that, the station must remain awake for the rest of the interval, i.e., $AW = BI$. Each nonquorum interval starts with an MTIM window. After that, the station can enter the sleep mode, i.e., $AW = MW$, if

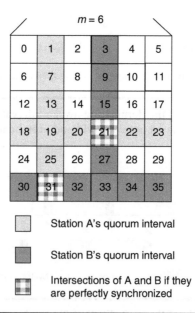

Figure 14.6 An example of the quorum-based protocol where $m = 6$.

there is no expected traffic. With such a design, the quorum-based protocol guarantees that a station always has at least two entire beacon windows that are fully covered by any station's active windows in every m^2 beacon intervals no matter how much their clocks drift away, as proved in Ref. [2]. The quorum-based protocol has an advantage in that it only contends to transmit beacons in $O\left(\frac{1}{m}\right)$ of the beacon intervals (on the contrary, the earlier two protocols have to contend to transmit a beacon in every interval). However, it also has the problem of some delay in perceiving a newly approaching neighbor.

Asynchronous protocols can guarantee network connectivity in highly dynamic networks since they trade power consumption for network connectivity. However, is it possible to find the minimum percentage of time that a station has to stay active given a desirable delay for neighbor discovery? Ref. [5] formulates this problem and proposes a wake-up mechanism. The problem statement is as follows. A multihop ad hoc network can be modeled by a directed graph $G(V,E)$, where V is the set of network stations and E the set of communication links. If a station v_i is within the transmission range of station v_j, then an edge (v_j, v_i) is in E. Bidirectional links are assumed. Therefore, if $(v_j, v_i) \in E$, then $(v_i, v_j) \in E$. $N(v)$ denotes the set of neighbors of a station v. The main purpose of the asynchronous wake-up mechanism is to maintain network connectivity despite the power modes which the stations may be in. Here the definition of connectivity is loose. Connectivity here implies that all stations are reachable from a station within a finite amount of time. In the proposed wake-up mechanism, each station follows its own wake-up schedule of length T, designated as the *wakeup schedule function* (WSF). The WSF of station v is represented by a polynomial of order $T-1$ as $f_v(x) = \sum_{i=0}^{T-1} a_i x^i$, where $a_i = 0$ or 1, and x is a place holder. If $a_i = 1$, the station should be active in slot i; otherwise, it can go to sleep in slot i. Therefore, $f_v(1)$ is the total number of slots in which station v is scheduled to be awake in every T slots. Thus, given a fixed value of T, the goal of power saving is to make $f_v(1)/T$ as small as possible subject to the constraint that the schedules of any two stations overlap by m slots. In the symmetric WSF design, let k be the desired value of $f_v(1)$ for all $v \in V$. Given T and m, a theoretical bound $k \geq \sqrt{mT}$ for the symmetric WSF design problem is derived in Ref. [5]. This lower bound can be achieved through a constructive method which maps the symmetric WSF design to the *symmetric block design problem* in combinatorics. (Further explanation about this method can be found in Ref. [5].)

14.2.3 Semiasynchronous Power-Saving Protocols

Although asynchronous protocols do not rely on clock synchronization, the energy cost is still relatively high when compared with synchronous protocols because stations need to keep awake for a longer time to discover neighbors. To conquer the deficiency of asynchronous protocols, Ref. [6] proposes several cluster-based semiasynchronous power-saving protocols for multihop MANETs. The basic idea is to cluster neighboring stations such that synchronous power-saving protocols can be adopted within individual clusters, and asynchronous power-saving protocols are only used at the boundaries of different clusters.

These semiasynchronous protocols work as follows: First, stations are clustered. The station with sufficient energy and the fastest clock among its direct neighbors will be chosen as the cluster head. Since there is only one hop from the cluster head to cluster members, it will be much easier for the cluster head to synchronize all member stations in the cluster. When all stations in the same cluster are synchronized, each station can simply adopt a synchronous power-saving protocol (e.g., the IEEE 802.11's power-saving mechanism) as its intracluster protocol. With clustering, the whole network is partitioned into several clusters whose clocks are not necessarily synchronized; so these clusters may have different wake-up patterns, and thus may be out of synchronization. To solve this problem, each cluster should also run an asynchronous power-saving protocol to detect its neighbors. Since each cluster may consist of several stations, it is not necessary for every cluster member to run the asynchronous protocol. Ref. [6] proposes that within each cluster, some stations should be delegated as *watchers* to run an asynchronous protocol for neighbor discovery. Members may also serve as watchers in turn. Therefore, the key challenge of the semiasynchronous protocol is how to delegate stations as watchers. Two schemes are proposed: *SNR-probability-based* and *location-based* schemes.

Before we explain these two schemes, we present the channel model used in Ref. [6]. To maintain the cluster structure and synchronize the stations in each individual cluster, the structure of beacon intervals is redesigned. Each beacon interval may contain four windows, namely *active window, synchronous window, beacon window,* and *MTIM window*. During the active window, stations have to be active to listen to frames and take proper actions. The synchronous window is for a cluster head to send its beacon, while the beacon window is for the other stations to send their beacons. The MTIM window is for stations to send their MTIM frames. To avoid collisions among beacons/MTIMs, a station should wait for a random number of slots before sending out its beacons/MTIMs.

Most stations will run the synchronous protocol. The structure of *synchronous intervals* is shown in Figure 14.7(a), where each interval starts with a synchronous window followed by an MTIM window. When serving as a watcher, the station will run an asynchronous power-saving protocol, such as the periodically-fully-awake-interval or the quorum-based protocol.[2] Suppose that the periodically-fully-awake-interval protocol is adopted. The beacon intervals for watchers are classified as *fully awake*

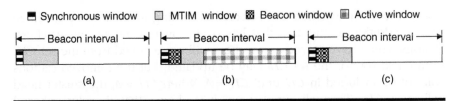

Figure 14.7 Three types of beacon intervals in the semiasynchronous power-saving protocol. (a) Synchronous interval. (b) Fully-awake interval. (c) Low-power interval.

(see Figure 14.7(b)) and *low-power* intervals (see Figure 14.7(c)). The fully awake intervals arrive periodically every p intervals and the rest of the intervals are low-power intervals.

In the semiasynchronous protocol, stations will decide their roles from round to round. Each round consists of q beacon intervals, where q is a multiple of p (the period of the adopted periodically-fully-awake-interval protocol). If a station is designated as a watcher, it will adopt the asynchronous protocol in this round; otherwise, it will adopt the synchronous protocol in this round. To detect the change in neighboring clusters, each cluster will have some stations to serve as watchers in each round. The formation of clusters has been studied extensively in many works, so we omit the details.

The SNR-probability-based scheme is designed for loosely coupled clustering (in such clustering, a cluster head does not try to maintain the structure of its own cluster). There is a probability associated with each cluster member, whose value is set according to the number of detected cluster members, the *signal noise ratio* (*SNR*) of the beacon frame sent by the cluster head, and the mobility of the station. The SNR is to estimate the station's distance to the cluster head. The greater the SNR the shorter the distance becomes. The station that has a longer distance to the cluster head will have a higher probability to serve as a watcher, since it is probably closer to the border of the cluster. For example, if the number of detected cluster members is n, the moving speed factor of the station is k, the estimated distance between the station and the cluster head is d, and the communication range is r, then the probability of the station to serve as a watcher can be set to $(d + r/2)k/nr$.

The *location-based* scheme is designed for tightly coupled clustering (in such clustering, the cluster head needs to record the physical locations of all its cluster members, which may be obtained via GPS receivers). The cluster head will dispatch k watchers in each round according to a greedy algorithm. Assume that the set of neighbors of the cluster head is CH. The cluster head first dispatches the station which is farthest from the cluster head as a watcher. Then the station is removed from CH. The rest of the $k-1$ watchers are selected according to their extra coverage area in addition to the area that has been covered by current watchers. Stations which have served as watchers will be removed from the set CH. In the next round, the set CH may be updated, but the removed stations will not be included in CH until $CH = \emptyset$. When $CH = \emptyset$, the cluster head will reset CH. Through simulations, it is shown that the cluster-based semiasynchronous power-saving protocols outperform the asynchronous power-saving protocols when applied to a multihop MANET.

14.3 Power Control

Varying transmit power level is another technique for reducing energy consumption. In addition, power control can also reduce interference and improve spatial reuse of wireless channels. In this section, we introduce how to incorporate the concept of power control in the MAC layer.

14.3.1 BASIC Protocol

Many works try to integrate power control into the IEEE 802.11 MAC protocol. The main idea is to use different power levels to transmit request to send (RTS), clear to send (CTS), DATA, and ACK frames. Among those schemes, a simple scheme[7] (referred to as the *BASIC* protocol here) shows that stations should use the maximum power to transmit RTS and CTS frames, and the minimum required power to transmit DATA and ACK frames. The RTS–CTS handshake is used to decide the minimum required power for the subsequent DATA–ACK transmissions. The detailed description is as follows. Let p_{max} be the maximum transmit power. Suppose that station v wants to send a data packet to station u. Station v should use a power level of p_{max} to send its RTS frame. When station u receives this RTS frame, it also replies a CTS frame at power level p_{max}. When station v receives the CTS frame, it calculates the minimum required power level, $p_{desired} = \frac{p_{max}}{p_r} \times p_{rmin} \times c$, based on the received power level p_r of the CTS frame, where p_{rmin} is the minimum necessary received signal strength and c is a constant. Then station v uses power level $p_{desired}$ to transmit its DATA frame. Similarly, station u calculates its power level $p_{desired}$ to transmit its ACK frame.

14.3.2 Power Control with Periodical Pulses

Ref. [8] points out that the BASIC protocol may degrade network throughput and even cause higher energy consumption. Before explaining the reasons, there are three terms to be defined: *transmission range, carrier sensing range*, and *carrier sensing zone*. When station v is within the transmission range of another station u, station v can receive and correctly decode frames from station u, while when station v is within the carrier sensing range of station u, station v can sense but not necessarily correctly decode station u's transmission. Usually, the carrier sensing range is larger than the transmission range (a typical assumption is that the radius of the former is twice larger than that of the latter). Note that the transmission range and the carrier sensing range also depend on the sender's transmit power level. The carrier sensing zone is defined as the area of the carrier sensing range excluding the transmission range. Thus, when a station is within the carrier sensing zone of a transmitter, it can only sense the signal but cannot decode the transmitted data correctly. These definitions are illustrated in Figure 14.8(a).

In the IEEE 802.11 MAC protocol, each station maintains a *network allocation vector* (*NAV*), which indicates the remaining time of ongoing

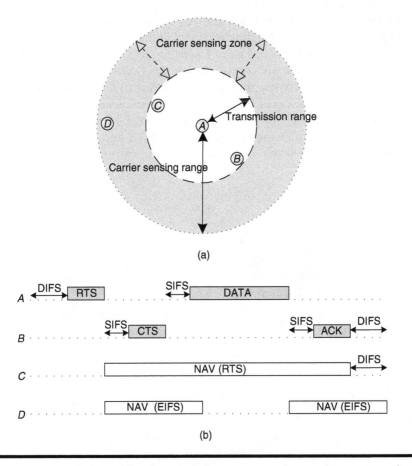

(a)

(b)

Figure 14.8 **(a) An example of transmission range, carrier sensing range, and carrier sensing zone. (b) Setup of NAVs by stations *C* and *D* when *A* and *B* exchange their RTS–CTS–DATA–ACK dialogue.**

transmission sessions. If a station hears an RTS or a CTS frame, it will update its NAV using the duration information specified in the frame. Thus, when source and destination stations transmit RTS, CTS, and DATA frames, stations in the corresponding transmission range can correctly receive these packets and update their NAVs. However, stations in the carrier sensing zones can only sense the signals, and cannot decode them correctly. So these stations only set their NAVs for an extended interframe space (EIFS) duration, whose purpose is to prevent possible collision with the ACK frame at the source side. So EIFS is equal to the sum of one short interframe space (SIFS), one distributed coordination function (DCF) interframe space (DIFS), and the time required to transmit an ACK frame at the lowest physical rate. Figure 14.8(b) shows an example, where A and B are exchanging RTS–CTS–DATA–ACK, C is within A's transmission range, and D is within A's carrier sensing zone.

In the BASIC scheme, since RTS–CTS frames are sent at the power level p_{max} and DATA–ACK frames are sent at the minimum required power level $p_{desired}$, stations in the carrier sensing zone of the RTS–CTS transmission may not sense any signal during the DATA–ACK transmission. If these nodes start a new transmission by sending an RTS at the power level p_{max}, a collision may occur and then trigger retransmissions. Consider the example in Figure 14.9. Suppose that A transmits a packet to B. When A and B reduce their transmission power for DATA and ACK transmissions, respectively, both their transmission ranges and carrier sensing ranges will

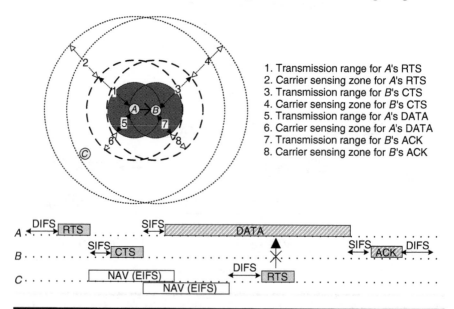

1. Transmission range for A's RTS
2. Carrier sensing zone for A's RTS
3. Transmission range for B's CTS
4. Carrier sensing zone for B's CTS
5. Transmission range for A's DATA
6. Carrier sensing zone for A's DATA
7. Transmission range for B's ACK
8. Carrier sensing zone for B's ACK

Figure 14.9 An example depicting that the power control mechanism in the BASIC protocol may result in collisions.

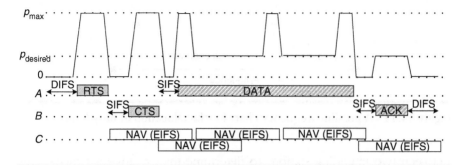

Figure 14.10 An example of power adjustment in Ref. [8] based on the scenario in Figure 14.9. To notify stations in the carrier sensing zone, the transmit power for the DATA frame is increased periodically to the power level p_{max}.

shrink. Therefore, C cannot sense these transmissions and may consider the channel to be idle. Once C starts transmitting an RTS or replying a CTS with the power level p_{max}, a collision occurs to the DATA–ACK transmission between A and B. The above discussion reveals that the BASIC protocol is liable to collisions, thereby degrading the network throughput, resulting in more energy consumption.

Ref. [8] proposes a new power control MAC protocol to prevent the potential collision problem in the BASIC protocol. In this protocol, the RTS and CTS frames are sent at power level p_{max}. Nodes in the carrier sensing zone set their NAVs for an EIFS duration when they sense some signals that cannot be decoded correctly. ACK frames are also sent at the minimum required power level $p_{desired}$. The main difference is that the power level for transmitting DATA frames is periodically increased to the power level p_{max} from the power level $p_{desired}$. That is, the power level to transmit DATA frames is alternated between p_{max} and $p_{desired}$ with a period of one EIFS. Figure 14.10 shows the changes of the transmit power levels during the RTS–CTS–DATA–ACK exchange. With such a modification, other stations that may cause collisions will periodically observe the existence of carriers and postpone their transmission. Since the transmit power for DATA frames is increased every EIFS duration, proper NAVs can be set at other stations. Also, the length of the duration to transmit at the power level p_{max} should be long enough for physical carrier sensing.

14.3.3 Power Control by Busy Tones

Ref. [9] also proposes a MAC protocol allowing nodes to use different transmission power levels to increase channel utilization. This MAC protocol combines the mechanisms of power control, RTS/CTS dialogue, and busy tones. The main idea is to use the exchange of RTS and CTS packets

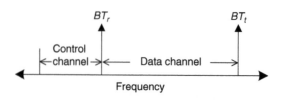

Figure 14.11 A possible frequency chart for the allocation of busy tones.

between two intending stations to determine their relative distances. This information is then utilized to constrain the power level at which a mobile host transmits its data packets. Using lower power levels can increase channel reuse, and thus channel utilization. It also saves the precious battery energy of portable devices and reduces cochannel interferences with other neighbor nodes.

Busy Tones, are introduced to prevent other mobile hosts, unaware of the earlier RTS/CTS dialogues, from destroying the ongoing transmission. The channel is split into two subchannels: a data channel and a control channel. The control channel is for transmitting RTS/CTS dialogues. Also, two narrow-band busy tones, called *transmit busy tone* BT_t and *receive busy tone* BT_r, are placed on the spectrum at different frequencies with enough separation. Figure 14.11 shows a possible spectrum allocation. BT_t indicates that some host is transmitting, while BT_r indicates that some host is receiving. A sending host must turn its BT_t on when transmitting a data packet and a receiving host must turn its BT_r on when replying the sender a CTS. When a host wants to send an RTS, it has to make sure that there is no BT_r around it. Conversely, to reply a CTS, a host must make sure that there is no BT_t around. In summary, a host should not send if it hears any BT_r, and should not consent to send if it hears any BT_t.

The MAC protocol in Ref. [9] intelligently incorporates power control information into RTS/CTS as follows: Consider the example in Figure 14.12(a), where a communication from C to D is ongoing. The communication from A to B cannot be granted because A can hear D's BT_r, and similarly that from E to F cannot be granted because F can hear C's BT_t. However, as the example shown in Figure 14.12(b), if transmitters' power levels are properly selected, all communication pairs can coexist without any interference. First, C uses a minimal power level to transmit its data packet and BT_t, but D keeps its BT_r at the normal (largest) power level. When E wants to communicate with F, E senses no BT_r, so it can send an RTS to F. At this moment, F hears no BT_t, so F can reply a CTS to E. Now if E appropriately adjusts its transmission power, the communication from E to F will not corrupt the ongoing transmission from C to D. Next, suppose that A wants to communicate with B. This case deserves more attention. At this time, A can sense D's BT_r. Based on the signal strength

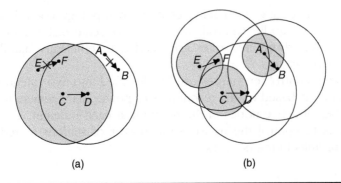

(a)   (b)

Figure 14.12 Transmission scenarios (a) without power control and (b) with power control. The areas of transmitting busy tones are shown in gray and those of receiving busy tones are shown in white.

of D's BT_r, A should send an RTS to invite B with a power level that is sufficiently large to reach B but not D. The basic idea is that A's yet-to-be-transmitted data packet should not corrupt D's reception. Host B, which must be closer to A than D is, will reply with a CTS. This causes no problem as B hears no BT_t. Then the communication from A to B can be started. To summarize, the rules in this protocol are: (i) data packet and BT_t are transmitted with power control, (ii) CTS and BT_r are transmitted at the normal (largest) power level, and (iii) RTS is transmitted at a power level to be determined based on how strong the BT_r tones are around the requesting host. Through simulations, this protocol[9] demonstrates that channel utilization can be significantly increased because more concurrent transmission pairs can be allowed.

14.4 Topology Control Protocols

In a dense wireless ad hoc network, topology control is an effective technique for achieving power saving. The topology of a wireless ad hoc network can be regarded as a graph with its nodes as vertices and communication links between node pairs as edges. Normally, the communication links are defined by assuming that all nodes use their maximum transmission power to send. As a result, the edge set is the largest possible one. However, in a dense ad hoc network, too many communication links are sometimes harmful for energy consumption, network throughput, and quality of service.

The topology of an ad hoc network can be affected by many factors.[10] Some of them are uncontrollable, such as node mobility, weather conditions, environmental interferences, and obstacles. Some of them, however, are controllable, such as transmission power, antenna direction, and duty-cycle scheduling. Generally, centralized topology control protocols can effectively handle controllable factors to achieve a suitable topology; however, distributed ones are needed to efficiently handle those uncontrollable factors.

Topology control techniques can be classified into two categories. The first category is based on the techniques of power management to control the number of active nodes in the network at each moment. The second category is to control the transmission power of stations to reduce the number of links in the network.

14.4.1 Topology Control by Power Management

Putting stations into the sleeping state can save significant energy. Our goal is to adjust the network topology by putting as many nodes without traffic into the sleep mode as possible while guaranteeing global network connectivity. In the following, two distributed topology control protocols which adopt different scheduling strategies are discussed.

14.4.1.1 ASCENT

Adaptive self-configuring sensor networks topologies (ASCENT)[11] is a distributed topology control protocol. In a multihop ad hoc network, to reduce cost, each node can only exchange information with local neighbors. Based on local observations, ASCENT is designed to help a node to find an equilibrium that saves power while preserving connectivity.

ASCENT adaptively selects some nodes to be active and allows the other nodes to remain passive. The active nodes will form a backbone to be responsible for transmitting and forwarding packets. The other passive nodes will operate in a low-duty cycle mode and periodically wake up to check if they should become active to join the backbone.

Each node will stay in one of the following states: *active* state, *test* state, *passive* state, or *sleep* state, as shown in Figure 14.13. Initially, only some nodes are active and the others will passively observe their surrounding status, including neighbor announcement messages and packet loss rate. While it is in the active state, a node will keep awake all the time and perform all necessary routing jobs until it runs out of energy. While it is in the sleep state, a node will turn its transceiver off and periodically wake up to enter the passive state. In the passive state, a node will gather local information to make sure that there are sufficient active neighbors surrounding it and that the current packet-loss rate is below a threshold. Note that this listening process will not cause interference to other nodes. According to the gathered network status, the node may enter the test state or go back to the sleep state. In the test state, a node will continuously conduct a similar check on the number of its active neighbors and the

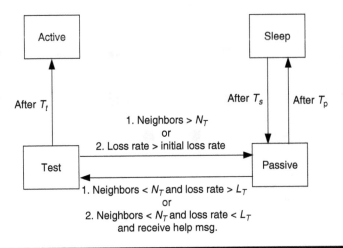

Figure 14.13 **The state transition diagram of the ASCENT protocol. T_s, T_p, and T_t express the timers in the sleep, passive, and test state, respectively. The threshold for the number of active neighbors is N_T and the threshold for the packet-loss rate is L_T.**

packet loss rate until a timer expiration. If it determines that it can improve the network connectivity, it will switch to the active state. Otherwise, it will go back to the passive state.

The example in Figure 14.14 shows how ASCENT works. The initial state is shown in Figure 14.14(a), where some nodes are active and some nodes are sleeping. Suppose that node A wants to transmit packets to node B. Under the current topology, the distance between nodes A and B is close to the limit of their transmission ranges, so node B will experience a high packet-loss rate. In response, node B will send *help* messages to ask surrounding passive neighbors to wake up. In Figure 14.14(b), some nodes in the sleep state wake up after their sleep timer T_s. Let us suppose that before their passive timers T_p expire, some of the passive nodes, say, D and F receive the *help* message from B and find that there are less than N_T active nodes in their surrounding area and that the packet-loss rate is less than L_T (a node can passively observe the packet-loss rate in its surrounding area by checking the sequence number attached in each packet). Then, D and F will transition to the test state. In addition, suppose that node E is far from node B; so it does not receive B's *help* messages. However, E finds that there are less than N_T active nodes in its surrounding area and the packet-loss rate is higher than L_T. So, it also changes to the test state. The passive node C also receives the *help* message. However, there are more than N_T active neighbors in its surrounding area, so it can switch back to the sleep state. Therefore, we see that in Figure 14.14(c), nodes D, E, and F go to in the test state, set their timers T_t, and broadcast neighbor announcement

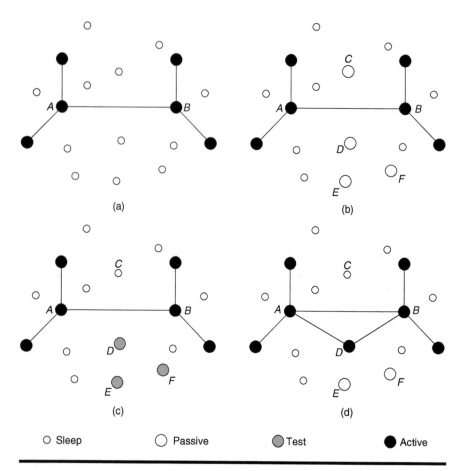

Figure 14.14 An example of the ASCENT protocol.

messages. Assume that the timer of node *D* expires before the timers of nodes *E* and *F*. Hence, node *D* will move to the active state first. Since nodes *E* and *F* observe that there are enough active nodes, they will change back to the passive state. Figure 14.14(d) shows the final topology.

In summary, ASCENT considers both connectivity degree and data loss rate to adjust the network topology. Each node decides whether or not it should join the network based only on local information. For denser network topologies, ASCENT can achieve longer network lifetime and increase the packet success rate.

14.4.1.2 Span

Span is a distributed power-saving protocol for wireless multihop ad hoc networks. Similar to ASCENT, Span allows nodes to go to sleep while maintaining performance and connectivity of the network. Span has the following features. First, it can guarantee network connectivity by ensuring that every node has at least one active node in its radio range. Second, fairness among nodes is based on two factors: the amount of energy remaining with of a node and the number of additional neighbor pairs that a node can connect if it becomes active. Hence, Span can balance both fairness and network lifetime.

In Span, active nodes will form a connected backbone. Each node broadcasts a HELLO message periodically. From received HELLOs, a node then decides whether it should participate in the backbone or not. In each HELLO message, a node should announce its current state (active or inactive), its current neighbors, and the status of its neighbors. From HELLO messages, a node can construct a table containing its neighbors and the neighboring active nodes of each neighbor. We then define an inactive node to be *eligible* if two of its neighbors cannot reach each other directly or through one or two active nodes. An eligible inactive node will contend to be an active node. Through this operation, each node ensures that any two of its neighbors will have a path to reach each other. If this rule is followed by all the nodes, it is guaranteed that active nodes must form a connected backbone and that each inactive node must be adjacent to an active node. This fact can be derived by induction as follows. First, the basis case is that every node i is guaranteed to be able to reach any two-hop neighbor through one active node or two connected active nodes. For the induction hypothesis, assume that there exists a $(k-1)$-hop path $i = n_0 \rightarrow n_1 \rightarrow \cdots \rightarrow n_{k-1} = j$ from each node i to each node j such that $n_1, n_2, \ldots, n_{k-2}$ are active. According this hypothesis, we can induce that there exists an active path $i = n_0 \rightarrow n_1 \rightarrow \cdots \rightarrow n_{k-2} \rightarrow n_{k-1} \rightarrow n_k = j$ because node n_{k-1} can guarantee an active path between node n_{k-2} and node n_k.

To achieve better fairness, eligible nodes will not become active immediately. Each eligible node will determine a backoff delay according to three factors: utility, remaining energy, and a random number. The utility of a node is the proportion of the number of additional neighbor pairs that can be connected if this node becomes active to the number of all neighbor pairs. The node which can provide more additional neighbor pairs should have a shorter backoff timer because it has more contributions to the network connectivity. Besides, to achieve equal energy consumption, the backoff mechanism should concern the remaining energy scaled to the maximum amount of energy available at each node. Nodes with more energies left should have shorter backoff delays. Finally, the random number is to prevent contending nodes from announcing that they have become active at the same time. After its backoff timer, an eligible node will reevaluate its eligibility again, based on the HELLO messages it received recently. If it is not eligible any more, it will go to sleep; otherwise, it will announce that it has become active.

On the other hand, an active node will periodically check if it can switch to the sleep state. Based on the above guideline, an active node may go to sleep if every pair of its neighbors can reach each other directly or through one or two active nodes. After being activated for a period of time, an active node will mark itself as *tentative*. A tentative active node still forwards packets as normal but it is treated as an inactive node by its neighbors. Hence, its nearby nodes will have chances to become active nodes.

In summary, Span is a distributed topology control protocol. It uses the connection relations of nodes' neighbors to guarantee network connectivity. Besides, a backoff mechanism is used to achieve better fairness among nodes.

14.4.2 Topology Control by Power Control

Power control is another way to adjust a network's topology. Instead of transmitting at the maximal power level, each node can choose a lower transmission power. In this section, we present two localized topology control algorithms proposed in Ref. [12]: *directed relative neighborhood graph (DRNG)* and *directed local spanning subgraph (DLSS)*. In both algorithms, each node w can adjust its transmission power to guarantee that there remains a path from any neighboring node u of w to any neighboring node v of w if there is a path from node u to node v in the original network topology (before power control). In both algorithms, it is assumed that each node knows its location. The notation r_u is the maximal transmission range of node u. The ordered pair (u, v) represents the directed edge from node u to node v, and $d(u, v)$ expresses the Euclidean distance between node u and node v.

14.4.2.1 DRNG

DRNG is composed of two phases: *information collection* and *topology construction*. The goal of the first phase is for each node u to obtain the set of nodes N_u that node u can reach directly using its maximal transmission power. To obtain the information, each node should periodically broadcast a *Hello* message using its maximal transmission power. Each *Hello* message contains the sender's *id*, location, and maximal transmission power. For a homogeneous network, this is sufficient for node u to calculate its neighbor set N_u, while for a heterogeneous network, this is not the case.

Therefore, for a general heterogeneous network, we will assume that the original network topology is dense enough such that after removing those unidirectional links from the original network, the network still remains strongly connected. Based on such an assumption, Ref. [12] shows that collecting neighbor information only from nodes whose transmission ranges cover a node u is sufficient for node u to construct a new topology which still preserves the network connectivity.

In the topology construction phase, each node will use the following definition of neighbor relation to construct its topology. Node u has a neighbor relation with node v (denoted by $u \xrightarrow{\text{DRNG}} v$) if and only if $d(u,v) \le r_u$ and there does not exist a third node p which can reach v (i.e., $d(p,v) \le r_p$) and also satisfy one of the following conditions:

1. $d(u,p) < d(u,v)$ and $d(p,v) < d(u,v)$ (see Figure 14.15(a)).
2. $d(u,p) = d(u,v), d(p,v) < d(u,v)$, and $id(p) < id(v)$ (see Figure 14.15(b)).
3. $d(u,p) < d(u,v), d(p,v) = d(u,v)$, and $id(p) < id(u)$ (see Figure 14.15(c)).
4. $d(u,p) = d(u,v), d(p,v) = d(u,v), id(p) < id(u)$, and $id(p) < id(v)$ (see Figure 14.15(d)).

These conditions avoid unnecessary links in the topology because node u does not need to reach v directly if there is a third node p which has a better situation, such as a shorter transmission range, to forward packets from u to v. Figure 14.15(a) illustrates the first condition that $v_3 \xrightarrow{\text{DRNG}} v_5$ can be deleted because of the existence of v_4, and so can $v_5 \xrightarrow{\text{DRNG}} v_3$. Figure 14.15(b) depicts the second condition that $d(u,v)$ is equal to $d(u,p)$. In this case, node u can select one of nodes p and v as its neighbor in DRNG because node u can reach one of them through the other. Here, node u will select the node with a smaller id as its neighbor in DRNG, i.e., $v_3 \xrightarrow{\text{DRNG}} v_5$ can be deleted. Figure 14.15(c) represents the third condition that node u can use a smaller power level to reach node v through node p and node p can use a smaller power level to reach node v through node u. In this case, either node u or node p can delete its neighbor relation with node v, but not both. Thus, nodes' ids are taken into consideration to allow the node with a larger id to delete its neighbor v, i.e., $v_3 \xrightarrow{\text{DRNG}} v_5$ can

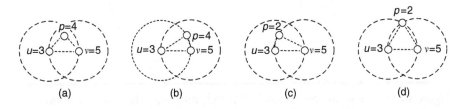

(a) (b) (c) (d)

Figure 14.15 Four conditions in the directed relative neighborhood graph. The dotted line denotes bidirectional connectivity.

be deleted, but $v_2 \overset{\text{DRNG}}{\longrightarrow} v_5$ cannot be deleted. Figure 14.15(d) portrays the fourth condition which can be seen as the combination of the second and the third conditions. In this case, *id* also plays an important role, resulting in the fact that $v_3 \overset{\text{DRNG}}{\longrightarrow} v_5$ and $v_5 \overset{\text{DRNG}}{\longrightarrow} v_3$ can be deleted, but $v_3 \overset{\text{DRNG}}{\longrightarrow} v_2$, $v_2 \overset{\text{DRNG}}{\longrightarrow} v_3$, $v_2 \overset{\text{DRNG}}{\longrightarrow} v_5$, and $v_5 \overset{\text{DRNG}}{\longrightarrow} v_2$ cannot be deleted.

Nodes' *id*s play an important role to avoid the network being partitioned caused by topology control. We use the following example to explain the importance of using *id*s in the second and the third conditions mentioned above. Consider the case in Figure 14.16(a). If we only consider distances between nodes, the neighbor relation $v_4 \overset{\text{DRNG}}{\longrightarrow} v_3$ can be deleted because of the existence of node v_5, and the relation $v_5 \overset{\text{DRNG}}{\longrightarrow} v_3$ can also be deleted because of the existence of node v_4. Unfortunately, this will result in node v_3 becoming unreachable from both v_4 and v_5, as shown in Figure 14.16(b). So, nodes' *id*s are used as a reference to decide which nodes have the right to reduce their transmission power.

After constructing all necessary neighbor relations, a node can determine its transmission power level. Specifically, a node should select the power level that can reach the farthest node which has a neighbor relation with it. Figure 14.17 shows an example. In this example, because of the existence of node v_6, node v_4 deletes the relation $v_4 \overset{\text{DRNG}}{\longrightarrow} v_5$, and so does node v_5 (deleting $v_5 \overset{\text{DRNG}}{\longrightarrow} v_4$). Similarly, because of the existence of node v_3, the neighbor relations $v_4 \overset{\text{DRNG}}{\longrightarrow} v_7$ and $v_7 \overset{\text{DRNG}}{\longrightarrow} v_4$ are deleted. Suppose that $d(v_3, v_7) = d(v_4, v_7)$. Hence, v_3 cannot delete the relation $v_3 \overset{\text{DRNG}}{\longrightarrow} v_7$ because node v_4's *id* is larger than node v_3's *id*. Also note that two directed links are shown in Figure 14.17(a) because of asymmetric transmission powers. These links are ignored by the DRNG algorithm. After the

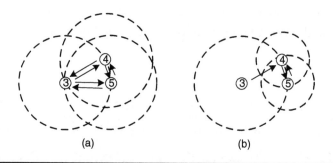

(a) (b)

Figure 14.16 An erroneous example if DRNG does not take nodes' *id*s into consideration. (a) The original network topology where $d(v_3, v_4) = d(v_3, v_5)$. (b) The result of DRNG without using *id* as a tiebreaker. Nodes v_4 and v_5 have reduced their transmission power, making v_3 unreachable from them.

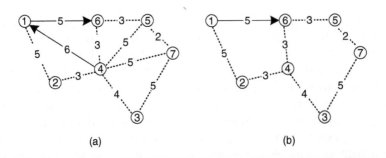

(a)   (b)

Figure 14.17 An example of the DRNG algorithm. The dotted line denotes bidirectional connectivity. (a) Original topology. (b) After topology control.

algorithm, some directed links may be deleted, as shown in Figure 14.17(b). The resulting network is guaranteed to be connected, disregarding those directed links.

14.4.2.2 DLSS

Similar to DRNG, the DLSS algorithm is also composed of two phases: information collection and topology construction. The main task of the first phase is for each node u to obtain its neighbor set N_u through exchanging *Hello* messages. As for the second phase, each node u will construct a *directed local spanning graph* S_u as follows. First, u builds a directed graph $G_u = (V(G_u), E(G_u))$, where $V(G_u) = \{N_u \cup u\}$ and $E(G_u) = \{(p, v): d(p, v) \le r_p, p, v \in V(G_u)\}$. Second, u sorts the edges in $E(G_u)$ in an ascending order according to the distances between edges' endpoints. If two edges have the same distance, then nodes' *id*s are used to break the tie. Specifically, the larger *id* is used first. Therefore, after sorting, the edge pairs (i, j) and (j, i) will be adjacent in the sorted sequence. From G_u, we will construct the graph S_u. The vertex set of S_u is also $V(G_u)$. The edge set of S_u is initially empty, and edges will be added to S_u as follows. From the above sorting result, each edge $(p, v) \in E(G_u)$ is visited sequentially. If node p already has a path to v in S_u, (p, v) will not be included in S_u; otherwise, the edge (p, v) will be included in S_u. After constructing S_u, we say that there is a neighbor relation $u \xrightarrow{\text{DLSS}} v$ if and only if $(u, v) \in E(S_u)$. Node u then selects the power level that can reach the farthest node which has a neighbor relation with it.

The DLSS algorithm guarantees that for any edge (u, v) in the original topology, there must exist a path from u to v in the final topology (denoted by G_{DLSS}) after each node reduces its transmission power. The proof is derived by an induction as follows. We first list all edges in the original topology in an ascending order according to their distances: $(u_1, v_1), (u_2, v_2), \ldots, (u_n, v_n)$, where n is the total number of edges in the original topology. We will show that for each (u_i, v_i), $1 \le i \le n$, there is a path in G_{DLSS} from u_i to v_i. The basis case is that the edge (u_1, v_1) with the minimum distance will be included in G_{DLSS} because for node u_1, (u_1, v_1) must be included in $E(S_{u_1})$. Then, assume the hypothesis that there is a path from u_i to v_i in G_{DLSS} for all edges (u_i, v_i), $1 \le i \le k - 1$. We need to prove that there exists a path from u_k to v_k in G_{DLSS}. If the edge (u_k, v_k) is in G_{DLSS}, then the proof is done. Otherwise, in the local topology construction for u_k, before edge (u_k, v_k) was selected into S_{u_k}, there must already exist a path $w_0 = u_k \rightarrow w_1 \rightarrow w_2 \rightarrow \cdots \rightarrow w_{m-1} \rightarrow w_m = v_k$, such that $(w_i, w_{i+1}) \in E(S_{u_k})$ and $d(w_i, w_{i+1}) \le d(u_k, v_k)$, $0 \le i \le m - 1$. The above claim is true because of the induction hypothesis.

An example of the DLSS algorithm is shown in Figure 14.18. Figure 14.18(a) presents the original network topology. Figure 14.18(b) shows the directed graph G_{v_3}. Based on the graph G_{v_3}, node v_3's directed local spanning graph S_{v_3} is shown in Figure 14.18(c). Note that the edge (v_4, v_7) is not included in S_{v_3} because (v_3, v_7) is inserted into S_{v_3} before (v_4, v_7). Thus, node v_3 only has two neighbor relations $v_3 \xrightarrow{\text{DLSS}} v_4$ and $v_3 \xrightarrow{\text{DLSS}} v_7$. According to these neighbor relations, node v_3 will control its transmission power to reach v_7 only. Figures 14.18(d) and (e) show the graphs G_{v_7} and S_{v_7} of v_7. Note that the edge (v_7, v_4) is not included in S_{v_7} because

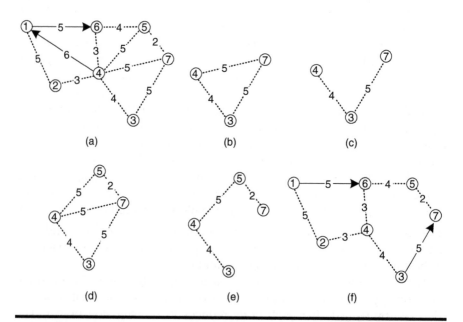

(a) (b) (c)

(d) (e) (f)

Figure 14.18 An example of the DLSS algorithm. The dotted line denotes bidirectional connectivity. (a) Original topology. (b) G_{v_3}. (c) S_{v_3}. (d) G_{v_7}. (e) S_{v_7}. (f) The final topology G_{DLSS}.

the edge (v_5, v_4) is inserted into S_{v_7} before (v_7, v_4). Node v_7 only has one neighbor relation $v_7 \xrightarrow{\text{DLSS}} v_5$. Therefore, node v_7 determines the required transmission power to reach node v_5 as its transmission power. Figure 14.18(f) shows the final network topology obtained by DLSS. Note that although there are unidirectional links, DLSS can guarantee that the final topology is connected.

14.5 Summary

This chapter discusses several techniques for energy conservation in wireless ad hoc networks. First, we review the basic power management mechanism in the IEEE 802.11 standard. Since the standard power management mechanism does not scale well in a multihop ad hoc network due to the difficulty of clock synchronization, several asynchronous and semiasynchronous power-saving protocols are discussed. Second, the techniques of power control in MAC layer are studied, in which nodes can decide proper power levels for their DATA–ACK transmission based on their RTS–CTS negotiation. Finally, we present two types of topology control protocols. The first type relies on power management to reduce the number of active nodes. The second type is based on power control to reduce the number of links in the network.

References

1. L. Huang and T.-H. Lai. On the scalability of IEEE 802.11 ad hoc networks. In *ACM International Symposium on Mobile Ad Hoc Networking and Computing (MobiHOC)*, pp. 173–182, 2002.
2. Y.-C. Tseng, C.-S. Hsu, and T.-Y. Hsieh. Power-saving protocols for IEEE 802.11-based multi-hop ad hoc networks. In *Infocom*, Vol. 43, no. 3, pp. 317–337, 2002.
3. D. Agrawal and A.E. Abbadi. An efficient and fault-tolerant solution for distributed mutual exclusion. In *ACM Transactions on Computer Systems*, Vol. 9, no. 1, pp. 1–20, 1991.
4. H. Garcia-Molina and D. Barbara. How to assign votes in a distributed system. In *Journal of ACM*, 32:841–860, 1985.
5. R. Zheng, J.C. Hou, and L. Sha. Asynchronous wakeup for ad hoc networks. In *MOBIHOC*, pp. 35–45, 2003.
6. C.-S. Hsu and Y.-C. Tseng. Cluster-based semi-asynchronous power-saving protocols for multi-hop ad hoc networks. In *ICC*, Vol. 5, pp. 3166–3170, 2005.
7. J. Gomez, A.T. Campbell, M. Naghshineh, and C. Bisdikian. Conserving transmission power in wireless ad hoc networks. *ICNP*, pp. 24–34, 2001.
8. E.-S. Jung and N.H. Vaidya. A power control MAC protocol for ad hoc networks. In *MOBICOM*, pp. 36–47, 2002.

9. S.-L. Wu, Y.-C. Tseng, and J.-P. Sheu. Intelligent medium access for mobile ad hoc networks with busy tones and power control. *IEEE Journal on Selected Areas in Communications*, Vol. 18, no. 9, pp. 1647–1657, 2000.

10. R. Ramanathan and R. Rosales-Hain. Topology control of multihop wireless networks using transmit power adjustment. In *IEEE INFOCOM*, vol. 2, pp. 404–413, 2000.

11. A. Cerpa and D. Estrin. Ascent: Adaptive self-configuring sensor network topologies. *IEEE Transactions on Mobile Computing*, 3(3):272–285, 2004.

12. N. Li and J.C. Hou. Localized topology control algorithms for heterogeneous wireless networks. In *IEEE/ACM Transactions on Networking*, Vol. 13, no. 6, pp. 1313–1324, 2005.

Chapter 15

Wireless LAN Security

Rong-Hong Jan, Huei-Ru Tseng, and Wuu Yang

The rapid growth of wireless communication means that issues in wireless local area network (LAN) security are of increasing practical importance. This chapter presents an overview of wireless LAN security for the IEEE 802.11 networks. First, we address the *wired equivalent privacy* protocol (WEP), the security method originally employed in IEEE 802.11 networks and show that it has numerous security weaknesses. Next, we introduce new security protocols, such as Wi-Fi protected access (WPA) and IEEE 802.11i, that replace WEP and provide real security for wireless LANs. Apart from reviewing these security designs, this chapter also illustrates several threats against these security mechanisms.

15.1 WEP and Its Security Weaknesses

The most significant difference between wireless and wired LANs is that wireless transmissions are not confined to a wire. That is, data in a wireless LAN is broadcast in the air. Anyone who knows how to intercept radio waves at the proper frequencies could access the data and become a potential security threat to the wireless LAN. Thus, the security problems in wireless networks are more complicated than those in wired networks.

In general, a secure communication network should satisfy three requirements: (1) data confidentiality, (2) data integrity, and (3) user authentication. In 1997, the IEEE 802.11 working group defined the WEP protocol. WEP is aimed at satisfying the requirements of data confidentiality and data integrity. Unfortunately, there are several security flaws in WEP.[1–4] WEP is found to be subject to several types of attacks. However, an understanding of WEP and its security weaknesses is helpful to learn which issues should be addressed for real security. In this section, we present the IEEE 802.11 security framework, WEP operations, and their security weaknesses.

15.1.1 The IEEE 802.11 Security Framework

IEEE 802.11[5] defines the authentication services and the WEP algorithm. The authentication services are used to establish the identity of a principle (a person or a workstation). There are two types of authentication services, open-system authentication and shared-key authentication. The open-system authentication, shown in Figure 15.1, involves two messages. The first message is the identity assertion and the request for authentication. The authentication result is the second message. The shared-key authentication requires the use of the WEP mechanism. The secret key for authentication is never transmitted over the communication channel. The authentication procedure is shown in Figure 15.2.

IEEE 802.11 also depends on WEP to protect authorized users against eavesdropping in wireless networks. The goal of WEP is to provide wireless users with security that is equivalent to what is available in wired networks. The encryption algorithm used by WEP is based on the RC4 algorithm. WEP cryptographic operations—encryption and decryption—are discussed in the next section.

15.1.2 WEP Cryptographic Operations

Originally, a message is a piece of plain text. The process of transforming a piece of text into a different form is called *encryption* (see Figure 15.3).

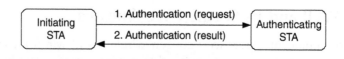

Figure 15.1 Open-system authentication in the IEEE 802.11 standard.

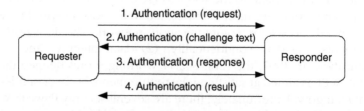

Figure 15.2 Shared-key authentication in the IEEE 802.11 standard.

The encrypted message is called the *ciphertext*.[6] The process of turning the ciphertext back into the original message is called *decryption* (see Figure 15.3). Encryption and decryption come in pairs. There are many different encryption algorithms, each with a different strength. The encryption algorithm used by WEP is based on the RC4 algorithm. To encrypt a message, the message is bitwise exclusive-ORed with a *keystream*, which is generated by the RC4 algorithm. Decryption is done similarly. The keystream is generated from a key, which is shared between the sender and the receiver. The encryption procedure is shown in Figure 15.4.

15.1.2.1 WEP Encryption Operation

The detailed WEP encryption operation is shown in Figure 15.5. The 40-bit WEP key concatenated with a 24-bit *initialization vector* (IV) forms the RC4 key. The RC4 key is fed into the RC4 algorithm to produce the required keystream. The keystream could be as long as is needed.

A 32-bit *integrity check value* (ICV) is produced from the message (i.e., the frame body) with the cyclic-redundancy-check (CRC-32) algorithm. The message concatenated with the ICV is bitwise exclusive-ORed with the keystream. The result is the ciphertext. Finally, a frame header, an IV header, and an *frame check sequence* (FCS) suffix is attached to the

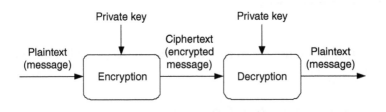

Figure 15.3   Encryption and decryption with a key.

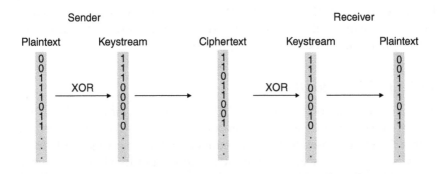

Figure 15.4   The procedure of encryption and decryption in WEP.

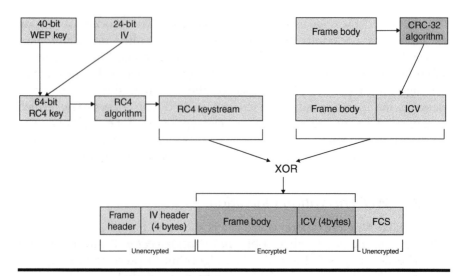

Figure 15.5   The encryption operation of WEP.

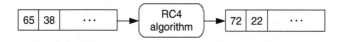

Figure 15.6   Stream cipher.

ciphertext, forming the complete encrypted message, which is sent to the receiver.

15.1.2.2 WEP Decryption Operation

The WEP key is shared between the sender and the receiver. The receiver can recover the 24-bit IV from the IV header in the encrypted message. From the shared WEP key and the IV, the receiver can generate exactly the same keystream as the sender. This keystream is bitwise exclusive-ORed with ciphertext, resulting in the original plaintext.

15.1.2.3 RC4 Algorithm

The encryption algorithm used by WEP is the RC4 algorithm. The fundamental function of RC4 is to generate a keystream. RC4 is a kind of a stream cipher. It acts as a pseudo-random number generator, taking in 1 byte from a key at a time as a seed and outputs a pseudo-random byte, as shown in Figure 15.6. The RC4 algorithm consists of two stages: S-box initialization and pseudo-random sequence generation.

Stage 1: S-box initialization

There are two 256-byte arrays, called the S-box and the K-box. Initially, the 256 bytes of S-box contains the values 0, 1, 2, . . . , 255, respectively. The K-box contains the key initially. If the key length is less than 256 bytes, the key repeats itself to fill up the K-box. For instance, the length of the key used by WEP is 8 bytes. Therefore, it has to repeat itself 32 times to fill up the K-box. The goal in this stage is to rearrange the bytes of S-box. The detailed algorithm is shown in Figure 15.7. For example, assume that we have S-box and K-box as shown in Figure 15.8a. When $i = 0$ and $j = 0$, $j = (0 + S_0 + K_0) \bmod 256 = (0 + 0 + 27) \bmod 256 = 27$. This means S_0 interchanges with S_{27}. We obtain the new S-box as shown in Figure 15.8b. This exchange process is performed on every byte of the S-box.

Stage 2: Pseudo-random sequence generation

After the S-box has been initialized, the RC4 algorithm goes to a pseudo-random sequence generation stage. The detailed algorithm for generating a pseudo-random sequence is shown in Figure 15.9. Each round of the

```
//Sₙ :  The value in S-box[n]
//Kₙ :  The value in K-box[n]
i = j = 0;
For i = 0 to 255 do
          j = ( j + Sᵢ + Kᵢ ) mod 256;
          Swap Sᵢ and Sⱼ
End;
```

Figure 15.7 S-box initialization.

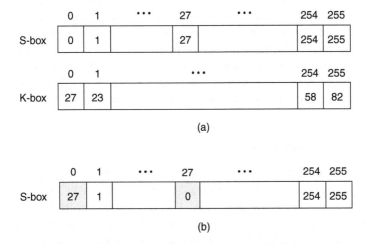

(a)

(b)

Figure 15.8 S-box and K-box.

```
//Sn :   The value in S-box[n]
//R   :   A pseudorandom value in one byte
i = ( i + 1 ) mod 256;
j = ( j + Si ) mod 256;
Swap Si and Sj
k =   ( Si + Sj ) mod 256;
R = Sk;
```

Figure 15.9   Pseudo-random sequence generation.

algorithm generates a 1-byte pseudo-random value (R). Repeating the algorithm results in the RC4 keystream.

15.1.3   WEP's Security Problems

As mentioned before, a secure communication network should satisfy three requirements: (1) data confidentiality, (2) data integrity, and (3) user authentication. However, the IEEE 802.11 standard fails to accomplish these security requirements. We look at each in detail.

15.1.3.1   Data Confidentiality

The objective of data confidentiality is to prevent the unauthorized disclosure of data. The problem associated with WEP lies in the limited length of the keys used in RC4 algorithm. The 40-bit WEP keys suffer from the dictionary attack, which becomes more and more feasible given today's extremely fast chip technology. In addition, the WEP key must be set up manually. This creates a potential hole in the overall security. Furthermore, key reuse is also a security problem in stream ciphers like RC4. For example, let P_1 and P_2 be two pieces of plaintext. Let C_1 and C_2 be P_1 and P_2, respectively, encrypted with the same key. Then $P_1 \oplus P_2 = C_1 \oplus C_2$. An attacker might take advantage of this property, say, if he already knows P_1 in advance. ($\oplus$ means the bitwise exclusive-OR operation.)

15.1.3.2   Data Integrity

WEP guarantees data integrity through the use of CRC-32. ICV produced by CRC-32 protects data against malicious modification. CRC-32 is originally intended for packet error detection. The mechanism suffers from several problems such as linearity. An attacker could predict which bits in the ICV will change if a single bit in the plaintext changes. It turns out that an attacker can intercept a message, modify it, and redirect it to the receiver, and the receiver will fail to detect the modification.

15.1.3.3   User Authentication

The most serious security problem for IEEE 802.11 is the lack of effective authentication services. Although IEEE 802.11 provides two kinds of authentication services (open-system and shared-key authentication) it still suffers from attacks. Actually, open-system authentication does not offer any access security, and shared-key authentication is vulnerable to the man-in-the-middle attacks.

15.2   802.1X Security Measures

WEP, the original 802.11 security mechanism, is vulnerable to various cryptographic attacks, such as dictionary attacks, linearity attacks, and man-in-the-middle attacks. To enhance the security level of 802.11, the industry developed authentication methods based on 802.1X specification,[7] which was originally designed for wired networks. The 802.1X standard defines a mechanism for port-based network access control that provides a means of attaching, authenticating, and authorizing devices to a LAN port having the point-to-point connection characteristics. Moreover, the mechanism prevents access to the port in case the authentication and authorization processes fails, and also provides a key distribution method.

802.1X relies on the extensible authentication protocol (EAP),[8] which supports various authentication mechanisms. The advantages of 802.1X authentication includes the following (1) every user in the network can be identified and authenticated; (2) it supports extensible authentication technologies, such as token cards, certificate/smart cards, and one-time passwords; and (3) it supports key managements, including key management and key reproduction.

15.2.1   The IEEE 802.1X Framework

In the 802.1X authentication system, there are three roles: supplicant, authenticator, and authentication server, as illustrated in Figure 15.10. The protocol between the supplicant and the authenticator is extensible authentication protocol over LANs (EAPOL). The protocol between the authenticator and the authentication server is the remote authentication dial in user service (RADIUS).[9]

The supplicant is a wireless client requesting to join the network. The entity that grants the network access is the authenticator, which is called an access point (AP) in 802.11 networks. The authentication server performs the authentication functions necessary to verify the supplicant. As illustrated in Figure 15.11, before being successfully authenticated by the authentication server, the supplicant is only allowed to send control messages to the server. (Messages containing data will be blocked.) After being authenticated by the authentication server, the supplicant is granted access to the whole network, as shown in Figure 15.12.

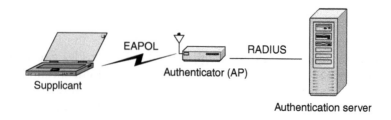

Figure 15.10   The IEEE 802.1X framework.

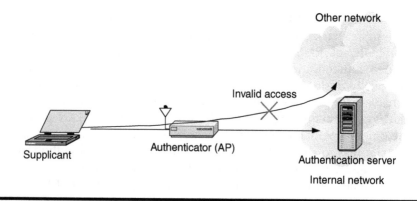

Figure 15.11   Invalid supplicant access before authentication.

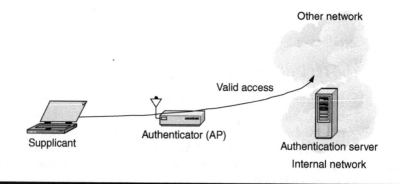

Figure 15.12   Valid supplicant access after authentication.

15.2.2   *The Extensible Authentication Protocol*

IETF RFC 2284[10] defines the point-to-point protocol (PPP), which provides a standard method for transporting multiprotocol datagrams over point-to-point links. In this protocol, it also defines an extensible link control protocol, called the *PPP EAP*, which allows negotiation for authenticating

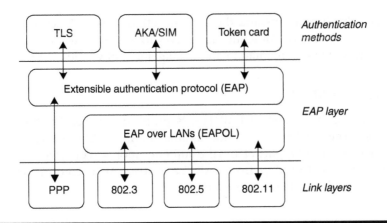

Figure 15.13   EAP architecture.

its peers before allowing network layer protocols to transmit data over the link. The EAP, shown in Figure 15.13, is based on the PPP EAP.[10]

Rather than specifying a fixed authentication mechanism, EAP provides an extensible authentication platform, allowing various types of authentication mechanisms, including EAP-MD-5, EAP-TLS (transport layer security), EAP-PEAP (protected extensible authentication protocol), EAP-TTLS (tunneled TLS), EAP-FAST (flexible authentication via secure tunneling), and Cisco LEAP (lightweight extensible authentication protocol), etc. We will discuss each of these in the following sections.

15.2.2.1   EAP-MD-5

EAP-MD-5 assumes that the supplicant and the authentication server share a common key. It uses the MD5 algorithm to protect the password of the supplicant. Since it only provides one-way authentication, EAP-MD-5 is typically not recommended for wireless LAN implementations.

15.2.2.2   EAP-TLS

EAP-TLS, proposed by Microsoft and Cisco, assumes that the supplicant and the authentication server have their own certificates for authentication. In addition, it can dynamically generate user- and session-based WEP keys to secure subsequent communications between the supplicant and the authenticator. EAP-TLS provides mutual authentication between the supplicant and the authentication server.

15.2.2.3   EAP-PEAP

EAP-PEAP, developed by Microsoft, Cisco, and RSA Security, can securely transmit authentication data via 802.11 wireless LANs. It uses *tunnels* between the supplicants and the authentication servers. Since PEAP only uses the certificate of authentication server to authenticate the supplicant, it simplifies the implementation and management of wireless LAN security.

15.2.2.4 EAP-TTLS

EAP-TTLS, an extension of EAP-TLS, is proposed by Funk Software and Certicom. The mechanism provides mutual authentication based on certificates between the supplicant and the authentication server via an encrypted channel. EAP-TTLS differs from EAP-TLS in that EAP-TTLS only needs the certificate of the authentication server.

15.2.2.5 EAP-FAST

In EAP-FAST, which is developed by Cisco, mutual authentication is achieved through protected access credentials (PAC), which are managed dynamically by the authentication server. EAP-FAST is available for enterprises currently. However, it cannot enforce a strong password policy.

15.2.2.6 Cisco-LEAP

Cisco-LEAP is typically used in Cisco Aironet WLANs. It uses WEP keys to encrypt the transmission data. Cisco-LEAP also provides mutual authentication.

15.2.3 The IEEE 802.1X Authentication Operations

802.1X is designed for user authentication.[11] This section presents the authentication procedures of the EAPOL protocol, which is shown in Figure 15.14. The steps in 802.1X authentication operations are:

1. The supplicant issues an association request to the authenticator.
2. In response to the association request, the authenticator sends an association response to the supplicant. At this stage, the supplicant is only allowed to send messages related to 802.1X authentication. All other traffic will be dropped by the authenticator.
3. The supplicant starts the 802.1X exchange by sending the EAP start frame to the authenticator.
4. The authenticator issues an EAP request/identity frame, requesting the supplicant to send his/her identity information.
5. After receiving the the EAP request/identity frame, the supplicant replies with an EAP response/identity frame, which is passed on to the RADIUS authentication server as a Radius-Access-Request packet.

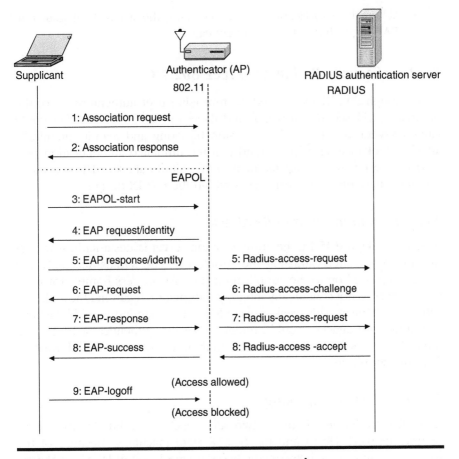

Figure 15.14   EAPOL operations on an 802.11 network.

6. The RADIUS authentication server replies to the authenticator with a Radius-Access-Challenge packet, which is passed back to the supplicant as an EAP request frame.

7. After receiving the EAP request frame, the supplicant extracts the challenge from the frame. The supplicant applies the MD5 algorithm to his password and the challenge, yielding an authentication string. The authentication string is packaged in an EAP response frame and sent to the authenticator. The authenticator translates this response frame into a Radius-Access-Request and sends it to the RADIUS authentication server.

8. The authentication server determines the legitimacy of the supplicant by checking if the authentication string is valid. If so, the authentication server will issue a Radius-Access-Accept packet, which is passed on to the supplicant as an EAP success frame.

9. When the supplicant is done accessing the network, it issues an EAP logoff frame to the authenticator.

15.2.4 Security Problems in IEEE 802.1X

Even though 802.1X was aimed at improving user authentication of the original 802.11 security, it was found flawed in 2002.[8] An attacker could launch several attacks, such as session hijacking and man-in-the-middle attacks. Asokan et al.[12] also found that the tunnelled authentication protocols were not secure against man-in-the-middle attacks. These security problems underline the weaknesses within the 802.1X design.

15.2.4.1 Man-in-the-middle Attack

In step 8 in Figure 15.14, the authentication server issues a Radius-Access-Accept packet after validating the supplicant. The Radius-Access-Accept packet is passed on to the supplicant as an EAP success frame. However, the frame does not contain any integrity information regarding the authentication mechanism. This weakness could invite a man-in-the-middle attack, as follows: An attacker could pretend to be the authenticator by forging the EAP success frame. The attacker can then request sensitive information from the supplicant.

15.2.4.2 Session Hijacking

According to Ref. [8], there are two state machines in 802.1X, the robust security network (RSN) and the 802.1X state machines. Because of the vague communication between the two state machines and the lack of message authenticity, an attacker could take advantage of the loose coupling of the state machines to initiate a session-hijacking attack.

15.3 ■ IEEE 802.11i Security

The IEEE 802.11i standard specifies advanced security mechanisms for IEEE 802.11 networks.[13] It emphasizes the improvements of the security protocols in the original 802.11 standard. In particular, 802.11i focuses on data confidentiality and key management.

802.11i defines two types of wireless networks, RSN and *transitional security network* (TSN).[14] RSN provides data confidentiality enhancements and key management protocols in wireless networks. RSN makes use of the *advanced encryption standard* (AES) and the CCMP (Counter mode with Cipher blocking chaining with Message authentication code Protocol) protocol to provide confidentiality, authentication, integrity, and replay protection. However, because it is practically impossible to ask *all* users to upgrade *all* existing equipment, the IEEE 802.11i defines TSN as an interim standard in which both RSN and WEP systems can cooperate.

Because the required AES–CCMP operations must be supported by the hardware, most existing 802.11 network interface cards cannot be upgraded to RSN. Although security is very important, switching to RSN means replacing all existing 802.11 equipment with new ones, which is deemed infeasible for most users, due to budgetary considerations. Software upgrade would be a much preferred approach. Therefore, 802.11i developed the *temporal key integrity protocol* (TKIP) based on the capabilities of existing 802.11 products. TKIP operates as an optional mode under RSN. Later, the Wi-Fi Alliance proposed a new security approach, called the WPA, which is based on the draft of RSN but only specifies TKIP.

The following sections review RSN protocols related to key hierarchy, operational phases, and data confidentiality.

15.3.1 Key Hierarchy of 802.11i

Note that WPA is a set of RSN. Both WPA and RSN share the same key hierarchy. As shown in Figure 15.15, there are three roles: an authentication server (AS), an AP, and a station (STA), in the 802.11i security framework. The authentication server is in charge of authenticating the stations. It maintains a database of legal users and related authentication information. The authentication server and the access point work independently. After a station is authenticated by the authentication server, the authentication server and the station will share a common *master key* (MK). Then the station and the authentication server derive a *pairwise master key* (PMK) from the MK. A PMK is 256 bits long. The authentication server distributes the PMK to the access point. The station and the access point use the PMK to enforce 802.11 channel access and to derive a *pairwise transient key* (PTK). A PTK

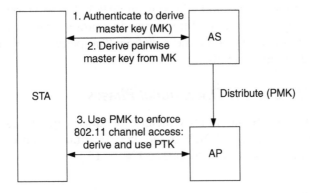

Figure 15.15 RSN key hierarchy.

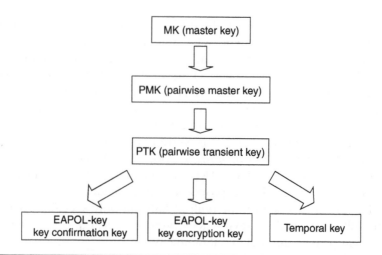

Figure 15.16 Pairwise key hierarchy.

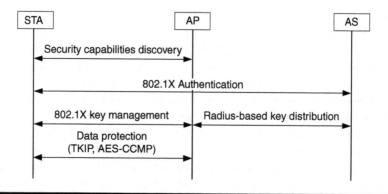

Figure 15.17 802.11i Operational phases.

consists of an EAPOL-Key integrity key, an EAPOL-Key encryption key, and a temporal key.* The temporal key is used to protect unicast communication between an access point and a station. The pairwise key hierarchy is summarized in Figure 15.16.

15.3.2 The 802.11i Operational Phases

The 802.11i operational phases include security capabilities discovery, 802.1X authentication, RADIUS-based key distribution and 802.1X key management, and data protection (see Figure 15.17).

---

* However, for TKIP, a PTK consists of four keys: EAPOL-Key integrity key, EAPOL-Key encryption key, a data encryption key, and a data integrity key.

15.3.2.1 Security Capabilities Discovery

First, AP advertises network security capabilities in *beacon frames* or *probe response frames* to STAs. 802.11i defines a new information element for RSN. The new RSN information element embedded in a beacon frame or a probe response frame includes information about all enabled authentication suites, all enabled unicast cipher suites, and/or multicast cipher suites. STA receives the beacon frame or probe response frame from AP. Then, STA responds to AP with an *association request frame*, the preferred authentication suite, and unicast cipher suite. In response to the association request frame, AP sends an *association response frame* to the STA. At this stage, STA and AP are said to be *associated*.

15.3.2.2 802.1X Authentication

After STA and AP are associated, STA and AS perform mutual authentication with the EAP-TLS protocol. An MK is generated after STA and AS are mutually authenticated. Note that the MK represents a decision to grant access based on the authentication. The MK is fresh and bound to this session between STA and AS. Then STA and AS derive a PMK from the MK. The PMK serves as an authorization token to enforce access control between STA and AP. Note that the MK is not equal to the PMK, for otherwise AP could make access control decisions.

15.3.2.3 RADIUS-Based Key Distribution and 802.1X Key Management

After the PMK is generated, AS uses the RADIUS protocol to send the PMK to AP. We assume that the channel used to distribute PMK from AS to AP is secure. Next, STA and AP mutually authenticate themselves by presenting their authorization token, i.e., the PMK. This can be done with a protocol called the *4-way handshake*, which makes use of 802.1X EAPOL-Key frames. In addition to authentication, the 4-way handshake generates the temporal keys for data confidentiality protocols.

As illustrated in Figure 15.18, four messages are exchanged between AP and STA. We will discuss the four messages in turn.

Message 1 After AP receives the PMK, it initiates the 4-way handshake. AP generates *ANonce*, which is a random bit string. ANonce is packaged in an EAPOL-Key frame and is sent to STA.

Message 2 STA generates *SNonce*, which is also a random bit string. STA then derives a PTK from ANonce, SNonce, medium access control (MAC) address of AP, and MAC address of STA. Finally, STA sends AP an EAPOL-Key frame containing SNonce, the RSN information element that is used during association, and a message integrity code (MIC). The MIC is

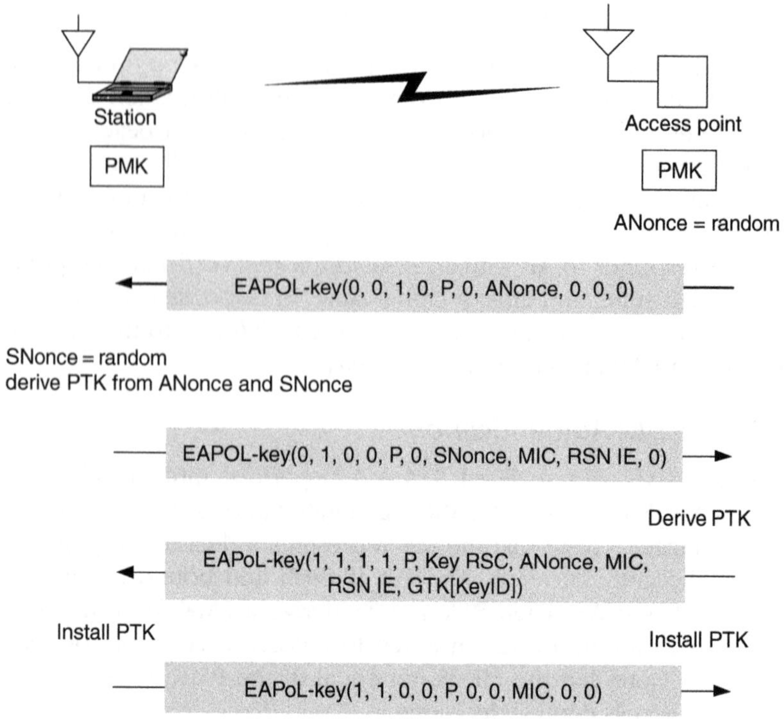

Figure 15.18 Four-way handshake in RSN.

computed by using the EAPOL-Key integrity key, which is a part of the PTK for preventing tampering. In addition, the MIC allows AP to verify that STA really knows the PMK.

Message 3 AP also derives the PTK from ANonce, SNonce, and MAC addresses of AP and STA. With PTK, AP verifies the MIC in the EAPOL-Key frame. Then AP sends an EAPOL-Key frame containing ANonce, the RSN information element, MIC, a flag indicating whether to install the temporal keys, and the encapsulated group transient key (GTK). The RSN information element is enclosed in the beacons and probe responses that are broadcast from AP.

Message 4 STA sends an EAPOL-Key frame, confirming that the temporal keys are installed and starts encryption.

15.3.3 Data Confidentiality Protocols in RSN

802.11i defines two RSN data confidentiality and integrity protocols: TKIP and CCMP. Since TKIP, which is based on the RC4 algorithm, is designed

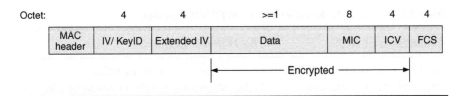

Figure 15.19 TKIP MPDU formats.

to improve WEP's security, an implementation of TKIP is optional for RSN. However, CCMP is mandatory for RSN compliance. CCMP is based on CCM of the AES encryption algorithm. It provides confidentiality, authentication, integrity, as well as replay protection.

15.3.3.1 Temporal Key Integrity Protocol

The TKIP is a data confidentiality protocol that is designed to enhance the WEP's security level. TKIP improves the security of WEP encapsulation with several functions, for example, the Michael algorithm[15] and a key mixing function.

WEP uses ICV for integrity check but ICV offers little real protection. However, ICV is still checked in TKIP. Besides, TKIP adopts the Michael algorithm as its data integrity mechanism. An 8-byte MIC is computed using the Michael algorithm and combines a data integrity key, which is a part of the PTK for TKIP. The frame format for TKIP is shown in Figure 15.19. Since the Michael algorithm is simple, it is vulnerable to brute-force attacks. TKIP adopts a series of countermeasures, such as changing the encryption key or stopping data transmission to make up for the vulnerability.

As mentioned earlier, the RC4 key in WEP is the concatenation of a 40-bit WEP key and a 24-bit IV. The IV is very short, so IV values are frequently reused in a busy network. TKIP increases the IV size to 48 bits. In addition, IV may serve as a sequence number, called the *TKIP sequence counter* (TSC), which could prevent replay attacks. TSC starts from 0 and is incremented by 1 for each frame sent. TKIP discards all frames whose TSCs are no greater than the last frame's. Next, TKIP applies a key-mixing function to generate an RC4 key for each frame. The key-mixing function consists two phases: Phase 1 uses the MAC address, the higher-order 32 bits of the IV, and the data encryption key (which is a part of the PTK) to calculate the phase-1 key. In phase 2, TKIP uses the phase-1 key, the lower-order 16 bits of IV, and the data encryption key to create the 128-bit RC4 encryption key.

All known security weaknesses of WEP have been solved in TKIP. The TKIP resolutions for WEP weaknesses are summarized in Table 15.1.

Table 15.1   The TKIP Resolutions for WEP Weaknesses

Weaknesses of WEP	The TKIP resolutions
The IV is too short	Increase the size of IV to 48 bit
Data integrity insufficient	Using Michael algorithm instead of CRC check sum, after generating 64-bit MIC, TKIP will encrypt the value
Using the master key directly	TKIP and Michael use a set of temporary keys, which derivatives from the master key and other values
Keys cannot be reset	WPA dynamically resets keys for deriving new temporary key set
No reperform data protection	TKIP uses IV as its frame counter for providing reperforming data protection

15.3.3.2 AES and the Counter-Mode with CBC-MAC Protocol (AES-CCMP)

802.11i allows for a variety of wireless network implementations and can use an optional TKIP. However, by default, RSN provides stronger and more scalable security protection with the mandatory AES-CCMP. AES is a block cipher and CCMP is the security protocol for RSN. CCMP defines a set of rules that uses AES to encrypt data frames. AES is to CCMP what RC4 is to TKIP.

For confidentiality, CCMP incorporates the *counter mode*. CCMP uses a 128-bit key. As shown in Figure 15.20, the payload is divided into 128-bit blocks. Each block is bitwise exclusive-ORed with the AES-encrypted counter value. Note that the counter may start from an arbitrary value. Whoever wants to decrypt the message must know the starting value of the counter. CCMP provides a 48-bit *packet number* (PN) in the frame header for the receiving party to derive the starting value of the counter.

Note that the counter mode does not provide any message authentication; only encryption is supported. Thus, CCMP uses the *cipher block chaining* to produce a 128-bit message authentication code, known as CBC-MAC. The idea of CBC-MAC method is quite simple (see Figure 15.20): The AES-encrypted block 0 serves as the first seed. The first seed is bitwise exclusive-ORed with block 1. The result is again encrypted with AES, producing a second seed. The second seed is bitwise exclusive-ORed with block 2 and then is encrypted with AES, producing a third seed, and vice versa for the other blocks. In general, the kth seed is bitwise exclusive-ORed with block k and is then encrypted with AES, producing the $(k+1)$th seed. The last seed is the desired MIC.

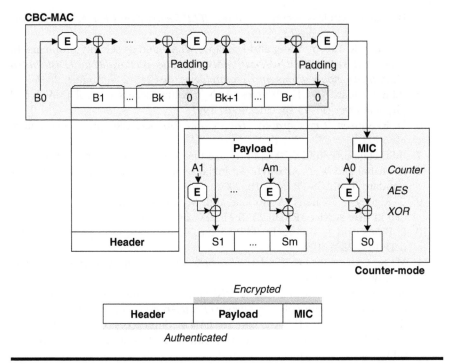

Figure 15.20   CCMP MPDU formats.

15.4   Summary

Wirelss LAN security has become more and more essential. This chapter presents several important frameworks and protocols, including IEEE 802.11, IEEE 802.1X, and authentication protocols, such as RADIUS and EAP. We also examine several existing threats against wireless networks.

At the end of this chapter, we cover the IEEE 802.11i standard, including the key management protocols and the data confidentiality protocols. Even though 802.11i provides stronger security protection than previous standards, there are still some weaknesses in 802.11i that may invite threats.[16]

References

1. J. C. Chen, M. C. Jiang and Y. W. Liu, Wireless LAN security and IEEE 802.11i, *IEEE Wireless Communications*, vol. 12, no. 1, pp. 27–36, 2005.
2. N. Chendeb, B. E. Hassan and H. Afifi, Performance evaluation of the security in wireless local area networks (WiFi), In *Proceedings of the IEEE International Conference on Information and Communication Technologies: From Theory to Applications*, pp. 215–216, April, 2002.

3. B. Potter, Wireless security's future, *IEEE Security and Privacy Magazine*, vol. 1, no. 4, pp. 68–72, 2003.
4. S. Wang, R. Tao, Y. Wang and J. Zhang, WLAN and its security problems, In *Proceedings of the IEEE International Conference on Parallel and Distributed Computing, Applications and Technologies*, pp. 241–244, August, 2003.
5. IEEE Standard 802.11, Wireless LAN medium access control (MAC) and physical layer (PHY) specifications, 1997.
6. D. R. Stinson, *Cryptography: Theory and Practice*, CRC press, Borakaton, FL, 1995.
7. IEEE Standard 802.1X-2004, Port-based network access control, 2004.
8. A. Mishra and W. A. Arbaugh, An initial security analysis of the IEEE 802.1X standard, UMIACS-TR-2002-10, 2002.
9. C. Rigney, S. Willens, A. Rubens and W. Simpson, Remote authentication dial in user service (RADIUS), IETF RFC 2865, 2000
10. L. Blunk and J. Vollbrecht, PPP extensible authentication protocol (EAP), IETF RFC 2284, 1998.
11. M. S. Gast, *802.11 Wireless Networks: The Definitive Guide*, O'Reilly, Sebastopol, CA, 2002.
12. N. Asokan, V. Niemi and K. Nyberg, Man-in-the-middle in tunneled authentication protocols, *Cryptology ePrint Archive*, Report 2002/163.
13. IEEE Standard 802.11i-2004, Wireless LAN medium access control (MAC) and physical layer (PHY) specifications: Amendment 6-meduim access control (MAC) security enhancements, 2004.
14. J. Edney and W. A. Arbaugh, *Real 802.11 Security Wi-Fi Protected Access and 802.11i*, Addison-Wesley, Boston, MA, 2004.
15. M. N. Ferguson, An Improved MIC for 802.11 WEP, 2002. Document number IEEE 802.11-02/020r0. Available on http://grouper.ieee.org/groups/802/11/ Documents/ DocumentHolder/2-020.zip
16. B. Brown, 801.11: The security difference between b and i, *IEEE Potentials*, vol. 22, no. 4, pp. 23–27, 2003.

Chapter 16

Temporal Key Integrity Protocol and Its Security Issues in IEEE 802.11i

Yang Xiao, Kamal Dass, Jun Zheng, and Kui Wu

16.1 Introduction

The IEEE 802.11i specification[1] specifies security mechanisms for the IEEE 802.11 wireless LANs to supersede the previous security specification, wired equivalent privacy (WEP), which was shown to have severe security vulnerabilities. Wi-Fi protected access (WPA), a subset of IEEE 802.11i, was introduced to improve WEP and served as a quick solution. Both WEP and WPA use Rivest Cipher 4 (RC4) stream cipher.

The IEEE 802.11i specification includes the following components:

- IEEE 802.1X for authentication using extensible authentication protocol (EAP) and an authentication server: IEEE 802.1X specification specifies a port-based network admission control for device authentications using EAP and an authentication server. EAP provides some common methods to negotiate a desired authentication mechanism. There are many EAP methods, e.g., EAP-TLS (transport layer security), EAP-SIM (GSM subscriber identity module), EAP-AKA (UMTS authentication and key agreement), PEAP (protected extensible authentication protocol), LEAP (lightweight extensible authentication protocol), EAP-TTLS (tunneled transport layer security), etc. An IEEE 802.11 access point (AP) serving as a network access server (NAS) with the IEEE 802.1x capability can invoke EAP to provide a secure authentication mechanism and negotiate a secure pairwise master key (PMK) between a station and the AP. The PMK is used for sessions under temporal key integrity protocol (TKIP) or advanced encryption standard (AES) encryption.

- Robust security network (RSN) for associations: RSN dynamically negotiates the authentication and encryption algorithms to be used for communications between the AP and stations, using AES+CCMP (cipher block chaining message authentication code protocol) up to 256 bits, 802.1x, and EAP.
- AES-based counter mode with CCMP for confidentiality, integrity, and authentication.

This chapter focuses on TKIP. Although TKIP is just a temporary solution, it is likely that TKIP will remain for many years since millions of IEEE 802.11 devices have been sold, and they can be fixed by using TKIP, but not by others based on AES in IEEE 802.11i, which need complete replacement of hardware, i.e., wireless LAN cards.[2] In other words, TKIP provides a software solution instead of a hardware solution to the devices already in the market and in usage, whereas other solutions in IEEE 802.11i require users to replace current wireless LAN cards with new ones that support the new security features.

In this chapter, we provide a survey on security issues for the IEEE 802.11i specification with a focus on TKIP. The chapter starts with WEP and its known vulnerabilities, and then moves toward WPA, which is used to improve WEP for better security. Then we introduce EAP and TKIP. Finally we present possible attacks on TKIP and relevant countermeasures.

16.2 Wired Equivalent Privacy and Its Weakness

WEP is an encryption algorithm designed to provide security to the IEEE wireless LANs. WEP encryption algorithm was used in IEEE 802.11a, IEEE 802.11b, and IEEE 802.11g. The main intention of the designers was to encrypt the data which are sent from one system to the other over wireless channels (usually data are transferred between a station and the AP). WEP was designed for this purpose and uses RC4 stream cipher. The implementation of RC4 in WEP requires a secret key (K) of 40 bits and an initialization vector (IV) of 24 bits. The secret key is preshared between the AP and the stations. During the authentication process, the AP sends a nonce (n) to a station to check whether the station has the right key to encrypt the nonce. After receiving the nonce from the AP, the station encrypts the nonce using the preshared key, and sends back the encrypted nonce $E_K(n)$ to the AP, which checks whether the decrypted nonce and the original nonce sent to the station are the same. If they are the same, the authentication is successful. Otherwise the AP sends a failure message back to the station.

WEP is designed to achieve three main goals as follows:

- Confidentiality—prevent eavesdropping
- Data integrity—prevent message modification
- Access control—prevent unauthorized access

Unfortunately, none of the security goals was achieved by WEP.[3–5] Some vulnerabilities of WEP are listed as follows[4,5]:

- *Key sequence reuse*: During the encryption process, WEP uses the same key but a different IV to encrypt the data. Since the IV is 24-bit long, it can produce only a maximum of 2^{24} different vectors. Eventually, the same IV will be used after some time, producing the same key stream for two different plaintexts to be encrypted. By constantly watching the network, the intruder will find the repeated IV and henceforth can intrude into the network with the help of the key stream.
- *Poor key management*: The 802.11 standard does not specify the way of sharing keys among the stations before the actual authentication process takes place. In practice, the same key is shared among all the stations. Therefore, the possibility of the key stream being reused increases, making the system vulnerable to the key stream reuse attack. This practice seriously affects the security of the system that depends on a single key for its entire network.
- *Message tampering*: CRC (32-bit long) in IEEE 802.11 is to ensure data integrity and not to ensure data security. Since the CRC checksum function is linear and stream ciphers such as RC4 are also linear, it is possible to use simple XOR operations to tamper with the message in transit without being detected.

16.3 Wi-Fi Protected Access

To design a better security mechanism, the IEEE 802.11 working group formed a Task Group, group "i", to develop the IEEE 802.11i standard, which was intended to address the problems faced by the previous security system.

16.3.1   802.1X EAP-Based Authentication

WPA uses an EAP-based authentication process to address the problems in user authentication in WEP. This is a mutual authentication process between a station and an authentication server in which the authentication server checks the station's ID and the station checks the authentication server's ID. The station (user) is the one who wants to get authenticated and enter the network. The authentication server (radius server) handles the actual authentications. An authenticator, usually the AP, acts as a middleman to carry out the authentication process. Figure 16.1 shows the EAP authentication process. The process of EAP-based authentication is described as follows[1]:

- The station which wants to get authenticated into the network initiates the connection with the AP.
- The AP asks for the station's ID. The station sends its ID to the AP.
- The AP now sends the station's ID to the radius server. The radius server checks for the validity of the station and sends an ACCEPT message back to the AP.
- The AP marks the station's state as the authorized state.
- The station requests the radius server's ID. The ID is passed to the station from the radius server through the AP. The station is now authenticated to the network and now will be able to transfer data securely.

This type of authentication is not suitable for home or small firms because of its complexity. The maintenance of the central radius server needs more resources and a higher budget.

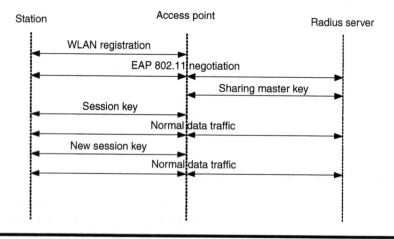

Figure 16.1   EAP authentication process.

A master secret key is calculated by the radius server and will be sent to the AP. Once the AP gets the master key, it makes a copy of the master key and sends the master key to the station. The AP generates the session key and encrypts the session key with the master key and sends it to the station. The session key is now shared between the station and the AP. The data is encrypted using the session key and is sent to the AP, i.e., the data is transmitted with the help of the session key.

16.4 Temporal Key Integrity Protocol

TKIP is proposed as a temporal solution until a strong encryption procedure is discovered. The main advantage of TKIP is that it uses the existing hardware to implement the new encryption procedure. TKIP is an enhanced version of WEP which solves most of the problems in WEP. Some of the features provided by TKIP are listed below[1]:

■ TKIP uses a key mixing function which mixes temporal address (TA), TKIP sequence counter (TSC), and temporal key (TK) to form the WEP seed. There are two phases in the key mixing function: key mixing phase 1 and key mixing phase 2. TSC is sent along with the MAC protocol data unit (MPDU) so that at the receiver end, the receiver can decrypt the message using TSC. The TSC here acts like an initialization vector in WEP.

■ The receiver will drop the MPDU at the other end if the received MPDU is not in sequence (ascending order). This avoids the replay attack.

■ IEEE 802.11i applies Michael key on MAC service data unit (MSDU) to produce message integrity code (MIC). The use of the Michael key ensures data integrity and keeps users from modifying the message.

16.4.1 TKIP Encryption Process

Encryption is the process of converting an originally intelligible message, called MSDU, into an encrypted message, called cipher text, using a temporal key which is shared among all the stations in the network. As we can see from Figure 16.2, there are two parts involved in this encryption process. One part generates the WEP seed and the other guarantees data integrity of the message.

The encryption process in the top part of Figure 16.2 calculates the WEP seed from the session or TK. TKIP uses a key mixing function to compute the WEP seed. The key mixing function has two phases: phase 1 and phase 2. Because of the mixture operations, the attacker does not get any conclusions about the key which is used in the generation of the key

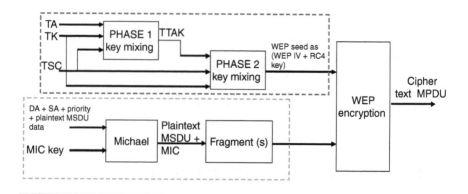

Figure 16.2   TKIP encryption process.

stream. In the generation of the WEP seed, TKIP uses transient address, TK, and TSC. Phase 1 mixes TA, lower part of TSC, and the TK. The output of the phase 1, i.e., TTAK, is given as an input to the next phase. Phase 2 mixes TTAK, the higher part of the TSC, and the TK to produce the desired output which is the WEP seed.[1]

The encryption process in the lower part of Figure 16.2 takes care of data integrity. The plaintext MSDU with its source address, destination address, and priority uses Michael key (64-bit long) to produce MIC. Michael key scheme is a keyed cryptographic hash algorithm which produces a 64-bit output. MIC is then applied to CRC (32 bits) to ensure that the sent data is the same as the one received at the other end. The resultant MIC is 64-bit long. The MSDU along with MIC can be fragmented into several fragments and then encrypted to form MPDUs. TKIP assigns an increasing TSC value to each fragmented MPDU. This is to make sure that all MPDUs obtained from the MSDU have the same extended IV. At the receiver end, MPDUs can be defragmented into the MSDU. The security of the Michael key relies on the assumption that the MIC values are always unknown to the attacker. This assumption is pretty unusual since the MIC should be designed in such a way that it should at least resist the known plaintext attack.

The WEP seed and the MSDU use XOR operation to form the encrypted data MPDUs. An MPDU is sent along with the TSC so that during the decryption process, the same TSC can be used to decrypt the MPDU at the receiver end.

16.4.2   TKIP Decryption Process

Decryption is the reverse process of encryption where the cipher text is converted into the plaintext. Figure 16.3 shows the decryption process. The cipher text is obtained along with the TSC (which is also called IV). TKIP

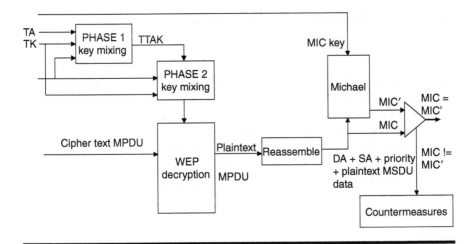

Figure 16.3 TKIP decryption process.

checks the sequence number of MPDU for its correctness. If the received MPDU violates the following rules, the TKIP rejects the MPDU. The rules for rejecting MPDUs are given below[1]:

- Each MPDU should have a unique TKIP TSC value.
- TSC values should be in an increasing order.
- Each transmitter will maintain separate TSC for each pairwise transient key (PTK) and group temporal key (GTK), where GTK is a random value used to protect broadcast/multicast MPDUs.

If an MPDU does not violate the above rules, the mixing function is applied on the MPDU to recover the original plaintext (MSDU). The key mixing phase 1 uses TK shared between the station and the AP after the authentication process, transient address, and the TSC to get the TTAK, which will be mixed in phase 2 with the TSC lower bits to provide the desired WEP seed. The MPDU uses Michael to get the MIC value. The MIC and WEP seed are applied on the decryption block to recover the original plaintext. Once the plaintext is obtained from that block, TKIP uses CRC (32 bits) to ensure that the received plaintext and the original plaintext sent to the station are the same. If both are the same, the authentication is successful; otherwise the AP sends a failure message back to the station.

The implementation of the encryption and decryption is very similar to that of the WEP encryption/decryption process with some minor changes. Most of the problems such as weak IVs, replay attacks, and modification attacks are avoided in the encryption and decryption processes of TKIP.

16.4.3 TKIP MPDU Message Format

Figure 16.4 depicts the layout of the encrypted message MPDU. The MAC header is the initial part of the MPDU. The IVs are split into IV with four octets containing TSC0, TSC1, and the extended IV with four octets containing TSC2–TSC5. The ExtIV bit in the IV octet indicates the presence of TKIP. For TKIP, the ExtIV is always set to 1, but for WEP, it is set to 0 since there does not exist any extended IV in the WEP. TSC5 is the most significant octet of TSC and TSC0 is the least significant octet of TSC. TSC2–TSC5 are used in the key mixing phase 1 along with the TA and TK. Octets TSC0–TSC1 are used in the key mixing phase 2 along with the output obtained from key mixing phase 1. The IVs are followed by the actual data or the encrypted message.[1]

16.4.4 TKIP MIC Processing Format

One of the most significant flaws of WEP is that it is vulnerable to active attacks. Michael key is used to generate the MIC to prevent active attacks. The solution given by Michael to prevent active attacks is not complete but it gives a better solution, compared to the WEP with limited resources. The TKIP MIC processing format, shown in Figure 16.5, contains MSDU

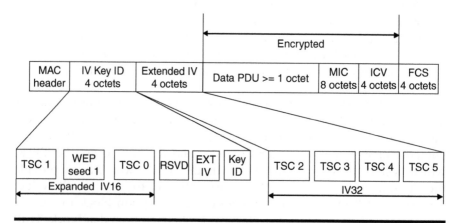

Figure 16.4 TKIP message format.

Figure 16.5 TKIP MIC processing format.

destination address (DA), MSDU source address (SA), MSDU priority, and unencrypted data (MSDU) along with the MIC. The priority octet will be 0 and will be reserved for future use.

16.4.5 TKIP Message Integrity Code

Michael is the most important part of TKIP's MIC. This is a new keyed cryptographic hash function. Michael is designed in such a way that the MIC value is always kept secret; furthermore, it is assumed that MIC values are unknown to attackers. If the MIC value is known to an attacker by some means, the attacker can get the Michael key because $m = Michael(K_{mic}, M)$, where $m = MIC$, $M = message$, and $K_{mic} = $ Michael key.

This assumption is weak since the MICs should at least be designed to prevent the known plaintext attack. The input given to the Michael block is a 64-bit key in the form of two 32 bits (K_0 and K_1) and a message (DA, SA, and the actual data), which is of 64 bits in length. The message is padded with a hexadecimal value 5a. This is to label the end of the message. The resultant MIC is 64 bits long. Each message word $[M_0, M_1, \ldots, M_{n-1}]$ is XOR-ed with one part of the key, K_0, and is applied on the block mixing function $b(\cdot)$. The block mixing function $b(\cdot)$ is an unkeyed four round Feistel structure, which uses shifting function, XOR function, and addition modulo. The Michael procedure for calculating the MIC is shown in Figure 16.6. The block procedure $b(\cdot)$ is given in Figure 16.7, where ROL indicates 32-bit left rotation and ROR indicates 32-bit right rotation. XSWAP is a function that swaps the position of the last 2 bits and the first 2 bits in a word.

16.4.6 TKIP Mixing Function

The TKIP mixing function has two phases. The first phase mixes the transmitter's address, TK, and TSC. The output of this phase, TTAK, is used as an input to the phase 2 mixing along with TSC and TK. The result of this

```
Procedure Michael ((K_0, K_1), (M_0, M_1, . . ., M_{n-1}))
Input given to the Michael block: Key (K_0, K_1) Message (M_0, M_1, M_2, . . ., M_{n-1})
Output: MIC value (V_0, V_1)
(l, r)←(K_0, K_1)
for i = 0 to N-1 do
     l← l ⊕ M_i
     (l, r)←b(l, r)
return (l ,r)
```

Figure 16.6 Michael.

Procedure b (l, r)

$r \leftarrow r \oplus$ (l ROL 17)

$l \leftarrow (l + r) \bmod 2^{32}$

$r \leftarrow r \oplus$ XSWAP (l)

$l \leftarrow (l + r) \bmod 2^{32}$

$r \leftarrow r \oplus$ (l ROL 3)

$l \leftarrow (l + r) \bmod 2^{32}$

$r \leftarrow r \oplus$ (l ROR 2)

$l \leftarrow (l + r) \bmod 2^{32}$

return (l, r)

Figure 16.7 $b(\cdot)$ in Michael.

Input : Transmit Address TA0-TA4 Temporal Key TK0-15 and TSC0-TSC5.
Output : Intermediate key TTAK0 - TTAK4
Phase 1-key mixing (TA0...TA5, TK0...TK15, TS0...TSC5)

PHASE1 Step_1:

TTAK0 $\leftarrow$ MK16 (TSC3, TSC2)

TTAK1 $\leftarrow$ MK16 (TSC5, TSC4)

TTAK2 $\leftarrow$ MK16 (TA1, TA0)

TTAK3 $\leftarrow$ MK16 (TA3, TA2)

TTAK4 $\leftarrow$ MK16 (TA5, TA4)

PHASE1 Step_2:

J $\leftarrow$ 2(j & 1)

TTAK0 $\leftarrow$ TTAK0 + S [TTAK4 XOR MK16(TK1+j, TK0 + j)].

TTAK1 $\leftarrow$ TTAK1 + S [TTAK0 XOR MK16(TK5+j, TK4 + j)].

TTAK2 $\leftarrow$ TTAK2 + S [TTAK1 XOR MK16(TK9+j, TK8 + j)].

TTAK3 $\leftarrow$ TTAK3 + S [TTAK2 XOR MK16(TK13+j, TK12 + j)].

TTAK4 $\leftarrow$ TTAK4 + S [TTAK3 XOR MK16(TK1+j, TK0 + j)]+i.

Figure 16.8   Phase 1 mixing.

phase produces a WEP seed, also called as per frame key. The two-phase process could be described as TTAK: = Phase1 (TK, TA, TSC), and WEP Seed: = Phase2 (TTAK, TK, TSC).

16.4.6.1 Key Mixing Phase 1

The phase 1 mixing algorithm is given in Figure 16.8. The inputs that are given to the key phase mixing phase 1 are TA, TK (128 bits), and the most significant 32 bits of the TSC. The output of the key mixing phase 1 is TTAK, which is 80 bits in length and is represented by an array of 16-bit values TTAK0, TTAK1, TTAK2, TTAK3, and TTAK4. XOR operation, bitwise AND operation, and the addition operation are used in phase 1. One of the functions used in key mixing phase 1 is MK16 that constructs a 16-bit value from two 8-bit inputs as $MK16(X, Y) = 256 \times X + Y$.

There are two steps in the key mixing phase 1 algorithm. The first step initializes TTAK from TSC and TA. The second step actually mixes all the information using XOR and MK16 functions, where *s* in Figure 16.8 is an S-box. The resultant output is TTAK, which is 80 bits in length.

16.4.6.2 Key Mixing Phase 2

The phase 2 mixing algorithm is given in Figure 16.9. The inputs to the phase 2 are TTAK0–5, which are the output of the key mixing phase 1. TK and the least significant bits of TSC are used in the key mixing phase 2. TTAK is 80 bits in length and they are split into five parts, i.e., TTAK0–TTAK4. The output from this phase is the WEP seed. The first 24 bits of the WEP seed will be transmitted in plaintext along with the encrypted message M. The mixing function uses XOR operation, addition operation, AND operation, OR operation, and the right bit shift operation ($>>$).

There are three steps involved in the key mixing phase 2. Step 1 makes a copy of the TTAK and reads in the TSC. Step 2 employs a 96-bit mixing

```
Input : intermediate Key TTAK0... TTAK4, TK, and TKIP sequence counter TSC
Output : WEP seed (WEPseed0 ... WEPseed15)
Phase2-key-mixing (TTAK0...TTAK4, TK0...TK15, TSC0...TSC5)
PHASE2_Step1:
     PPAK0 ←TTAK0
     PPAK1←TTAK1
     PPAK2←TTAK2
     PPAK3←TTAK3
     PPAK4←TTAK4
     PPAK5←TTAK4 + Mk16 (TSC, TSC0)
PHASE2_Step2:
     PPAK0 ←PPAK0 + S [PPAK5 XOR MK16 (TK1, TK0)]
     PPAK1 ←PPAK1 + S [PPAK0 XOR MK16 (TK3, TK2)]
     PPAK2 ←PPAK2 + S [PPAK1 XOR MK16 (TK5, TK4)]
     PPAK3 ←PPAK3 + S [PPAK2 XOR MK16 (TK7, TK6)]
     PPAK4 ←PPAK4 + S [PPAK3 XOR MK16 (TK9, TK8)]
     PPAK5 ←PPAK5 + S [PPAK4 XOR MK16(TK11, TK10)]
     PPAK0← PPAK0 + RotR1 [PPAK5 XOR MK16 (TK13, TK12)]
     PPAK1← PPAK1 + RotR1 [PPAK0 XOR MK16 (TK15, TK14)]
     PPAK2← PPAK2 + RotR1 (PPAK1)
     PPAK3← PPAK3 + RotR1 (PPAK2)
     PPAK4← PPAK4 + RotR1 (PPAK3)
     PPAK5← PPAK5 + RotR1 (PPAK4)
PHASE2_Step3:
     WEPSeed0 ←TSC1
     WEPSeed1 ←TSC1
     WEPSeed2 ←TSC0
     WEPSeed3 ←Lo8 ((PPK5 XOR MK16 (TK1, TK0)) >>1)
     For I = 0 to 5
          WEP Seed 4 + (2-i)← Lo8 (PPKi)
          WEPSeed4 + (2-1)← Hi8 (PPKi)
     End
Return WEPSeed0 ... WEP Seed15
```

Figure 16.9 Phase 2 mixing.

operation, and Step 3 reads in the last of the TK bits and assigns the 24-bit WEP IV value.

16.5 Fragility of Michael

16.5.1 Inverse Function of Michael

There exists a simple function InvMichael, which can recover the secret MIC key (K_{mic}) when a single known message M and its matching MIC value (m) are known.[2] This is because the mixing function does not depend on the key. The inverse Michael $b^{-1}(\cdot)$ works in exactly the opposite way as the actual Michael block function $b(\cdot)$ by altering between XOR and subtraction modulo 2^{32} operations. The block procedure for the inverse Michael function is given in Figure 16.10, and the pseudo procedure for the $b^{-1}(\cdot)$ function is given in Figure 16.11.

By this point, it is clear that by hacking MIC and message M, we can calculate the Michael key using the $b^{-1}(\cdot)$ function. This is a definite threat to the new security system.

$$
\begin{aligned}
&\text{Procedure InvMichael }((V_0, V_1), (M_0, M_1, \ldots, M_{n-1})) \\
&\text{Input given to the Inverse Michael block:} \\
&\qquad \text{MIC Value }(V_0, V_1), \\
&\qquad\qquad \text{Message }(M_0, M_1, M_2, \ldots, M_{n-1}) \\
&\text{Output: Key }(K_0, K_1) \\
&(l, r) \leftarrow (V_0, V_1) \\
&\text{for } i = 0 \text{ to N-1 do} \\
&\qquad (l, r) \leftarrow b^{-1}(l, r) \\
&\qquad l \leftarrow l \oplus M_i \\
&\qquad \text{return } (l, r)
\end{aligned}
$$

Figure 16.10 Inverse Michael.

$$
\begin{aligned}
&\text{Procedure } b^{-1}(l, r) \\
&\qquad l \leftarrow (l - r) \bmod 2^{32} \\
&\qquad r \leftarrow r \oplus (l \text{ ROR } 2) \\
&\qquad l \leftarrow (l - r) \bmod 2^{32} \\
&\qquad r \leftarrow r \oplus (l \text{ ROL } 3) \\
&\qquad l \leftarrow (l - r) \bmod 2^{32} \\
&\qquad r \leftarrow r \oplus \text{XSWAP }(l) \\
&\qquad l \leftarrow (l - r) \bmod 2^{32} \\
&\qquad r \leftarrow r \oplus (l \text{ ROL } 17) \\
&\qquad \text{return } (l, r)
\end{aligned}
$$

Figure 16.11 $b^{-1}(\cdot)$ in inverse Michael.

16.5.2 An Attack

Consider that an attacker has obtained two MPDUs such as M_1 and $M_2{}^2$ such that

- The same IVs are used to encrypt both the messages M_1 and M_2.
- M_2 is not longer than M_1.
- The plaintext P_1 of M_1 is known to the attacker.

By knowing the above information, an attacker can calculate K_{mic}. The IVs used to calculate the key stream are same for both the messages M_1 and M_2. M_1 and the plaintext 1 (P_1) are known to the attacker, but the MIC value is unknown to the attacker. The attacker can recover the key stream by using the formula $K[i] = M_1[i] \oplus P_1[i]$ for $i = 1, 2, 3, \ldots,$ length (M_1). By calculating the key stream, we can calculate the MIC value by $m = K \oplus M_1$, where K is the key stream. By finding the MIC value and the plaintext 2, we can calculate the Michael key by applying InvMichael to the MIC value and the message.

16.6 TKIP Countermeasures

As we have seen in the previous section, Michael provides a weak protection against active attacks. A successful attack against the Michael MIC means that the attacker could inject unwanted data frames and perform attacks like data modification against the encryption key. TKIP has come up with countermeasures,[1] if TKIP detects an active attack on the network. The countermeasures are given as follows. When TKIP finds two MIC failures within 60 s, the station and the authenticator which detect the MIC failures must be disabled for a period of 60 s. By doing this, it will become difficult for the attackers to inject a large number of forged data frames in a short period of time.

The PTK and the GTK must be changed once the station or the AP detects the MIC failure.

To calculate the time, a timer is used to log MIC failures. The supplicant (wireless device), when it finds an MIC failure, sends an MIC failure report back to the AP. A Michael MIC failure report is a key frame with the MIC bits, the error bits, the request bit, and the secure bit all set to 1. MLME-MICHAELMIC FAILURE.indication is used to indicate Michael MIC failure to a local supplicant and the authenticator. When the system meets an MIC failure, it relies on the log sheet and a timer to check whether there will be another MIC attack in 60 s seconds. If there is no MIC failure in another 60 s, the process continues; otherwise the AP and the station connected to the authenticator are disconnected for about 60 s. After 60 s, the station gets reassociated to the network and hence carries out the data transfer through the authenticator.

■ *TKIP countermeasures for the authenticator.* When the authenticator receives a frame with a Michael MIC error, it discards the affected frame, increases the Michael MIC failure counter by one, generates the MLME-MICHAELMIC FAILURE.indication primitive, starts the timer, and checks if the Michael MIC failure is reported again. If it is reported again within 60 s, the AP deauthenticates, deletes all PTK for all stations using TKIP, and discards the GTK, and after 60 s, new PTK and GTK are constructed. TKIP countermeasures for the authenticator are shown in Figure 16.12.

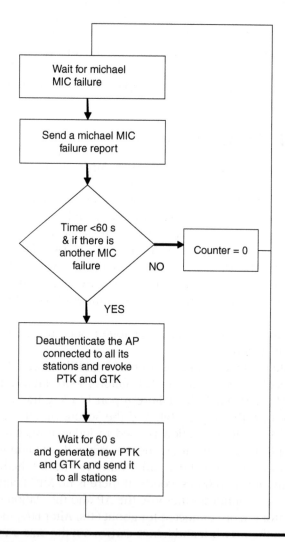

Figure 16.12 TKIP countermeasures for authenticator.

■ *TKIP countermeasures for the supplicant:* When the supplicant receives a Michael MIC error frame, it discards the affected frame, increments the Michael MIC failure counter by one, and generates an MLME-MICHAELMIC FAILURE.indication primitive. For the supplicant which receives a MLME-MICHAELMICFAILURE.indication primitive from the station, it sends a Michael MIC failure report frame to the AP. If in less than 60 s of the previous detected failure, another MIC failure occurs, the supplicant deletes the PTK and GTK, deauthenticates the station from the authenticator, and waits for 60 s before getting reassociated with the same authenticator.

16.7   Key Handshake Procedure

The second layer of the authentication process is involved between a station and the AP.[1,6] The key handshake process is shown in Figure 16.13. The station has already obtained the preshared master key from the AP through the EAP authentication process. The station that wishes to get authenticated to the AP sends a request to the AP. The AP sends an ANonce to the station to check whether the station has the right master key to enter into the network, as shown in Step 1 in Figure 16.13. After receiving the ANonce from the AP, the station encrypts the ANonce with the master key and sends

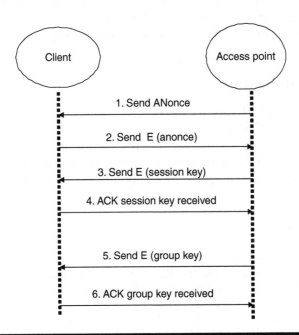

Figure 16.13   Key handshake process.

it back to the AP, as shown in Step 2 in Figure 16.13. Once the AP receives the encrypted ANonce, it decrypts the message and checks whether the original message sent to the station and the message received from the station are the same. If both the messages are the same, the AP allows the station into the network; otherwise the AP rejects the station. If the AP accepts the station, it encrypts the session key or the temporal key using the master key and sends it to the station, as shown in Step 3 in Figure 16.13. On receiving the session key, the station sends an acknowledgment back to the AP, as shown in Step 4 in Figure 16.13.

This session key is valid only for a specified session time. Once the session key expires, the AP sends a new session key to the station by encrypting it using the master key. The station uses the new session key to transfer the message to the other stations through the AP. The session key is used to defend against the key stream reuse attack. The disconnected user problem is also solved in this process. If a user goes out of the network for a period of time and then comes back to the network later, the user might get disconnected from the network since he/she might be using an expired session key. In the key handshake process, if the user is disconnected from the network, he/she can come into the network by requesting for the session key from the AP. The AP encrypts the session key with the master key and sends it to the station. The station gets back the session key through this process. The AP also sends the encrypted group key to the station, as shown in Step 5 in Figure 16.13. The group key is used to broadcast the message to all the stations connected to that AP. An acknowledgment is sent back to the AP from the station, as shown in Step 6 in Figure 16.13.

16.8 Conclusions

TKIP uses the same hardware to come up with a better security solution than the previous WEP security system. Although TKIP is a temporary solution for the wireless LAN security, it will exist for many years since other solutions in the IEEE 802.11i need replacements of hardware, i.e., the old wireless LANs cards. Using TKIP, users do not need to buy new cards.

References

1. IEEE 802.11i standard, 2004.
2. A. Wool, "A note on the fragility of the 'Michael' message integrity code," *IEEE Transactions on Wireless Communications*, Vol. 3, No. 5, 2004, pp. 1459–1462.
3. W. Arbaugh, N. Shankar, and J. Wang, "Your 802.11 network has no clothes," *Proceedings of the 1st IEEE International Conference on Wireless LANs and Home Networks*, December 2001.

4. N. Borisov, I. Goldberg, and D. Wagner, "Intercepting mobile communications: The insecurity of 802.11," *Proceedings of the 7th Annual International Conference on Mobile Computing and Networking*, pp. 180–188, 2001.
5. Y. Xiao, C. Bandela, X. Du, Y. Pan, and K. Dass, "Security mechanisms, attacks, and security enhancements for the IEEE 802.11 WLANs," *International Journal of Wireless and Mobile Computing* (accepted).
6. C. He and J. Mitchell, "Analysis of the 802.11i 4-way handshake," *Proceedings of ACM WiSe'04*, pp. 43–50, 2004.

INTEGRATED SYSTEMS

Chapter 17

Wireless Mesh Networks: Design Principles

Wai-Hong Tam, Chih-Yu Lin, and Yu-Chee Tseng

17.1 Introduction

IEEE 802.11 wireless local-area networks (WLANs) have been widely accepted recently. However, to cover a larger-scale area, multiple access points (APs) are required and each AP has to be connected to a wired network. Thus, the cost is not well justified. The *wireless mesh network* (WMN)[1] provides a good solution to this problem. WMNs have been receiving much attention from both the academia and the industry. WMNs can provide broadband wireless access for wider areas without costly infrastructures. Therefore, WMNs have great potential to solve the last-mile problem and to provide ubiquitous broadband wireless access. Moreover, A WMN benefits from easy deployment and easy maintenance because of its self-configuring and self-healing capabilities. Owing to the advantages mentioned above, the WMN technology may be used to construct a wireless backbone for metropolitan or rural areas. In this chapter, we will present some design principles of networking technologies for building and maintaining a WMN.

17.2 Generic Architecture and Basic Requirements of Wireless Mesh Networks

To begin with, we describe the general architecture of WMNs. The architecture is shown in Figure 17.1, where dash lines indicate wireless links.

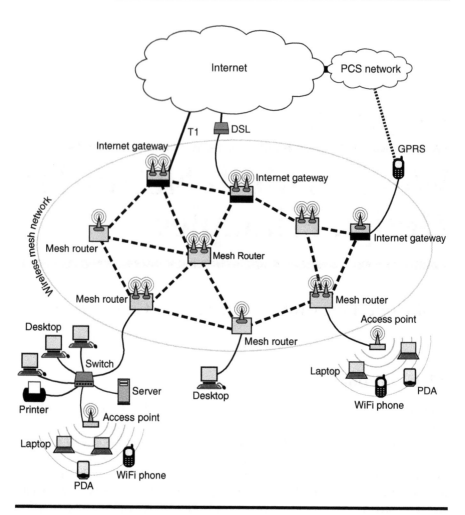

Figure 17.1   The generic WMN architecture.

A WMN consists of multiple *mesh routers* (or simply called *nodes*), which can forward packets on behalf of other mesh routers through wireless communication. To further improve the performance, a mesh router may be equipped with more than one radio interface. Mesh routers that are connected to the Internet are called *Internet gateways* (or simply called *gateways*). Mesh routers usually have minimal mobility and are usually fixed, such as being deployed on roofs or streetlight poles. Like any access point, a mesh router can serve local mobile stations in its area. The stations may be desktop/laptop PCs, PDAs, WiFi phones, etc. The mesh routers wirelessly relay packets of stations to other routers until they reach a gateway that connects to the Internet. More generally, instead of serving as a

WiFi access point, a mesh router can also interoperate with a local-area network such as an Ethernet, Bluetooth, or sensor network.

WMNs can be applied to many application scenarios. For example, the WMN architecture can be used as a residential network.[2,3] Traditionally, each family in a community may use cable or digital subscriber line (DSL) to connect to the Internet. The Internet service provider may need to send an engineer to each family to install and set up the network. In addition, information sharing inside the community always has to go through the Internet. With the help of a WMN, the cost can be reduced. WMN can also be used in a metropolitan area, as shown in Figure 17.2. The mesh routers form a low-cost wireless backbone to provide network access in a metropolitan area. Since WMNs usually have a higher data rate than cellular networks, many advanced broadband services can be provided (e.g., multimedia services and voice-over-IP services).

A WMN is conceptually similar to a mobile ad hoc network (MANET), in which mobile stations can communicate with each other in a multi-hop fashion. However, there are some differences between a WMN and a MANET. First, the mesh routers in a WMN are usually stationary. On the contrary, each node in a MANET may be capable of mobility. This feature may influence the protocol design. For example, some routing protocols are designed for recovering broken routes efficiently in MANETs. The same protocol, if applied to a WMN, may not be efficient. Thus, the protocol

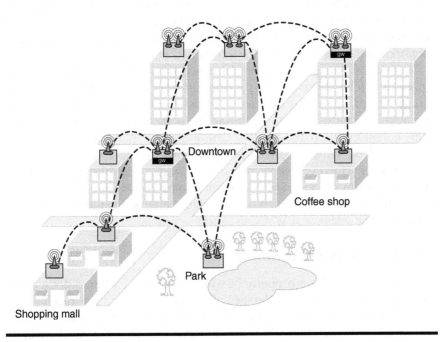

Figure 17.2   A WMN as a metropolitan-area network.

design should be reconsidered. Second, network throughput should be the major concern in WMNs, because a WMN may serve hundreds or even thousands of stations. The scalability problem will be an important issue in WMNs. Third, a WMN should be autoconfigurable, because the deployment area may be wide. Otherwise, the configuration and maintenance costs do not justify its benefits. To achieve this goal, autoconfiguring technologies are needed. Fourth, power consumption is not a critical issue in WMNs. Typically, mesh routers are plugged to electric outlets and a mesh router may concurrently serve many traffic flows. Therefore, it is reasonable to use more advanced hardware in mesh routers.

17.3 Network-Planning Techniques

WMNs are generally constructed in two ways. The first way is to carefully construct a network by placing mesh routers in preferred locations, equipping them with appropriate numbers of antennas, and attaching some of them to the Internet.[3–5] These jobs are usually done by wireless Internet service providers (WISPs) with design goals of high throughput and good connectivity. The other way is to adopt unplanned autonomous mesh routers, which have the ability to form networks without the use of any centralized infrastructure.[2,6,7] Such a network allows nodes to configure themselves automatically and attach themselves to the existing mesh network dynamically and is thus easy to be deployed. However, the lack of planning might reduce the network performance. This section and the next section will cover these techniques: *network planning* for planned networks and *self-configuring* for unplanned networks.

Network planning is typically concerned with the cost-efficient deployment of a wireless backbone to provide sufficient connectivity, throughput, and quality of service for end users. To overcome these constraints, some researchers suggest using multiple radios in each mesh router to provide more communication bandwidth. One key issue is the *channel assignment* problem, which involves assigning channels to radio interfaces of mesh routers. However, some researchers suggest establishing additional Internet gateways to increase the Internet portal bandwidth. This raises the *gateway placement* problem, which involves determining the exact number of Internet gateways in a WMN and their locations.

17.3.1 Channel Assignment

Owing to the broadcast property of wireless medium, when a node transmits a packet, the other nodes near the transmitter and the receiver are forbidden to initiate any transmission. A WMN may suffer from dramatic performance degradation when the network size grows. Use of multiple nonoverlapping channels instead of a single channel is an efficient method to overcome this problem. The IEEE 802.11b standard and IEEE 802.11a/g standard offer 3 and 12 nonoverlapping channels, respectively. Each radio interface can be switched to one of the available channels appropriately. When a pair of radio interfaces are on the same channel and are within each other's transmission range, they can communicate with each other. Two transmissions on different channels in a neighborhood are allowed to happen at the same time without any interference. Enabling multiple simultaneous transmissions within a neighborhood, which can improve the network throughput, is the goal of channel assignment.

To benefit from using multiple channels, each node can be equipped with either multiple interfaces, each of which is associated with a fix channel, or one single interface, which is capable of switching (or hopping) among multiple channels frequently. Both these approaches involve a channel assignment issue. We divide the channel assignment problem into four classes: (i) static assignment for a long interval of time (such as every time when the topology changes), (ii) dynamic assignment for a shorter interval of time (such as about 100 ms), (iii) dynamic assignment per packet, and (iv) dynamic assignment per flow. Class (i) assignment strategies usually require equipping each node with multiple interfaces, such as inexpensive commodity 802.11 network interface cards (NICs), without any hardware or device driver modification. After being assigned with a channel, each radio interface no longer switches its channel again, unless the network topology or the traffic loads change. Class (ii) and (iii) assignment strategies may affect the behavior of the medium access control (MAC) protocol. They always require some modification of device drivers or some special hardware. Class (iv) assignment strategies are mainly combined with the path selection protocol being used. In this subsection, we will focus on class (i) channel assignment strategies and introduce two representative methods in detail.

17.3.1.1 Load-Aware Channel Assignment

In Ref. [8], assuming that each node may have multiple interfaces, Raniwala et al. propose a centralized channel assignment algorithm to satisfy the traffic load requirement on each link. The channel assignment depends on the expected load on each link, which in turn depends on the routing protocol being used. However, routing also depends on the link capacity, which is determined by the channel assignment. Thus, there is a circular dependency between radio channel assignment and packet routing. In Ref. [8], to break this circularity, the expected load of each link without regard to the link capacity is used as an initial estimation, that is, the load of each link is considered as the input to the scheme. Then the algorithm performs channel assignment and routing steps iteratively until the available bandwidth of each link satisfies its expected load as much as possible.

Initially, given the estimated traffic loads for all communicating node pairs, a WMN topology, and the number of nonoverlapping channels, the algorithm first estimates the *initial link capacity* of each link, say (A,B), to be the total aggregated channel capacity divided by the total number of links in node A's and node B's interfering ranges. For example, given that the capacity per channel is 11 Mbps and that the number of nonoverlapping channels is 3, if there are 6 links within node A's and node B's interfering ranges (included link (A,B)), the initial link capacity of link (A,B) is $\frac{3*11}{6}$ Mbps. The algorithm then performs a routing based on these initial link capacities to obtain the initial routing paths between all communication pairs, and thus the *initial expected load* on each link.

Afterward, based on the expected load on each link and the number of network interfaces available on each node, the goal of this channel assignment algorithm is to assign each network interface to an appropriate channel such that the resulting available bandwidth of each link is at least equal to its expected traffic load. Raniwala et al.[8] show that this problem is *NP-hard* and present a greedy approach. In the algorithm, links in the WMN are visited in the decreasing order of the expected loads of links. When a link (A,B) is traversed, it is assigned a channel based on the current channel assignment of interfaces of nodes A and B. A concept called *degree of interference* is defined, which is the sum of expected loads on links in the interference region of nodes A and B that are also using the same channel as link (A,B). More specifically, when conducting channel assignment for link (A,B), there are three possible cases:

- Both node A and node B have one or more free interfaces (i.e., they have not been assigned with channels). We will assign a channel to link (A,B) that has the least degree of interference.
- One of the nodes, say node A, has one or more free interfaces, but the other node B has no free interface. We will choose a channel that has been assigned to an interface of node B and assign it to the link. The selected channel is the one which incurs the least degree of interference. This implies that an interface of A will be assigned to this selected channel, but node B will use the interface that has been assigned to this channel to communicate with A.
- All interfaces of node A and node B are occupied. If there are common channels shared by interfaces of node A and node B, we will pick the common channel that minimizes the degree of interference, and assign it to link (A,B). Otherwise, we will pick one channel from node A and one channel from node B, merge them into one common channel, and assign this merged channel to the link. Again, the choice of the two channels to be merged is the one such that the combined degree of interference of the two channels is minimized.

After the current iteration, if some of the link loads exceed their capacities, the algorithm will go back to the routing step to recalculate all routing paths. This will give a new expected load for each link. Then the algorithm will repeat again (based on the new link loads) to find a better channel assignment, and compare the new link loads with the new link capacities. This iterative process keeps on going until all link loads are satisfied or no further improvement is possible.

In summary, the algorithm performs the following sequence of steps:

1. Estimate the initial link capacity.
2. Perform routing to obtain the initial expected load.
3. Determine a channel assignment based on the expected load.
4. Calculate the new link capacity.
5. Perform routing to obtain the new expected load.
6. If all link loads are satisfied, exit the algorithm; otherwise, go to Step 3.

17.3.1.2 Connectivity-Preserving Channel Assignment

With the previous approach, it is possible that the length of the routes between nodes may increase. Also, the network may be partitioned if there is no communication demand between some network partitions. Therefore, Marina and Das[9] propose a connectivity-preserving channel assignment algorithm to solve this problem. Connectivity preserving means all links that are connected if all interfaces are set to a single common channel remain connected after the channel assignment. This implies that the shortest path between any two network nodes remains available after the channel assignment and there is no network partitioning concern. However, this approach is not concerned with the traffic demand. The goal of the algorithm is to provide a connected and low-interference topology in a *traffic-independent* manner.

The channel assignment algorithm is based on a greedy heuristic. At the beginning, each node is given an initial priority based on some criterions. The algorithm assigns channels to nodes in the order of their priorities. However, during the process, it may raise the priorities of some nodes to a value greater than the current maximum priority for those nodes which lack flexibility in assigning their channels. In other words, a node whose interfaces are all occupied will receive the highest priority. In addition, when facing with a decision to pick a channel, it always makes a locally optimal choice from the feasible set of channels.

Consider the example shown in Figure 17.3. There are two interfaces in each of the nodes A, C, and D, and there is only one interface in each of the nodes B and E. The initial order of priorities is A, E, B, C, and D. Based on the priority, the algorithm starts at node A to assign channels to

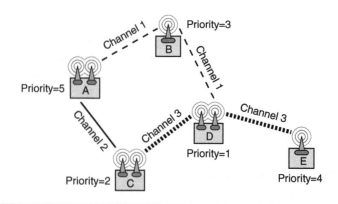

Figure 17.3 An example of the connectivity-preserving channel assignment.

its incident links. Suppose that it chooses channel 1 for link (A,B). This assignment causes node B to lose its flexibility in choosing channels for its other incident links. So the algorithm will raise node B's priority to the highest. This will allow node B to start assigning channels to its incident links, which will result in node B reusing channel 1 for link (B,D). Then, going back to node A, suppose that it chooses channel 2, which will cause the least interference to its neighborhood, for link (A,C). Next, node E has the next highest priority and can assign channels. Suppose that node E chooses channel 3 for link (E,D). Because of the lack of flexibility of node D, it will raise its priority to the highest and assign channel 3 to the link (D,C). Finally, all network interfaces have been assigned with channels.

17.3.1.3 Other Approaches for Channel Assignment

Besides the above two channel assignment schemes, there are some channel assignment algorithms that consider different metrics. In Ref. [10], the objective is to find a channel assignment, which maximizes the number of simultaneous transmissions to increase the network throughput. Das et al.[10] formulate this problem as an optimization problem, and use two mixed integer linear programming methods to solve it. Fujiwara and Matsumoto[11] argue that the main reason for the TCP throughput degradation in a multihop environment is the exposed terminal problem. A channel allocation principle is proposed to alleviate this degradation. The channel assignment problem is formulated as a node-coloring problem by transforming the carrier sensing condition into a chromatic graph.

17.3.2 Gateway Placement

Another network-planning issue is the gateway placement problem. In a WMN, gateways serve as portals to the Internet and thus their locations are critical to the performance of the WMN. Since traffics may be congested nearby the gateway areas, a poor placement strategy may lead to performance degradation. Also, the number of gateways reflects the wiring cost. For these reasons, the solution to the gateway placement problem is to deploy the minimum number of gateways such that the traffic demands of users can be satisfied.

In Ref. [12], the placement decisions are based on the network topology, user demands, and wireless link characteristics. Suppose that the locations of all nodes and the possible positions of gateways are known. Each node in the WMN also knows its traffic demand. The goal is to determine whether a gateway needs to be opened or not, such that all nodes can obtain their required bandwidth, without violating the wireless link capacity constraints. Qiu et al.[12] use three coarse-grained interference models to capture the throughput degradation of wireless links due to interference among competing nodes. Consider the traffic flows of multiple source–destination pairs in the network. The first model, which is an ideal link model, assumes that the throughputs of all links constituting a flow are all the same, regardless of the length of the flow. This model is appropriate for environments with very efficient use of spectrum, such as using directional antennas, power control, multiple radios, and multiple-channel technologies. The second model, which is a bounded hop-count model, assumes that the throughputs of all links constituting a flow are all the same if the length of the flow is less than a threshold; otherwise, the flow throughput is zero. The third model, which is a smooth throughput degradation model, assumes that the throughputs of all links constituting a flow are inversely proportional to the length of the flow. Note that the above models are to approximate the throughputs of flows by taking interference and contention among nodes into account.

With these simplified interference models, Qiu et al.[12] develop some polynomial-time greedy placement algorithms. For the first model, the placement problem is formulated as a network flow problem. The algorithm uses a greedy heuristic to determine which gateways need to be opened. The main idea is to iteratively pick a new gateway that maximizes the total satisfied user demand when the gateways chosen in the previous iterations are all opened. Computing the total satisfied demand relies on a maximum-flow algorithm.

Consider the network in Figure 17.4(a). Nodes A to H are hosts, and nodes X, Y, and Z are possible gateways. The algorithm first transforms the network architecture into a flow network as shown in Figure 17.4(b). Initially, all the gateways are closed, such that link (X, sink), link (Y, sink), and link (Z, sink) are all disconnected. In the first iteration, we will try to open gateway X and perform a maximum-flow algorithm to find the total flow f_X that can be supported from the source to the sink. Also, we can try to open each of gateways Y and Z and find the total flows f_Y and f_Z, respectively, from the source to the sink. We then choose to open the gateway

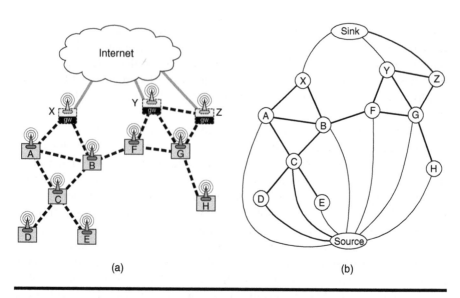

Figure 17.4 An example of the gateway placement algorithm. (a) The network architecture. (b) The corresponding flow network.

that produces the maximal flow. This is called as iteration. This process will be repeated until all demands are satisfied or all gateways are opened.

For the second and third models, the placement is formulated as a linear programming problem. Similarly, it iteratively selects gateways to maximize the total user demands, but computing the total satisfied demand relies on a linear-programming algorithm. Although the interference models are not accurate, Qiu et al.[12] demonstrate that the greedy algorithms give close-to-optimal solutions over a variety of scenarios.

17.4 Self-Configuring Techniques

In an unplanned environment, a network provider may like to spend minimal setup and administration overheads to build a WMN. Because of this, adding a new node or replacing an existing node should be as simple as plugging it in and turning it on. It is desirable that a new node can automatically join the network and dynamically configure itself to adapt to the current state of the network. Then, it can freely communicate with other nodes.

Before connecting to the Internet, a node typically needs to perform a sequence of operations. First, it has to acquire an IP address so that the other nodes in the network can identify it. Additionally, it shall switch its radios to appropriate channels if the multiple-channel model is used. Moreover, it needs to select the best gateway to access the Internet. Also, a robust path-selection protocol and a media access control protocol are required.

17.4.1 Address Autoconfiguration

Currently, IP address configuration can be classified into two types: global (public) addressing and local (private) addressing. In the former case, each node is assigned a globally unique IP address, which is valid in the Internet and has been properly registered. In the latter case, however, private unregistered addresses are introduced. This is allowed because the address realm in a WMN is isolated from the global Internet. So IP addresses can be reused among different WMN subnetworks and this approach greatly extends the scalability. However, mapping of a local address to a global address is necessary, which can be resolved using network address translation (NAT).

In general, the address of a new node can be manually configured in advance, dynamically assigned by dynamic host configuration protocol (DHCP), or dynamically assigned by an autoconfiguration method. Pre-configuration is not always possible and may take a number of operational overheads. Since it is not desirable to rely on a central component in a WMN, the traditional DHCP approach may not work properly (for example, the DHCP server itself may be disconnected from the network). Hence, address autoconfiguration is essentially required to enable dynamic address assignment in a WMN. In the following, we will introduce some address autoconfiguration protocols that have been proposed for wireless ad hoc networks or WMNs.

17.4.1.1 Hardware-Based Solution

We first describe an obvious solution to the address autoconfiguration issue.[2] This method relies on the hardware address (usually MAC address), which uniquely identifies every piece of network interface card (NIC). A new node can choose a local 32-bit IP address whose lower 24 bits are the same as the lower 24 bits of the NIC's MAC address and whose higher 8 bits are the private class-A network prefix of the WMN. For example, suppose that the NIC's MAC address is 01:02:03:04:05:06 and the network prefix is 10.0.0.0/8, the selected IP address is 10.4.5.6. Addresses obtained in this way are likely to be unique only inside the WMN. The gateways must provide NAT functionality for private/public address translation. This autoconfiguration method is easy to implement, but sometimes it may lead to address conflict.

17.4.1.2 Ad Hoc Address Autoconfiguration

In Ref. [13], the *duplicate address detection* (DAD) is introduced. Initially, a new node selects a temporary address randomly from the address range 169.254.0.0/16, which is registered with the Internet Assigned Numbers Authority (IANA) for this purpose. Then, a query-based DAD is performed

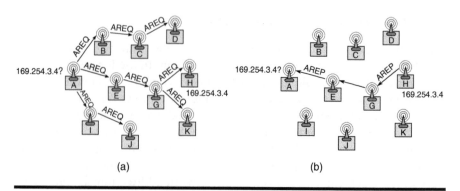

Figure 17.5 An example of the query-based DAD. Node A happens to select the same address as node H's. (a) Address request. (b) Address reply.

to guarantee the uniqueness of the selected address. During this process, the new node broadcasts an *address request* (AREQ) message containing its selected address. If the address is already in use, the node that uses the same address replies with an *address reply* (AREP) message to the requesting node. Unicasting this AREP back relies on establishing a reverse route in the earlier broadcast process. The absence of an AREP after a certain time interval implies the uniqueness of the requested address. However, this method may fail when two network partitions are merged together. An example is shown in Figure 17.5, which shows the process how a new node A selects an address same as node H's.

17.4.1.3 Weak DAD

Contrary to the above approach, the weak DAD[14] only ensures that packets are routed to the correct destination when duplicate addresses exist. Vaidya[14] argues that the "strong" DAD, which guarantees every address is unique, is sometimes infeasible if the network can be merged and partitioned dynamically. In the weak DAD, at the initialization time, a node will generate a key (e.g., the MAC address or a random number) and distribute this key along with its address in all routing packets. All other nodes will learn and store both the address and the key in their routing tables, and distribute further. Once a node detects two routing entries with the same address but different keys, a duplicate address is discovered. The node then marks the entry as invalid to prevent misrouting of data packets and notifies other nodes about the presence of the duplicate address. The same address and key generated by two different nodes can also cause address conflict, but the possibility is relatively low and this situation can be alleviated by increasing the key length. Figure 17.6 shows an example how node E detects that an address conflict happens to node A and node B.

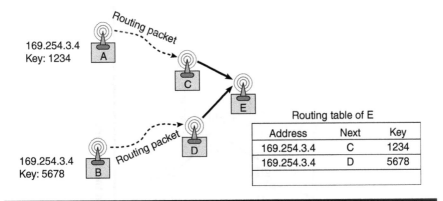

Figure 17.6 An example of the weak DAD. Node E detects that node A and node B use the same address.

17.4.1.4 MANETconf

In MANETconf,[15] each node maintains an address allocation table that contains all addresses either occupied or available in the network. A new node first broadcasts a *neighbor query* message to its direct neighbors. From the replies, one of its neighbors will be picked as the *initiator*. The initiator chooses an address that is available according to its allocation table, and floods an *initiator request* message containing this selected address and asks for a positive *initiator response* message from each known node to indicate that the address is available for use. If the initiator receives a positive reply from each node, the allocation is successful. The initiator assigns the proposed address to the requester using a *requester reply* message and informs the network about this allocated address by flooding an *address announce* message. Figure 17.7 shows the whole process how node A successfully obtains a new address. In addition, MANETconf handles the case of network merging and partitioning by using the idea of partition ID. However, the address allocation time of MANETconf may be much longer depending on the number of requests.

17.4.2 Dynamic Channel Selection

At powering up, a mesh router should dynamically determine to which channels its radio interfaces should switch. Unlike the channel assignments discussed in Section 17.3.1, it does not need a control center to coordinate the channel assignment, and, thus, it is more suitable for unplanned networks. In the following, we will describe four dynamic channel selection techniques. The first two are aimed at the single-interface model, and the last two are aimed at the multiple-interface model.

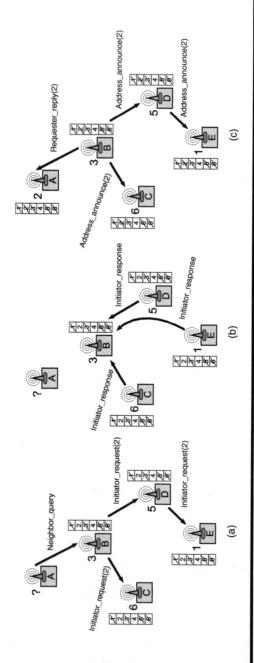

Figure 17.7   An example of MANETconf. Node A obtains an address through the existing nodes in the network.

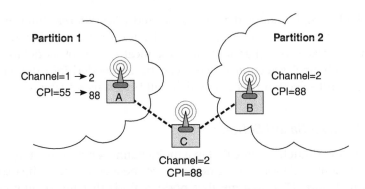

Figure 17.8   An example of the IEEE 802.11s SEE-mesh's channel unification protocol. Node C is a newly joining node, which will trigger partition 1 to choose channel 2 so as to join partition 2.

17.4.2.1   Indicator-Based Scheme

The IEEE 802.11s SEE-Mesh proposal[16] specifies a simple channel-unification protocol. If devices are not powered on at the same time, and nodes are allowed to choose their channels randomly, the network may be divided into several partitions. The SEE-Mesh protocol provides a con-trolled way for a mesh router to change its channel such that the radio interfaces of all nodes are merged into the same channel and hence the network is fully connected. This is achieved by using *channel precedence indicator (CPI)*, which is a random number, in each node. Each node maintains the highest known CPI in the network. When two or more dis-joint network partitions discover each other, the one with a lower CPI will change its channel to merge with the one with a higher CPI. Nevertheless, this protocol is only feasible for the single-interface environment.

In the example shown in Figure 17.8, suppose partition 1 and partition 2 are turned on at different time and choose channels 1 and 2, respectively. Also, suppose that nodes A and B are too far away and they do not realize the presence of each other. The CPIs of partitions 1 and 2 are 55 and 88, respectively. When a new node C that can hear both nodes A and B joins the network, it will choose to join the partition with the highest CPI, that is, partition 2. Then, it will inform partition 1 to merge with partition 2 by changing its channel from 1 to 2. This information will be propagated along partition 1 and all nodes will be merged into the same channel.

17.4.2.2   Gateway-Based Scheme

The above indicator-based protocol is very simple, but the whole WMN merely uses one channel. To reduce interference, So and Vaidya[17] suggest that neighboring partitions that are associated with different Internet gateways can operate on different channels. Therefore, a node should switch its radio interface to a channel that is identical to that of its default gateway. This channel selection process can be embedded into the gateway discovery and selection process, which will be discussed later.

17.4.2.3 Tree-Based Scheme

Raniwala and Chiueh[18] argue that in a multichannel environment, the channel dependency relation among nodes might be a concern. In the example shown in Figure 17.9, assume that node A finds that the quality of link (A,C) becomes worse and thus decides to switch one of its radio interfaces from channel 1 to a better channel 5. This implies that node C also needs to change one of its interfaces from channel 1 to channel 5. This in turn implies that link (C,F) also needs to change from channel 1 to channel 5. This ripple effect further propagates to link (F,G) and link (G,I). As can be seen, node I may set both of its interfaces to channel 5, which is inefficient. The channel dependency relationship makes it difficult for an individual node to predict the effect of a local channel change.

To solve this problem, Raniwala and Chiueh[18] propose a tree architecture for a multichannel WMN. Under this tree structure, the interfaces of each node are divided into two subsets. Some interfaces are used to communicate with the node's parent and thus are assigned the same channels as its parent's channels. The other interfaces are used to communicate with its children and the node is free to choose the least interfered channels according to the node's neighborhood condition. This selection is started from the root node and is gradually propagated to its child nodes. Whenever there is some modification of channels at a node, it only affects its

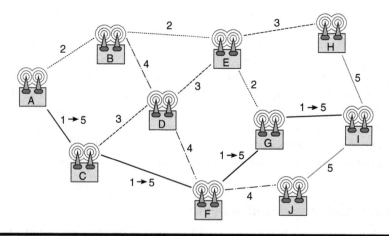

Figure 17.9 An example to illustrate the channel dependency problem.

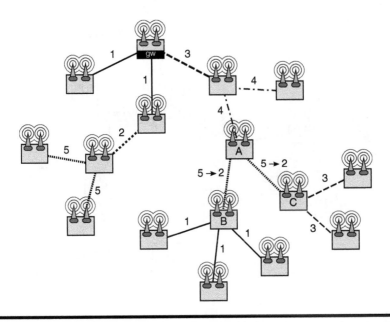

Figure 17.10 An example of the tree-based channel selection algorithm. The change of a channel at node A is only rippled to nodes B and C.

child nodes. In the example shown in Figure 17.10, when node A changes its channel from channel 5 to channel 2, only its child nodes B and C are affected. Therefore, this distributed channel selection protocol effectively reduces the channel dependency problem.

17.4.2.4 Cluster-Based Scheme

Another method to prevent the channel dependency problem is based on clustering.[19] All nodes in the same cluster use the same channel to communicate. This channel is selected by the cluster head based on neighborhood condition. Different clusters might use different channels to reduce interference. The intercluster communication is achieved by using a common default channel. Figure 17.11 shows an example of the cluster-based channel selection.

17.4.3 Gateway Discovery

In a WMN, since a large amount of traffic may be intended to access the Internet, each mesh router should discover all reachable gateways and then select one of them as its default gateway. Because of the dynamic nature of WMNs, gateway discovery is nontrivial.

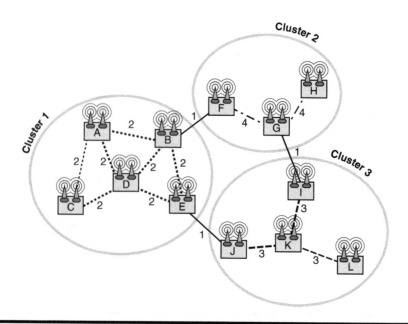

Figure 17.11   An example of the cluster-based channel selection.

Generally, there exist three types of gateway discovery approaches: active discovery, passive discovery, and a hybrid method. In active discovery, the search is initiated by the mesh router. Upon initialization, a mesh router will send out a *gateway request* message. This request is broadcasted or multicasted via a specific *Gateway Multicast Address*, which represents the group of all Internet gateways in the WMN. When a gateway receives this request, it replies a *gateway response* message via unicasting. This gateway response message contains the gateway's address, hop count to the requesting node, current traffic load, quality of links, etc.

In passive discovery,[2,18,20,21] each gateway periodically broadcasts *gateway advertisement* messages to announce its presence. All nodes within its transmission range are responsible for forwarding the advertisements further in the network. A new node can passively listen to the advertisements from its neighbors to collect all gateways' information. The gateway discovery time is highly dependent on the period of advertisements.

The hybrid method[17] involves both active and passive discovery methods. Similar to the passive discovery method, each gateway periodically broadcasts advertisements to the network and this information is stored by all nodes. When a new node sends an active gateway request message, an intermediate node with stored gateway information can directly reply a gateway response message. Hence, this method not only reduces the signaling traffic but also decreases the latency of gateway discovery.

17.4.4 Gateway Selection

When a node detects its reachability to more than one gateway, it should select the best gateway as its default gateway according to a certain metric. This decision is called *gateway selection*. The goal is to select a default gateway to provide a better quality of Internet access. The quality can be mainly affected by two factors: the link quality on the route and the traffic load on the route. The former is concerned with the bandwidth and the bit error rate (BER) of each wireless link. The latter is concerned with network congestion. Below, we describe several approaches.

17.4.4.1 Link-Quality-Based Scheme

In the current implementation of WLAN, each mobile station associates with the AP that has the strongest received signal strength indicator (RSSI). The RSSI is a simple indication of the link quality. However, in a multihop environment, the estimation of link quality is more complicated.

Routing metrics are widely used for route selection. Thus, most of the gateway selection algorithms are directly based on routing metrics. As a simple solution, a node can select a gateway with the minimal hop count.[21] Other metrics, such as estimated transmission time (ETT)[2] or weighted cumulative expected transmission time (WCETT),[22] can also be used.

17.4.4.2 Traffic-Load-Based Scheme

The goal of traffic-load-based gateway selection algorithms is to balance the loads among gateways to avoid bottlenecks and to utilize the network resource more efficiently. Below, we present two gateway selection algorithms proposed in Ref. [18].

The first algorithm is aimed to alleviate the bottleneck problem. It is based on the *gateway link capacity*, which represents the residual capacity of the gateway link that connects the gateway to the Internet. Once a new node associates with a gateway, the residual capacity of the gateway link is recomputed by subtracting the current usage of all associated nodes from its overall capacity. Each node will select the gateway with the maximal gateway link capacity as its default gateway, and then sends a *gateway association request* message containing its demanded traffic load. If the gateway has enough capacity to serve the demand, it updates its residual capacity and sends a *gateway association reply* message to the associated node.

Being more general, the second algorithm considers that bottlenecks can take place in any link of a path, instead of always in gateway links. In a real-world network, because of interference from other radio sources, intermediate wireless links could become bottlenecks too. Therefore, this algorithm is based on the *path capacity*, which indicates the minimum residual bandwidth of all constituting wireless links on the path. The residual capacity of a wireless link is approximated by subtracting the aggregate usage of its interfering sources on the same channel from its raw capacity. Since the selection is determined by the residual capacity of the path, this algorithm can balance the traffic loads among all paths. However, more control overhead is required. After adding a new node, not only the associated gateway needs to update its residual capacity, but also all nodes interfered by the path need to update their residual capacities.

17.5 Conclusions

In this chapter, we have presented a variety of design principles for WMNs. From the perspective of deployment, the way to efficiently utilize the limited resources (e.g., wireless channels, radio interfaces, and Internet bandwidth) is an interesting and challenging issue. We introduce two critical problems, the channel assignment problem and the gateway placement problem, and discuss some solutions. From the perspective of maintenance, minimizing the setup cost and administration overhead is always desired. We have given a comprehensive overview of self-configuring techniques from address autoconfiguration to gateway selection. With the help of these techniques, WMNs can become more friendly to both network providers and subscribers.

References

1. I.F. Akyildiz, X. Wang, and W. Wang. Wireless mesh networks: A survey. *Comput. Networks ISDN Syst.*, 47(4):445–487, 2005.
2. J. Bicket, D. Aguayo, S. Biswas, and R. Morris. Architecture and evaluation of an unplanned 802.11b mesh network. In *Proc. MobiCom*, August 2005.
3. R. van Drunen, J. Koolhaas, H. Schuurmans, and M. Vijn. Building a wireless community network in the Netherlands. In *USENIX/Freenix Conference*, June 2003.
4. P. Bhagwat, B. Raman, and D. Sanghi. Turning 802.11 inside-out. In *Proc. HotNets-II*, November 2003.
5. B. Raman and K. Chebrolu. Revisiting MAC design for an 802.11-based mesh network. In *Proc. HotNets-III*, November 2004.
6. Microsoft Mesh Networking. http://research.microsoft.com/mesh/.
7. Seattle Wireless. http://www.seattlewireless.net/.
8. A. Raniwala, K. Gopalan, and T.-C. Chiueh. Centralized channel assignment and routing algorithms for multi-channel wireless mesh networks. In *Mobile Computing and Communications Review*, pp. 50–65, April 2004.
9. M.K. Marina and S.R. Das. A topology control approach for utilizing multiple channels in multi-radio wireless mesh networks. In *Proc. BROADNETS*, October 2005.

10. A.K. Das, H.M.K. Alazemi, R. Vijayakumar, and S. Roy. Optimization models for fixed channel assignment in wireless mesh networks with multiple radios. In *Proc. SECON*, September 2005.

11. A. Fujiwara and Y. Matsumoto. Centralized channel allocation technique to alleviate exposed terminal problem in CSMA/CA-based mesh networks. *IEICE Trans. Commun.*, 8(3):958–964, 2005.

12. L. Qiu, R. Chandra, K. Jain, and M. Mahdian. Optimizing the placement of integration points in multi-hop wireless networks. In *Proc. ICNP*, October 2004.

13. C.E. Perkins, J.T. Malinen, and R. Wakikawa. IP address autoconfiguration for ad hoc networks. *IETF Internet Draft*, November 2001.

14. N. Vaidya. Weak duplicate address detection in mobile ad hoc networks. In *Proc. MobiHoc*, June 2002.

15. S. Nesargi and R. Prakash. MANETconf: Configuration of hosts in a mobile ad hoc network. In *Proc. INFOCOM*, June 2002.

16. IEEE 802.11 TGs Simple efficient extensible mesh (SEE-Mesh) proposal.

17. J. So and N.H. Vaidya. Routing and channel assignment in multi-channel multi-hop wireless networks with single network interface. In *Proc. QShine*, August 2005.

18. A. Raniwala and T.C. Chiueh. Architecture and algorithms for an IEEE 802.11-based multi-channel wireless mesh network. In *Proc. INFOCOM*, March 2005.

19. J. Zhu and S. Roy. 802.11 mesh networks with two radio access points. In *Proc. ICC*, May 2005.

20. C.-F. Huang, H.-W. Lee, and Y.-C. Tseng. A two-tier heterogeneous mobile ad hoc network architecture and its load-balance routing problem. *Mob. Network Appl.*, 9(4):379–391, 2004.

21. J. Shin, H. Lee, J. Na, A. Park, and S. Kim. Gateway discovery and routing in ad hoc networks with NAT-based Internet connectivity. In *Proc. VTC*, September 2004.

22. R. Draves, J. Padhye, and B. Zill. Routing in multi-radio, multi-hop wireless mesh networks. In *Proc. ACM MobiCom*, September 2004.

Chapter 18

Wireless Mesh Networks: Multichannel Protocols and Standard Activities

Chih-Yu Lin, Wai-Hong Tam, and Yu-Chee Tseng

18.1   Introduction

We have investigated the architecture of wireless mesh networks (WMNs) and some planning/self-configuring techniques for WMNs in the previous chapter. With those techniques, a WMN can be formed. However, protocols are needed for communication among mesh routers. Mesh routers themselves form a network that is similar to a multihop ad hoc network. Therefore, we can consider to apply the communication protocols designed for ad hoc networks to WMNs. However, there are still some differences between ad hoc networks and WMNs. WMNs are typically stationary, and most of the traffics are intended for the Internet. Besides, power consumption and hardware cost are not major concerns in WMNs.

In addition, many standard activities for mesh networks have been conducted by industrial groups. For example, IEEE 802.15,[1] IEEE 802.16,[2] and IEEE 802.11[3] have all set up subworking groups to investigate standards for mesh networks. The mesh networks mentioned in the IEEE 802.15 subworking group are aimed at home networking. The mesh networks mentioned in the IEEE 802.11 and 802.16 subworking groups are aimed at extending the reach of the network to areas inaccessible to fixed base

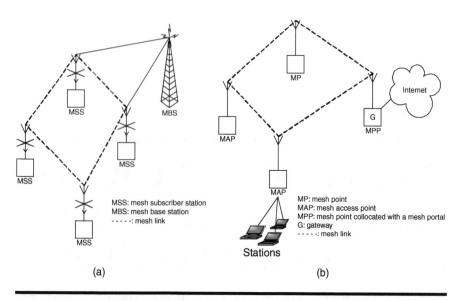

MSS: mesh subscriber station
MBS: mesh base station
- - - -: mesh link

MP: mesh point
MAP: mesh access point
MPP: mesh point collocated with a mesh portal
G: gateway
- - - -: mesh link

(a)

(b)

Figure 18.1 Architectures of 802.16 and 802.11 mesh networks. (a) An 802.16-based mesh network. (b) An 802.11-based mesh network.

stations. Figure 18.1 compares the architectures of the 802.16- and 802.11-based mesh networks. Note that a mesh network can also be used as a backbone of some local-area networks. For example, in an 802.11-based mesh network, each mesh node can be connected to a wireless LAN with an 802.11 access point. This access point can serve some stations under the access point's coverage.

In this chapter, we will focus on the medium access control (MAC) protocols and the routing protocols to be used among mesh routers for an 802.11-based mesh network. Note that the MAC used in IEEE 802.16 is based on TDM/TDMA, so the results discussed in this chapter may not be applicable to IEEE 802.16 mesh networks. Finally, we will also give a brief survey of some standard activities.

18.2 Multichannel MAC Protocols

MAC protocols are used to solve the contention and collision problems. However, because of the interference problem, MAC is critical to the performance of WMNs. One way to improve the throughput of a WMN is using multiple channels to reduce interference. Conceptually, when a radio interface experiences interferences, it is allowed to switch to a different channel, instead of waiting in the same channel. For example, three nonoverlapping channels are available in IEEE 802.11b.[4] However, the MAC protocol in IEEE 802.11b only allows a radio interface to operate on one channel at a time. Therefore, it deserves to investigate the design of multichannel

MAC for WMNs. In this section, we will review some multichannel MAC protocols that have been proposed for WMNs.

18.2.1 Multichannel MAC Protocols with a Single Transceiver

Consider the radio interfaces that are used for communication among mesh routers. Here we assume that each mesh router has only one such interface. We will look at the MAC issue of allowing one interface to operate on multiple channels. In the next subsection, we will further investigate allowing multiple transceivers in a mesh router. In a single-transceiver model, two problems should be addressed:

- *Channel switching problem.* Because each node has only one transceiver and the transceiver can only operate on one channel at a time, channel switching is required to support multiple channels. For example, if two nodes intend to communicate with each other and these two nodes operate on different channels, then at least one of them has to switch its current channel. Channel switching may lead to other problems such as the *multichannel hidden terminal* problem and the *time synchronization* problem, which will be discussed later.
- *Channel selection/scheduling problem.* When a node with a transceiver decides to change its operational channel, it also has to decide which channel it intends to switch to. There exist several possible approaches to this problem. For example, the sender and the receiver can negotiate a channel by using some mechanisms, after which they can switch to the negotiated channel to transmit packets. Another way is using a schedule. First, time is divided into slots and each node will maintain a channel usage schedule that defines which channel should be used in each slot. Based on its schedule, a node can "hop" among different channels.

Below, we will introduce two multichannel MAC protocols under the single-transceiver model: *multichannel MAC* (MMAC)[5] and *slotted seeded channel hopping* (SSCH).[6] Before introducing these two protocols, we will discuss the multichannel hidden terminal problem that may occur in a multichannel environment.

18.2.1.1 The Multichannel Hidden Terminal Problem

To illustrate the multichannel hidden terminal problem,[5] we make the following assumptions. We assume that each node will use a specific channel for a fixed period and it will use another channel when the period expires.

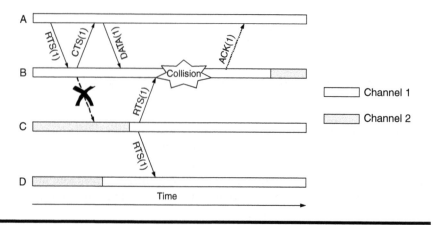

Figure 18.2 The multichannel hidden terminal problem.

However, nodes are not time synchronized. In each channel, the IEEE 802.11 MAC protocol is used. Consider the scenario in Figure 18.2. In the beginning, node A and node B exchange their RTS and CTS using channel 1, and node C listens on channel 2, so it does not hear the CTS transmitted by node B. Then, when node A starts to send its DATA packet to node B, node C switches its used channel to channel 1 and transmits an RTS to node D on channel 1. This will cause a collision at node B, which is referred to as the *multichannel hidden terminal problem*. Note that if only one channel is used, node C can hear the CTS sent by node B and thus the collision can be avoided. We will explain how MMAC and SSCH solve this problem.

18.2.1.2 MMAC

MMAC[5] is designed based on the IEEE 802.11 *power-saving mechanism (PSM)* used in ad hoc mode. First, we give a brief introduction to IEEE 802.11 PSM used in ad hoc mode. In this mechanism, time is divided into beacon intervals. Beacon intervals of nodes are synchronized by periodic beacon transmissions. A beacon interval is further divided into two parts. The first part is called the *ad hoc traffic indication message (ATIM) window*. The second part can be used for data transmission as in typical 802.11 MAC or for energy conservation by switching the node to the doze mode. Figure 18.3 illustrates how the PSM works. If node A intends to send a packet to node B, it will send an ATIM packet to B during the ATIM window. When B receives the ATIM packet, it replies an ATIM–ACK to A. Then, both A and B can exchange packets after the ATIM window. On the contrary, if a node does not receive or send any ATIM packet, it can go to sleep after the ATIM window to save energy.

MMAC uses a mechanism similar to that used in IEEE 802.11 PSM. The basic idea is that a common channel will be used for all nodes during the

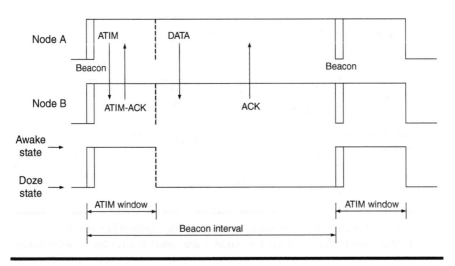

Figure 18.3   The IEEE 802.11 power-saving mechanism.

ATIM window. After the ATIM window, nodes can select different channels to transmit or receive packets. The packet transmission procedure is described as follows. If node A intends to transmit a packet to node B, it will send an ATIM packet to node B with A's *preferable channel list* (*PCL*) that indicates which channels are preferable to use for node A. When node B receives the ATIM packet, it will choose a channel based on its own PCL and A's PCL. Generally, the less used (or interfered) channel will be selected. (The detailed channel selection rules will be described later.) Then node B can reply an ATIM–ACK packet that includes the selected channel. When node A receives the ATIM–ACK packet, it will check whether it can select the specified channel. If A cannot select the specified channel, it can try again in the next beacon interval. Otherwise, node A will reply an ATIM–RES packet to node B. After the ATIM window, nodes A and B will switch to the selected channel to communicate. Figure 18.4 shows the procedure mentioned above. In this example, the common channel used during ATIM windows is channel 1 and node B selects channel 2 to communicate with node A.

In a node's PCL, each channel is categorized as *high preference* (HIGH), *medium preference* (MID), or *low preference* (LOW). For each node, at most one channel will be in the HIGH state. In the beginning, each channel will be associated with a counter initialized to zero. The channel states will be changed according to the following rules:

- All channels are reset to the MID state at the beginning of each beacon interval.
- If two nodes agree on a channel, the channel will be set to the HIGH state by both nodes.

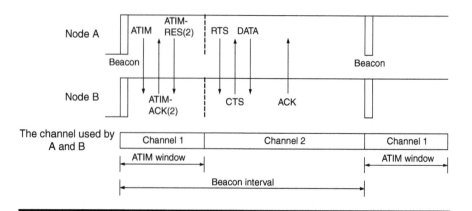

Figure 18.4   A transmission example of MMAC. The number in parentheses denotes the channel that will used by node A and node B after the ATIM window.

■ When a node x overhears an ATIM–ACK or ATIM–RES sent by another node, three cases are considered. (1) If the state of the channel carried by the ATIM–ACK/ATIM–RES packet is MID, x will change the state of the channel to LOW and increment the counter for that channel by one. (2) If the state of the channel is LOW, the counter for that channel will be incremented by one. (3) If the state of the channel is HIGH, its state will remain the same.

Now we return to the above example and describe how node B selects a channel based on its PCL and A's PCL. The selection procedure used by node B is as follows: (1) If there is a HIGH state channel in B's PCL, this channel will be selected. (2) Otherwise, if there is a HIGH state channel in A's PCL, this channel will be selected. (3) Otherwise, if there is a MID state channel in both A's and B's PCLs, this channel will be selected. (4) Otherwise, if there is a MID state channel in A's or B's PCL, this channel will be selected. (5) Otherwise, this means that all of the channels are in the LOW state. For each channel, the corresponding counters of the A's and B's PCL will be summed together. Then the channel with the least count will be selected.

To summarize, MMAC solves the multichannel hidden terminal problem by time synchronization and by prohibiting nodes from switching channels after the ATIM window of each beacon interval. Ref. [5] discusses a possible scenario in which a node is allowed to switch its channel after an ATIM window. When a node switches its channel after an ATIM window, it should listen on that channel for a period that is long enough before transmitting a packet to avoid the multichannel hidden terminal problem.

18.2.1.3 SSCH

In MMAC, a sender and a receiver negotiate a channel using an on-demand approach. An alternative approach is the *match-based* scheme. For example, the *slotted seeded channel hopping (SSCH)* is a link-layer protocol using such an approach. SSCH addresses the packet and channel scheduling issues. For the MAC part, it can adopt the IEEE 802.11 MAC protocol.

In SSCH, time is divided into fixed-length slots. Each node will maintain its own channel schedule for slots. Packet scheduling is to address how a node transmits buffered packets in a slot. Channel scheduling is to address how a node chooses channels such that (1) the number of slots that it can use to communicate with its intended receivers should be as many as possible (a slot can be used if the channel selected by the sender in that slot matches the channel selected by the receiver), and (2) for each neighboring node, there always exist some slots such that the node uses the same channel as the neighboring node. The second condition is to avoid the network being partitioned.

To begin with, we introduce the packet scheduling mechanism in SSCH. Each node will maintain a FIFO queue for each neighbor. These queues are ordered based on certain priorities. At the beginning of a slot, one buffered packet will be drawn for transmission from the first queue. If the transmission completes successfully, then one packet will be drawn from the second queue and the same procedure will continue until all packets in queues are transmitted or the current slot expires. However, if the transmission of any queue's packet fails, then the priority of that queue will be reduced. For example, when a node A intends to transmit a packet to its neighbor node B. If node A fails to transmit a packet to node B, then the priority of the queue for B will be reduced. This is because node B may use a channel that is different from that used by node A. Thus, to reduce the waste of bandwidth, the priority of the queue for B should be reduced. More details about the packet scheduling can be found in Ref. [6].

For channel scheduling, each node will maintain m pairs of numbers, $(x_1, a_1), (x_2, a_2), \ldots, (x_m, a_m)$, where m is a fixed integer, x_i represents a channel, and a_i represents a seed. Note that x_i is an integer in the range $[0, n-1]$ and a_i is an integer in the range $[1, n-1]$, where n is the total number of available channels and it should be a prime number. Each pair will be used once per m slots. In the first m slots, the channels used by the node are $x_1, x_2, \ldots, x_m$, respectively. In the next m slots, x_i will be set to $((x_i + a_i) \bmod n)$, $i = 1, \ldots, m$, and the channels used by the node are the new $x_1, x_2, \ldots, x_m$, respectively. This process is repeated for each set of the forthcoming m slots. It is not hard to observe that when the process is repeated n times, the channels used by the node will return to the original $x_1, x_2, \ldots, x_m$. Thus, $n \times m$ slots can be referred to as a *cycle*. However, to avoid the network partitioning problem mentioned earlier, one additional

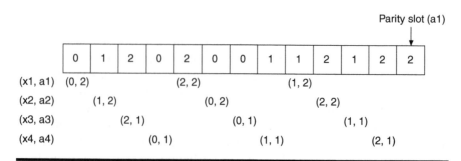

Figure 18.5   An example of channel scheduling in SSCH.

slot called *parity slot* is inserted after every $n \times m$ slots. For parity slots, we always use channel a_1. Therefore, the length of a cycle is $(m \times n) + 1$ slots. Figure 18.5 shows an example where $m = 4$ and $n = 3$. Now we explain why the parity slot can be used to prevent the logical partition problem. Three scenarios should be considered. For simplicity, we assume that $m = 1$. (1) If the pairs used by two nodes are identical, then obviously they can communicate with each other in each slot. (2) If the pairs used by the two nodes are different and their seeds are different too, it can be shown that these two nodes will meet exactly once in each cycle. To see this point, suppose that in a cycle, the two nodes use pairs (x, a) and (x', a'), respectively, such that $a \neq a'$. There always exists exactly one k, where k is an integer in the range $[0, n-1]$, such that

$$(x + ka) \equiv (x' + ka')(\mathrm{mod}\ n)$$
$$\Leftrightarrow (x - x') \equiv ((a' - a)k)(\mathrm{mod}\ n)$$

because $(a - a') \neq 0$ and n is a prime number. (3) If the pairs used by the two nodes are different and the seeds are identical, then these two nodes cannot communicate with each other without parity slots. Thus, the parity slots are used to prevent the network partitioning problem.

One important issue in SSCH is the selection of the pairs (x_i, a_i), $i = 1, \ldots, m$. In WMNs, most traffic flows are multihop flows. To support multihop flows, SSCH introduces a *partial synchronization* technique. For example, in Figure 18.6, there is a flow from node A to node B and then to node C. Here we assume $m = 2$ and $n = 3$ in this scenario. Node A can "synchronize" with node B using its first pair (x_1, a_1). That is, node A will set its (x_1, a_1) to be the (x_1, a_1) currently used by node B. Because node B has to relay A's packets to node C, node B can also "synchronize" with node C using its second pair (x_2, a_2). Since neighboring nodes' parameter pairs are not completely synchronized, this is called partial synchronization. More issues about the synchronization can be found in Ref. [6]. Finally, note

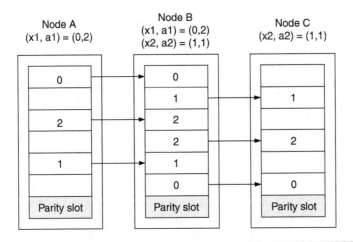

Figure 18.6 An example of the partial synchronization technique in SSCH.

that a transceiver must wait for the maximum length of a packet transmission after switching to a new channel in SSCH. Therefore, the multichannel hidden terminal problem is avoided.

18.2.2 Multichannel MAC Protocols with Multiple Transceivers

Using multiple transceivers is another possibility to improve the performance of WMNs. Under the multitransceiver model, solutions can be divided into two classes. In the first class, transceivers are given different functionalities or missions; for example, one transceiver is used to transmit control packets and others are used to transmit data packets. In the second class, transceivers are independent of each other and each transceiver may use an existing single-channel MAC protocol. However, a virtual MAC protocol designed in the link layer is required to integrate all transceivers. We will introduce one protocol, called DCA,[7] belonging to the first class and two protocols belonging to the second class.

18.2.2.1 DCA

Dynamic channel assignment (DCA)[7] is a multichannel MAC protocol using two transceivers, one dedicated to transmitting control packets and the other dedicated to transmitting data packets. The first transceiver always operates on a dedicated channel, and the second transceiver may switch among several available data channels.

Now we introduce how the DCA works. Each node x will maintain a *channel usage list (CUL)* that records the channel usage states of x's neighbors. A *free channel list* (*FCL*) will be dynamically computed from *x*'s CUL. Suppose that a node, say A, intends to transmit a packet to another node, say B. First, node A will check (1) whether the data transceiver of node B will be free to receive data and (2) whether node A itself has a free data channel to be used for communicating with node B. If both answers are affirmative, then A will send an RTS with A's FCL on its control transceiver to B. When node B receives A's RTS, it will check its CUL and A's FCL to determine whether there exists a free channel that can be used for communicating with node A. If there is no free channel, then node B will reply a CTS to node A indicating when A can try again (note that this is possible because control packets will carry the durations of nodes' reservations and this information is recorded in nodes' CULs). Otherwise, node B will choose a free channel and reply a CTS indicating the selected channel. When node A receives the CTS, it will switch its data transceiver to the selected channel and send its DATA packet to B. At the same time, node A will send an RES (reservation) packet to its neighbors on its control transceiver. Note that nodes who overhear the CTS sent by node B or the RES sent by node A will update their own CULs. Figure 18.7 shows the operation of DCA, where it is assumed that channel 1 is the control channel. Because the control transceiver is always listening on the control channel, the multichannel hidden terminal problem will not happen.

18.2.2.2 MUP

An alternative way to utilize multiple transceivers is to allow transceivers of a node to work independently of each other. In this approach, each transceiver operates on a specific channel. However, we will need a link layer protocol (also called *virtual MAC protocol*) to unify these transceivers.

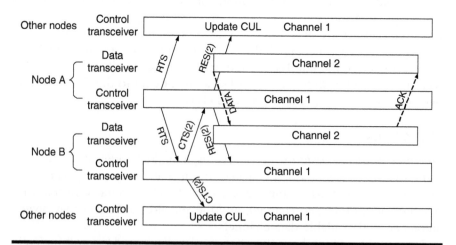

Figure 18.7 An example of the operation of DCA.

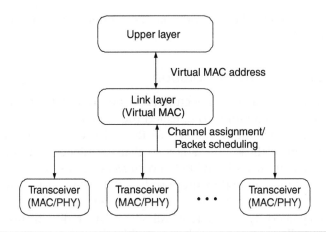

Figure 18.8 The high-level architecture diagram of the virtual MAC protocols.

Figure 18.8 shows the high-level architecture of the protocol. Two tasks should be done by the virtual MAC protocols. The first one is channel assignment. The second one is to choose a transceiver for each packet coming from the upper layer. Note that the upper layer only sees the whole system as a single NIC. The *multiradio unification protocol (MUP)*[8] is designed based on this philosophy.

MUP is a link layer protocol. In MUP, each node is assumed to have multiple radios, each of which uses the standard-compliant IEEE 802.11 hardware and operates on a specific channel. MUP is responsible for coordinating multiple radios operating on different channels. In MUP, once a channel is assigned to a transceiver, it will not be changed. However, each MUP-enabled node can dynamically decide which transceiver is used for communicating with each of its neighbors. The decision is based on the channel quality, which is estimated as follows. A node will send a *channel select (CS)* message over each transceiver periodically. When its neighbor receives the CS message, a *CS–ACK* message will be replied immediately. Then the node which sends the CS message will calculate the round-trip time (RTT), and it will incorporate the RTT measurement into a weighted average called *smoothed RTT* as follows[8]:

$$SRTT = \alpha RTT_{new} + (1 - \alpha)SRTT \qquad (18.1)$$

The weighted average is used as the channel quality estimate. When a randomized time interval (typical values for this interval are in the range of 10–20 s) expires, a node is allowed to change the selected transceiver for each neighbor. The decision is based on the SRTT mentioned above. However, the channel assignment problem is not addressed in Ref. [8]. Also, when the number of radios of a node is smaller than the number

of available channels, the network may be partitioned if channels are not
chosen properly.

18.2.2.3 A Cluster-Based Solution

A cluster-based two-radio architecture is proposed in Ref. [9]. Each node is
assumed to have two radios: one called the *primary radio* and the other
called the *secondary radio*. (Similar to Ref. [8] a virtual MAC protocol is
used to coordinate these two radios so that the higher layer only sees a
single NIC.) The network is divided into *clusters* (for example, the *highest-
connectivity cluster* algorithm proposed in Ref. [10] can be used for the
clustering). All primary radios of nodes in the network will use a common
channel designated by the system. All secondary radios of nodes in the
same cluster will use a common channel, which is chosen using a *minimum
interference channel selection (MIX)* algorithm, by which a clusterhead will
select the secondary radio channel (denoted as k) with the minimum energy
on air. Let $\overline{E}_{ij}$ denote the average energy on the ith channel sensed by node
j. The MIX algorithm will choose the k such that

$$\overline{E}_{kj} = \min\{\overline{E}_{1j}, \overline{E}_{2j}, \ldots, \overline{E}_{nj}\} \qquad (18.2)$$

where n is the number of available channels in the system. Figure 18.9
shows an example of the proposed channel assignment scheme. With this

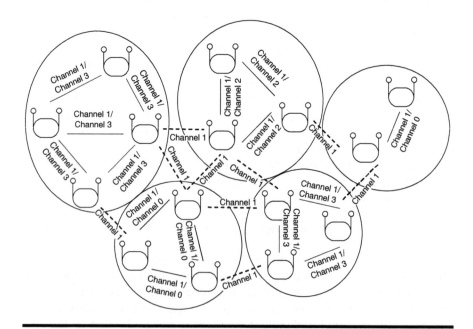

Figure 18.9 An example of the two-radio architecture.

two-radio architecture, intercluster traffics and control/management traffics are transmitted on the primary radio, and intracluster traffics are transmitted on the secondary radio. Note that the logical partition problem will not happen, because each node has a primary radio operating on a common channel used by all nodes in the network.

18.3 Multichannel Routing Protocols

In this section, we will discuss the design of routing protocols that adopt multiple channels. Four protocols will be reviewed.

18.3.1 Hyacinth

To begin with, we introduce a tree-based routing protocol. We assume that most of the user traffic is directed to/from gateways. A tree is constructed for each gateway (i.e., the tree is rooted at the gateway) to perform routing. Based on such a tree-based routing, Ref. [11] proposes an IEEE 802.11-based multichannel WMN architecture, called the *Hyacinth* architecture. Both channel assignment and routing are addressed. First, we introduce the channel assignment mechanism in Ref. [11]. Transceivers on a node are classified into two classes: *UP–NICs* and *DOWN–NICs*. UP–NICs of a node are used to communicate with the node's parent and the channels used by the UP–NICs are determined by its parent. However, DOWN–NICs of a node are used to communicate with the node's children and the channels used by DOWN–NICs are determined by the node itself. To assign channels to its DOWN–NICs, a node has to estimate the channel usage status within its interference neighborhood. To achieve this, each node will periodically exchanges a CHNL_USAGE packet that carries its channel usage information with all its $(k + 1)$-hop neighbors, where k is the ratio of the interference range and the transmission range (k is typically between 2 and 3). Then, each node will estimate the aggregate traffic load of a particular channel by summing up the loads contributed by all its interfering neighbors that use this channel. To account for the MAC overhead, the *total load of a channel* is a weighted combination of the aggregated traffic load and the number of nodes that use the channel. Based on this information, a node can determine the channels to be used by its DOWN–NICs.

For packet routing, the routing tree is constructed in a top-down manner from the gateway to leaves. The gateway will broadcast ADVERTISE packets periodically. When a node, say x, receives an ADVERTISE packet from a node, say y, the following actions will be taken by x. If x does not have a path to the gateway or it finds that the metric of the path through y to the gateway is better than the metric of its current path, then x will set y as its parent by sending a JOIN packet to y. Upon receiving the JOIN packet, node y will send an ACCEPT packet to node x. Note that if the old parent of x, say y', exists, then x will also send a LEAVE message to y'. After finishing these steps, x will have a path to the gateway and x will also start to broadcast ADVERTISE packets periodically. Note that when node x builds a new path to the gateway, the routing tables maintained by some nodes may need to be updated. To perform these updates, node y and y' will send RT_ADD and RT_DEL packets, respectively, along the tree paths to the root. Figure 18.10 shows the procedure mentioned above. Note that the metric used by nodes to determine its parent should consider the

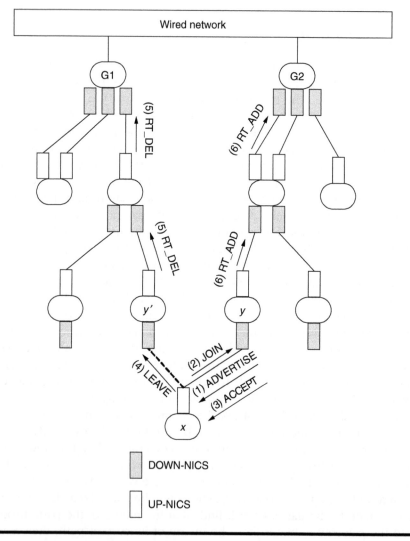

Figure 18.10 The routing procedure proposed in the Hyacinth architecture.

factor of load balancing. To achieve this, Ref. [11] proposes two load-aware metrics: the gateway link and the path capacity.

18.3.2 WCETT Metric

A general multichannel routing protocol can be designed by combining an existing single-channel routing protocol with a new routing metric by taking multichannel effects into consideration. The WCETT metric proposed in Ref. [12] is such a metric for routing in multi-radio, multihop WMNs. The routing protocol proposed in Ref. [12] is called *multi-radio link-quality source routing (MR-LQSR)*. MR-LQSR is a routing protocol that combines the WCETT metric with the LQSR protocol.[13] In MR-LQSR, each link will be assigned a weight *ETT* (*the expected transmission time*). ETT_i denotes the expected amount of time that link i will take to transmit a packet of some fixed size S. When an n-hop path is found, the path metric can be computed by simply summing the ETTs of all hops together:

$$WCETT = \sum_{i=1}^{n} ETT_i \qquad (18.3)$$

where ETT_i is the ETT of the ith link. We then choose a path with the least WCETT as the routing path. However, Eq. (18.3) does not consider the impact of channel diversity. More specifically, when two interfering links try to transmit simultaneously, both their ETTs may increase significantly. On the contrary, if they switch to different channels, then their ETTs may remain the same. Therefore, Ref. [12] introduces another term X_j to reflect the impact of channel diversity:

$$X_j = \sum_{\text{Hop } i \text{ is on channel } j} ETT_i \qquad (18.4)$$

Then, the WCETT can be redefined as follows:

$$WCETT = \max_{1 \le j \le k} X_j \qquad (18.5)$$

where k is the total number of channels in the system. Therefore, a path that is more channel-diversed may have a lower value of the WCETT metric. However, it is also observed that the WCETT in Eq. (18.5) may ignore the impact of path length. To take path length into consideration, Ref. [12] further combines Eq. (18.3) and Eq. (18.5) together by taking a weighted average:

$$WCETT' = (1 - \beta) \sum_{i=1}^{n} ETT_i + \beta \max_{1 \le j \le k} X_j \qquad (18.6)$$

where β is a weighting factor, $0 \le \beta \le 1$. Eq. (18.6) provides a tradeoff between delay (which is considered in the first term) and throughput (which is considered in the second term). The channel assignment issue is not explicitly specified in Ref. [12], but we may apply some existing schemes.

18.3.3 Fixed and Switchable Channels

In Ref. [14], an interface assignment scheme is proposed first. The proposed scheme does not require modifications to the IEEE 802.11 standard, and it allows the number of interfaces on a node to be less than the total number of channels. Suppose that a node has m interfaces. k of these m interfaces will be dedicated to *fixed* interfaces, and the remaining $m - k$ interfaces are dedicated to *switchable* interfaces. The former will operate on more static channels, and the latter will operate on more dynamic channels. When a node x intends to communicate with another node y, one of x's switchable interfaces will switch to the operational channel of one of y's fixed interfaces. Figure 18.11 shows a three-node example.

To assign channels to fixed interfaces, Ref. [14] proposes that each node adopts a function f known to all other nodes. Alternatively, one may

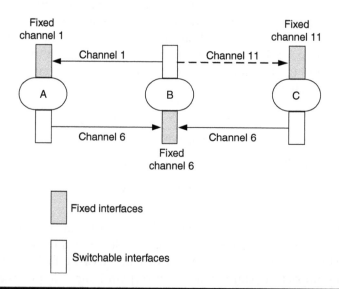

Figure 18.11 An example of the interface/channel assignment scheme used in Ref. [14]. When node B intends to communicate with node A, the switchable interface of node B will change its operational channel to channel 1. Later, when node B wants to communicate with node C, its switchable interface will operate on channel 11.

consider to allow a node to collect neighborhood information and choose those less interfered channels.

Ref. [14] also identifies routing heuristics to complement its interface assignment strategy. Because the switchable interfaces may switch their operational channels, not only the channel diversity but also the cost of interface switching should be taken into consideration. The path cost should be computed as the weighted combination of the switching cost, the diversity cost, and the global resource usage cost. Similar to the WCETT, the proposed routing heuristics can be incorporated into the existing routing protocols.

18.3.4 CA-AODV

Ref. [15] proposes an approach that combines channel assignment with the AODV (ad hoc on demand distance vector) routing protocol,[16] called CA-AODV. It assumes one control channel and several data channels. All control packets are transmitted on the control channel, but nodes may use different channels to exchange data packets. Senders are the ones who determine the data channels to be used. Initially, a source node will select a channel to use. If the source is not an active node that belongs to any routing path (i.e., the source node does not have a channel yet), then it can pick a channel randomly. Then, the source node will broadcast a route request (RREQ) that includes the channel index choosen by itself. When a node, say x, receives the RREQ packet, x will also select a channel to use. If it has been an active node of another route, it will use the channel that it selected for that route. Otherwise, it will check the path in the RREQ packet, and then choose a channel that is not used by any node on the path within k-hops (usually, k is equal to 2 or 3). Then, it will check whether it has a path to the destination. If the answer is no, it will rebroadcast the RREQ packet by appending its information. Otherwise, it will send a route reply (RREP) packet to the source along the path included in the RREQ packet. When nodes send the RREP packet back to the source, some nodes may further change their selected channels. This will be discussed in the following example.

In Figure 18.12, we assume that an active route from node E to node G has been established. Suppose that node A intends to find a route to node D. Node A will broadcast a RREQ packet that carries the channel index used by itself (let it be channel 5). When node B receives the RREQ packet, suppose that it chooses channel 3 as its operational channel. Assume that node B has no path to node D, so it will rebroadcast the RREQ that carries the channel indices used by node A and itself. When node C receives the RREQ packet, C can only choose channel 3, because it is an active node of an existing route using channel 3. When node D receives the RREQ, it will reply a RREP that carries the channel index used by itself (let it

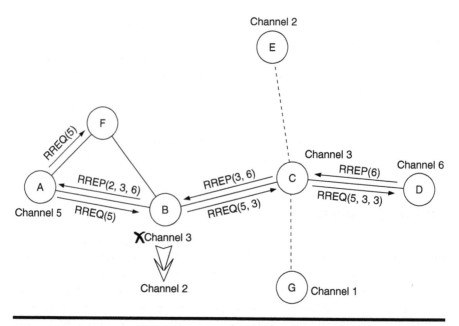

Figure 18.12 An example of the routing protocol CA-AODV.[15]

be channel 6). Similarly, node C will send a RREP back that carries the channel indices used by node D and itself. When the RREP arrives at node B, it finds that node C has chosen the same channel 3 as it has. Therefore, it decides to change its channel from channel 3 to channel 2. When node A receives the RREP, the channels used by all nodes are determined. To summarize, Ref. [15] is a sender-based solution. So each sender will choose the best channel to transmit according to its observation. A multichannel MAC protocol that divides bandwidth into one control channel and several data channels can be used (e.g., DCA[7]). However, the sender chooses a data channel to use based on routing information rather than the channel usage state around the sender.

18.4 Standard Activities of Mesh Networks

In this section, we will introduce some standard activities related to mesh networks: the working group 802.11s of IEEE 802.11,[3] the working group IEEE 802.15.5,[1] the IEEE 802.16 WiMax standard.[2]

18.4.1 IEEE 802.11s Mesh Networks

The mesh network architecture proposed by the drafts of IEEE 802.11s are close to what have been addressed in this chapter. The 802.11s working group is formed to standardize the extended service set (ESS). Below, we review the 802.11 TGs Simple Efficient Extensible Mesh (SEE-Mesh),[17] which is one of the 802.11s proposals that have been proposed. The proposal is targeted at unplanned mesh networks. To build a mesh network or add a new node to an existing WMN, some procedures are required:

- *Neighbor Discovery.* The neighbor discovery procedure relies on the profiles maintained by each node. A profile describes a specific mesh network that consists of a mesh ID, a path selection protocol identifier, and a path selection metric identifier. A node may have more than one profile. A mesh node that is not a member of any WLAN mesh performs passive or active scanning to discover neighboring mesh points (MPs). A node, say y, is considered a neighboring MP of a new node, say x, if the used profile of y matches at least one of x's profiles. For example, if a mesh node, say y, uses a path selection protocol that is not supported by the mesh node x that performs neighbor discovery, then y will not be a neighboring MP of x. Furthermore, the SEE-Mesh proposal limits the number of neighbors of a node to a fixed value. Therefore, a neighboring MP will be considered a candidate peer if and only if the neighboring MP has nonzero peer link value.

- *Link Establishment.* When a node, say x, intends to create a link with another node, say y, it will transmit an association request frame to y. When y receives the association request frame, it may accept or reject the request by replying an association reply frame. If y accepts the request, y will become the *superordinate node* of that link and x will become the *subordinate node* of that link. To ensure that the measured link state is symmetric, the superordinate is responsible for measuring the link quality and it will send a local link state announcement frame to the subordinate node. Then, the subordinate node can update its neighbor MP table.

- *Channel Selection.* The channel selection mode for each interface on a mesh node can be specified as either *simple unification mode* or *advanced mode.* An interface that uses the advanced mode will select its operational channel based on a more complicated channel assignment algorithm (which is beyond the scope of the SEE-Mesh proposal). On the contrary, an interface that adopts the simple unification mode will determine its operational channel by following the subnetwork with the larger CPI (channel precedence indicator). The SEE-Mesh also proposes an optional *common channel framework (CCF)* as follows. When the simple channel unification protocol ends, all interfaces in the same mesh network will have a common channel to use. However, it allows a communication pair to exchange a request-to-switch frame and a clear-to-switch frame on the common channel and then switch to a different channel to exchange data packets. After the packet exchange, the pair can switch back to the common channel for other activities.

Currently, the draft proposes to use EDCA as defined in 802.11e[18] as the basis for the 802.11s media access mechanism. As to the path selection problem, the draft proposes a protocol called the *hybrid wireless mesh protocol* (*HWMP*) that combines on-demand routing and proactive tree-based routing. Other related issues (e.g., power management and security) are also addressed in the SEE-Mesh proposal.

18.4.2 IEEE 802.15 Mesh Networks

IEEE 802.15.5 is established to determine the necessary mechanisms in the PHY and MAC layers to enable mesh networking in wireless personal area networks (WPANs). Figure 18.13 shows an example. End devices must be connected to a coordinator to form a star topology. Coordinators then can form a mesh network. One issue in WPAN meshes is the beacon alignment

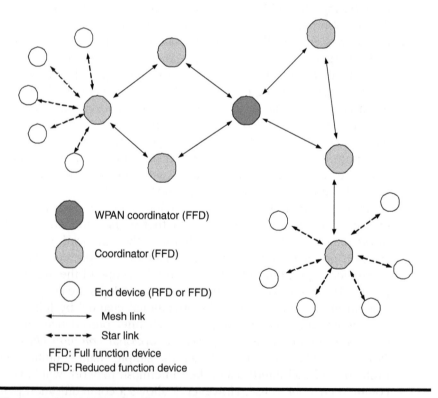

Figure 18.13 A WPAN mesh network example.

problem. Each coordinator will send out beacons periodically. Thus, how to arrange beacon transmission time is a difficult problem. The WPAN mesh network is normally aimed at home networking. Besides, the mesh nodes in a WPAN may be capable of mobility.

18.4.3 IEEE 802.16 Mesh Networks

Initially, IEEE 802.16 was created to address point-to-multipoint (PMP) communications. The standard IEEE 802.16a further defined the mesh mode. A mesh network is managed by a *mesh base station* (*MBS*) that controls one or more *mesh subscriber stations* (*MSS*). In the PMP mode, traffic only occurs between the BS and SSs. In the mesh mode, traffic can be routed through other SSs and can occur directly between SSs.

18.5 Conclusions

In this chapter, we have presented some communication protocols and some standard activities for WMNs. For WMNs, throughput and scalability are major concerns. One way to improve throughput and solve the scalability problem is using multiple channels. We have studied some MAC protocols and routing protocols that use multiple channels. However, there are still more research issues to be addressed for multichannel WMNs, such as clock synchronization, power management, load balance of gateways, etc.

References

1. IEEE 802.15 Standard Group Web Site. http://www.ieee802.org/15/.
2. IEEE 802.16 Standard Group Web Site. http://www.ieee802.org/16/.
3. IEEE 802.11 Standard Group Web Site. http://www.ieee802.org/11/.
4. IEEE Std 802.11b-1999. Supplement To IEEE Standard For Information Technology-Telecommunications And Information Exchange Between Systems- Local And Metropolitan Area Networks- Specific Requirements- Part 11: Wireless LAN Medium Access Control (MAC) And Physical Layer (PHY) Specifications: Higher-speed Physical Layer Extension In The 2.4 GHz Band.
5. J. So and N. H. Vaidya. Multichannel MAC for ad hoc networks: handling multichannel hidden terminals using a single transceiver. In *Proc. MobiHoc*, May 2004.
6. P. Bahl, R. Chandra, and J. Dunagan. SSCH: Slotted seeded channel hopping for capacity improvement in IEEE 802.11 ad hoc wireless networks. In *Proc. MobiCom*, Sept. 2004.
7. S.-L. Wu, C.-Y. Lin, Y.-C. Tseng, and J.-P. Sheu. A new multichannel MAC protocol with on-demand channel assignment for multihop mobile ad hoc networks. In *Proc. ISPAN*, Dec. 2000.

8. A. Adya, P. Bahl, J. Padhye, A. Wolman, and L. Zhou. A multi-radio unification protocol for IEEE 802.11 wireless networks. In *Proc. BROADNETS*, Oct. 2004.

9. J. Zhu, and S. Roy. 802.11 mesh networks with two radio access points. In *Proc. ICC*, May 2005.

10. M. Gerla, and J. T.-C. Tsai. Multicluster, mobile, multimedia radio network. *Wireless. Network*, 1(3):255–265, 1995.

11. A. Raniwala, and T. cker Chiueh. Architecture and algorithms for an IEEE 802.11-based multichannel wireless mesh network. In *Proc. INFOCOM*, Mar. 2005.

12. R. Draves, J. Padhye, and B. Zill. Routing in multi-radio, multihop wireless mesh networks. In *Proc. ACM MobiCom*, Sept. 2004.

13. R. Draves, J. Padhye, and B. Zill. The architecture of the link quality source routing protocol. Technical Report MSR-TR-2004-57, Microsoft Research, 2004.

14. P. Kyasanur and N. H. Vaidya. Routing and interface assignment in multichannel multiinterface wireless networks. In *Proc. WCNC*, New Orleans, U.S.A., Mar. 2005.

15. M. X. Gong, and S. F. Midkiff. Distributed channel assignment protocols: A cross-layer approach. In *Proc. WCNC*, New Orleans, U.S.A., Mar. 2005.

16. C. E. Perkins, E. M. Belding-Royer, and S. R. Das. Ad hoc on-demand distance vector (AODV) routing. RFC Experimental 3561, Internet Engineering Task Force, July 2003.

17. IEEE 802.11 TGs Simple Efficient Extensible Mesh (SEE-Mesh) Proposal. http://www.802wirelessworld.com/.

18. IEEE Std 802.11e-2005 (Amendment to IEEE Std 802.11, 1999 Edition (Reaff 2003). IEEE Standard for Information technology – Telecommunications and information exchange between systems – Local and metropolitan area networks – Specific requirements Part 11: Wireless LAN Medium Access Control (MAC) and Physical Layer (PHY) specifications Amendment 8: Medium Access Control (MAC) Quality of Service Enhancements.

Chapter 19

Integrated Heterogeneous Wireless Networks

Dave Cavalcanti, Anup Kumar, and
Dharma P. Agrawal

19.1 Introduction

The proliferation and fast deployment of a variety of radio communication technologies clearly indicate that future environments will be heterogeneous in terms of wireless networks, devices, and applications. While some wireless technologies are already an integral part of our everyday lives, such as wireless WANs (e.g., 2G and 2.5G cellular systems), WLANs (e.g., IEEE 802.11a/b/e/g), and wireless PANs (e.g., Bluetooth), upcoming wireless standards, for example, 3G and Beyond cellular systems, IEEE 802.16 WMANs, and high-rate/low-rate IEEE 802.15.3/4 WPANs are expected to provide even more exciting and efficient services. Such technological diversity is also impacting the design of mobile computing devices, which are being equipped with multiple communication interfaces, for instance, multimode devices with WLAN and Bluetooth access capabilities in addition to the traditional cellular mode. Furthermore, advances in the software defined radio (SDR) technology[1] will also make multimode terminals common. The availability of a multitude of wireless technologies has motivated research efforts toward the integration of heterogeneous wireless networks to allow users to experience "Always Best Connectivity" anytime, anywhere, and with any device. Although there are different visions of next-generation communication systems,[2–4] it is possible to identify three key characteristics that must be integrated to achieve the desired ubiquitous connectivity, namely,

- *Multiservices*: Users must be able to access a wide range of services, including data and multimedia, with seamless mobility and quality of service (QoS) guarantees. The available services would include voice, multimedia, messaging, e-mail, information services (e.g., news, stocks, weather, and travel), M-commerce, entertainment, location-based public utility and health-care services, and so on.
- *Heterogeneous multinetworks*: The cooperation and integration of networks with distinguishing characteristics of data rate, coverage range, power, mobility, and price is paramount to support multiple services with different QoS requirements. Legacy (e.g., cellular and WLANs single-hop networks) and new network paradigms, such as the mobile ad hoc networks (MANETs)[5] and moving networks[6] are also expected to coexist and converge.[3] Moreover, heterogeneous wireless networks must cooperate to provide an underlying *data transport service* that is transparent to the applications and users.
- *Multimode terminals*: Multiple communication capabilities must be opportunistically exploited to ensure that the user will use always the best connectivity[2] opportunity available at any given time and location for each service (or application). Multimode terminals may include cellular/WLAN, cellular/Bluetooth, cellular/satellite, WLAN/WPAN, and so on. As pointed out in Ref. [7], multimode or multiinterface terminals are expected to play a fundamental role in future networks.

The technology evolution toward an integrated heterogeneous network with these characteristics is not simple, however. Integrating networks with distinguishing characteristics of data rate, coverage range, power, mobility, and administrative domains will involve not only technical issues in several layers of the protocol stack,[8] but also economical and political issues. Integration of heterogeneous networks has gained more and more attention from both industry and academia in the recent years. Industry standardization groups such as the Third Generation Partnership Project (3GPP), 3GPP, and the IEEE 802 are already developing the interworking of heterogeneous infrastructure-based wireless networks. For instance, the 3GPP has been working on an integrated architecture for interconnection of cellular systems and WLANs.[9] On the other hand, link-layer support for enabling interworking of IEEE 802-based networks with external networks (e.g., cellular networks) is been considered under the IEEE 802.21 work group.[10] The IEEE 802.11 Task Group u (TGu)[11] is also considering integration with cellular systems, but it only addresses issues related to 802.11 WLANs. A common aspect in all these works toward integration of heterogeneous networks is the fact that they consider only the single-hop operation mode, i.e., direct link connections with a cellular base station (BS) or a WLAN/WMAN access point (AP). Despite the increasing standardization efforts, there are more open problems than standardized solutions, which indicate that the work is only at the initial stages. Examples of open problems include support to seamless mobility across networks and QoS provisioning across different networks.

More recently, a broader vision of the integration of heterogeneous networks has been considered, which integrates heterogeneous infrastructure-based wireless networks with the new paradigm of multihop communications used in MANETs.[8] In such a system, the mobile stations (MSs) could access the network directly through base stations (BSs) and access points (APs) or through other MSs by employing a multihop routing protocol. This type of scenario has been referred[12] to as a heterogeneous wireless multihop network (HWMN), and it could represent another step toward the desired ubiquitous "Always Best Connectivity." However, several challenges need to be addressed to provide an integrated architecture as well as transparent and self-configurable protocols for HWMNs. The important issues include design of standard interfaces for connecting networks of different operators and technologies, integrated routing, mobility management, QoS support, network management, load balancing, and security.

In this chapter, we provide an overview of the standardization efforts toward integrated heterogeneous wireless networks, and discuss the state-of-the-art research on HWMNs, which integrate infrastructure-based and infrastructure-less networks. We review and compare state-of-the-art integrated architectures and routing protocols for heterogeneous wireless networks. Finally, we highlight some of the important open research issues in this area.

19.2 Integration of Infrastructure-Based Heterogeneous Wireless Networks

In infrastructure-based wireless networks, the users, or MSs access the network through a direct (or single-hop) link-layer connection with a network attachment point. The network attachment point can be called BS or AP depending on the specific technology. In the remaining part of this chapter, we use the term BS to refer to the cellular network attachment points, while AP is used for all other networks, such as WLANs (e.g., 802.11) and WMANs (e.g., 802.16). One of the motivations for the integration of such networks is to enable multimode-enabled MSs to use the network that better satisfies the users' preferences as well as the application requirements.

Currently, the main application scenario driving research and standardization work towards the integration of heterogeneous wireless networks is the interworking of 3G cellular systems with WLAN hot-spot areas. Cellular systems and WLAN are viewed as complementary technologies. While cellular system provides long-range coverage at relatively low data rates, for example, up to 2 MB/s in 3G UMTS systems, WLANs can provide hot-spot coverage at much higher transmission rates, for example, an 802.11a WLAN interface can support up to 54 MB/s physical layer data rate. Therefore, WLANs are viewed as an alternative to opportunistically off-load some of the traffic from the cellular system, and as a new possibility for the cellular service providers to enhance services provided to the users. Figure 19.1 shows an example of an integrated 3G/WLAN network. As can be seen, the dual-mode MSs would have the possibility to select between the two technologies in several hot-spot areas.

Several standardization groups are considering scenarios similar to the one shown in Figure 19.1. Next, we provide an overview of the work under several related groups including 3GPP, 3GPP2, IEEE 802.21, and 802.11.

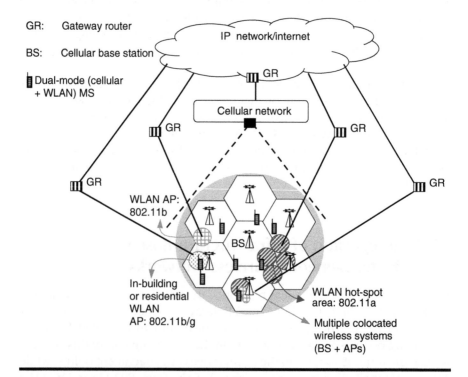

Figure 19.1   3G/WLAN interworking scenario. (From D. Cavalcanti et al. Ed. *IEEE Wireless Communications Magazine,* **June 2005.)**

In addition, we highlight the main underlying principles and research challenges.

19.2.1 3GPP/WLAN Interworking Architecture

The 3GPP has been developing an internetworking architecture for 3G cellular systems and WLANs with the aim of enhancing the services provided to subscribers by 3G operators.[9] The proposed 3G/WLAN architecture has been discussed in several papers.[13–15] The basic internetworking aspects considered in the 3GPP architecture are network selection, authentication, authorization and accounting (AAA), and routing in the fixed infrastructure connecting the WLAN APs and the 3G networks, and security. The underlying principle is to consider WLANs as an independent access network, which is connected to the core cellular infrastructure via a gateway router. Although, a final version of the standard has not yet been defined, the basic components of the proposed architecture for integration of 3GPP networks and WLANs have been defined in Ref. [9].

The 3GPP2 group is also working on 3G/WLAN internetworking, but it mostly addresses the WLAN/CDMA2000 interoperation,[16] whose basic underlying principles are in line with the work of 3GPP. Therefore, without loss of generality, in the remaining part of this section, we focus on the 3GPP/WLAN interworking model.

Five basic integration levels for 3GPP and WLAN have been identified in Ref. [15]:

- Level 1: common billing and customer care integration.
- Level 2: 3GPP system-based access control and charging.
- Level 3: access to 3GPP system IP multimedia services (IMS) from the WLAN without mobility support.
- Level 4: access to 3GPP system IP multimedia services (IMS) from the WLAN with mobility support for service continuity.
- Level 5: access to 3GPP system IMS from the WLAN with seamless mobility support.

The integration at level 1 can be viewed as only administrative, as it does not require actual integration of the network infrastructures. The practical realization of each of the remaining levels of integration (level 2–5) involves several technical challenges. The ultimate goal is to enable seamless mobility (level 5) between networks with no performance degradation for the users, which is not a simple task, however. In fact, only levels 2 and 3 are being considered in the current 3GPP draft.[9] The problems related to levels 4 and 5 have been addressed at research level, but not yet considered in the standard draft.

The 3GPP specification[9] defines two new procedures in the 3GPP system for interworking with WLANs, namely,

- WLAN Direct IP Access: It enables access to the WLAN and the locally connected IP network (e.g. Internet) to be authenticated and authorized through the 3GPP system. In this type of access, 3GPP subscribers could use their 3GPP credentials to access a local WLAN hot spot.
- WLAN 3GPP IP Access: The objective of this option is to enable WLAN terminals, or WLAN user equipments (UEs) to establish connectivity with external IP networks via the 3GPP system. The external networks would include 3G operator networks, corporate intranets, or the Internet.

Figure 19.2 illustrates the simplified reference model for 3GPP–WLAN interworking[9] that would enable WLAN direct IP access and WLAN 3GPP IP access when the WLAN access network is directly connected with the 3GPP home network of the WLAN user equipment. The WLAN access network includes WLAN access points and one or more gateway routers that would also host intermediate AAA elements. In the WLAN direct IP access, only authentication and accounting information would be exchanged between the WLAN access network and the 3GPP AAA server in the 3GPP network.

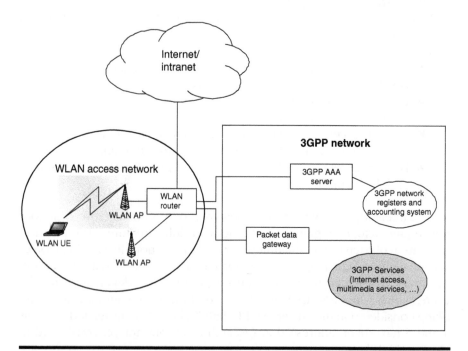

Figure 19.2   3GPP/WLAN simplified integration view.

On the other hand, the packet data gateway would support the WLAN 3GPP IP access to external IP networks by forwarding between the WLAN and the 3GPP services network. As 3GPP–WLAN internetworking concentrates on the interfaces between 3GPP elements and the interface between the 3GPP system and the WLAN, the internal operation of the WLAN is only considered to assess the impact of architecture options/requirements on the WLAN.

Some of the requirements and principles identified in Ref. [9] as important points to be considered during the systems' design include:

- Minimal impact on existing WLAN and 3GPP networks.
- Legacy WLAN terminals should be supported, although software upgrades may be required to access a universal subscriber identifier module (USIM), which is the module in 3GPP devices that stores users' profiles, and authentication information.
- WLAN authentication signaling is carried between WLAN UE and WLAN access network by WLAN-specific protocols.
- WLAN authentication signaling should be transported between any WLAN and 3GPP network by a standard protocol.
- Data flows must be able to be routed to the home or visitor 3GPP network.
- Users should experience no degradation in QoS, when accessing these services via WLAN.

The 3GPP/WLAN interworking standard also considers the possibility of roaming users to access both types of services. In this case, the WLAN access network would be directly connected to a 3GPP visitor's network, which would be interconnected via standard interface to the user's 3GPP home network. The home network would be responsible for access control, while charging records could be generated in the visited or the home 3GPP networks. The 3GPP AAA server in the visited network would act as a proxy relaying access control signals and accounting information to the home 3GPP AAA service using a standard interoperator interface. The reader can refer to Ref. [9] for more details on the 3GPP/WLAN interworking architecture.

It must be noted that most of the standardization work done at 3GPP and 3GPP2 address mostly issues related to the interfaces and protocols interconnecting the wired infrastructure, while issues related to the air interface of each specific technology are not taken into account. Issues related to the air interface are critical in two main situations, namely, network selection and mobility management. As it was pointed out in Ref. [4], the selection of the best link interface for a given application is an issue for the MS, which would be out of the current scope of 3GPP UMTS/WLAN interworking architecture. Furthermore, although some network advertisement and selection procedures have been proposed for

3GPP UMTS/WLAN integrated environment,[9] they do not consider all the relevant parameters for the mobility and the QoS support.

19.2.2 The IEEE 802.21 Media-Independent Handover Function

Among several standardization groups that are developing solutions for heterogeneous networks, the IEEE 802.21 is particularly related to the network selection and mobility management problems. While other groups, such as 3GPP and 3GPP2, are also looking at the integration problem, their solutions do not address the network selection and mobility from the mobile user's perspective. On the other hand, the IEEE 802.21 working group is developing a standard to enable handoffs and interoperability between heterogeneous networks, including both 802-based and non-802 networks (e.g., cellular networks). The 802.21 standard does not focus only on the network side, but it also considers the MS as an important component. In fact, the main goal is to provide link-layer intelligence and information to the upper layers to optimize the network selection and handoff decisions. Although the standard has not yet been finalized at the time of this writing, the underlying framework and some of the basic requirements are well defined in the current draft.[10]

A simplified reference model for multimode devices based on the 802.21 standard[10] is depicted in Figure 19.3. As can be seen, the main component is the media-independent handover (MIH) function, which can be seen as a sublayer between multiple link layers (layer 2 and below) and the upper layers (L3 and above). The purpose of the MIH function is to provide services to the upper layers. According to the 802.21 standard, the upper layers may register with the MIH_SAP (MIH service access point) to receive events, commands, and access information about the lower layers, which can be used for network selection and handoff decision. On the other hand, the MIH function interacts with the multiple communication interfaces available through media-specific SAPs. It is through the media-specific SAPs that the MIH function, in the mobile device, can access link/medium access control (MAC)/physical layer information, issue commands (e.g., trigger a L2 handoff), and exchange information with a MIH function located in the access network, i.e., in the BS or AP. One of the goals of the 802.21 working group is to specify the services that will be available through the media-specific SAPs and MIH_SAP, as well as to suggest amendments to various link/MAC/physical layer supported standards. It is worth mentioning that the MIH function is available at both ends of the link-layer connection, i.e., the mobile station and the network attachment point, which can be an AP or BS, depending of the type of interface. The main idea is to enable the MS to cooperate with the access network to enhance the performance during handoffs. An underlying assumption in the 802.21 standard is that the

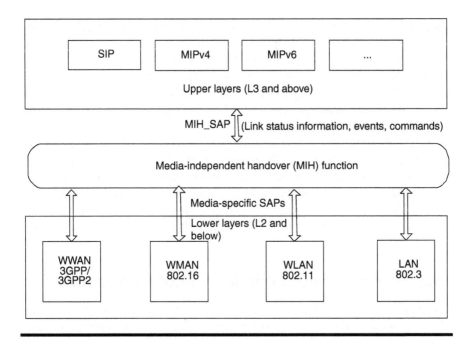

Figure 19.3 Simplified 802.21 reference model for multimode devices.

link-layer connections can only be established directly with the network attachment points, i.e., single-hop mode.

19.2.3 *The IEEE 802.11u Integration with External Networks*

The IEEE 802.11 Task Group u (TGu)[11] has been recently formed to enhance the IEEE 802.11 standard for supporting efficient integration with external networks. The main motivation for the 802.11u is the increasing interest for interconnection of 802.11 WLANs and cellular networks. Some of the application scenarios considered by 802.11u are similar to the ones considered by 3GPP and 3GPP2 interworking architectures. The problems that have been addressed in 802.11u include:

Network selection: The goal is to enable a WLAN user to select an external network, or subscription service provider network (SSPN), to authenticate to gain access to the local WLAN as well as to services provided by the SSPN. As several SSPN may be available at a single WLAN access network, the WLAN has to advertise information about the supported SSPNs, such that the users can make an optimized selection.[17] The task of the 802.11u is to specify what information related to network selection will be advertised and how this information will be advertised using the 802.11 protocol.

User Protection: To enable an efficient network selection procedure, the users may have to exchange information with the WLAN, even before authentication and protecting the users against possible security threats, such as hijacking of MAC address, is critical.

QoS Mapping: Another important issue is the provisioning of QoS in the WLAN and the mapping of the users' QoS profiles supported in the SSPN to the WLAN. Owing to the distinct protocols used in different networks, the parameters that specify QoS in the SSPN may not have a direct correspondence in the 802.11 MAC.

Although some overlapping areas may exist between the 802.11u and the 802.21 groups, the main differences are the facts that 802.11 considers solutions for the 802.11 networks only, and mainly layer 2 solutions, while the 802.21 standard considers a broader range of technologies and can be considered a solution above the layer 2, as shown in Figure 19.3. The 802.11u is in its early stages and is expected to be completed by end of 2008.

19.3   Heterogeneous Wireless Multihop Networks

The main idea of multihop communication paradigm is to exploit the use of every node in the network to forward data packets toward their destination. This is an underlying principle of the MANETs, which can operate without any infrastructure. Despite a great deal of research on MANETs, their deployment in commercial scale has not fully happened yet. Recently, another application of the multihop communication paradigm has gained increasing interest, which is the integration of multihop communications with existing infrastructure-based networks, as an alternative to improve capacity and achieve greater flexibility in existing single-hop networks.[8] In such an integrated environment, users would be able to access the network directly via infrastructure-based networks (e.g., cellular systems, WMANs, and WLANs), or through multihop communication, as in an infrastructure-less ad hoc network. This integrated network environment has been referred in Ref. [12], as HWMN and it would integrate multiple networks (multinetworks), with multimode communication paradigms, and multimode (or multistandard) mobile devices to provide multiservices to mobile users. The multinetworks available may include heterogeneous wireless technologies with distinguishing characteristics of coverage range, data rate, reliability, mobility support, and cost. Figure 19.4 shows a possible example of a HWMN, integrating 3G and WLAN technologies, but it could also include other types of networks, such as WMANs and WPANs. In a HWMN, the MS may be equipped with multiple wireless communication capabilities, including any combination of WWAN, WMAN, WLAN, or WPAN interfaces.

A HWMN shown in Figure 19.4, would allow the mobile users to be connected "anytime, anywhere," and with any device. This integrated environment would also provide the required networking infrastructure

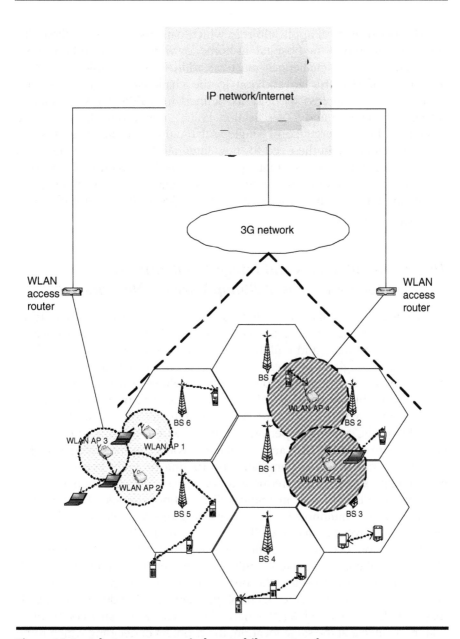

Figure 19.4   A heterogeneous wireless multihop network.

and enable new applications in several areas, including entertainment, health care, information services, location services, etc. For instance, new health care applications could be developed where the patients and medical staff could access medical services anytime and everywhere through multimode devices. The seamless connectivity through multimode devices could support medical applications in which patients could have their vital signals monitored in the hospital, at home, or while moving in between.

Several research challenges must be addressed to enable a practical realization of HWMNs. In addition to the challenges to enable service continuity and seamless mobility as defined in 3GPP/WLAN integration scenarios at levels 4 and 5 (see Section 19.2.1), new protocols are needed to enable the convergence of single-and multihop communications. All these factors highlight the need for new integrated and adaptable network architectures and protocols that operate in any heterogeneous networking environment, hiding the underlying heterogeneity from upper layers and taking into account all wireless technologies available for end-to-end service provisioning.

19.3.1 Architectures and Integrated Routing for Heterogeneous Multihop Wireless Network

Routing is a major challenge in a heterogeneous scenario that includes technologies providing different service levels, and in particular in a HWMN, that integrates single- and multihop communication paradigms. In today's isolated and single-hop-based networks, such as cellular systems and WLAN hot spots, the routing problem is simplified, since in most of the cases, the mobile terminal usually forwards the data packets to a default access router, which can be reached through the BS or the AP. Then, after this point, routing is performed in the wired infrastructure. But, as infrastructure-based WLAN and cellular networks are integrated with the multihop communication paradigm, the routing problem becomes much more complex due to the frequent topology changes and higher control overhead to discover and maintain valid multihop routes.[8] For instance, in an integrated multihop network consisting of only cellular and WLAN technologies, two dual-mode (cellular/WLAN) terminals could establish an end-to-end connection in 10 different ways, as illustrated in Figure 19.5. The main challenge is how to identify the different end-to-end routes available, and then how to select the best route. However, in a more generic environment, including other technologies and multimode user terminals (e.g., cellular/WLAN and WPAN-enabled devices), with many connectivity options available at the same location, the problem of deciding which route meets the QoS requirements of the application with the lowest cost is not simple.

An overview and comparison of several architectures and hybrid routing protocols that have been proposed to integrate infrastructure-based and multihop networks is provided in Ref. [8]. As pointed out in Ref. [8], most existing architectures and integrated routing solutions do not fully exploit the connectivity opportunities that will be available in future heterogeneous

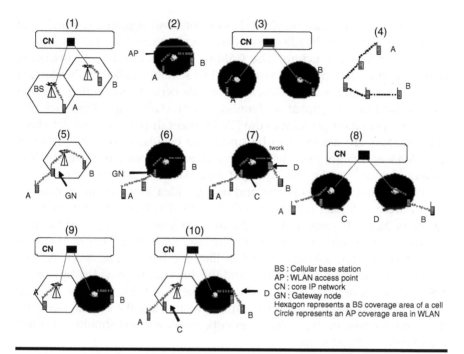

Figure 19.5   Connection alternatives between two dual-mode MSs.

networking environments, where users will be equipped with multistandard or multimode devices. In fact, most solutions discussed in Ref. [8] consider only cellular or hot-spot WLAN users and try to exploit the MANET mode only to improve the performance for those users. Next, we describe some recent works on architecture and routing for integrating infrastructure and multihop networks that were not considered in Ref. [8]. One important feature of the solutions discussed next is the fact that they consider the multihop access at the same level of importance as the single-hop (or infrastructure-based) access. In other words, the multihop connections are not only used to improve the performance of the infrastructure-based networks, but multihop connections are considered just as another way for the users to connect to the network, which should also provide QoS guarantee.

19.3.2   CAMA

With the aid of the existing cellular infrastructure, the cellular aided mobile ad hoc network (CAMA) architecture[18] has an underlying goal to enhance the performance of MANETs. This concept is slightly different from most integrated solutions discussed so far, since the cellular network is used only to control the operation of the MANET by providing authentication, routing, and security. Only control data is sent to the cellular network, i.e., the cellular channels can be viewed as out-of-band signaling channels that are to be used by multiradio MSs to connect to the CAMA agents located in the cellular infrastructure. The CAMA agents perform route discovery using a centralized position-based routing scheme, called multiselection greedy positioning routing (MSGPR). In this centralized routing, all MSs are assumed to have global positioning system (GPS) capability, such that MSs can report their position to the CAMA agent through the cellular channels. All MSs are assumed to be under the cellular coverage, such that the CAMA agents are provided with information about the entire MANET. Bhargava et al.[18] claim that low-cost high-data-rate ad hoc channels are suitable for multimedia services, and use this idea as the prime motivation in forwarding all data traffic through the MANET. However, the dynamic nature of MANETs cannot always assure QoS guarantee required by multimedia services. For example, while the data rates are higher in MANETs, delay and jitter are affected by factors like interference and mobility. The connectivity between the MANET and the Internet is provided by special MSs (gateways or APs) connected to the fixed network. Another point to be noted is that although no data traffic flows through the cellular network, the control traffic in the cellular network increases and should not affect the QoS for cellular users.

As can be noted, CAMA is not a generic architecture for heterogeneous wireless networks, but it uses the cellular infrastructure to improve the performance of MANETs. As suggested in Ref. [18], this concept could be extended to provide other control operations in the MANET, such as topology control and power management. Although CAMA is not designed to provide services to cellular users, it could be applied only to a scenario similar to (4) in Figure 19.5, if all MSs are inside the coverage of a cellular network.

19.3.3 Two-Tier Heterogeneous MANET Architecture

The two-tier heterogeneous MANET[19] considers WLAN MSs operating as a MANET and dual-mode MSs with WLAN and cellular capabilities, which are able to operate as gateways between the MANET and the Internet. The basic problem considered is how to efficiently share the gateways among the MSs in the MANET. The MSs operating in the MANETs form the lower tier, while the dual-mode gateways form the second tier of the architecture. Load-balancing issues come into picture as the gateways use limited bandwidth of the cellular channels to relay and provide Internet connectivity to multiple MSs. Several gateway selection schemes have been proposed in Ref. [19] that partition the network into clusters by associating MSs to gateways in a dynamic way as per different metrics.

Similar to CAMA, the two-tier heterogeneous MANET architecture does not provide services to cellular users, as the main objective is to connect the

MANETs with the Internet through the cellular network. The destination or the source node of a connection is always in the fixed network. Therefore, the architecture supports only a variation of the scenario (5) shown in Figure 19.5.

19.3.4 Hierarchical Cellular Multihop Networks

The hierarchical cellular multihop network (HCMN)[20] integrates a 3G cellular system with WLAN hot spots, whose coverage is extended through multihop capable nodes (MHN). The cellular system is at the top level of the HCMN architecture, and it is considered as the primary network that can provide full coverage and can always be used as a fall back solution in case that a mobile node (MN) loses connection to any other network. The WLAN hot spots are the next level of the architecture and they provide broadband access to the Internet. Furthermore, the MHNs are special nodes (static or mobile) introduced to increase the coverage of the WLAN APs through multihop routing. Moreover, a cellular network is used to provide signaling and routing services to the MN. When a MN wants to access the Internet it sends a route request to the cellular network. Thus, the cellular system is assumed to be always available to the MNs and MHNs. Also, it is assumed that the MNs' locations can be derived by a satellite system to be used for geographic cellular-based routing. Simulation results[20] indicate that the proposed cellular-based routing performs better than traditional multihop routing protocols (dynamic source routing [DSR] and ad hoc on demand distance vector [AODV]), which was anticipated, even though the simulation environment is not realistic, as it assumes a collision-free MAC protocol for the multihop links.

As in CAMA and the two-tier heterogeneous MANET architecture, the cellular network is used to provide signaling and routing information to support multihop connections. But, the HCMN goes a little beyond CAMA and the two-tier heterogeneous MANET architecture, as it assumes that the cellular network can always be used as an alternate second lower-bandwidth connectivity option. Additionally, the multihop connections are used only to reach WLAN AP and not to access the cellular network. Therefore, HCMN is clearly not a generic solution for heterogeneous wireless environments. HCMN could support variations of scenarios 2–4, 6, 8, and 9, if it is assumed that the cellular and satellite coverage is always available.

19.3.5 The Integrated Routing Protocol

The integrated routing solutions proposed in Refs. [21,22] consider generic heterogeneous wireless networks, and can provide connectivity to not only cellular and WLAN hot-spot users, but also to mobile users operating in a MANET. In Ref. [21], Cavalcanti et al. proposed and evaluated the integrated routing protocol (IRP) with two different gateway discovery approaches, namely, reactive gateway discovery (IRP-RD), and proactive gateway discovery (IRP-PD), whereas in Ref. [22], a new mechanism based on the IRP with a new integrated routing metric for heterogeneous wireless networks is described. The new routing metric used in Ref. [22] takes into account throughput, delay, and retransmissions on heterogeneous wireless links.

IRP includes a topology discovery process at the BSs and the APs, and a gateway discovery procedure to provide connectivity to MSs outside radio coverage of these networks through MSs that act as gateway nodes (GNs). The topology discovery is implemented by allowing every WLAN-capable MS to periodically transmit "Hello" packets on their WLAN interface to discover direct links with other WLAN-capable MSs. Then, these MSs maintain a 1-hop neighbors' table, which is sent to their current BS/AP every time it is updated. Furthermore, the neighbors' table maintains the costs of the direct links between MSs, which will define the routing metric used in the route discovery by the BSs and APs. Thus, each MS i associates a generic link cost c_{ij} to its neighbor j. The BSs and APs also maintain the quality (cost) of downlinks and uplinks for the MSs within their coverage. Then, BSs and APs use the neighbors' tables to build the integrated network topology, and each time they receive a RREQ, they execute a shortest path algorithm (e.g., Dijkstra's algorithm) to find the minimal cost route.

The Hello packets are also used to propagate information about GNs and allow the MS outside of the coverage of BSs and AP to proactively discover routes to the GN. As the MSs send the Hello packets they also include the estimated cost metric to reach the gateway to allow other out-of-coverage MS to select the best GN. Any cost metric can be used in this process, such as the integrated metric proposed in Ref. [22]. The IRP with a reactive gateway discovery scheme (IRP-RD) has also been considered in Ref. [21], and in this case, there is no defined association between MSs and GNs. However, owing to space constraints here, we only describe the proactive approach.

In addition, IRP differentiates the route discovery procedure when the MS is within the coverage of a BS and AP from the case when no direct connectivity to BS or AP is available. When it is inside of coverage of an infrastructure-based network (cellular BS or WLAN AP), the MS tries to explore the topology information stored at the BS/AP to discover an end-to-end route. On the other hand, if no direct coverage to a BS or AP is available, the MS uses a GN to discover an end-to-end route and to provide connectivity to the infrastructure-based network. With such strategy, IRP can be used in any of the scenarios depicted in Figure 19.5. In fact, it can even be used as a multihop routing protocol for an isolated MANET, as it provides route discovery through broadcast over the MANET in case the infrastructure or GN is not available.

A comparison of several characteristics of the integrated architectures described above is given in Table 19.1

Table 19.1 Comparison of Integrated Architectures

Architecture/ Integrated Protocol	Operation Modes Considered	Main Optimization Goal	Mobile User Stations	Supported Scenarios or Its Variation in Figure 19.2	Support of Out of Coverage MSs	Connection Mode/ Gateway Discovery
CAMA	Cellular–MANET	Improve MANET performance	Single and dual mode	(4), but with all MSs under cellular coverage	No	Gateways are used only to provide access to the CAMA agents in the cellular network
Two-Tier Het. MANET	Cellular–MANET	Connect MANETs with the Internet	Single and dual mode	(5)	No	Several gateway selection schemes provide load balance among the gateways
Hierarchical Multihop Cellular Network	Cellular–WLAN–MANET	Enhance coverage of WLAN hot spots	Dual mode	(2), (3), (4), (6), (8), and (9), but with all MSs under cellular coverage	Yes, but only out of coverage from APs	Routing is performed by the cellular network, which is assumed to location information
Integrated Routing Protocol	Cellular–WLAN–MANET	Explore all connectivity options to discover end-to-end routes	Single and dual mode	(1)–(10)	Yes	Two gateway discovery options can be used: reactive or proactive

19.4 Research Issues for Heterogeneous Wireless Networks

The ultimate goal in a HWMN is to exploit multinetworks and multimode terminals to provide seamless connectivity to multiservices with guaranteed QoS. This is not a simple problem, in fact, it could be divided into four major subproblems: network availability detection, network selection, connection establishment, and connection management. First, the networks, or link-layer connections, available at a given location for a multimode MS need to be identified. Second, this information is used as input of a network selection algorithm, which can also use the applications' QoS requirements to choose the option that bests matches the application needs and user's preferences. Then, the connection has to be established by setting up an end-to-end route between the source MS and the destination host. Finally, the ongoing connection must be seamlessly maintained to avoid disconnections due to mobility or to other changes in the network conditions that may degrade the QoS experienced by the application. Such reconfiguration task is usually performed by mobility management protocols that work at link (L2), network (L3), or application layer.

Although the integration of heterogeneous wireless networks has been the subject of intensive research, most works have focused mainly on a subset of the features and the scenarios we envision for HWMNs. The main research and standardization focus has been on interworking of cellular systems and WLANs. Also, the existing integrated architectures and hybrid routing protocols either have not fully explored all the possibilities of using multihop communication in HWMNs, or do not provide the required reliability and QoS support for real-time applications. Last, but not the least, the selection of the best connectivity in heterogeneous networks has to take into account a broader range of performance-affecting factors, in addition to QoS requirements and network conditions (e.g., minimal throughput, delay bound, and maximal burst size), such as mobility profile and specific traffic requirements.[12]

In summary, some of the open research issues toward future integrated heterogeneous wireless networks are

New Architectural Components: One important open issue toward our envisioned HWMNs is to identify the main architectural components required to enable the convergence of heterogeneous networks, such as the 3G cellular systems, WMANs (e.g., the IEEE 802.16), WLANs (e.g., the IEEE 802.11a/b/e/g), and WPANs (e.g., Bluetooth and IEEE 802.15). This task would require the extention of existing standard solutions with new components to support multiservices with QoS and seamless mobility as well as to exploit multihop communications. It is important to remark that to enable realistic and practical solutions, one has to consider existing standard solutions (e.g., the 3GPP/WLAN internetworking architecture[9]) as the basis for developing the solutions for HWMNs, and then, introduce new components as needed. Nevertheless, completely innovative approaches may also be required for supporting integration scenarios not yet considered in current architectures. For instance, new approaches must be explored to opportunistically exploit multihop communications and multimode devices.

New Protocol Stack and Cross-Layer Design: It is also important to revisit the traditional protocol stack, based on the OSI model, to introduce new cross-layer features that allow multimode devices to opportunistically exploit the multiple networks available. These new features may include network availability detection, network state estimation, network selection, horizontal and vertical handoff decision, and connection management. Some of these features have been discussed in the context of the IEEE 802.21 standard, but they are far from solving the problems in HWMNs, as the network scenarios considered in the IEEE 802.21 standard are much simpler (e.g., include only single-hop or infrastructure networks). The main idea behind these features is to use information from different layers, such as physical, MAC, link, and application layers, to decide which connectivity alternative best matches the requirements of an application. This may involve selecting one specific communication interface or even combining multiple interfaces.

Integrated Routing and Mobility Management: The routing layer and mobility support is particularly important for the successful integration of heterogeneous networks. The integrated solutions have to support multimode devices and take this information into account when discovering and establishing routes, while preserving the end-to-end semantics of the applications. As a matter of fact, in an HWMN, multiple routes for a given source–destination pair may be available at a given location, and the integrated network layer should be able to identify the route that best satisfies the QoS requirements of the application with minimal cost for the system.[21] Furthermore, the best end-to-end route may be the one that spans multiple networks with different characteristics, or it may be the case where a single network cannot satisfy the QoS requirements of the application. In the latter case, using multiple physical end-to-end connections for a single application may be the best solution. Also, to provide seamless mobility in the HWMNs, new location and mobility management must be investigated. We refer the reader to Ref. [23] for a detailed description of the specific mobility management issues and existing solutions in HWMNs.

Performance Evaluation and Experimental Platforms: Last, but not the least, another important challenge is to allow the effective evaluation of the proposed integrated solutions. Most of the existing work addresses the evaluation of integrated heterogeneous networks through simulations and analytical models. Obviously, the evident complexity and resource constraints avoid the set up of large-scale experiments, including all available wireless technologies. However, experiments involving different flavors of WLANs and WPANs and multiradio devices (e.g., cellular + WLAN) can provide valuable information to drive the design of future HWMNs.

19.5 Conclusions

Enabling cost-effective data services across heterogeneous networks is a challenging task. In this chapter, we have described some of the standardization work toward integrated heterogeneous wireless networks and discussed the state-of-the-art architectures and protocols for heterogeneous wireless multihop networks. One of the main issues is to efficiently use multihop connections and multi-interface capabilities of mobile stations to improve the systems' capacity and to provide ubiquitous connectivity to the users. Mobility management across heterogeneous networks is a real challenging task as distinct levels of mobility support could be provided by different technologies.

Acknowledgment

This work has been partially supported by the Ohio Board of Regents, Doctoral Enhancement Funds.

References

1. J. Mitola, Ed., Special Issue on Software Radio, in *IEEE Communications Magazine*, May 1995.
2. E. Gustafsson and A. Jonsson, Always Best Connected, *IEEE Wireless Communications*, vol. 10, pp. 49–55, February 2003.
3. C. Polits, T. Oda, S. Dixit, A. Sieder, H. Lanch, M. Smirnov, S. Uskela, and R. Tafazolli, Cooperative Networks for the Future Wireless World, *IEEE Communications Magazine*, September 2004.
4. V. Gazis, N. Alonistioti, and L. Merakos, Toward a Generic Always Best Connected Capability in Integrated WLAN/UMTS Cellular Mobile Networks, *IEEE Wireless Communications*, June 2005.
5. C.M. Cordeiro and D.P. Agrawal, *Ad Hoc and Sensor Networks*, World Scientific Press, New Jersey, USA, 2006.
6. H.-Y. Lach, C. Janneteau, and A. Petrescu, Network Mobility in Beyond-3G Systems, *IEEE Communications Magazine*, July 2003.
7. P. Bahl, A. Adya, J. Padhye, and A. Wolman, Reconsidering Wireless Systems with Multiple Radios, *ACM SIGCOMM Computer Communication Review*, vol. 34, issue 5, pp. 39–46, 2004.
8. D. Cavalcanti, C.M. Cordeiro, D.P. Agrawal, B. Xie, and A. Kumar, Issues in Integrating Cellular Networks, WLANs, and MANETs: A Futuristic

Heterogeneous Wireless Network, in *IEEE Wireless Communications Magazine*, Special Issue on Toward Seamless Internetworking of Wireless LAN and Cellular Networks, June 2005.

9. 3GPP TS 23.234, GPP System to WLAN Interworking, V.6.5.0, June 2005.

10. IEEE 802.21 Working Group, Draft IEEE Standard for Local and Metropolitan Area Networks: Media Independent Handover Services, October 2005. http://www.ieee802.org/21/index.html.

11. IEEE 802.11 Task Group u (TGu), http://www.ieee802.org/11/

12. D. Cavalcanti, Integrated Architecture and Routing Protocols for Heterogeneous Wireless Networks, Ph.D. dissertation, University of Cincinnati, January 2006.

13. K. Ahmavaara, H. Haverinen, and R. Pichna, Interworking Architecture between 3GPP and WLAN Systems, *IEEE Communications Magazine*, 41(11), November 2003.

14. A.K. Salkintzis, Interworking Techniques and Architectures for WLAN/3G Integration Toward 4G Mobile Data Networks, *IEEE Wireless Communications*, June 2004.

15. F.G. Marquez, M.G. Rodriguez, T.R. Valladates, T. De Miguel, and L.A. Galindo, Interworking of IP Multimedia Core Networks between 3GPP and WLAN, *IEEE Wireless Communications*, June 2005.

16. M. Buddihikot, G. Chandranmenon, S. Han, Y. Lee, S. Miller, and L. Salgarelli, Design and Implementation of a WLAN/CDMA2000 Interworking Architecture, *IEEE Communications Magazine*, 41(11), November 2003.

17. Q. Song and A. Jamalipour, Network Selection in an Integrated Wireless LAN and UMTS Environment Using Mathmatical Modeling and Computing Techniques, in *IEEE Wireless Communications*, June 2005.

18. B. Bhargava, X. Wu, Y. Lu, and W. Wang, Integrating Heterogeneous Wireless Technologies: A Cellular Aided Mobile Ad Hoc Network (CAMA), *Mobile Networks and Applications*, Kluwer Academic Publishers, pp. 393–408, 2004.

19. C. Huang, H. Lee, and Y. Tseng, A Two-Tier Heterogeneous Mobile Ad Hoc Network Architecture and Its Load-Balance Routing Problem, *Mobile Networks and Applications*, Kluwer Academic Publishers, pp. 379–391, 2004.

20. M. Lott, M. Weckerle, W. Zirwas, H. Li, and E. Schulz, Hierarchical Cellular Multihop Networks, in *Proceedings of the 5th European Personal Mobile Communications Conference (EPMCC2003)*, Glascow, Scotland, April 2003.

21. D. Cavalcanti, C.M. Cordeiro, A. Kumar, and D.P. Agrawal, Self-Adaptive Routing Protocols for Integrating Cellular Networks, WLANs and MANETs, *Journal of Wireless Communications and Mobile Computing*, to appear.

22. D. Cavalcanti, C.M. Cordeiro, A. Kumar, and D.P. Agrawal, A New Routing Mechanism for Integrating Cellular Networks, WLAN Hot Spots and MANETs, in *Proceedings of the 16th IEEE PIMRC*, 2005.

23. A. George, A. Kumar, D. Cavalcanti, and D. Agrawal, A Survey of Mobility Management Protocols for Heterogeneous Multi-Hop Wireless Networks, *Journal of Pervasive and Mobile Computing*, 2006.

Chapter 20

Intrusion Detection for Wireless Network

Cynthia Kersey, Zhenwei Yu, and Jeffrey J.P. Tsai

20.1 Introduction

Recently, there has been a rapid proliferation of wireless devices and the use of wireless networking. Two types of wireless networks, mobile ad -hoc networks (MANETs) and wireless sensor networks (WSNs), have attracted many researchers' attention and have been demonstrated with various applications in many areas. While these networks provide immense flexibility they have an increased vulnerability to attacks and intrusions. Because of this increased vulnerability, intrusion prevention alone does not solve the problem and intrusion detection takes on an enhanced role in these wireless networks. In this chapter, we survey the current research in the area of intrusion detection for these two types of wireless networks in two parts. The first part will discuss intrusion detection for MANETs, while the second part will focus on WSNs.

20.2 Background on Intrusion Detection

An intrusion is defined as an action that attempts to compromise the confidentiality, integrity, or availability of a resource.[1] As the name states, an intrusion detection system (IDS) is a system that detects a network intrusion. It is often termed a second line of defense because it is only activated when the intrusion prevention system has failed. Ideally, such a system can detect, identify, and eject an intruder before any damage is done. In this way, an IDS can also serve as a deterrent because intruders recognize that even if they can gain access they are likely to be expelled by the IDS.

IDSs for traditional networks function under the assumption that normal activity and intrusion activity have distinct behavior.[2–4] Additionally, to implement an IDS, users and program activities must be observable, for example, via a system auditing mechanism,[5] so that deviations from the norm can be recognized. Based on the type of audit data collected, an IDS can be classified as network- or host-based. Network-based IDS operate by passively or actively monitoring the network itself. Packets are collected from network traffic and analyzed to identify an intrusion. Network-based IDS often requires a dedicated host or special equipment, which makes them vulnerable to attack. Host-based IDS monitors activity on each individual node. Data is collected from the system's audit trails, system and application logs, or audit data generated by a model that intercepts system calls.[6]

IDS can be further classified on the basis of detection techniques. Intrusion detection techniques can be categorized into misuse detection and anomaly detection. Misuse detection uses the signature of known attacks to identify an intrusion. The advantage of this technique is that instances of known attacks can be quickly and accurately identified. However, misuse detection lacks the ability to detect newly invented attacks leaving the network vulnerable. In anomaly detection, a profile of normal activity is created and is used to classify any unreasonable deviations from the established norm as a potential attack. Data mining technology is often used in the profile creation because it is beneficial to automatically construct models due to the large amount of data collected. The advantage of anomaly-based detection is that no prior knowledge of intrusions are required, so novel attacks can be detected. However, this technique may suffer from high false-positive rates and additionally may not be able to accurately describe the attack that is occurring.

20.3   Intrusion Detection for Mobile Ad Hoc Networks

20.3.1   Introduction

A wireless ad hoc network provides communication between various devices (nodes) via a shared wireless channel. However, unlike a more conventional wireless network, nodes in an ad hoc network communicate without the assistance of a fixed network infrastructure. Nodes within one another's radio range can communicate through wireless links and dynamically form networks.[7] Additionally, nodes must cooperate by forwarding packets so that nodes not directly connected or beyond radio ranges can communicate with each other. Often the nodes in an ad hoc network are mobile. These networks are called MANETs.

Ad hoc networks are suited for situations where rapid network deployment is required or it is prohibitively costly to deploy and manage a network infrastructure. Some examples include military soldiers in the field, emergency services in a disaster area, attendees in a conference room, sensors scattered throughout a city for biological detection, space exploration, forestry or lumber industry, and temporary offices such as campaign headquarters.[8]

While there has been much work in IDS for traditional wired networks, it is difficult to apply much of this research to wireless ad hoc networks because of key architectural differences, most notably the lack of a fixed infrastructure. The lack of centralized audit points, such as switches, routers, and gateways, makes it difficult to collect audit data for the entire network. Data collection, in a wireless ad hoc network, is limited to activities taking place with radio range, so IDSs must work with localized partial information. Also, without a centralized authority, the algorithms used for intrusion detection must be distributed in nature, yet it must be kept in mind that attacks may be made from nodes inside the network. This means that one of the nodes participating in a collaborative intrusion detection algorithm may be a malevolent node. Additionally, while misuse detection can be applied successfully in traditional networks, this is not the case for wireless ad hoc networks. Since they are relatively new, not many specific attacks have emerged for wireless ad hoc networks. Therefore more emphasis should be given to anomaly-based detection. Anomaly-based IDSs detect patterns based on long-term modeling and the classification of normal and abnormal activity. Since wireless ad hoc networks are very dynamic in structure, this can be very challenging. And owing to mobility and power constraints, there is not always a clear separation between normalcy and anomaly in an ad hoc network. Constrained battery power also affects the detection algorithms used, since a limited power supply requires that intrusion detection algorithms be highly efficient.

20.3.2 Requirements for an IDS for Mobile Ad Hoc Networks

An IDS in a wireless ad hoc environment must be effective and efficient. An effective IDS correctly classifies normal and malicious activities. It must be fault-tolerant and resist subversion and it cannot introduce a new weakness into the network. An efficient IDS is cost-effective and uses little system resources, since to be effective an IDS must run continuously. An IDS in an ad hoc environment must work collaboratively to identify intrusions. And lastly, all IDSs must initiate a proper response when an intrusion is detected. In an ad hoc environment, these responses include reinitializing communication channels, identifying a compromised node and reorganizing the network to exclude that node, notifying the end user to take action, and even launching a counterattack.

20.3.3 Security Vulnerabilities

The lack of centralized control and infrastructure of an ad hoc network increases its vulnerability and exposure to attacks. Unlike its fixed wired counterpart where an attacker must gain physical access through several lines of defense at firewalls and gateways, attacks on a wireless network can come from all directions including nodes thought to be participating in the network, as the absence of authorization facilities impedes the usual practice of distinguishing nodes as trusted and nontrusted. Additionally, since the nodes are often mobile, the topology of the network may be constantly changing as nodes join in and move out of the network as they move in and out of radio range. Also, nodes may operate in a disconnected state to preserve a limited power supply, which also affects the network topology. This dynamically changing topology makes it difficult for nodes in a network to recognize a malicious node.

20.3.4 Attacks against Mobile Ad Hoc Networks

Attacks against wireless networks fall into two categories: passive attacks and active attacks. Passive attacks, such as eavesdropping, can be devastating to security critical areas such as military applications. Active attacks, on the other hand, involve replication, modification, and deletion of data.[7] And since nodes without adequate protection in a wireless ad hoc network are prone to being captured, compromised, or hijacked, these networks are particularly vulnerable to attacks that come from inside. Internal attacks are far more damaging and difficult to detect. A malicious node can disrupt the network by deleting or modifying messages or even attacking the routing protocol by refusing to forward messages or advertising incorrect paths. This can be difficult to detect, because false routing messages could be benign, just the result of an outdated routing table. Other active attacks include energy exhaustion attacks, referred to as sleep deprivation torture,[6] and denial-of-service (DoS) attacks. DoS attacks can be launched by a rogue node by sending a large number of route requests or by a node spoofing its IP and sending route requests with a fake ID to the same destination, causing a DoS at the destination node.

20.3.5 Intrusion Prevention for Mobile Ad Hoc Networks

The prevention of intrusions in wireless ad hoc networks would require the development of new secured protocols or modification of the logic of existing protocols to enhance their security. Traditional security solutions that require trusted authorities or certificate repositories are not well suited for securing wireless ad hoc networks as these networks exhibit frequent partitioning due to node mobility and disconnection. Several solutions have been presented to deal with these issues using either a partially distributed certificate authority[9] or a self-organized public-key management system.[10] A self-organized key management system allows users to generate public–private key pairs, issue certificates, and to perform authentication regardless of the network partitions and without any centralized services or trusted authority.

While these intrusion prevention techniques can be used to reduce intrusions, none are completely foolproof. History has shown us that regardless of the number and types of prevention measures that are inserted into a network, there are always some weak links through which attackers can gain access. As a second line of a defense, an IDS can be used to identify an intrusion and eject an intruder potentially before any damage is done. Given its inherent weaknesses, such a system is a necessity for a wireless ad hoc network.

20.3.6 *Intrusion Detection for Mobile Ad Hoc Networks*

In this section, we present an overview of the current research in intrusion detection for wireless ad hoc networks, including architecture, data sources for detection models, and detection algorithms.

20.3.6.1 *Distributed Intrusion Detection*

Zhang et al.[5] propose a distributed and cooperative IDS for wireless ad hoc networks. In their system, every node participates in intrusion detection and response via an IDS agent placed on it. The IDS agent is divided into six pieces: data collection module, local detection agent, cooperative detection agent, local response module, global response module, and a secure communication module, which provide a high-confidence communication channel among nodes in the network.

The data collection module gathers streams of audit data from various sources including system activity within the node, communication activities by the node or observable by (within radio range of) the node. This data can be integrated and used in a multilayer intrusion detection method.

The data collected in the collection module is analyzed by the local detection agent for signs of an intrusion. Traditional IDSs use data only from the lower layer, as the application level can be protected through application layer firewalls and application-specific modules. But in wireless networks there are no firewalls to protect the application layer, so intrusion detection in this layer becomes necessary. Also, certain attacks, for example, a DoS attack, may be more quickly identified in the application layer. Therefore, this IDS uses modules from the lower layer as well as the application level. Detection at each layer can be initiated or aided by evidence from other layers. If a node considers the evidence of the intrusion as "strong," it can independently determine that there is a network intrusion and initiate the proper response. However, if the node considers the evidence of intrusion as weak, it can start the cooperative detection agent by propagating state information among neighboring nodes. This information could include only the level of confidence of an intrusion or it could include the identity of the suspected malicious node along with the confidence level. On receiving an anomaly state request, each node, including the initiator, sends its state information to its immediate neighbors. Each of the nodes then decides whether the majority of the reports received reflect an anomaly. If so, any node can conclude that the network is under attack. The node that makes such a conclusion can initiate an appropriate response.

The intrusion response can be either local or global. In a local response, a node initiates actions local to itself, while in a global response, a node coordinates actions among neighboring nodes in the network. The actions taken are based on the network, applications, and confidence in the evidence. Some possible responses include forcing a re-key or identifying the compromised node or nodes and reorganizing the network to exclude such nodes.

This system uses anomaly-based intrusion detection by creating a model that can be used to classify an action as normal or a potential intrusion. The model is constructed by defining a set of features, which can be used to classify a system state. Because the set of features that could potentially identify a system state is quite large, an unsupervised method is used to determine the set to be used in classification that is called the essential feature set. A classifier is then used to compute rules to partition the data into the two classes. Intrusion reports are created by examining the current state of the essential feature set and using this information to classify the system (network) state as normal or abnormal.

The system was tested by creating four separate models, using two different feature sets with information available from the routing protocol, which collect data only from the local node. Two different classifiers were used, RIPPER, a decision-tree-equivalent classifier, and SVM Light that partitions the data with a hyperplane. Simulation data was then run using three wireless ad hoc protocols, dynamic source routing (DSR), ad hoc on-demand distance vector routing (AODV), and destination-sequence distance-vector routing (DSDV). In general, good results were obtained, particularly using the SVM Light classifier and the DSR protocol, which showed anomaly detection rates of approximately 99% and a false alarm rate of less than 0.05%.

20.3.6.2 Hierarchical Cooperative Intrusion Detection

Sterne et al.[1] take Zhang et al.'s idea of a cooperative IDS and augment it with a dynamic hierarchical structure. While cooperative IDSs may be successful in detecting malicious behavior with respect to routing protocols, such systems have not shown that they are applicable to more conventional attacks. Additionally, a hierarchical structure is traditionally more amenable to growth. In the fully cooperative, distributed IDS, such as the one discussed above, communication overhead can rise very quickly, in the order of the square of the number of nodes. A hierarchical model, on the other hand, allows data sharing without such a rapid increase in communication overhead. The proposed architecture was designed for military applications and as such mimics the structure found in such organizations in the manner in which intrusion detection data is passed up the hierarchy while intrusion response directives flow down to the lower levels.

Ad hoc networks typically construct routes using topology-based clustering. Nodes create neighborhoods based on proximity. Such clusters can then select a node to be a neighborhood representative called a cluster head. The cluster heads then organize into a second level of clusters and select representatives who join in a third level of clusters and so on until all the nodes in the network are interconnected. In a dynamic hierarchical structure, the cluster head is selected based on a variety of attributes including connectivity, hardiness, power and storage capacity, and bandwidth capabilities.

In the proposed architecture, the nodes cooperate to protect the network but remain responsible for intrusion detection mechanisms to protect themselves. The nodes share tasks such as monitoring, logging, analyzing, and reporting data at various layers of the network. Monitoring is both promiscuous and direct. Promiscuous monitoring is monitoring the communication of neighboring nodes even when a node is not involved in the transmission of a message. Direct monitoring involves reporting of a node of its activity. In a fully cooperative IDS, all nodes monitor the traffic that flows through it. In the hierarchical model, monitoring responsibilities are given to the two nodes that are the first and last hop between each pair of nodes. The responsible nodes are automatically updated when a route changes as a node would be aware of the path a packet is taking and what its position is in the routing of the packet. This simple strategy can dramatically reduce the amount of communication overhead and duplicated effort. Additionally, this sort of monitoring is suitable for detecting conventional attacks on the network.

Nodes at the lowest level are responsible for collecting certain data as well as intrusion detection and reporting. The key principle in this system is that intrusion detection should occur at the lowest level of the hierarchy at which data is available to make an accurate decision. Since leaf nodes do not aggregate data they generally do not analyze intrusion information since this typically requires large amount of data. This analysis is performed by the cluster heads that collect from their cluster members and perform detection computations on the consolidated data.

A cluster head may query members of its clusters or its peers for additional information. Additionally, a cluster head sends consolidated data to its superior. Nodes at the top of the hierarchy have responsibility for managing the IDS through activities such as distribution decision rules and signatures of known attacks. A node's authority increases as it moves toward the top of the hierarchy, thus mimicking the structure found in many organizations.

20.3.6.3 Mobile Agent-Based Intrusion Detection Systems

Karchirski and Guha[11] prose a distributed IDS for wireless ad hoc networks based on mobile agent technology. Mobile agents are autonomous software entities that can halt themselves, ship themselves to another agent-enabled host on the network, and continue execution, deciding where to go and what to do along the way.[11] Agents are dynamically updateable and have a specific functionality.

The proposed system uses a modular architecture with several types of mobile agents that perform functions such as network monitoring, host monitoring, decision making, and action. Only certain nodes will have agents for network packet monitoring while every node in the network will have an agent to monitor system- and application-level activities. In the cooperative decision-making process, every node will decide on an intrusion threat level at the host level while only those nodes containing a network monitoring agent will participate in making decisions on a network-based intrusion. All nodes will contain an action host that is responsible for responding to an intrusion. By distributing the functions of the IDS into separate modules represented by a lightweight mobile agent, the workload of intrusion detection is spread across the nodes of the network to minimize power consumption and reduce processing time. There are three agent classes: action, decision, and monitoring. The monitoring class is further divided into agents that monitor packet-level data, user (application) data, and system-level data.

Since the agents that monitor network packets and make network intrusion detection decisions are located on a subset of the nodes of the network, a distributed algorithm is used to select the nodes to host these agents. The algorithm used logically divides a mobile network into clusters with a single cluster head for each. The cluster head then hosts the network-monitoring sensor, which collects all packets within radio range and analyzes them for known patterns of attacks.

The cluster heads monitor packets sent by every member of its cluster, while ignoring those sent by nodes outside of its cluster. This prevents duplicate processing of packets by two different cluster heads. The packet information is inserted into a fixed-sized queue, which is used by the decision agent to analyze the state of the network and its nodes. Local detection agents monitor local activity looking for suspicious activities. If an anomaly is detected with strong evidence, action is taken to terminate the suspicious activity. If an anomaly is detected with less confidence, the node reports its status to the decision-making agent on the cluster head.

The proposed system uses a decision-making process, where individual nodes make decisions on their local state while the global decision-making agent, located on the cluster head, collects information from the network and all the nodes within its cluster. The agent can then conclude with some confidence whether a node has been compromised. When such a determination is made the agent instructs the local node to take action. This action should result in a decreased threat level. If that does not occur, the node can be excluded from the network. The authors propose the use of an anomaly detection model to identify potential intrusions into the network.

The mobile agent approach creates an IDS that minimizes the use of scare computational and power resources. However, at the same time, it creates points of failures that could be exploited by an attacker. The authors recognize this limitation and propose additional research into an effective means of defense.

20.3.6.4 Cross-Feature Analysis for Intrusion Detection

As mentioned above, while misuse detection can be effectively used to identify intrusions in a wired network this is not the case for ad hoc networks, given their relative infancy. Therefore, anomaly detection is currently the preferred methodology. Anomaly detection generally involves mining historical data to detect patterns related to normal and abnormal activities and then building a classifier based on these patterns. One method for building such a classifier is suggested in Ref. [12], using a technique for identifying anomalies called cross-feature analysis.

A basic assumption for a network of any kind is that there exists a set of features that can unambiguously identify whether a network is in a normal or abnormal state. The set of features can be stored in a feature vector and often there are a set of such feature vectors related to a normal network state. Cross-feature analysis attempts to explore the relationship between the values in the feature vector and the state of the network. Using all normal feature vectors, a classifier is built that predicts the value of a given feature based on the values of the other features and a normal system state. During the training process a classifier is built for each of the features, f_i, in the feature vector of the form C_i: $\{f_1, f_2, \ldots, f_{i-1}, f_{i+1}, \ldots, f_n\} \rightarrow f_i$.

This classifier contains a set of rules or a decision tree that can predict the value of a feature given the other features. The assumption made is that if the predicted value of a feature does not match the actual value of the feature, it can be assumed that there is anomaly. At the end of the training process there exists a set of classifiers, one for each of the features in the feature vector.

These classifiers are then used to analyze the network logs and identify anomalies. Two different algorithms are suggested. The simplest one is called average match count. When an event is analyzed, the classifiers are used to predict each of the features in the feature vector and a count is kept of the number of matches that occur. A simple average is taken and if the average number of matches is less than a designated threshold, the network is assumed to be in an abnormal state. A second algorithm is suggested that uses probabilities instead of the simple binary matching classification. Most classifiers can return the probability that the labeled feature contains a certain value, given the values of the remaining features. The classifiers are used to estimate the probability of each value in the feature vector. These probabilities are averaged and again the network is assumed to be in an abnormal state if the average probability is less than a given threshold.

Cross-feature analysis was tested using a feature vector designed to identify routing anomalies. Using a network simulator and four different routing protocols, two routing anomalies were generated: black hole attacks and selected packet dropping. Three different classifiers were built using three different classification algorithms: decision tree, class association rules, and naïve Bayes. Near-perfect results were obtained using the decision tree classifier and average probability detection.

20.3.6.5 SVM-Based Intrusion Detection Systems

An alternate method for using a set of features to classify a network state as normal or abnormal is to use a support vector machine (SVM).[13] SVMs are classifiers that identify a hyperplane to separate two classes of data: positive and negative. Data is mapped to very high-dimensional space using a special type of function called a kernel function. Then a hyperplane is defined that works as a decision boundary between the two classes of data. This idea works on the heuristic that data that appear nonseparable in lower dimensions is separable in higher dimensions.

The proposed SVM-based intrusion detection module has two components. An unsupervised SVM detection module, 1-SVMDM, which can be used when no training data is available and a supervised SVM detection module, 2-SVMDM, which can be trained using available attack data. 1-SVMDM can be used until a system has a history that can be used to train 2-SVMDM. Unsupervised intrusion detection can be modeled as outlier detection, using the assumption that an abnormal state is sufficiently different from a normal network state. Once the system has been used for a period of time, the abnormal data, outliers, can be labeled as normal or abnormal and be used to train the model and derive a new decision boundary. This revised decision model can then be used to classify a network state as normal or abnormal.

The proposed IDS consists of four components: local data collection, SVM-based intrusion detection, local response, and global response. Data is collected locally from various network audit streams and is passed to the SVMDM. The SVMDM classifies the network state as normal or a possible intrusion in which case it also identifies the source node. The local response module distributes local detection results based on the data collected locally while the global response module consolidates other nodes' locally collected data and makes a decision based on this consolidated data. The method of sharing data is dependent on the IDS architecture and this type of detection is conducive to either a fully distributed architecture such as the system proposed by Zhang et al.[5] or a hierarchical architecture as in Ref. [1].

The system was tested using a network simulator, which created simulations of two different DoS attacks against the AODV routing protocol, black hole attacks, and frequent false routing requesting (FFRR). The detection rate for 2-SVMDM was approximately 96% for both the fully distributed and the hierarchical system architectures, with a slightly higher false alarm rate in the fully distributed system. 1-SVMDM was able to detect both types of attacks with a detection rate of approximately 85% but with a false alarm rate approaching 20%. While the system was tested on only a single routing protocol and a specific set of routing-based attacks, the authors believe that the system can be extended to other routing protocols and attack types with the appropriate parameter selection.

20.3.6.6 A Game Theory Approach to Intrusion Detection

Game theory has been used extensively to model a variety of problems, such as routing behavior and distributed power control, in wireless ad hoc networks. Patcha and Park[14] present the use of game theory to model the interaction between an attacker and IDS. This scenario is modeled as a two-player game. The key to such a model is the interaction of the players such that the actions of one player affect the other player in either a positive or negative way. This is obviously the case in an IDS, as an intrusion negatively impacts the node being attacked, while stopping an intrusion has a negative impact on the attacker. Additionally, in game theory, a player always takes actions that are in that player's best interest. This, again, is the case with IDS in wireless ad hoc networks.

In the proposed game model, the objective of the attacker is to send a malicious message with the intention of attacking the other player, which is another node in the network. The intrusion is considered successful if the malicious message reaches the target without being detected while the IDS is successful if it detects the intrusion and the intruding node is blocked. In the game theory model presented, the attacker is considered the sender and the host-based IDS is the receiver. The host-based IDS has a prior belief regarding the probability that a node is an attacker or a regular node. The IDS uses this probability to calculate the expected payoff from blocking the sender's transmission.

$$\text{Payoff}_{\text{receiver}} = (s\gamma_{\text{miss}}) + (t\gamma_{\text{falseAlarm}}) - (st(\gamma_{\text{detect}} + \gamma_{\text{falseAlarm}} + \gamma_{\text{miss}}))$$
(20.1)

where γ_{miss} is the cost of missing an intrusion, $\gamma_{\text{falseAlarm}}$ the cost of a false alarm, γ_{detect} the gain of detection, s the probability that the sender is an attacker, and t the probability of detecting the intrusion.

The payoff for the attacker is found using,

$$\text{Payoff}_{\text{sender}} = (t\delta_{\text{caught}}) + ((1 - t)\delta_{\text{intrude}})$$
(20.2)

where δ_{caught} is the cost of being detected and blocked and δ_{intrude} the gain of a successful intrusion.

The strategy for the sending node is to decide whether to send a message based on the strategy of the IDS and to send a message if it maximizes its expected payoff. The choice of strategy by the IDS is based on the receiver's prior belief, calculated using Bayes rules, so that it is able to maximize the effective payoff by minimizing the cost due to false alarms and missed attacks. Since Bayes theorem is recursive in nature, these probabilities will be recalculated regularly and this should reduce the number of false alarms and missed intrusions.

20.3.6.7 Combining Misuse Detection with Anomaly Detection

The idea of combining anomaly detection with misuse detection is presented by Nadkarni and Mishra.[18] The idea behind this approach is that while anomaly detection leads to a high degree of false positives and misuse detection can miss some attacks, the combination of the two methods is superior to using either separately. Additionally, this proposed IDS is adaptive in adjusting its thresholds to abnormal activities, effective with an average accuracy rate of over 90%, efficient in conserving resources and power consumption, and protocol-independent.

The proposed IDS can be broken into three stages: initialization, audit data analysis, and threshold adjustment. During the initialization phase, a node analyzes network traffic and gathers information about the normal behavior of the network. Using this information, initial threshold values are created for each of the "normal" occurrence of attack-like actions in the network. Two arrays are maintained, one with predefined threshold values that store the number of maximum number of symptoms for each attack that occur under normal operating conditions and another array of maximum time intervals during which such symptoms occur. During the operating stage, audit data consisting of routing updates and packet headers is used for analysis and identification of abnormal behavior. At each node, a three-dimensional array of abnormal event counters is maintained. It consists of a counter related to each attack for each of the other nodes in the network and for each variation in the type of abnormal behavior for the attack. Neither one incident of abnormal behavior nor a series of widely dispersed incidents signals an attack. However, a series of abnormal behaviors all associated with the same attack symptoms occurring at higher-than-normal frequency may signify an attack. Therefore this set of counters is maintained, one for each type of attack. A counter is incremented when a related incident occurs and after a single incident of abnormal behavior, the suspicious status of the related node is noted and the activity of the node is monitored for a possible intrusion. If the suspicious node continues to display abnormal behavior that can be interpreted as some symptom of the attack, or the variation of such an attack, during a specific time frame the IDS identifies that there is an intrusion and initiates an appropriate response.

The adaptive properties of this IDS are noted in the threshold adjustment stage. After regular time intervals without an intrusion, threshold values are adjusted. This is to prevent the possibility that malicious nodes are operating just under a threshold level. Therefore the threshold for each attack, or variation of such an attack, is increased by a fixed percentage. If an attack does occur, the threshold is adjusted to take into account the properties of the attack. It is revised to the difference in the detected rate of abnormal behaviors and the "normal" rate of abnormal behaviors multiplied by time interval of the attack.

The proposed IDS was tested using the user datagram protocol (UDP) and mobile nodes with the testing focusing on the detection accuracy when varying node mobility and density. Preliminary results showed that the IDS detected malicious nodes over 90% of the time with a low false alarm rate of approximately 2%.

20.3.6.8 Watchdog and Pathrater

One of the earliest proposals for an IDS for wireless ad hoc networks is by Marti et al.[7] While the goal of their proposal is to increase throughput in the network, it focuses on intrusion prevention methods by introducing two overlays to the DSR algorithm, in which every packet has a route path that consists of nodes that have agreed to forward the packet. The proposed system consists of two tools to detect and mitigate abnormal routing behavior. The Watchdog tool identifies misbehaving nodes, while Pathrater aids the routing protocol in avoiding such nodes.

Watchdog works by using promiscuous listening. Each node in a routing path verifies that its successor appropriately forwards the message to the next node in the path. For example, node S wishes to send a message to node D using the routing path S-A-B-C-D. When node B forwards packet to D through C, A can listen to B's transmission and verify that B has attempted to pass the packet to C. Additionally, if encryption is performed separately for each link, A can also tell if B has tampered with the header or the message itself. Since failing to forward a single packet is not indicative of a malicious node, each node maintains statistics for the routing behavior of its neighbors. This is accomplished by maintaining a buffer of recently sent packets. Each time a node monitors a message, it compares it with the packets in the buffer to see if there is a match. If there is a match the packet is removed from the buffer. And if a packet remains in the buffer for longer than a specified time period, the Watchdog increments a failure counter for the node responsible for forwarding the packet. After the counter exceeds a certain threshold the node is identified as misbehaving and a message is sent to the source identifying the misbehaving node.

The information collected by Watchdog can be used by the Pathrater to determine an efficient route that avoids routing packets through misbehaving nodes. Each node maintains a rating for all other nodes in the network. A newly discovered node receives a neutral score. For every time interval where a node acts appropriately in forwarding a message, its score is increased. However, if Watchdog notes that a node failed to forward a packet, the node's score is decremented. If a node is designated as misbehaving it receives a high negative score. When computing a route, each potential path for a message receives a score, which is the average rating of the nodes in the path. If there are multiple paths to the same node, the path with the highest score is chosen. This guarantees that messages are routed through the most reliable nodes.

Even though the combination of Watchdog and Pathrater increases the overhead at a node, testing showed that overall network throughput was increased. Additionally, simulations showed that network throughput was not adversely affected by false detection.

20.4 Intrusion Detection for Wireless Sensor Networks

20.4.1 Introduction

A WSN consists of a large set of tiny sensor nodes. Sensor nodes can perform sensing, data processing, and communicating but with limited power, computational capacities, small memory size, and low bandwidth. Unlike MANETs, the senor nodes in WSNs are usually static after deployment, and communicate mainly through broadcast instead of point-to-point communication. Sensor networks have been used in a variety of domains, such as military sensing in a battlefield, perimeter defense in critical area such as airport, intrusion detection for traditional communication network, disasters monitoring, and home healthcare. Obviously, some applications are security-critical, which attract many researchers' attention to secure a sensor network. Some security protocols or mechanisms have been designed for sensor network. For example, SPINS, a set of protocols, provides secure data confidentiality, two-party data authentication, and data freshness and authenticated broadcast for sensor network.[15] LEAP, a localized encryption and authentication protocol, is designed to support in-network processing based on the different security requirements for different types of message exchange.[16] INSENS is an intrusion-tolerant routing protocol for WSNs.[17] A lightweight security protocol relying solely on broadcasts of end-to-end encrypted packets was reported in Ref. [18]. However, in a sensor network, as a complicated system, there are always some vulnerabilities to be attacked. In this part, we will investigate intrusion detection approach in sensor network and its related topics.

20.4.2 Challenges on Intrusion Detection in Wireless Sensor Networks

Security properties or the challenges of the WSNs have been reported in various literatures.[19–22] We summarized the challenges on designing IDS in WSN from the limited resources of the sensor nodes, the wireless communication, the dynamic topology of the network, and the hostile working environment.

Sensor network nodes usually have severe constraint in computational power, memory size, and energy. A representative processor used in well-known Crossbow MICA2/MICAz series sensor nodes is an 8 MHz 8-bit Atmel ATMEGA128L CPU. This CPU has only 128 kB of instruction memory, 4 kB of RAM for data, and 512 kB of flash memory.[23] The sensor node power is usually provided by two AA batteries. With these limited resources, some effective security defense techniques for traditional LAN/WAN/Internet are no longer suitable for WSN. For example, asymmetric cryptography is often too expensive for many WSN applications. Intrusion detection, as another layer of security, plays a more important role to secure WSNs. However, the low computational power and the insufficient available memory pose big challenges to the design of an IDS for WSN: the intrusion detection components should optimize resource consumption, and it might sacrifice its performance to fit the resource constraints. Another challenge is the available data; owing to low available storage (if it has), only limited log/audit data could be used for intrusion detection.

Sensor nodes use wireless communication in WSNs. Any information over the radio can be intercepted and the private or sensitive information could be captured by a passive attacker. An aggressive attacker can easily inject malicious messages into the wireless network to perform varied attacks. Unlike wireless local area networks (LANs), whose available bandwidths could be 54 Mbps, the data rate for WSN is likely far less than 1 Mbps. For example, the Crossbow MICA2 series motes feature a multichannel radio from Chipcon, delivering up to 38.4 kbps data rate up to 1000 ft.[24] The MICAz series motes offer 250 Kbps data rate up to 100 m.[25] The low bandwidth prevents some analysis on suspicious data being executed promptly in the powerful remote base station. On the other hand, communication is a very energy-hungry task in sensor node; transmitting with maximal power could consume about 3–4 times the power a processor does in active mode. Most of the communication ability should be reserved for target-sensed information. Only limited amount of security-related data could be sent to the powerful base station for further comprehensive analysis to detect intrusions.

Knowledge of the network is a very useful information to detect intrusions. In a WSN, the topology of the network is usually not a priori. Even after the deployment, the network is always evolving due to frequent failure of sensor nodes, and new added sensor nodes. It could be a big challenge to build a base profile in such a dynamic network for an IDS.

WSNs may be deployed in hostile environments such as battlefields, where sensor nodes are susceptible to physical capture. Security information (e.g., shared key) might be exposed by compromised nodes. The development of tamper-proof nodes is one possible approach to security in a hostile environment, but the complicated hardware and high cost keep it away from WSN applications. An IDS for WSN has to handle physical attacks as well as network attacks.

20.4.3 Attacks against Wireless Sensor Networks

Researchers have identified some of the attacks that could be performed against WSNs. In this section, we will discuss those attacks.

One of the most challenging issues facing sensor networks is how to resist *physical attacks*. In a traditional computer communication network, physical security is often taken for granted. Attackers are simply denied physical access to network nodes. Since WSNs might be deployed in a hostile environment or densely populated areas, it is very hard to prevent sensor nodes from being accessed and captured physically. A captured sensor could be simply disabled to destruct the network, or the memory in the captured sensor node could be analyzed to expose its data or cryptographic keys. The exposed key could be used to perform further attacks. A captured sensor node could be reprogrammed by the attacker and be rejoined to the network in the form of a subverted sensor node,[20] which is potentially undetectable to neighboring nodes.

Jamming is a well-known, DoS attack on wireless communication. Different layers could be targeted by jamming attack. At the physical layer, the attacker can send out interfering RF signals to impede communication. The jamming attacker can also inject irrelevant data to waste/drain battery energy on the receiving node for radio reception. Link-layer jamming exploits properties of the medium access control protocol employed. For instance, the attack can induce malicious collisions or attempt to get an unfair share of the radio resource.[26] At the network/routing layer, an attacker injects malicious routing information that causes other nodes to form a routing loop to waste precious communication and battery resources.

Misdirection is an attack against routing algorithms that can be performed by spoofed, altered, replayed routing information. By forwarding messages along wrong paths, an attacker misdirects them, perhaps by advertising false routing updates.[26] An attacker could inflict the attack on a particular sender by diverting only traffic originating from the victim node.

A *sinkhole attack* is also against the routing algorithms in a WSN. It typically works by making a compromised node to allure nearly all the traffic from a particular area through the node, forming a routing hole.[27] Through a sinkhole attack, the compromised node on the path that packets follow can enable many other attacks, like selective forwarding.

In a *selective forwarding attack*, the compromised nodes simply drop certain messages to ensure that those messages are not forwarded any further, but forward the other traffic as normal to reduce the risk of being detected.[27] An extreme form of selective forwarding attack, called *black hole attack*,[26] is that a malicious node refuses to forward every packet. However, such black hole attack runs the risk of its neighboring nodes concluding that the malicious has failed and deciding to seek another route.

The *Sybil attack* is where a malicious node illegitimately claims multiple identities and works as if it were many nodes.[28] The identities could be stolen or fabricated. The Sybil attack can be exploited at different layers to cause DoS in sensor networks.[26] At the MAC layer, the malicious node can claim a dominating fraction of the shared radio resource by Sybil attack. At the routing layer, by claiming a large number of identities, the Sybil node (the node performing Sybil attack) will be selected as the next hop node with a high probability to create a "sinkhole." Any system whose correct behavior is based on the assumption that most nodes will behave properly may be at risk for Sybil attacks. For example, in WSN, usually aggregation of sensor readings rather than individual sensor reading will be sent to base station to conserve energy. By using the Sybil attack, one malicious node may be able to contribute to the aggregate many times to alter the aggregate reading.

In the *wormhole attack,* an adversary tunnels messages received in one part of the network over a low latency link and replays them in a different part.[29] The simplest instance of this attack is a single node situated between two other nodes forwarding messages between the two of them. However, wormhole attacks will more commonly involve two distant malicious nodes colluding to understate their distance from each other by relaying packets along an out-of-bound channel available only to the attacker.

HELLO *flooding attack* is against many node discovery protocols. Those protocols require nodes to broadcast HELLO packets to announce their existences to their neighbors, and a node receiving such a packet may recognize the sender as its neighbor according to the normal radio range. This assumption could be attacked by using a laptop-class node that can broadcast HELLO packets with large enough transmission power that could convince every node in the network that the adversary was its neighbor.[27]

Rushing attack is against many flooding-based broadcast algorithms, which employ duplicate suppression on incoming messages where a node only forwards a message once and drops the duplicate message with the same ID. All nodes under the rushing attack will suppress the legitimate broadcast message if a bogus message with the same message ID has been broadcast by an attacker.[30]

A *stealthy attack* is to deliberately deceive the network with a false data value.[31] This attack is mainly against the data aggregation scenario. There are several methods to alter the normal aggregation result. For instance, a compromised sensor/aggregator can report significantly biased false values. A compromised node can also perform a Sybil attack to have greater impact on the aggregated result. The attacker can also perform DoS attacks on legitimate nodes to suppress their impact on the aggregation result. A stealthy attack can also be executed to disseminate false timing information to desynchronize nodes in the network by intercepting and delaying synchronization messages, or spreading false synchronization messages.[20]

20.4.4   Current Defenses against Specific Attack

20.4.4.1   Packet Leashes

Packet leashes is a general mechanism proposed in Ref. [29] for detecting and defending against wormhole attacks. The key idea is that information (called leash) is added to a packet to restrict the packet's maximum allowed transmission distance. Two leashes, geographical leashes and temporal leashes, were introduced in Ref. [29]. A geographical leash ensures that the recipient of the packet is within a certain distance from the sender. A temporal leash ensures that the packet has an upper bound on its lifetime, which restricts the maximum travel distance. Either type of leash can prevent the wormhole attack, because it allows the receiver of a packet to detect if the packet traveled further than the leash allows.

A geographical leash includes location and timestamp information. When a node sends a packet, its own location and sending timestamp are added to the packet as leash. After a node receives a leashed packet, the receiving node compares these leashes' values to its own location and the receiving time. The upper bounds on distance and delay between the sender and receiver are used to detect the wormhole attack.

A temporal leash is a sending timestamp or an expiration time. When a node sends a packet, the sending timestamp or an expiration time is included in the packet. After a node receives a packet, the receiving node compares this value to the time at which it received the packet. The receiver is thus able to detect if the packet traveled too far.

A geographical leash builds on correct location and synchronized clocks. A temporal leash requires more tightly synchronized clocks than a geographical leash does. It requires supports not only from the hardware but also from other detection and defense mechanism against the attacks to location discovery schemes and time-synchronization protocols.

20.4.4.2   Localization Anomaly Detection

A scheme taking advantage of the deployment knowledge to detect localization anomalies caused by adversaries was proposed in Ref. [32]. Detecting localization anomalies was formulated as an anomaly intrusion detection problem. Deployment points to be deployed are stored in the sensors' memories prior to the sensors' deployment as a baseline profile. After sensors derive their locations, the real profile is compared with the baseline profile. The level of inconsistency above certain threshold indicates malicious attacks on localization.

A model called group-based deployment model was presented in Ref. [32], where all sensor nodes to be deployed were divided into equal-size groups. The deployment points (the location to be deployed) were arranged in a grid. After deployment, the resident point (the final actual location) of the nodes in one group follows a two-dimensional Gaussian distribution function.

After the sensors are deployed, each sensor broadcasts its group ID to its neighbors, and each sensor can count the number of neighbors from each group, which was called the actual observation of the sensor. However, an expected observation of the sensor and the likelihood of its actual observations could be computed based on the derived location using certain localization scheme and deployment model. If the expected observations are too different from its actual observations, or if the likelihood of the actual observation is too low, a sensor can claim that the expected observation is inconsistent with the actual observations, which indicates an anomaly. Three metrics, the difference metric, the add-all metric, and the probability metric, were proposed to measure how different these two observations are in Ref. [32].

20.4.4.3 Localized Encryption and Authentication Protocol

LEAP is designed to support in-network processing bases on the different security requirements for different types of message exchange.[16] LEAP supports the establishment of four types of keys for each sensor node. An individual key shared with the base station for secure communication between the node and the base station. A pairwise key shared with each of its immediate neighbors for securing communications that require privacy or source authentication. A cluster key shared with multiple neighboring nodes for securing locally broadcast messages. And a group key is globally shared by all the nodes in the network for encrypting messages that are broadcast to the whole group by the base station. In LEAP, every node only accepts packets from its authenticated neighbors, which can prevent HELLO flooding attack. The neighbor discovery phase of the pairwise key establishment process is the only chance for an outside attacker to launch wormhole attack in LEAP. After that phase, a node knows all its neighbors. Thus the adversary cannot later convince two distant nodes that they are neighbors to launch wormhole attack.[16]

20.4.4.4 Radio Resource Testing

Radio resource testing was proposed to defend against Sybil attack in Ref. [33]. This approach relies on two assumptions. The first assumption is that any physical device has only one radio. The second assumption is that a radio is incapable of simultaneously sending or receiving on more than one channel.

During the test, the testing node assigns each of its neighbors a different channel to broadcast some message and then chooses a channel randomly to listen. If the neighbor that was assigned that channel is legitimate, it should hear the message or the channel was assigned to a Sybil node. To increase the probability of detecting the Sybil node, a sensor node can repeat this test multiple times.[33]

20.4.4.5 Random Key Predistribution

In random key predistribution, a random set of keys or key-related information are assigned to each sensor node, so that in the key setup phase, each node can discover or compute the common keys it shares with its neighbors; the common keys will be used as a shared secret session key to ensure node-to-node secrecy. Extension on existing random key pre-distribution techniques is proposed to defend again the Sybil attack by associating the node identity with the keys assigned to the node randomly and key validation in Ref. [33]. Even if an attacker captures a limited set of keys, because the keys associated with a random identity are not likely to have a significant intersection with the compromised key set, which makes it hard for the fabricated identity to pass the key validation, there is little probability that an arbitrarily generated identity is going to work.

20.4.5 Intrusion Detection for Wireless Sensor Networks

20.4.5.1 Decentralized High-Level Rule-Based IDS Model

A decentralized high-level rule-based IDS model was proposed in Ref. [34]. The proposed system is decentralized. All IDS functions, from data acquiescing to analyzing, are implemented in monitor nodes. Only intrusion alerts are sent to the base station. This IDS performs analysis on data message listened by the monitor node that is not addressed to it and message collision when the monitor node tries to send a message.

The proposed system is a rule-based IDS model. Seven high-level rules were defined in Ref. [34]. An *interval rule* is to check the time interval on receptions of two consecutive messages to detect abnormal message sending. A *retransmission rule* is to monitor the selected routing node on its neighbors to forward the received message to detect black hole and selective forwarding attack. An *integrity rule* is to check the message payload along the path from its origin to a destination to be modified. A *delay rule* is to check the retransmission of a message by a monitor's neighbor within a defined timeout. A *repetition rule* is to count the same retransmitted message by the same neighbor to detect a DoS attack. A *radio transmission range rule* is to check the direct sender of all monitored messages to detect wormhole and hello flooding attack. A *jamming rule* is to count the number of collisions associated with a message sent by the monitor to detect jamming attack.

After messages are collected in a promiscuous mode and the important information is filtered and stored, a sequent rule-matching procedure is executed on every message. The order of rules depends on the message type. When a rule fires on a message, the rule-matching procedure will stop and the message will be discarded to save the storage space. Instead of reporting an alarm on attack, a failure counter is incremented when a rule fires on a message. An attack is alerted only if the counting failure number is greater than an expected value by the monitor node during the analysis of messages transmitted on its neighborhood in a round. This expected number is calculated dynamically by the monitor node according to the failure history for each node in its neighborhood.

20.4.5.2 Noncooperative Game Approach

The intrusion detection problem in WSN was formulated as a noncooperative two-player nonzero-sum game between the IDS and the attacker in Refs. [35, 36]. The basis is that in noncooperative games there exist sets of optimal strategies (the so-called Nash equilibrium) used by the players in a game such that no player can benefit by unilaterally changing his or her strategy if the strategies of the other players remain unchanged. The relationship between an attacker and the IDS is noncooperative in nature because no outside authority could assure any agreement between an attacker and the IDS. The proposed IDS is able to monitor all sensor nodes, but due to system limitations it can only protect one sensor node at each time slot, and based on a game theoretic framework it will choose such a sensor node (called cluster head) for protection. So the task of IDS is to find a cluster head to achieve equilibrium where the security of the sensor network is maximized.

Owing to limited available resource and shortage of global identification in WSN, all sensor nodes are divided into clusters. In each cluster one node was chosen as the cluster head, and each cluster head has a unique ID. Weight clustering approach (WCA) proposed by Mainak Chatterjee et al. was used to choose a cluster head because their cluster head election procedure is not periodic and is invoked as rarely as possible, which reduces system updates and hence computation and communication costs.[35] In this IDS, it is assumed that to protect, one cluster head the IDS can only do one inspection in one time slot, and when IDS chooses to protect the cluster head, it is impossible for an attacker to attack that cluster.

All of the three possible attacker strategies (AS) in one time slot with respect to one given cluster are AS1, which attacks the given cluster; AS2, which does not attack any cluster; and AS3, which attacks another cluster. Meanwhile, for the IDS, the two system strategies (SS) are SS1, which defends the given cluster and SS2, which defends another cluster.[35] In a two-player game, each player has a payoff matrix to represent the payoffs under all strategy combinations. The rows in the payoff matrices represent one player's strategy and the columns represent the other player's strategy. The authors defined the IDS' payoff matrix based on four parameters: the utility of the sensor network's ongoing sessions ($U_{ij}(t)$), the average loss by loosing cluster k (AL_k), the average cost of defending cluster k (C_k), and the number of nodes in a cluster k (N_k). The Attack's payoff matrix is calculated on the three parameters: the cost of waiting (CW) and deciding to attack in the future, the cost of intrusion (CI) for the attacker, and the average profit of each attack (PI(t)). Four DoS attacks (homing, misdirection, desynchronization, and jamming) are investigated to calculate the attacker's payoff matrix. The authors reduced the size of the payoff matrices through finding dominant strategies and concluded that strategy pair (AS1, SS1), constituted the Nash equilibrium to this game. So the IDS is to choose the cluster with the highest value of $U(t) - C_k$ to protect to maximize its payoff.[35,36]

20.4.5.3 MDP-Based Intrusion Detection Scheme

A scheme using Markov decision process (MDP) to predict the most vulnerable sensor node to defend was also presented in Ref. [36]. In this scheme, the past behavior of the attacker and so the past states of the system are assumed to be known, and the task is predicting the most vulnerable node, which attacker most probably will attack. A MDP model is a four tuple (S,A,R,tr), where S is a set of states, A a set of actions, R a reward function, and tr the state-transition function. The actions change the states and the effect of the actions on the states is captured by the transition function. The transition function assigns a probability distribution to every (state, action) pair. The reward function assigns a real value to each (state, action) pair, which describes the immediate reward (or cost) of executing this action in that state. Each action of the MDP corresponds to one intrusion detection of a sensor node. When IDS detects an intrusion on a node, the MDP can either accept this detection and thus the state will be transferred, or it selects another node. Only if IDS chooses the right cluster for protection, it will gain that reward. The rewards in the MDP encode the utilities of detecting an intrusion. The criterion used in selecting the action in every state is the maximization of its future reward. The objective is to maximize the expected value of received reward over time.[36]

20.4.5.4 Combination of the Intrusion Detection and
Failure Recovery

Some research works that target on the attackers who try to capture nodes by taking control of the code they are executing were presented in Ref. [37]. The authors proposed to combine the IDS and failure recovery to raise the cost for a potential attacker and provide a chance to detect and recover a compromised node. Some canonical conditions on the basic behavior of a sensor node, including correct forwarding of messages, correspondence between current incoming and outgoing traffic, conformance to communication schedules, and integrity of application code[37] are identified to build the system base profile to detect intrusion. Some available features in the microcontrollers, including *boot-loader* and *memory protection,* provide the hardware support for the proposed failure recovery procedure. A boot-loader is a piece of software that supports reprogramming "on board" without the help of external devices. Memory protection allows the boot-loader to be protected from the application code.[37] The system works in two phases. After the system detects a compromised node at the detection phase, the boot-loader accepts application code from the base station and writes it directly to node memory to recover the compromised node in recovery phase.

20.4.5.5 Hierarchical-Based Intrusion Detection

Unlike the decentralized IDS model in Ref. [34], where all IDS functions were embedded in the monitor node, a hierarchical-based IDS was reported to be built in Ref. [38], where the functions of sensing, computation, and data delivery are unequally divided among the different level nodes. In their proposed IDS, sensor nodes will collect application data and monitor behavior of neighboring nodes. The aggregation points and base station aggregate the application data from lower level nodes nearby, monitor behavior of the network or individual nodes, and then identify the intrusions by analyzing the aggregated data and the network behavior by χ^2 distance measure, EWMA forecasting, or Markov model.[38]

20.4.6 Wireless Sensor Networks for Physical Intrusion Detection

Although WSNs for physical IDSs are designed to protect critical areas such as airport or militarized zone rather than the sensor network itself, we think some variation of such systems could be used to defend against the physical attack on the sensor network in some applications. This kind of sensor nodes for physical intrusion detection can be mixed with the general sensor nodes for sensing interest to defend against the physical attacks on the sensor nodes.

20.4.6.1 Reconfigurable Sensor Network for Intrusion Detection

The development of a WSN with a two-level structure to detect intruder in a monitored area was reported in Ref. [39]. The first-level sensor nodes have basic sensing and wireless transmission capabilities. The second-level sensor nodes are built around a high-performance FPGA controlling an array of cameras. The first-level nodes acquire the indication of the presence of an intruder and send the warning message to the second-level FPGA node to activate the cameras. The FPGA can be dynamically configured to perform three types of intrusion detection algorithms on the captured images by the cameras. Static intrusion detection only extracts the shape of the intruder from a captured image and classifies the intruder by its shape. Dynamic intrusion detection extracts the shape variations of a moving intruder from a sequence of images and classifies the intruder by the type of its mobility. Intrusion visualization extracts the full image of the intruder from the scene and sends them to a distant human operator for a visual verification.[39]

20.4.6.2 SOC- and HMM-Based Intrusion Detection

An IDS-based self-organized criticality (SOC) and hidden Markov models (HMM) for WSN were presented in Ref. [40]. SOC is a concept to understand the internal interactions of complex systems. Research shows that large interactive systems always self-organize into a critical state governed by a power law. Inspired by SOC, Doumit and Agrawal[40] observed the rate of temperature change of real sites and discovered that the temperature dynamics in the real world are not just cyclic but also follow a power-law distribution that is distinct for every locality. Therefore the temperature in every location could be deduced after all the local natural dynamics such as the wind, foliage, and clouds have played their roles in affecting the overall temperature to reach its SOC. A sensor network was designed to monitor the change of temperature on an area. A HMM model was built and trained on the data (temperature) accumulated from previous reading, based on the SOC of that area. In this system, an activity that can alter the temperature values beyond a certain threshold (whether increase or decrease) from the expected values that the HMM has predicted was considered as an intrusion.

20.4.6.3 A Wireless Sensor Network for Target Detection

An application of sensor networks in the intrusion detection problem, called "a Line in the Sand," was reported in Ref. [41]. The system consists of a large number of sensor nodes distributed over an extended geographic area that is to be monitored. The objective of this system is to identify a breach along a perimeter or within a region. The intruding object, or target, may be an unarmed person, a soldier carrying a ferrous weapon, or a vehicle. A set of essential features for the three types of targets in six fundamental energy domains are identified. The six energy domains include optical, mechanical, thermal, electrical, magnetic, and chemical.[41]

The system can detect the presence of the targets by monitoring the disturbances on these energy domains that the targets cause. The authors carefully select a set of sense nodes appropriate for sensing the disturbances on these energy domains. The system detects a target's presence by detecting a signal's presence. Specifically, an analytical model for each target and sensor type was derived and used to determine the sensor output that would result if a target were present.[41]

20.5 Conclusion

Wireless networks are increasingly implemented for situations where fixed infrastructure networks are not practical. However, with this flexibility comes an additional security burden. Intrusion prevention is not always practical, so intrusion detection becomes an important second line of defense. Because of this, there has recently been a significant amount of research on this topic. Several studies deal more extensively with the architecture of such IDS. While the architectures of such systems are inherently distributed, they can vary from fully distributed to hierarchical and cooperative. Also, different methods can be used to identify intrusions. In the topic of intrusion detection for MANETs, two of the proposed methods, cross-feature analysis and SVM-based analysis use data mining techniques to identify anomalous network states. Other wireless ad hoc IDS approaches were suggested including a game theory-based IDS, an IDS that uses the combination of misuse detection and anomaly detection, and an IDS that is an overlay to the DSR protocol. While none of the methodologies are foolproof, and it is unlikely that any will ever be, they all present an improvement in the security situation of ad hoc networks. In the area of WSNs, the IDS could be designed to cooperate with those prevention techniques if they are adopted by the sensor network, which could keep the IDS efficient and reduce the resource requirement. We present current research on design IDS in WSN and the systems using WSN to detect physical intrusion. We suggest that this kind of sensor nodes for physical intrusion detection can be mixed with the general sensor nodes for sensing interest to defend against the physical attacks on the sensor nodes.

References

1. Sterne, D., Balasubramanyam, P., Carman, D., Wilson, B., Ko, C., Balupari, R., Tseng, C.-Y., Bowen, T., Levitt, K. and Rowe, J., A general cooperative intrusion detection architecture for MANETs, in *The 3rd IEEE International Workshop on Information Assurance*, March 2005.
2. Ma, L. and Tsai, J.J.P., *Security Modeling and Analysis of Mobile Agent Systems*, Imperial College Press, London, 2006.

3. Ma, L. and Tsai, J.J.P., Attacks and countermeasure in software system security, *Handbook of Software Engineering and Knowledge Engineering*, Vol. III, World Scientific Publisher, Singapore, 2005.

4. Yu, Z. and Tsai, J.J.P., An efficient intrusion detection system using a boosting-based learning algorithm, *International Journal of Computer Applications in Technology*, 27(4):223–231, 2006.

5. Zhang, Y., Lee, W. and Huang, Y., Intrusion detection techniques for mobile wireless networks, *ACM Wireless Networks*, 9(5):545–556, 2003.

6. Burtch, P. and Ko, C., Challenges in intrusion detection for wireless ad-hoc networks, *Symposium on Applications and the Internet Workshops (SAINT'03 Workshops)*, 2003.

7. Marti, S., Giuli, T., Lai, K. and Baker, M., Mitigating routing misbehavior in mobile ad hoc networks, in *Proceedings of the 6th Annual International Conference on Mobile Computing and Networking*, pp. 255–265, August 2000.

8. Nadkarni, K. and Mishra, A., A novel intrusion detection approach for wireless ad hoc networks, *IEEE Wireless Communications and Networking Conference*, 2:831–836, March 2004.

9. Zhou, L. and Hass, Z.J., Securing ad hoc Networks, *IEEE Network*, 13(6): 24–30, 1999.

10. Capkun, S., Buttyan, L. and Hubaux, J.P., Self-organized public-key management for mobile ad hoc networks, *IEEE Transactions on Mobile Computing*, 02(1):52–64, 2003.

11. Karchirski, O. and Guha, R., Effective intrusion detection using multiple sensors in wireless ad hoc networks, in *Proceedings of the 36th Annual Hawaii International Conference on System Sciences*, 2(2):57, 2003.

12. Huang, Y.A., Fan, W., Lee, W. and Yu, P.S., Cross-feature analysis for detecting ad-hoc routing anomalies, in *Proceedings of the 23rd International Conference on Distributed Computing Systems*, 478, September 2003.

13. Deng, H., Zeng, Q.A. and Agrawal, D.P., SVM-based intrusion detection system for wireless ad hoc networks, in *Proceedings of the 58th IEEE Vehicular Technology Conference*, Vol. 3, pp. 2147–2151, October 2003.

14. Patcha, A. and Park, J.M., A game theoretic approach to modeling intrusion detection in mobile ad hoc networks, in *Proceedings of the 5th Annual IEEE Information Assurance Workshop*, pp. 280–284, June 2004.

15. Perrig, A., Szewczyk, R., Tygar, J.D., Wen, V. and Culler, D., SPINS: Security protocols for sensor networks, *Wireless Networks*, 8(5):521–534, 2002.

16. Zhu, S., Setia, S. and Jajodia, S., LEAP: Efficient security mechanisms for large-scale distributed sensor networks, in *Proceedings of the 10th ACM Conference on Computer and Communications Security (CCS '03)*, October 2003.

17. Deng, J., Han, R. and Mishra, S., A performance evaluation of intrusion-tolerant routing in wireless sensor networks, in *Proceedings of the 2nd International IEEE Workshop on Information Processing in Sensor Networks (IPSN'03)*, April 2003.

18. Undercoffer, J. et al., Security for sensor networks. *CADIP Research Symposium*, 2002.

19. Chong, C.Y. and Kumar, S.P., Sensor networks: Evolution, opportunities and challenges, in *Proceedings of the IEEE*, 91(8):1247–1256, 2003.

20. Shi, E. and Perrig, A., Designing secure sensor networks, *IEEE Wireless Communications*, 11(6):38–43, 2004.

21. Akyildiz, I.F. , Weilian, S., Sankarasubramaniam, Y. and Cayirci, E., A survey on sensor networks, *IEEE Communications Magazine*, 40(8):102–114, 2002.

22. Tubaishat, M. and Madria, S., Sensor networks: An overview, *IEEE Potentials*, 22(2):20–23, 2003.

23. Online ATMEGA128L datasheet, http://www.atmel.com/dyn/resources/prod_documents/doc2467.pdf, January, 2006.

24. Online MICA2 datasheet, http://www.xbow.com/Products/Product_pdf_files/Wireless_pdf/MICA2_Datasheet.pdf, January, 2006.

25. Online MICAz datasheet, http://www.xbow.com/Products/Product_pdf_files/Wireless_pdf/MICAz_Datasheet.pdf, January, 2006.

26. Wood, A.D. and Stankovic, J.A., Denial of service in sensor networks. *IEEE Computer,* 35(10):54–62, 2002.

27. Karlof, C. and Wagner, D., Secure routing in wireless sensor networks: Attacks and countermeasures, *Ad Hoc Networks, Special Issue on Sensor Network Applications and Protocols*, 1(2–3):293–315, 2003.

28. Douceur, J.R., The Sybil attack, in *Proceedings of the 1st International Workshop on Peer-to-Peer Systems (IPTPS '02)*, March 2002.

29. Hu, Y.C., Perring, A. and Johnson, D., Packet leashes: A defense against wormhole attacks in wireless networks, *IEEE Infocom, 22nd Annual Joint Conference of the IEEE Computer and Communications Societies*, 3:1976–1986, March 2003.

30. Hu, Y.C., Perring, A. and Johnson, D., Rushing attacks and defense in wireless ad hoc network routing protocols, in *Proceedings of the 2nd ACM Workshop on Wireless Security*, 30–40, 2003.

31. Przydatek, B., Song, D. and Perrig, A., SIA: Secure information aggregation in sensor networks, in *Proceedings of the 1st International Conference on Embedded Networked Sensor Systems*, 255–265, 2003.

32. Du, W., Fang, L. and Ning, P., LAD: Localization anomaly detection for wireless sensor networks, in *Proceedings of the 19th International Parallel and Distributed Processing Symposium (IPDPS)*, April 2005.

33. Newsome, J., Shi, E., Song, D. and Perrig, A., The Sybil attack in sensor networks: Analysis and defenses, in *Proceedings of the 3rd International Symposium on Information Processing in Sensor Networks*, 259–268, 2004.

34. Silva, A.P., Martins, M., Rocha, B. and Loureiro, A., Decentralized intrusion detection in wireless sensor networks, in *Proceedings of the 1st ACM International Workshop on Quality of Service and Security in Wireless and Mobile Networks*, 16–23, October 2005.

35. Agah, A., Das, S.K. and Basu, K., A non-cooperative game approach for intrusion detection in sensor networks, *IEEE Vehicular Technology Conference (VTC)*, Fall 2004.

36. Agah, A., Das, S.K. and Basu, K., Intrusion detection in sensor networks: A non-cooperative game approach, in *The 3rd IEEE International Symposium on Network Computing and Applications, (NCA'04)*, 343–346, 2004.

37. Vogt, H., Ringwald, M. and Strasser, M., Intrusion detection and failure recovery in sensor nodes, in *Proceedings of the Workshop on Tagungsband INFORMATIK 2005, Lecture Notes in Informatics, Vol. P-68, Gesellschaft für Informatik*, 161–163, September 2005.

38. Ngai C.H. and Lyu, M.R., Intrusion detection for wireless sensor networks, Technical Report, http://www.cse.cuhk.edu.hk/~lyu/student/phd/edith/edith_term2.pdf, Spring 2005.

39. Sluzek, A. and Annamalai, P., Development of a reconfigurable sensor network for intrusion detection, in *The 8th Annual MAPLD International Conference*, September 2005.

40. Doumit, S.S. and Agrawal, D.P., Self-organized criticality and stochastic learning based intrusion detection system for wireless sensor networks, in *Proceedings of IEEE Military Communications Conference (MILCOM 2003)*, October 2003.

41. Arora, A., Dutta, P., Bapat, S., Kulathumani, V., Zhang, H., Naik, V., Mittal, V., Cao, H., Demirbas, M., Gouda, M., Choi, Y.R., Herman, T., Kulkarni, S., Arumugam, U., Nesterenko, M., Vora, A. and Miyashita, M., A line in the sand: A wireless sensor network for target detection, classification, and tracking, *Computer Networks*, 46(5) 605–634, 2004.

Chapter 21

Security Issues in an Integrated Cellular Network—WLAN and MANET

Bin Xie, Anup Kumar, and Dharma P. Agrawal

21.1 Introduction

It is widely recognized that many new wireless technologies have been introduced to cater the ever-growing demands for diversified services.[1] Various fast-growing services provided by different networks include voice, multimedia, messaging, e-mail, information services (e.g., news, stocks, weather, and travel), M-commerce, entertainment, location-based public utility and health-care services, and so on. A multiinterface mobile device may allow a user to access the Internet by a WLAN interface for a data service (e.g., checking e-mail, news, and even a multimedia service). In addition, by using the same device, the user may simultaneously start another application, such as receiving a call from its cellular interface. While wireless networks are becoming ubiquitous, integration of ad hoc network devices with the Internet is the fundamental to providing mobile users with Internet accessibility anytime and everywhere, with quality of service (QoS) and security guarantees.[2]

Wireless LANs (e.g., IEEE 802.11a/b/e/g and HiperLAN/2), MANs (e.g., IEEE 802.16), and WANs (e.g., 1G, 2G, 2.5G, 3G, GSM, and the proposed

IEEE 802.20) employ different radio spectrums to access to the Internet. In this manner, the IP-based network environment has become heterogeneous in terms of services, devices, and networks. This phenomenon is totally different from a pure mobile ad hoc network (MANET), where a self-configurable network capability is provided by communication among ad hoc mobile stations (MSs) without any Internet-based infrastructure support.

In the last few years, increasingly more attention is being focused on integration of cellular network, WLAN and MANET.[2] Once attached to the Internet, the MANET communication is not isolated anymore, but is an integral part of the infrastructure-based networks for Internet access. An MS in MANET can access the services provided by the Internet using direct link or MANET relay. A multitude of architectures and protocols[3–21] has been proposed to extend traditional cellular networks and WLANs to multihop communication using the MANET. Recent research has shown that the adoption of multihop communication from WLANs to cellular networks improves the wireless radio coverage and the network robustness against propagation phenomena resulting from multipaths, radio interfaces, fading, and obstacles.[10–12] To access the Internet data and multimedia applications, such integration allows an MS, which is outside the radio coverage of the base station (BS*), to establish a communication to a BS with a multihop connection.[13–21] The multihop communication can be used to increase the utilization and capacity of a BS by decreasing the cochannel interference via lowering the transmission power either of the BS or of the MSs.[4] In addition, the integration can be useful in achieving load balancing by forwarding a part of the traffic from an overloaded BS to a neighboring BS.[5–6]

Secure communication in such heterogeneous networks is inhibiting the commercial acceptance of these networks.[3] A wide variety of security threats are possible in integrated networks due to the vulnerability of heterogeneous communication and multihop relaying in the MANET.[22–27] The security design and implementation should take the heterogeneous communication path into account, which spans the Internet, cellular networks, WLAN, and MANET. The possible attacks occurring in either the Internet, cellular, WLAN, or MANET may compromise the security of communication session in the integrated network.

Many security issues in the integrated networks should be reconsidered in a hybrid environment rather than only as an individual cellular network, WLAN or MANET. At first, it is a fact that current authentication protocols of UMTS, IEEE 802.16 and IEEE 802.11 security are not enough for the integrated networks because these protocols are developed based on

---

* In this chapter, a BS refers an access point for mobile station, e.g., a cellular base station or an IEEE 802.11 access point.

the assumption that the MS can connect to a BS directly and exchange authentication packets. Second, a secure multihop route is required before communicating with the Internet which does not exist in single-hop cellular or WLAN networks. Since the Internet mobility protocol (i.e., mobile IP) in the conventional cellular and WLAN lacks security support for a multihop MS, new security features should be considered for the multihop MS to access the network with mutual authentication by multihop. Third, in the integrated networks, possible noncooperative behavior in MANET may significantly and adversely affect the performance of network in the case of multihop packet relaying. Therefore, collaborative protocols are needed in the integrated network to encourage cooperation during multihop packet relay.

The rest of this chapter is structured as follows: The architecture and communication scenarios of the integrated cellular, WLAN, and MANET are discussed in Section 21.2. Then, the security impacts from the unique characteristics of the integrated network are studied in Section 21.3. Section 21.4 describes the potential security threats with many examples of security attacks. A detailed investigation of security solutions for the integrated networks is addressed in Section 21.5, which is followed by the open issues and the design challenges of security protocols for the integrated network in Section 21.6. Finally, concluding remarks are provided in Section 21.7.

21.2   Architecture of the Integrated Network

Before investigating security threats in integrated networks, it is necessary to know the basic network architecture of the integrated cellular, WLAN, and MANET. The architecture in Figure 21.1 depicts a network that integrates cellular network, WLAN, and MANET.[2–21] The core components of this network are MSs, BSs, home agent (HA)/foreign agent (FA), and the core IP network. The BS provides one or more types of wireless radio access interfaces (i.e., cellular, IEEE 802.11, and IEEE 802.16) to MSs and serves as the communication bridge between MSs to access a wireless or wired network. Within the same domain (or subnet), cellular BSs, IEEE 802.11 access points (APs) as well as other types of BSs can be interconnected with universal gateway[2,27] that constitutes the heart of a heterogeneous network. Some of the BSs in a domain can be connected to an internet gateway (IGW) directly through a wired or wireless connection. Finally, IGW connects to the Internet. A cellular BS and an IEEE 802.11 a/b/g AP can be colocated in the hot-spot area where the traffic is high, as illustrated in Figure 21.1. A multimode MS possesses multiple radio interfaces, i.e., a separate cellular interface and a WLAN interface. The MS can connect to an available BS with a single- or multihop path, using an appropriate radio interface. In the multihop communication, an MS operates in

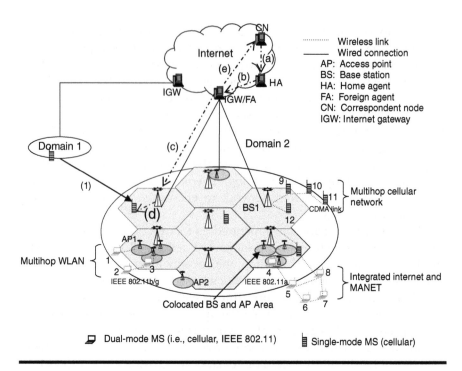

Figure 21.1   A simplified integrated network architecture.

MANET communication mode. At a given instance, a particular MS can be either located within or outside the coverage of BSs. In a dual coverage area (where BS and AP are colocated), a multimode MS may have more than one alternative to access the Internet. If an MS moves outside the coverage of direct transmission from BSs, multihop relaying is needed for the MS to obtain services from a BS to communicate with the Internet.

21.2.1   Communication and Mobility

To support the communication in a visiting network for an MS, mobile IP[27–30] is needed to provide continuous mobility management. As shown in Figure 21.1, mobile IP has two main entities, namely HA and FA. An HA is the server on the mobile host's home network that maintains the information about the MS's current location, as identified as care-of-address (CoA), billing, account, and security credentials. However, an FA is the server on the visiting network providing the CoA and security administration of the visiting network. By a process of registration, an MS updates its mobility binding at the HA, which is a mapping between the CoA and its permanent home IP address. When the MS moves to a new network domain, it receives a CoA with a beacon (i.e., advertisement) from the visiting FA, then, the MS registers with the FA. The registration request is delivered to the HA through the FA. After the HA updates the mobility binding for the MS, it sends a registration reply to MS. Two possible ways of communication between correspondent node (CN) and MS are illustrated in the architecture of Figure 21.1. Data packets sent by a CN toward the MS are intercepted by the HA, and the HA encapsulates the packets with the CoA as shown in steps (a) and (b) of Figure 21.1. Upon receiving data packets sent by the HA, the FA delivers the data packets to the destination MS through the previously established path (during registration) as shown in steps (c) and (d) of Figure 21.1. However, after receiving data packets sent by the MS, the FA delivers them to the CN using IP routing in the infrastructure network as shown in step (e) of Figure 21.1.

21.2.2 Multihop Routing Protocols

In an integrated cellular, WLAN and MANET, multihop route discovery protocols are required to support multihop wireless communication. A multihop MS located outside the coverage of the BS should detect the availability of a BS by a process of a route discovery. The route discovery protocol further enables the MS with capability to reconfigure a new route to the BS when current path is broken due to intermediary node mobility. In general, the integrated multihop networks can be divided into three main categories:

- Multihop cellular networks.[3–9]
- Multihop WLAN networks.[10–12]
- Integrated Internet and MANET networks.[13–21]

The multihop cellular networks provide the Internet connectivity to MSs through WAN/MAN BSs, such as GSM, IEEE.802.20, and IEEE 802.16. For example, as shown in Figure 21.1, MS 11 connects to the Internet with the path MS11–MS10–MS9–BS1. The routing protocols for multihop cellular networks include A-GSM,[3] MCN,[4] integrated cellular and ad hoc relay (iCAR) system,[5] MADF,[6] unified cellular and ad hoc network (UCAN),[7] opportunity-driven multiple access (ODMA),[8] and SOPRANO.[9] For example, ODMA[8] breaks a single CDMA transmission from an MS to a cellular BS into several multiple wireless hops. Thus, ODMA reduces the transmission powers and cochannel interferences.

The multihop WLAN networks allow the MSs to obtain services from the Internet through WLAN.[10–12] A WLAN AP provides a higher communication speed to MSs at the expense of smaller radio coverage as compared to a cellular BS. For example, MS2 in Figure 21.1 accesses WLAN AP and obtains services from the Internet with the help of MS3 using the IEEE

802.11g link. The routing protocols for multihop WLAN networks include two-hop relay,[10] hybrid wireless network (HWN),[11] 1- and 2-hops direct transmission.[12] For instance, the protocol 1- and 2-hops direct transmission combines the single- and the multihop operations in a WLAN environment to solve the problems, such as any weak WLAN connection of a single hop, handoff procedures in single-hop mode, and AP failures.

In integrated Internet and MANET networks,[13–21] each ad hoc MS runs a MANET routing protocol that is used to construct the communication path between two network stations. The process of an ad hoc routing discovery, such as ad hoc on-demand distance vector (AODV) and dynamic source routing (DSR), enables the ad hoc MS with the capability to reach an FA. Through the established route to the FA, a separate protocol[13–21] coordinates mobile IP and the MANET routing protocol to obtain the Internet connectivity for ad hoc MSs.

21.3 Security Impacts from the Unique Network Characteristics

Similar to other wireless systems (single-hop cellular, WLAN, and MANET), an integrated network has the same basic characteristics, such as open wireless medium, mobility, and constrained terminal power capability. Because of the open wireless medium, passive/active link attacks, such as eavesdropping and spoofing, are possible. Owing to mobility of MSs, the network topology changes frequently and poses a challenge to the security issues within a nomadic environment. In this part, we analyze the security-related features of the integrated networks compared to pure MANETs, conventional cellular networks, and WLANs.

21.3.1 Internet and Infrastructure Support

A pure MANET is a self-configurable network with the capability of communication among ad hoc MSs without any infrastructure or any centralized administration. The multihop communication path for traffic in MANETs between the source and destination nodes can be any pair of MSs through any intermediary node. The integrated networks share similarities with MANETs: multihop networking and multihop communication. However, it has a clear distinction between the integrated networks and MANETs: Internet with infrastructure support in the integrated networks. The wireless infrastructure such as a BS provides the Internet accessibility to MSs either by single- or multihop. The Internet infrastructure such as FA provides centralized administration for the MSs in the integrated networks. Owing to lack of any centralized administrative infrastructure support in pure MANETs, it is difficult to establish any distributed trust relationships among ad hoc MSs. In a pure MANET, an adversary within a nomadic environment can easily execute various attacks with different identities. Without any central authority in pure MANETs, the adversary has the ability to forge identity. The malicious MS can use a forged ad hoc identity and then makes feigned trust relations with other MS to attack the network internally. However, in an integrated network, an MS can access the Internet by way of single- or multihop connectivity through a BS. The wireless and Internet infrastructure (i.e., AAA server) can serve as an authentication authority and an administrative center for security. The Internet and infrastructure-supported security deployment has a fundamental impact on the security implementation of the integrated networks as compared to pure ad hoc security. In an integrated network, it is very important to protect and authenticate the identity as well as its associated credentials of an MS with the help of the Internet and wireless infrastructure.

Besides the infrastructure-based authentication, the integrated network can take advantage of the Internet and wireless infrastructure for secure key management. In general, security goals for wireless networks are achieved through cryptographic schemes, such as symmetric or asymmetric keys for encryption/decryption or signature/verification. The goals of key management in a MANET are difficult to achieve in which MSs randomly move and continuous connections are not maintained. The proposed mechanisms for key management, such as secret-key cryptography, public-key cryptography, or third-party authentication in MANET, are vulnerable. However in the integrated networks, mobile IP protocol maintains the continuous connectivity for MSs, and the Internet infrastructure (e.g., AAA home/foreign server[28–31]—authentication, authorization, and accounting) may be used for key creations and distribution center with a stronger basis than a self-configurable key scheme deployed in MANETs.

21.3.2   *Wireless Multihop Communication*

An MS in an integrated network communicates with a BS toward the Internet or another MS by using single- or multihop path with the help of intermediary nodes. On the contrary, an MS in a traditional cellular network or a WLAN connects to the Internet through BS directly. All the packets including control and data packets are exchanged between the MS and the BS directly without any intermediary wireless device. Although a malicious MS can eavesdrop on the exchanged packets with a wireless receiver, it cannot modify these packets during transmission. Compared to single-hop networks, the multihop wireless communication in the integrated networks imposes many new challenges, such as collaboration of packet forwarding and dynamic network topology that have a significant impact on security. All possible attacks found in MANETs can be easily mounted on integrated networks due to the common characteristic of multihop route. The routing protocols used for the integrated networks, such as AODV, iCAR, UCAN, and A-GSM[21] may be attacked in the integrated networks.

In single-hop cellular networks or WLANs, security solutions only provide protection for one-hop connectivity between an MS and BS by securing MAC/link-layer protocols. On the contrary, in the integrated cellular network, WLAN and MANET, the security deployment should be extended beyond the single- to multihop routing security at the network layer. Therefore, traditional cellular and WLAN security protocols are not enough in terms of multihop authentication and multihop wireless communication. Although the security protocols for MANETs provide routing security, unfortunately, they also cannot be adopted for the integrated networks directly because these protocols are based on the assumption that the infrastructure-supported authentication is not present as illustrated in Section 21.3.1.

21.3.3 Multimode with Multiple Radio Interfaces

Another distinguishing characteristic of the integrated networks is multiple radio interfaces in MSs. A dual-mode MS equipped with two radio interfaces may switch from one radio interface to another (e.g., redirecting a flow from a cellular radio interface to WLAN radio interface when a dual-mode MS moves to the WLAN area) as network accessibility or topology changes. Cellular, WLAN and MANET security protocols are based on a single radio interface and do not have any provision for protection when communication migrates from one radio interface to another. It is necessary to develop integrated schemes for security interworking between multimode radios. For example, a 3rd generation partnership project (3GPP) subscriber MS, which is equipped with cellular–WLAN radio interfaces, may handoff its service from a 3GPP network to a public WLAN network. Here, to provide security features to the MS while accessing the public WLAN network, it is necessary for the public WLAN network to reuse the 3GPP subscription and 3GPP-based authentication/authorization as well as 3GPP-based security key agreement using SIM/USIM card. In the case of multihop route, the infrastructure security protocols (e.g., 3GPP cellular security, mobile IP security for WLAN) must coordinate with multihop routing protocols for multihop MSs. At the same time, it is important for the security protocols to minimize the authentication latency induced by the multiple radio interfaces (e.g., networking selection) and multihop route.

21.4 Potential Security Threats

Security has always been an important issue for mobile communication. The security attacks can coarsely be classified into two categories: Internet

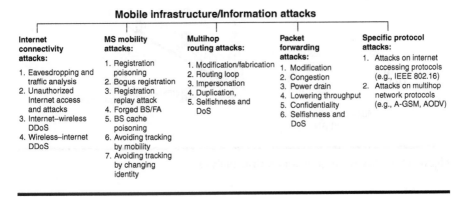

Figure 21.2   Types of security attacks on the integrated cellular, WLAN and MANET.

infrastructure/information attacks, and mobile infrastructure/information attacks. The Internet infrastructure attacks can be further divided into four subcategories: domain name service (DNS) "hacking," routing table "poisoning," packet "mistreatment," and denial-of-service (DoS) attacks. To provide a way to locate a host on the Internet, DNS maintains the hierarchical distributed database on the Internet to provide the translation between an IP address and its host name, as well as the mapping of a name server to an Internet domain. The DNS "hacking" may result in cache poison by modifying domain registration information at an insecure name server. Numerous approaches have been developed for defending against these possible attacks. Chakrabarti and Manimaran[32] provide a comprehensive survey of various types of attacks and existing solutions on the Internet infrastructure attacks.

In this chapter, we focus on the mobile infrastructure/information attacks. Figure 21.2 gives the detailed classification of security attacks on the integrated cellular, WLAN, and MANET. In the following section we provide a general discussion on these types of attacks and emphasize the specific attacks that differ from an individual wireless network (e.g., pure MANET).

21.4.1   Internet Connectivity Attacks

The Internet connectivity-related attacks are being focused on this part. Attacks of this type have illustrated the lack of protection of Internet infrastructure or its resources.

21.4.1.1   Eavesdropping and Traffic Analysis

The broadcast nature of the wireless transmission medium renders radio links insecure in the integrated networks. It is very easy for an attacker to

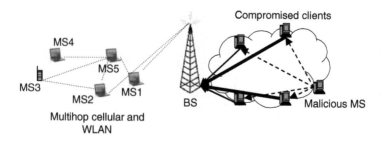

Figure 21.3   Eavesdropping and traffic analysis.

eavesdrop on an ongoing communication by residing on the transmission range of the intended wireless device. In addition, an attacker in an integrated network may reside on the multihop path and present itself as an intermediate relaying MS. The data packets can be copied and distributed by the attacker when it forwards packets to the next hop (e.g., MS5 is the intermediary relaying MS for MS3 and MS4 in Figure 21.3).

Although the data packets over the wireless link may be encrypted against eavesdropping from reading transmitted data, it is possible for an attacker to gather information by examining the monitored packets such as address, size, number, and time of transmission. The attacker can seize the information of BS, such as location and IP address. Also, the attacker can know the critical MSs, which provide the Internet connectivity for other MSs (e.g., MS1 and MS5 in Figure 21.3). The information obtained is useful for many attacks. As an example, in Figure 21.3, by the analysis of the traffic that flow through MS1, the MS2 (an attacker or attacker conspirer) does know that MS1 is the critical MS for providing the Internet connectivity for MS3, MS4, and MS5. Then, MS2 can launch the attack toward MS1 for the purpose of turning down the Internet connection of MS3, MS4, and MS5.

21.4.1.2   Unauthorized Internet Access and Attacks

A malicious MS may access the integrated network or the Internet and enjoy free network usage by way of single- or multihop communication. Wireless Internet service providers (ISPs) may be accessed by an unauthorized MS because of the lack of a correct ISP configuration. As for the network devices without having adequate security measures are concerned, the security threats may come from within the network itself. A registered MS of the network may access, read, copy, and distribute an unauthorized data file. A malicious MS may access an enterprise network by remote wireless access and then destroy data on the network itself.

When the Internet is accessed by way of a multihop communication, the malicious MS consumes precious resources like power and bandwidth.

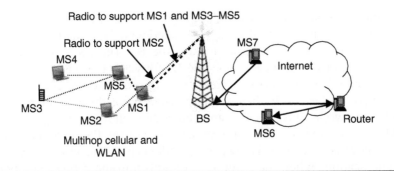

Figure 21.4 Unauthorized network access and man-in-the-middle attack.

In the multihop network as shown in Figure 21.4, MS1 should be allocated more radio resource to support the throughput needed by MS3–MS5.

Although the free network usage may not be a significant threat to the Internet, an authorized access is the first step for the MS to control the Internet infrastructure and attack these components. After entering the Internet, the attacker may use some techniques like medium access control address spoofing to gain access to the network infrastructure. For instance, in Figure 21.4, MS1 implements the man-in-the-middle attack between MS1 and its default network router (router shown in Figure 21.4) from which MS1 sees all the traffic between MS7 and router. First, MS1 connects to the Internet and sends a malicious address resolution protocol (ARP) reply to router, associating MS1's MAC address with MS7's IP address. With such access, the router assumes MS1 to be MS7. Next, MS1 sends a malicious ARP reply to MS7 associating MS1's MAC address with the router's address. In this case, MS7 believes MS1 is its router (router). Finally, MS1 can access the session between MS7 and router. Thus, all data packets from MS7 will be delivered to MS1 first and then MS1 forwards the received data packet to router. Finally, the router sends the packets to the destination (e.g., MS6 as shown in Figure 21.4). In the opposite direction, all the packets from MS6 will be forwarded to MS1 by router and MS1 sends the packets to MS7. In this manner, MS1 intercepts the traffic between MS7 and MS6.

21.4.1.3 Internet–Wireless DDoS

A multihop cellular and WLAN network may be crippled by the attacks from the Internet, referred to as internet–wireless DDoS. Compromising a BS or wireless router can disable or congest the wireless communication in a domain. As a typical example, as shown in Figure 21.3, the malicious MS first initiates the control over one or several computers (i.e., compromised clients as seen in Figure 21.3) on the Internet by using some kind of automatic intrusion software to hack them. After gaining control of the computers, the attacker synchronizes them to send traffic in bulk toward one or more multihop MSs that are associated with the same BS. The traffic first travels through the Internet toward the BS. The BS is not aware of the illegality of the packets. Then, the BS forwards the traffic toward the destination MSs (e.g., MS4 in Figure 21.3) according to the address given in each packet header. In the end, the packets travel through the multihop wireless network. This process can achieve the following two goals.

First, it may exhaust the wireless resources of the BS: limited availability of BS radio spectrum is always a bottleneck in the Internet–wireless communication. When the compromised MSs transmit a huge number of packets to a BS, the BS may exhaust its radio resource by forwarding the packets on its air interface. Therefore, the attack can immediately block a large number of MSs that are communicating through the BS.

Second, it may deplete the wireless resource of MSs (power, radio/bandwidth): an MS has limited communication capability compared to fixed devices in terms of available power and bandwidth. The packets transmitted by compromised computers traveling through the multihop wireless network definitely cause the power and bandwidth consumption at each intermediate MS. The multihop transmissions can simply disable or slow down other communications along the multihop path. It can be observed from Figure 21.3 that the traffic from the compromised clients consumes the bandwidth of the intermediary relaying MS1. The attack at least slows down the communication between MS5 and CN because MS5 uses the same intermediary relaying MS1 as the attack. Moreover, if the power of MS1 is exhausted, then the Internet connections for MS3–MS5 will also be turned down because MS1 is the only host to reach BS from MS3–MS5.

21.4.1.4 Wireless–Internet DDoS

An attacker can initiate an attack from the wireless side to the Internet, referred as wireless–Internet DDoS. It can attack the BS radio spectrum when a number of wireless devices (attackers) around a BS send packets simultaneously in bulk to the BS by single- or multihop routes. In this case, at a given instance the BS may be disabled due to the unavailability of the communication channel.

By single- or multihop communication, the attacker may access the Internet through a BS. In this manner, the attacker can impair the Internet infrastructure and its communication. As shown in Figure 21.5, the attacker, which operates in an ad hoc mode and runs the ad hoc routing protocol, connects to BS via the MANET/cellular/WLAN. Moreover, the attacker can collude with other attackers or compromised Internet components for implementing more sophisticated attack on the Internet or wireless network. For example, to execute DDoS, an attacker from the wireless network can compromise several computers on the Internet as attacking "agents."

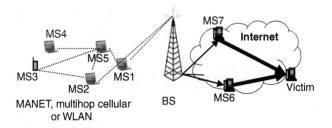

Figure 21.5 Wireless-internet DDoS.

Then, the attacker directs the "agents" to send a huge number of packets (i.e., UDP, TCP, or ICMP packets) to disable a target (e.g., a server). The compromised "agents" may also launch the "Internet–wireless DDoS" to attack the wireless network as discussed in Section 21.4.1.3. Since the attacker has accessed to the Internet by multihop path and a temporary ID, it is very difficult for any victim to trace back the source of the attack (see Section 21.4.2).

21.4.2 MS Mobility Attacks

In this section, we focus our attention to the mobility-related attacks in the integrated network. The mobile IP protocol enables an MS to be located by using its home IP and CoA addresses. The data packets from the Internet can travel toward the current attacked network (steps (a) and (b) as illustrated in Section 21.2.1). In the multihop network, the mobile IP has been extended to support the Internet accessibility for multihop MS. The basic procedure of a mobile IP with multihop support[14–21,27] is illustrated in Figure 21.6 using the following steps:

- **BS/FA discovery:** Each BS/FA periodically advertises its presence with beacons. The registration occurs immediately after the MS moves to a visiting network. In case when an MS is single hop away from the BS (e.g., MS1 and MS2 in Figure 21.5), the MS obtains the connection to the BS directly and can hear the beacon from the BS directly. On the other hand, if the MS is outside of the BS's transmission range (e.g., MS3–MS5), it has the knowledge of BS/FA with a process of BS/FA discovery, from which the MS establishs a multihop route to the BS/FA. For example, MS4 in Figure 21.5 knows the existence of BS/FA with the path MS4–MS5–MS1 after a BS/FA discovery.
- **Advertisement with CoA:** While obtaining the connection to the Internet, the MS solicits the identity (i.e., CoA) of the visiting network. The CoA is used for registration with the HA through the BS/FA.

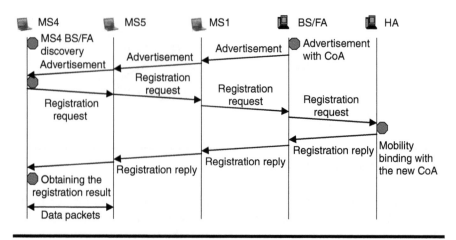

Figure 21.6   Mobile IP with multihop extension.

- **Registration request:** The MS sends a registration request to the FA, carrying out its home IP address, FA address, HA address, and registration lifetime. The registration request is forwarded to the FA by single hop or multihop For example, in Figure 21.5, path MS4–MS5–MS1 is used by MS4 to send its registration request to the BS. Upon receipt of the registration request, the foreign network (i.e., FA) forwards the request to the MS's home network.
- **Mobility binding:** The HA updates the mobility binding for the MS by creating a mapping between the home IP address with the CoA.
- **Registration reply:** After the mobility update for the MS, the home network replies the foreign network with a registration reply message carrying the registration result. Then, the foreign network forwards the registration reply to the MS with the reversed multihop route. In the end, the MS can send or receive data packets to or from the Internet from the visiting network.

The registration procedure shown in Figure 21.6 provides the desirable property of mobility with Internet connectivity, but it may results in several security threats. The typical types of registration attacks[23–26,29–31] include registration poisoning, bogus registration, replay attack, forged FA, BS cache poisoning, avoiding tracking by mobility, or by changing identity.

21.4.2.1 *Registration Poisoning*

In registration poisoning, the adversary corrupts a registration procedure of a multihop MS, and consequently keeps the multihop MS from obtaining services from the wired networks.[27] In a registration procedure, a multihop

MS registers with the foreign network through some intermediary MSs. The adversary (e.g., MS2 in Figure 21.5) can entice a multihop MS (e.g., MS3) to choose it as an intermediate MS by claiming to have a short or fast route to a BS. When the adversary is selected as the intermediary MS for a registering MS, it can modify or drop the MS's registration request/reply. Once the adversary modifies the registration request, the MS cannot correctly register with the foreign network. If the adversary modifies the registration result, the MS cannot access the Internet as if its registration request is rejected.

21.4.2.2 Bogus Registration

This occurs when a malicious MS does a fake registration by spoofing an IP address to masquerade itself as someone else.[27] A bogus registration may cause a wrong mobility binding so that all packets are tunneled to the malicious MS rather than the correct MS. By a bogus registration, the attacker obtains the right to access the Internet, and can implement further attack on the Internet, such as wireless–Internet DDoS. For instance, in Figure 21.5, MS2 does a forged registration by masquerading itself as MS3. Thus, if the bogus registration accomplishes, then all packets coming from the Internet for MS3 are forwarded to MS2. In this case, MS3 cannot receive any packet from the Internet.

21.4.2.3 Registration Replay Attack

The replay attack[27] happens when the attacker obtains a copy of a legitimate registration message sent by an MS and then replays the message later to lead to a false authentication. In Figure 21.5, MS5 may repeatedly forward a copy of the registration request message from MS3 and cause FA and HA to initiate the process of authentication for MS3 again and again. In the multihop paths, replay attacks may be present in the uplink or downlink of a multihop path, which is not possible in the single-hop wireless networks. A replayed registration request will lead all intermediate MSs to forward the replayed message along the uplink of a multihop path to the BS. However, a replayed registration reply will force all intermediate MSs to forward the replayed message along the downlink of a multihop path until reaching the destination MS.

Moreover, as can be seen from Figure 21.7, an adversary may spoof an FA and access the Internet by replay attack as follows.[23–25,29–31] At first, the adversary (i.e., M1 in Figure 21.7) intercepts the *request* and *reply* messages used in an MS registration process. The *request* message is issued by the MS, and the *reply* message is sent by the HA in the registration. The *request* and *reply* have the nonce of MS and HA, respectively. Then, the attacker, M1 in Figure 21.7, replays the intercepted request of the MS which will be delivered to FA (step 1 in Figure 21.7) by single hop or multihop. The

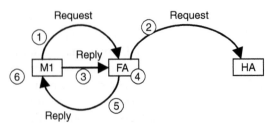

Request and Reply are two old messages of an MS and intercepted by M1
from the MS and the HA. M1 then replays the two messages to spoof FA.

Figure 21.7　Replay attack by single hop or multihop.

request will be forwarded to the MS's HA (step 2). HA ignores the request
checking the nonce of MS. However, the FA is not aware of the stale request
because the two messages lack the nonce of FA. M1 replays the intercepted
reply message to FA (step 3) as if it is HA. Then, FA considers M1 has been
authenticated by the HA and trusts M1 as a legitimate MS (step 4). In the
end, M1 can access the Internet via FA (step 6) after receiving the reply
from FA (step 5).

21.4.2.4　Forged BS/FA

In an integrated network, an attacker can attack the multihop wireless
network by advertising itself as a genuine BS using some forged messages or duplicate beacons recorded from a correct BS procured by
eavesdropping.[27] When an MS hears the fraudulent beacons from the malicious MS, it assumes that it is within the radio coverage of a genuine BS and
then initiates a registration procedure. A registration request is issued from
the MS to the forged BS. The forged BS replies with a bogus registration
reply carrying the acceptance of the registration request. After receiving the
registration result, the MS assumes that it has obtained the Internet connection through the forged BS and disconnects its communication from the
genuine BS. One by one, the forged BS could entice a number of MSs to
disconnect from the genuine BSs and establish connections with the forged
MS either by single- or multihop route. However, the MSs cannot obtain
any Internet service correctly from the forged BS. This attack is valid in
cases where the BS is not authenticated by the MS. For instance, in Figure 21.5, malicious MS2 advertises a high-speed connectivity to the Internet
by sending bogus beacons to its neighbor. After hearing the beacons from
the forged BS (i.e., MS2) but without realizing the fraud, MS3 and MS5
register with MS2 by a single hop. After registration, MS3 and MS5 believe
that they are connected to a BS with a higher speed, and thus disconnect
with genuine BS.

21.4.2.5   BS Cache Poisoning

In a single-hop wireless network, it is difficult for a malicious MS to modify the radio mappings from which MSs enter a BS because each MS and the BS interact with each other directly. On the contrary, the BS in the integrated network suffers from possible BS cache poisoning as multihop communication is now allowed.[27] To support multihop communication, the routing cache is needed at each BS for the purpose of recording the multihop routes between the BS and each multihop MS. The multihop routing cache of an MS may be poisoned in several ways. For instance, when a multihop MS sends a multihop route update packet for creating or updating its multihop paging cache at a BS, the malicious MS may modify the packet that could result in multihop routing cache poisoning. Also, a malicious MS may send a wrong route update packet on behalf of a genuine one. And the BS updates the routing information for the genuine MS with the wrong information sent by the malicious MS. To locate a multihop MS, the BS finds the first hop MSs in routing cache that can reach the destination MS. The data packet from the BS is forwarded hop by hop to the destination, in accordance with the multihop route. When a multihop route is poisoned, the BS is unable to locate the destination MS by following the multihop route provided in the BS cache. For instance in Figure 21.5, MS2 sends a multihop route update packet to the BS on behalf of MS3, and the BS updates the multihop routing cache from MS3–MS5–MS1–BS to MS2–MS1–BS. In this case, the packets of MS3 from the Internet will be lost due to incorrect routing information.

21.4.2.6   Avoiding Tracing by Mobility

An adversary always prefers to hide itself from being located while attacking a victim. Many technologies, including link testing, logging, ICMP trace back, and IP trace back[33] have been developed to enable the victim to trace back the source of the attacker. For example, link testing iteratively checks the upstream link until the source is reached. In the multihop environment, the malicious MS may take advantage of the integrated network to hide itself from being traced more easily than in fixed network. At first, due to mobility, malicious MS can change its multihop route to access the BS that makes the victim hard to trace the actual location of the malicious MS. Secondly, the adversary may also change its attached network (i.e., BS). The existing tracking-back technologies based on the fixed network may be invalid in the multihop networks because of the dynamic topology of multihop network. As for the mobile IP mobility protocol, an MS has two IP addresses: the home address and the CoA. The CoA is temporarily assigned by foreign network and used as the current address for communication. The non disclosure method (NDM)[33] prevents the tractability of network connections in mobile environment by hiding the source and the destination addresses of an IP packet from every forwarding device except the packet destination.

21.4.2.7 Avoiding Tracing by Changing Identity

Another important issue for an adversary to avoid being tracked is to hide itself from being identified while attacking the network. In a traditional attack, the adversary uses a forged or spoofed address as its identity for this purpose. A new method can be used for the attacker to hide itself in an integrated Internet and MANET. MANET protocols may use different addressing solutions[15–21,34]; AODV, DSR, and temporally-ordered routing algorithm (TORA) ad hoc protocols use Node ID; hierarchical state routing (HSR)[34] protocol has a hierarchical addressing solution; zone-based hierarchical link state routing (ZHLS)[34] protocol uses <zone id + node id> as MS's ad hoc address. When an MS enters a MANET, the MS will be automatically configured with an identity by the ad hoc protocol, using one of the above address solutions. For example, a DSR network will assign an MS with a Node ID (e.g., 001) when the MS enters the DSR network. If the MS leaves the DSR network and enters the DSR network again, the MS will be assigned a new available Node ID (e.g., 012).

In an integrated Internet and MANET, a malicious MS can participate in a MANET and establish its Internet connectivity by using its ad hoc identification. In this case, the malicious MS can easily implement its attacks to the Internet or wireless network over and over again by masquerading itself. When a malicious MS enters the integrated Internet and MANET, the network configures the MS with an ID. If the malicious MS leaves the network, and enters the network again, the network will automatically configure the malicious MS with a new ID without knowing its last ID. The network cannot track and monitor the history of the malicious MS because each MS does not have a unique and consistent ID while leaving and entering the network. When a malicious MS exhibits illegal actions on the integrated Internet and MANET, the malicious MS can clear its bad record by reentering the network and reinitiating a trust relation with the network by using a new ID. As shown in Figure 21.8, the attacker exhibits its wireless–Internet attack by using ID_1. If the network finds the attack from ID A, the malicious MS reenters the network with a new ID_2, and could attack again. The ability of the malicious MS to modify packets to spoof MAC addresses prevents a detection mechanism to quickly identity the malicious MS.

21.4.3 Multihop Routing Attacks

As illustrated in Section 21.2.2, multihop route discovery is responsible for detecting the multihop routes between a pair of MSs or between an MS and

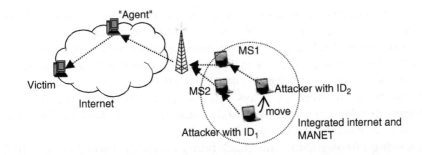

Figure 21.8   Avoiding tracing by changing identity.

a BS. Almost all routing attacks in MANET may happen in the integrated network. In this section, we give a brief introduction of these possible attacks[34,35]:

Modification/fabrication: Modification/fabrication violates the integrity of the routing packets. The malicious MSs modify, inject, or delete some fields of a routing packet, and then forward the packet with falsified values in the packet fields. These fields may include the source or destination address, hop count, sequence number, etc.

Routing loop: A malicious node creates a routing loop by modifying the routing packet in such a manner that a loop is created. The routing packet may cycle in the loop many times, but never reach the destination. Routing loop results in the consumption of power and bandwidth at the traveled MSs.

Impersonation: To hide its identity, a malicious node misrepresents itself with the identity of an existing node or a fictitious identification, which is achieved by modifying its medium access address and IP address in the outgoing packets. Also, the malicious node may use the broadcast address in the network to send packets.

Duplication: A malicious node sends a legitimate routing packet repeatedly in the name of the source node of this routing packet. These duplicated messages cause duplicated receptions and processing overhead on adjacent MSs.

Selfishness and DoS: An adversary drops the routing packets. The selfishness is the action that a node avoids participating in routing process of other's traffic for the purpose of conserving its own energy.[36,37]

21.4.4   Packet Forwarding Attacks

During the data transmission phase, an adversary that acts as the intermediary relay node, gets hold of the data packets and mistreats them. In a multihop network, the protection for routing security cannot guarantee each intermediary MS to forward the data packet in accordance with the routing table. The most probable attacks[34] of this type include:

Modification: While acting as an intermediary node and receiving the data packet from a source node, a malicious MS intentionally modifies, drops, or injects data packets before forwarding to the next hop.

Congestion: An adversary relays data packets to heavily loaded MSs.

Power Drain: An adversary sends unnecessary packets or broadcast packets to drain the battery of other MSs.

Lowering Throughput: This attack brings down the throughput of a multihop route and causes higher packet delay by reducing its forwarding speed.

Confidentiality: An attacker may violate the privacy of the source node by copying and then revealing the data packets to a third party.

Selfishness and DoS: It occurs when a node does not perform its expected functions by refusing to collaborate.[35,36] A selfish node does not execute packet forwarding function for the purpose of saving its battery. *Black hole* attack drops all the packets while *grey hole* is the attack that selectively drops some packets and relays others.

21.4.5 Specific Wireless Protocol Attacks

An adversary may focus its attack on the Internet accessing protocols (i.e., IEEE 802.11 and IEEE 802.16), or multihop network protocols (i.e., A-GSM and UCAN). IEEE 802.16 is a standard for constructing wireless metropolitan area networks (MANs). The IEEE 802.16 security is implemented as a privacy sublayer below the MAC protocol. It provides protection in terms of one-hop connection between an MS and a BS. However, the IEEE 802.16e security cannot prevent an MS from the BS forgery or replay attacks due to lack of a BS certificate. Like IEEE 802.16 security, most of the wireless protocols suffer from various security threats. For instance, the GSM security protocol may suffer from the attacks on authentication algorithm like cloning, confidentiality attacks like brute-force attacks, crypt analytical attacks, and attacks using loopholes. It is well known that IEEE 802.11 WEP protocol is vulnerable to many attacks.

21.5 An Investigation and Analysis of Security Protocols

There are many security protocols developed on the basis of single-hop wireless networks including GSM security, UMTS security, Mobile IP security, HiperLan/2, IEEE 802.15, IEEE 802.11, IEEE802.16, and IEEE 802.20. In addition, many security protocols[34] have been proposed for the

MANET security. As discussed in Section 21.3, the above protocols are not enough to provide security protection for the integrated cellular, WLAN and MANET. In this section, we provide a survey of the existing security protocols for the integrated cellular, WLAN and MANET. Some of the security protocols are based on a specific type of multihop network, such as UCAN[7] for multihop cellular network, two-hop relay[10,22] for multihop cellular/WLAN, and MCIP[27] for multihop cellular IP networks. Other security protocols are developed based on a generalized type of multihop network, e.g., a charging and rewarding scheme[33] and GMSP[23] for multihop cellular and multihop WLAN. The details of the above protocols are discussed as follows. To simplify the description, the area inside of a dotted circle in Figures 20.9 to 20.12 represents the coverage of a BS in this chapter.

21.5.1 Unified Cellular and Ad Hoc Network

Figure 21.9 depicts the architecture of a UCAN[7] system in which each MS is equipped with two interfaces: high data rate (HDR) and IEEE 802.11b. As shown in Figure 21.9, for all MSs associated with the HDR BS, some of them (e.g., in Figure 21.9, A, B, C) are actively sending/receiving packets while other MSs (e.g., in Figure 21.9, MSs D) are in the idle state and have their HDR interfaces in the dormant mode. When a destination MS experiences the low HDR downlink channel rate (e.g., in Figure 21.9, A) through its direct link to the HDR BS, the HDR BS transmits data packets to the destination through a multihop path (e.g., in Figure 21.9, HDR BS->proxy client C -> Relay client B -> Destination A). It is because proxy client C has a better HDR downlink channel than MS A. To support the multihop communication in UCAN system, a routing discovery protocol[7] has the capability to find a high-speed proxy client as well as the corresponding multihop route.

To protect the multihop relay path and to motivate each intermediary client to actively participate in packet forwarding, Luo et al.[7] proposed a security scheme, called *secure crediting*, as one of the design goals of

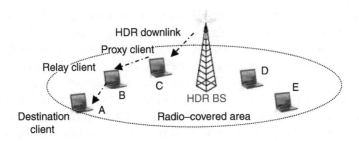

Figure 21.9   UCAN system.

(1) Mutual authentication and generation of a temporal secret key
(2) Verification of each relay client

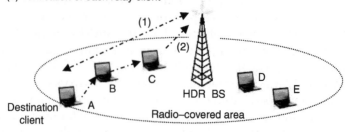

Figure 21.10 UCAN secure crediting.

the UCAN system (Figure 21.10). In the beginning, each MS must register with its HA through the visiting network (i.e., BS) by following the 3G security specification. The registration prevents mobility-related attacks by implementing a process of the mutual authentication between each MS and the BS. At the same time, the BS validates the credentials of MS by HA. If the MS is validated, a temporal secret key with the same period of the registration lifetime is negotiated for future security authentication to preclude packet forwarding attacks. The secret key can be derived from a pre-established secret key between the MS and its HA, by which the system saves the overhead for key management.

In essence, any secure crediting scheme stimulates packet forwarding by awarding credits to all the intermediate nodes. Two problems need to be solved in the crediting scheme: *deletion of legitimate clients* and *addition of extra clients*. In the deletion of legitimate clients, a node (e.g., B) on an ad hoc relay path may intentionally remove another legitimate intermediary node (e.g., A) that has actually forwarded the data packets on the relay path. It causes the collaborating client to lose its credits. On the other hand, a malicious node may add another intermediary node to earn credits. However, the intermediary node has no contribution to the packet relaying. Addition of extra clients discourages the network operators from enabling UCAN relay.

To avoid the deletion of legitimate clients and addition of extra clients, the UCAN security scheme requires each routing request (RTREQ) to be first formed with a signature by the source node using the secret key. The signature piggybacks a message authentication code in the RTREQ. After receiving a RTREQ from a neighboring node, an intermediary node creates its own message authentication code and appends it on the RTREQ before propagating the RTREQ again. When the RTREQ propagates to the BS through intermediary clients, each client has a signature on the RTREQ. Thus, the BS can validate the multihop path and can also keep track of each registered client. By verifying the signature on each data packet, the

BS can precisely record the number of data packets that are forwarded by each intermediary node.

The shortcoming of the UCAN system is that it does not support multihop communication for an MS which is outside the coverage of the BS. The registration and authentication of an MS are based on the assumption that the MS is under the coverage of the BS. The BS cannot authenticate an MS outside its transmission range.

21.5.2 Two-Hop Relay

Wei and Gitlin[10] proposed an integrated two-hop-relaying WMAN/WLAN system to increase the performance and coverage of WMAN and WLAN system. Figure 21.11 illustrates a two-hop-relay communication scenario that integrates the cellular network paradigm with WLAN relay.

As shown in Figure 21.11, the relay gateway (RG) is a dual-mode MS with both WWAN (i.e., cellular) and WLAN radio interfaces. Therefore, it can act as the intermediary node between the BS and the destination MS. When the destination MS, a dual-mode or WLAN-only terminal, suffers a low-speed link toward the BS because of poor radio, it establishes a two-hop-relay connection to the BS. The data packets from the BS are first sent to the dual-mode MS with the WMAN interface, and then forwarded to the destination with the WLAN interface.

Before the traffic between the destination and the BS can be forwarded by the intermediate RG, the two-hop-relay system carries out three security-related steps (Figure 21.12). At first, the RG, which has subscribed the integrated service, has to register with the BS using a process of mutual authentication provided by the cellular security protocol (e.g., UMTS security). After authentication, the RG periodically advertises its existence with a *relay advertisement*. Meanwhile, in the second step, the destination MS and the BS are also mutually authenticated when an MS enters the cellular

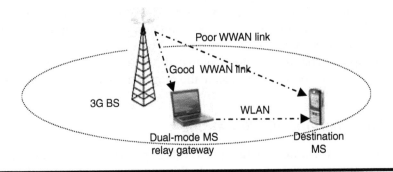

Figure 21.11   Two-hop-relay system.

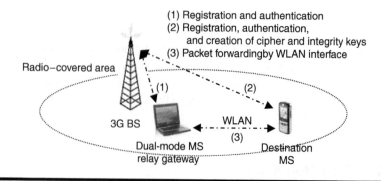

Figure 21.12 Security authentication in a two-hop-relay network.

network. In case of a successful registration, the destination MS negotiates a pair of session keys (cipher key and integrity key) with the BS. Finally, the destination MS exchanges traffic with the BS through the intermediary RG while enduring a low-speed communication with the direct link toward the BS and hearing a *relay advertisement*. The session keys of the destination maintain the privacy and integrity of the forwarded packets.

Wei and Gitlin[22] assume that all dual-mode MSs are selfish and are not willing to relay traffic for other MSs without any specific benefits. In other words, every MS only cares about the utility function itself rather than others. Under this assumption, Wei et al.[22] take the advantage of the *game theory* to model the behaviors of selfish nodes and design an incentive scheme to encourage relay. In the proposed incentive scheme, the BS allocates radio resource to MSs in such a way that a collaborative MS receives a higher priority during the time slot allocation process, and obtains a higher throughput from the BS. The service of quality (e.g., throughput) of the selfish node will degrade in case the collaborative MSs have the higher priority to take up more radio resource for services. To maximize its utility function, even if a selfish node, each MS should act cooperatively and expect to improve its own performance to a greater degree.

Similar to UCAN systems, the two-hop-relay system has the shortcoming that it does not support the multihop communication for an MS, which is outside the coverage of the BS. The application of game theory for the proposed approach has the requirement that all MSs are under the coverage the BS, with poor or good link to execute the mutual authentication with the BS.

21.5.3 Charging and Rewarding

Salem et al.[38] advocate a charging/rewarding approach to encourage packet relay in multihop cellular networks. In this approach, an initiator of a communication pays for the services provided and all intermediary nodes are

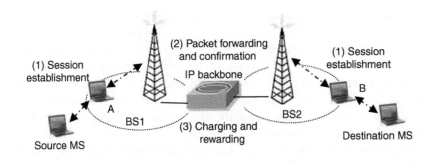

Figure 21.13   Charging and rewarding.

rewarded after relaying packets.[36] Figure 21.13 illustrates a multihop cellular network, which can be A-GSM,[3] MCN,[4] or MCIP.[27] In these multihop networks, an MS outside the coverage of the BS can communicate with the Internet by a multihop connection. As shown in Figure 21.13, the source MS communicates with the Internet using intermediary node A.

The charging/rewarding scheme is developed based on three assumptions. First, it assumes that in multihop cellular networks all traffic travels through the BS, which means all data packets initiated by the source MS are forwarded to a BS by a single- or multihop path, and then pass through the Internet toward the destination. Second, the communication path between the source and the destination is secure and the BS maintains the routing tables for the MSs in the cell without being corrupted (e.g., cache poisoning). Third, the approach assumes that the network operator needs to be fully trusted by all the MSs, with the capability of manipulating the account of each MS.

The charging/rewarding protocol requires each node to register with the operator before using the system for communication. When an MS is registered with the system, it receives a long-term symmetric secret key, which is shared with the operator and used to authenticate the MS. In a multihop communication, the incentive mechanism, as shown in Figure 21.13, includes three phases: (i) session establishment, (ii) packet forwarding and confirmation, and (iii) charging and rewarding. In the session establishment phase, the path for communication (e.g., source MS–A–BS1–IP backbone–BS2–B–destination MS as shown in Figure 21.13) is constructed. This path can be reconfigured if an intermediary node moves away and the link is broken. During this period of path establishment, the operator authenticates the source and destination MSs as well as each intermediary node (e.g., A and B) on the path.

In the second stage, the data packets are forwarded between the source node and the destination where each intermediary node on the path accepts to forward packets (e.g., A and B in Figure 21.13). The integrity of each packet is protected by the source MS which creates a message authentication code on the packet by using its secret key. When the BS (e.g., BS1 in Figure 21.13) receives the data packet, it verifies the message authentication code of the source and checks the freshness of the packet. Then, the packet is forwarded to travel through the Internet with IP security protocol. The BS on the other side, forwards the packet to the destination after appending a new message authentication code by using the secret key associated with the destination. In the end, the destination receives the packet and validates the message authentication code of the BS. The destination acknowledges the reception of every packet either on per packet basis or batch basis. The destination reports to the BS every time when it receives a data packet from the source. To save resources, the destination acknowledges all received packets in a single batch at the end of packet transmission. The BS can maintain the record of all the activities of packet forwarding for each participating MS. Finally, based on the reports from multihop network, the network (BS) determines which accounts should be charged or credited. To prevent misbehavior from a malicious node, the proposed protocol is designed in such a manner so as to resist cheating during the process of charging and rewarding: refusal to pay, incorrect reward claims, free-riding, and invasive adversary.[38]

21.5.4 Multihop Cellular IP

MCIP[27] divides a cellular access network into separate local domains (e.g., domains 1 and 2 in Figure 21.1). Within a local domain, the BSs are interconnected by wired cables, and each MS connects one of the BSs in a domain by single hop or multihop. When an MS moves from one domain to another, it involves a global mobility (macro-mobility) as described in Sections 21.2.1 and 21.4.2. If an MS moves from one BS to another BS within the same domain, it is called local mobility (micro-mobility). During the micro-mobility for an MS, it updates its new route to its new attacked BS at the FA, without the requirement of a macro-mobility mobile IP registration.

SM^{3}P (secured multihop macro-micro-mobility protocol)[23] has been designed to address macro/micro-mobility security in the MCIP network. In the MCIP network, each BS has a pair of public and private key (e.g., K_{BS1}/K_{BS1}^{-1} for BS1 in Figure 21.14) and a domain secret key ($S_{BS1-BS2}$ for BS1 and BS2 in Figure 21.14) with each neighboring BS. When a control packet is transmitted between two BSs in the local domain, the secret keys protects the integrity of the packets as shown in Figure 21.14. Also, each MS has a secret key that is a priori associated with its HA.

Upon entering the local domain of a network, the security protocol requires each MS to present its authentication information to the FA. The single-hop MS (e.g., MS1 in Figure 21.14) registers with BS directly (e.g., step 1 of MS1 in Figure 21.14). The key challenge is how to authenticate

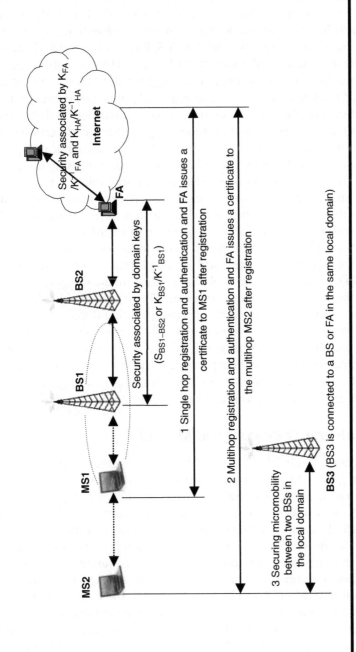

Figure 21.14 Security associations (SAs) in an MCIP local network.

a multihop MS (e.g., MS2) outside the coverage of the BS and prevent the mobility-related attacks as illustrated in Section 21.4.2. To provide a secure multihop authentication and registration, a multihop MS (e.g., step 2 of MS2 in Figure 21.14) has to go through the following activities: (i) the FA first validates each intermediary node that relays the registration request from the registering MS, (ii) then, the multihop MS, FA, and HA authenticate one another, (iii) as a successful authenticated MS, its HA computes a pair of temporary public and private keys (K_{MS}/K_{MS}^{-1}) for the MS, (iii) the private key (i.e., K_{MS}^{-1}) is delivered to the MS secretly and public key is encapsulated into a certificate by the FA. The certificate has a valid period that is from issue time (T_{issue}) to the expiration time (T_{expire}). Each certificate has the signature of the FA. Since each registered MS has an authenticated public key (i.e., K_{FA}) of the FA, they can validate the certificate using the public key of FA. The certificate can be used as the security token for the micro-mobility security, and multi-hop routing discovery between two MSs. Each MS discards the multihop control and data packets from a neighboring MS without a valid certificate. Before the expiration of a certificate, MS requests to update its certificate if the MS is still in the MCIP local domain.

As for the security for the local mobility (e.g., step 3 of MS2 in Figure 21.14), a secure micro-mobility process maintains the multihop paging/routing cache from being poisoned. This is achieved by the certificates of MSs to protect the integrity of paging/routing-update packets and to prevent multihop paging/routing caches from poisoning. Figure 21.14 shows the security associations in the multihop network, in which each BS, the FA and the HA have their public and private keys (e.g., K_{HA}/K_{HA}^{-1} for HA and K_{FA}/K_{FA}^{-1} for FA). SM^{3}P implements the goal that any pair of components in the local domain can be mutually authenticated.

21.5.5 Integrated Internet and MANET

In the integrated Internet and MANET,[13–21] mobile IP and ad hoc routing protocols (e.g., AODV, DSR) coordinate with each other to build the connectivity across the heterogeneous networks as shown in Figure 21.15. In Figure 21.15, MSs B, G, and I are located in the coverage of the BS. Other MSs (e.g. MSs E, F) are located outside the coverage of the BS and may connect to the BS via MANET. There are two kinds of possible communication in the integrated network. Intra-MANET communication involves interaction between hosts within the ad hoc network, e.g., in Figure 21.15, communication between A and D. Inter-MANET is the communication that involves infrastructure nodes (like CN) and ad hoc network nodes.

To support the secure inter-intra-MANET communication, a security protocol[25] is proposed for the MANET routing where the FA serves as the authentication center to check the credentials for MANET MSs. Before having the capability to communicate with other MANET MSs or the wired

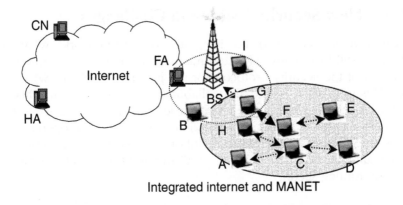

Integrated internet and MANET

Figure 21.15 Communications in integrated internet and MANET.

network, a MANET MS has to perform the following operations to establish a trust relationship with the integrated network. Otherwise its data packets will be ignored by other MANET MSs. The security protocol[25] includes the following steps:

1. *Key establishment*: The MS generates a pair of private and public keys.
2. *FA discovery and FA advertisement*: The MS finds a route to a FA, sends its public key to FA, and obtains FA's advertisement with CoA.
3. *MS registration and authentication*: The MS follows an authentication protocol to register with the FA. The security protocol ensures that the MS's registration request has not been changed during the forwarding from the MANET MS to the FA/HA.
4. *Validation*: The HA or a trusted authority validates the identity of the MS as well as its credential, and then gives the result back to the FA. By trusting HA for authentication, the security protocol ensures that the MS's registration request is legitimate.
5. *Identity Binding at FA*: MS's home IP address, which is the permanent address of the MS, MS's temporal ad hoc identifier, and its public key are uniquely bound by FA. Then, FA issues the MS a certificate.

Only authenticated MS has a certificate from the registered FA. It guarantees that only the authenticated nodes have the capability to participate in ad hoc routing discovery. All unauthorized nodes cannot obtain services from either the Internet or MANET because the packets issued by unauthorized nodes are ignored by other nodes. The routing security for multihop communication includes ad hoc routing discovery, routing cache, and routing maintenance.

21.6   New Security Issues and Challenges

Research in the security of the integrated cellular network, WLAN and MANET is still in its early stage and many issues remain unexplored. The final goal of the security protocols for the integrated cellular, WLAN and MANET is to provide secure and collaborative communication. This goal cannot be achieved easily due to intricacies of communication protocols and specific network traits as illustrated in Section 21.3. The open research issues range from the Internet to multihop wireless network security and can be summarized in two key issues: distributed trusted relationship and multihop routing security.

21.6.1   *Distributed Trusted Relationship*

First of all, the fundamental vulnerability of the integrated networks comes from its distributed environment, open wireless medium, and heterogeneous multihop wireless communication, and this makes the establishment of the distributed trust relationship difficult. The distributed trust relationship is the basis for the security framework in the integrated networks against the connectivity and mobility-related attacks in addition to routing-related threats. These integrated networks require four types of distributed trust relationships:

(1) An MS (single- or multihop) ↔ Home network
(2) An MS (single- or multihop) ↔ Visiting network
(3) Home network ↔ Visiting network
(4) An MS (single- or multihop) ↔ Another MS (intermediate or destination)

The distributed trust relationship addresses two important problems: who can be trusted and how can the trust be established. A plethora of security protocols focus on the above trust establishments based on various single-hop wireless networks. These security protocols include GSM security, UMTS security, standard Mobile IP security, HiperLan/2, IEEE 802.15, IEEE 802.11, IEEE802.16, and IEEE 802.20. All these protocols provide procedures for creating trust relationship between a single-hop MS and the visiting network.

The distributed trust relationship in the integrated networks is a more challenging issue than in single-hop wireless networks.[23] The existing mechanisms in the literature to establish distributed trust relationship for a multihop MS have three steps[22–27,34,39]: (i) the multihop MS sends its credentials to the network (BS) so that the BS can authenticate the MS, (ii) the multihop MS validates the BS so that the MS can trust the BS for accessing the network, and (iii) after the mutual authentication, the MS can create a security binding with the integrated network. The security binding provides keys to be used for multihop routing security and packet forwarding security. However, the procedure for trust establishment is still largely open for further investigation:

■ It is a critical issue to develop a security protocol in a way that a multihop MS can effectively validate the authenticity of any other entity when required in a multihop WLAN or a multihop cellular network.[34]

■ The private/public key-based protocols cause heavy computation delay compared to secret key-based cryptosystem. However, it is hard for the secret key-based scheme to achieve scalability due to key management. And it is an important issue how to facilitate the process of expensive trust establishment with high efficiency and scalability.

21.6.2 Multihop Routing Security

Multihop routing security in the integrated networks is another critical issue. If the multihop routing is corrupted by a malicious intermediary node, the MS cannot get a correct Internet connection with services. There are three types of routes in the integrated networks:[24] the route from a BS to an MS, the route from an MS to a BS, and the route between two MSs without BS. Corresponding to the three types of routes, multihop routing security should provide security to all the above types of routes. In the process of route discovery, it is necessary to execute the required principles to enforce registered MSs to participate in honest route discovery and maintenance and to exclude the malicious nodes from the routing paths. Owing to infrastructure-supported multihop routing security, the home network has the capability to manipulate MS's billing and credential when an MS has any malicious action in the process of route discovery. Thus, it is possible to have a scheme that the Internet maintain a set of metrics to record the past misbehavers of an MS and a multihop selects well-behaved MSs as the intermediary MS for packet forwarding. Also, the infrastructure-based scheme for detecting various misbehaviors is an important issue in the integrated networks but has been neglected in current security designs. A lot of issues about securing multihop routing and packet forwarding remain unexplored:

■ How to enforce service availability and cooperation in the integrated network with a secure mechanism to stimulate MS to participate in packet forwarding, to refrain from overloading the network, to thwart the "selfish" MS, and to deter malicious behaviors.

■ How to implement fair charging and rewarding for the cooperation between MSs in packet-forwarding protocol and a reasonable fine for misbehavers.[40-46]

21.7 Conclusion

The integrated cellular network, WLAN and MANET provide the Internet connectivity and ad hoc communication. In this chapter, we consider the security issues including security threats and solutions. After the analysis of the security characteristics, the possible security threats against the integrated networks have been presented. The security attacks may seriously degrade the performance of Internet services for multihop MSs. However, the current security solutions for the integrated networks are only the first step toward tackling various security attacks. For the practicality, robust and efficient security schemes have to be designed across heterogeneous multihop wireless networks and the Internet. Compared to the security solutions for pure MANETs, the security protocols for the integrated networks should takes advantage of infrastructure-based Internet authentication to establish the trustworthiness among MSs. In this chapter, the security challenges of such schemes are discussed for providing ubiquitous and secure Internet services for mobile users in support of heterogeneous multihop communication.

References

1. D.P. Agrawal and Q.A. Zeng, *Introduction to Wireless and Mobile Systems.* Brooks/Cole Publishing, pp. 436, 2003.
2. D. Cavalcanti, C. Cordeiro, D. Agrawal, B. Xie, and A. Kumar, "Issues in integrating networks, WLANs and MANET: A futuristic heterogeneous wireless network," *IEEE Wireless Communications Magazine*, Vol. 12, No. 3, 2005, pp. 30–41.
3. N. Aggelou and R. Tafazolli, "On the relaying capability of next-generation GSM cellular networks," *IEEE Personal Communications*, Vol. 8, 2001, pp. 40–47.
4. Y.-D. Lin and Y.-C. Hsu, "Multihop cellular: A new architecture for wireless communications," in *Proc. INFOCOM*, Vol. 3, 2000, pp. 1273–1282.
5. H. Wu, C. Qiao, S. De, and Q. Tonguz, "Integrated cellular and ad hoc relaying systems: iCAR," *IEEE Journal on Selected Areas in Communications*, Vol. 19, No. 10, 2001, pp. 2105–2115.
6. X. Wu, S.-H. Chan, and B. Mukherjee, "MADF: A novel approach to add an ad-hoc overlay on a fixed cellular infrastructure," in *Proc. IEEE WCNC*, Vol. 2, 2000, pp. 549–554.
7. H. Luo, R. Ramjee, P. Sinha, L. Li, and S. Lu, "UCAN: A unified cellular and ad-hoc network architecture," in *Proc. ACM MOBICOM'03*, September 2003, pp. 353–367.

8. T. Rouse, I. Band, and S. McLaughlin, "Capacity and power investigation of opportunity driven multiple access (ODMA) networks in TDD-CDMA based systems," in *Proc. IEEE ICC*, Vol. 5, 2002, pp. 3202–3206.

9. A. Zadeh, B. Jabbari, R. Pickholtz, and B. Vojcic, "Self-organizing packet radio MANETs with overlay (SOPRANO)," *IEEE Communications Magazine*, Vol. 40, No. 6, 2002, pp. 149–157.

10. H.-Y. Wei and R. Gitlin, "Two-hop-relay architecture for next-generation WWAN/WLAN integration," *IEEE Wireless Communications*, Vol. 11, No. 2, April 2004, pp. 24–30.

11. H. Hsieh, K.-H. Kim, Y. Zhu, and R. Sivakumar, "A receiver-centric transport protocol for mobile hosts with heterogeneous wireless interfaces," in *Proc. ACM MobiCom*, September 2003, pp. 1–15.

12. R. Chang, W. Yeh, and Y. Wen, "Hybrid wireless network protocols," *IEEE Transactions on Vehicular Technology*, Vol. 52, No. 4, 2003, pp. 1099–1109.

13. A. Sleem, A. Kumar, and K. Kamel, "Using cell topography in predicting intra-cell mobility in wireless," in *Proc. SCI2003*, Orlando, 2003.

14. S. Sharma, N. Zhu, and T.C. Chiueh, "Low latency mobile-IP handoff for infrastructure-mode WLANs," *IEEE Selected Areas in Communications*, Vol. 22, No. 4, 2004, pp. 643–652.

15. Y. Sun, E.M. Belding-Royer, and C.E. Perkins, "Internet connectivity for ad hoc mobile networks," *International Journal of Wireless Information Networks, Special Issue on Mobile MANETs*, Vol. 9, No. 2, 2002, pp. 75–88.

16. H. Lei and C.E. Perkins, "Ad hoc networking with mobile IP," in *Proc. Second European Personal Mobile Communication Conference*, 1997, pp. 197–202.

17. D.B. Johnson and D.A. Maltz. "Dynamic source routing in ad hoc wireless networks," in *Mobile Computing*, Kluwer, chap. 5, 1996, pp. 1369–1379.

18. U. Jonsson, F. Alriksson, T. Larsson, P. Johansson, and G. Maguire, Jr., "MIPMANET-mobile IP for mobile MANETs," in *Proc. IEEE/ACM Workshop on Mobile and MANETing and Computing*, 1999.

19. P. Ratanchandani and R. Kravets, "A hybrid approach to Internet connectivity for mobile MANETs," in *Proc. IEEE WCNC*, Vol. 3. March 2003, pp. 1522–1527.

20. J. Broch, D.A. Maltz, and D.B. Johnson, "Supporting hierarchy and heterogeneous interfaces in multi-hop wireless MANETs," in *Proc. I-SPAN*, June 1999, pp. 370–375.

21. B. Xie, and A. Kumar, "A protocol for efficient bi-directional connectivity between ad hoc networks and Internet," *Journal of Internet Technology, Special Issue on Wireless Ad Hoc and Sensor Networks*, Vol. 6, No. 1, 2005, pp. 101–109.

22. H.-Yu Wei and R.D. Gitlin, "Incentive scheduling for cooperative relay in WWAN/WLAN two-hop-relay network," in *Proc. IEEE WCNC*, 2005, pp. 1696–1701.

23. B. Xie, A. Kumar, D.P. Agrawal, and S. Srinivasan, "Securing macro/micro mobility for multi-hop cellular IP," *Journal on Security in Wireless Mobile Computing System, Elsevier Special Issue of Pervasive and Mobile Computing*, 2006.

24. B. Xie, A. Kumar, D.P. Agrawal, and S. Srinivasan, "GMSP: A generalized security multi-hop protocol for heterogeneous multi-hop wireless network," in *Proc. IEEE WCNC*, 2006.

25. B. Xie and A. Kumar, "A framework for integrated Internet and ad hoc network security," in Proc. *IEEE Symposium on Computers and Communications*, Vol. 1, June 2004, pp. 318–324.

26. M. Shi, X. Shen, and J.W. Mark, "IEEE 802 Roaming and Authentication in WLAN/Cellular Mobile Networks," *IEEE Communication Magazine*, Vol. 11, No. 4, August 2004, pp. 66–75.

27. B. Xie, A. Kumar, D. Cavalcanti, and D.P. Agrawal, "Multi-hop cellular IP: A new approach to heterogeneous wireless networks," *International Journal of Pervasive Computing and Communications*, 2006.

28. C.E. Perkins, "IP mobility support for IPV4," IETF Internet Draft, draft-ietf-mobile-rfc2002-bis-08.txt, 2001.

29. J. Zao, S. Kent, J. Gahm, G. Troxel, M. Condell, P. Helinek, N. Yuan, and I. Castineyra, "A public-key based secure mobile IP," *Wireless Networks*, Vol. 5, No. 5, 1999, pp. 373–390.

30. S. Jacobs, "Mobile IP public key based authentication," http://search/ietf.org/internet-drafts/dreaft-jacobs-mobileip-pki-auth-02.txt, 1999.

31. Sufatrio and K.Y. Lam, "Mobile-IP registration protocol: A security attack and new secure minimal public-key based authentication," in *Proc. Int. Symp. on Parallel Architectures*, 1999.

32. A. Chakrabarti and G. Manimaran, "Internet infrastructure security: A taxonomy," *IEEE Network*, Vol. 15, No. 6, November–December 2002.

33. A. Fasbender, D. Kesdogan, and O. Kubitz, "Analysis of security and privacy in mobile IP," in *Proc. Int. Conf. on Telecommunication Systems*, March 1996.

34. H. Yang, H. Luo, F. Ye, S. Lu, and L. Zhang, "Security in mobile ad hoc networks: Challenges and solutions," *IEEE Wireless Communication*, Vol. 11, No. 1, 2004, pp. 38–47.

35. H. Deng, W. Li, and D.P. Agrawal, "Routing security in ad hoc networks," *IEEE Communications Magazine*, Special Topics on Security in Telecommunication Networks, Vol. 40, No. 10, October 2002, pp. 70–75.

36. Y. Yoo and D.P. Agrawal, "Why it pays to be selfish in a MANET?" *IEEE Wireless Communications*.

37. Y. Yoo, S. Ahn, and D.P. Agrawal, "A credit-payment scheme for packet forwarding fairness in mobile ad hoc networks," *Proc. IEEE ICC*, May 2005, pp. 3005–3009.

38. N. Ben Salem, L. Buttyan, J.-P. Hubaux, and M. Jakobsson, "Cooperation in multi-hop cellular networks with extended security analysis," in *Proc. MobiHoc*, June 2003.

39. G.M. Koien and T. Haslestad, "Security aspects of 3G-WLAN interworking," *IEEE Communication Magazine*, Vol. 41, No. 11, November 2003, pp. 82–88.

40. L. Buttyan and J.P. Hubaux, "Enforcing service availability in mobile ad-hoc WANs," in *Proc. MobiHoc*, August 2000, pp. 87–96.

41. S. Marti, T.J. Giuli, K. Lai, and M. Baker, "Mitigating routing misbehavior in mobile ad hoc networks," in *Proc. ACM MobiCom*, 2000, pp. 255–265.

42. L. Butty'an and J.-P. Hubaux, "Stimulating cooperation in self-organizing mobile ad hoc networks," *Mobile Networks and Applications*, Vol. 8, No. 5, 2003, pp. 579–592.
43. B. Raghavan and A.C. Snoeren, "Priority forwarding in ad hoc networks with self-interested parties," in *Proc. Workshop on Economics of Peer-to-Peer Systems*, 2003.
44. S. Zhong, J. Chen, and Y.R. Yang, "Sprite: A simple, cheat-proof, credit-based system for mobile ad-hoc Networks," in *Proc. IEEE Infocom*, 2003, pp. 1987–1997.
45. P.-W. Yau and C.J. Mitchell, "Reputation methods for routing security for mobile ad hoc networks," in *Proc. Mobile Future and Symposium on Trends in Communications*, 2003, pp. 130–137.
46. E. Huang, J. Crowcroft, and I. Wassell, "Rethinking incentives for mobile ad hoc networks," in *Proc. ACM SIGCOMM*, 2004, pp. 191–196.

Chapter 22

Fieldbus for Distributed Control Applications

N.P. Mahalik and Kiseon Kim

22.1 Introduction

The fields of wireless communication and digital electronics have been encouraging industrial automation and control system designers to develop a cost-effective and reliable networked control architecture. With regard to industrial machine and process control systems, described by the features presented in Table 22.1, the individual sensors, actuators, valves, or drives at the field level must be able to interact with each other via controllers and other intelligent devices such as personal computer (PC) platforms. A fundamental enabling technology for realising an intelligent and responsive process is that of distributed real-time control, sometimes referred to as industrial networking system (INS) or *fieldbus technology.* Distributed control using fieldbus technology is considered as the first step in the application of power processing to field devices. Fieldbus technology enables intelligence to be distributed to the device level in a way that greatly improves monitoring, and control is possible at reduced cost. It is apparent that the fieldbus systems will be the preferred choice for the design of many industrial machines and process control systems. Wireless communication plays a potential role in fieldbus systems. To consider fieldbus as the basic enabling technology for designing distributed control system (DCS), it is important to comprehend the fundamental details about it. This chapter deals with aspects of wireless fieldbus systems. The rest of the chapter is organized as follows. The next section describes the origin of fieldbus technology. Section 22.2 presents the correlation between

Table 22.1 Communication Requirement[1]

Features	Necessity (%)
Time synchronous communication	Required: 64
	Not required: 36
Sampling mode for network node	Cycling sampling: 73
	Event-driven sampling: 27
Cycle periodicity	1 ms: 26
	10 ms: 48
	100 ms: 26
Signal response time	1 ms: 40
	10 ms: 34
	100 ms: 26
Length of fieldbus system	<100 m: 66
	<1000 m: 31
	>1000 m: 3

Source: From Lawrenz, W., in *2nd International CAN Conference, ICC-95*, London, 3–4 October, 1995.

the OSI model and the fieldbus layers. The fundamental aspects of DCS is described in Section 22.3. Some commonly used standards and frequency bands are highlighted in Section 22.4. Section 22.5 describes some of the important and popular wired and wireless fieldbuses currently available in the market. In particular, Profibus, R-fieldbus, DeviceNet, LON, Foundation Fieldbus, and HART fieldbuses are described. Section 22.6 discusses how to select a fieldbus.

22.1.1 Origin of Fieldbus Technology

In the early 1980s, Honeywell introduced the concept of superimposing a digital signal upon the 4–20 mA current loop system for their field devices (sensors, actuators, etc.). This is believed to be the birth of the smart field devices. In manufacturing and process industries, the number of point-to-point connections between master (supervisory) and field devices has increased.[2] Each new device has its own principles of communication in terms of transmission rate, messaging format, and medium of communication. Handling of interoperative communications has been more complex. The problem was then tackled by borrowing ideas from computer networking. There was a need to introduce one serial communication link to which the existing field devices could be connected. This was the concept behind the INS that led to the fieldbus standard in the mid-1980s.[3]

The concept of a fieldbus makes it possible to interconnect sensors, actuators, valves, motor starters, motor controllers, operator interfaces, and other I/O devices as well as control devices such as programmable logic controllers (PLCs) and supervisory computers. It integrates in a uniform and simple way to collect, send, and control data/signals through a single communication channel. Instead of an extensive number of physical point-to-point connections between a master and each field device, all devices are connected to one single physical communication link. The data system at the field level must be able to communicate under real-time conditions. Further, digital communication has replaced the analog connections. The media used for fieldbus can be twisted pair cables, optical fiber, radio frequency (RF), power line cables, etc.

22.1.2 OSI Model and Fieldbus Layers

International Organization for Standardization (ISO) is a world-wide federation of national standard bodies. The work of maintaining international standards is normally carried out through technical committees. ISO collaborates with the International Electrotechnical Commission (IEC) on electrotechnical standardization. In 1978, ISO issued a recommendation for a standard network architecture which is in the form of a seven-layer model known as OSI reference model, (OSI RM). The purpose of this model was to provide a common basis for the coordination of the development of communication system standards and to allow existing standards to be placed into perspective. The ISO/OSI RM was developed as the architecture model for open data communications. In production automation sectors, the manufacturing automation protocol (MAP) also complies with the ISO/OSI RM. MAP contains communication standards in all seven layers. The data system at the sensor–actuator level (called field level) does not need the services of all the seven layers of the ISO/OSI RM. What is important is that with the increasing overhead of the defined layers from the ISO/OSI RM, the system becomes more complex and the real-time response can be affected.

Fieldbus protocols are developed in line with the ISO/OSI model (Figure 22.1). It is believed that there was no recommended guideline to design a standard fieldbus protocol to be used in real-time industrial process control applications. The OSI model was only a reference model used for computer networking and was apparently not an implementation specification. As a matter of fact, in 1985, both IEC and Industry Standard Architecture (ISA) published guidelines for the standardization of a fieldbus suitable for use in industrial applications. Companies and individuals active in the fieldbus standard areas formed the International Fieldbus Consortium (IFC) in 1990. They carried out field trials of the evolving fieldbus standards to demonstrate the performance of the fieldbuses. They also provided information feedback to the appropriate standard bodies. The IFC could not succeed in its objectives as the member body split into groups. With the failure to finalize internationally accepted standards, we can still

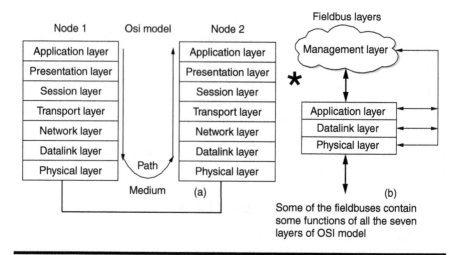

Figure 22.1 (a) ISO OSI reference model. (b) Fieldbus layers.

find many fieldbus vendors and suppliers in the technology market. Out of these, the standard protocols are LonWorks Technology, CANbus, FIP, WorldFIP, ISA, IEC, ISP, HART, P-NET, BACNet, DeviceNet, Bitbus, EHS, Profibus, SP50, Foundation Fieldbus, and many others. It is expected that in future no single fieldbus protocol will predominate. On the other hand, a single fieldbus cannot provide the solution for all the applications. From the literature review it has been found that, CANbuses are mainly used in automotive engine applications and machine control industries, Profibus in textile industry, and LonWorks and Bitbus in building automation and industrial control environment.[4] However, the fieldbus technology is best suited for most industrial applications. An abstract level comparison of some important fieldbuses is presented in Figure 22.2(a) and (b). Fieldbus specifications are partitioned into layers. The important layers are described below, although some fieldbuses claim the services of all the seven layers.

- *Physical Layer:* This layer describes the physical medium used to carry fieldbus messages encoded as appropriate for the signaling media.
- *Datalink Layer:* This layer manages the routing of messages to the correct destination using device and data address references. It also manages access rights enabling multiple devices to share the usage of a common physical medium. Accurate time distribution and the bridging of multiple segments into a single fieldbus network are also part of data link functionality.
- *Application Layer:* It supports user functions like device identification and data structures.

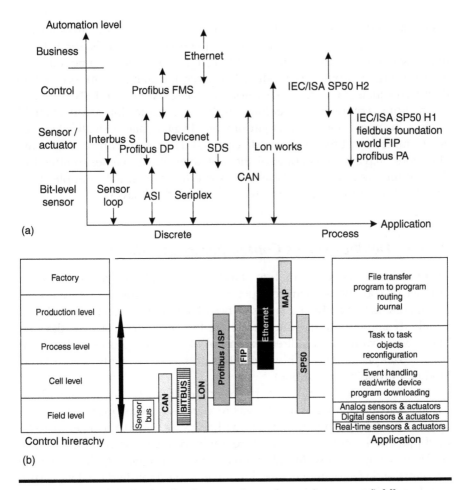

Figure 22.2 **(a) An abstract-level comparison of some important fieldbuses. (b) Scope of different buses.**

■ *Management Layer:* This layer provides various monitoring, config-uration, and initialization services for all the layers, together with diagnostic support and fault-recovery functions.

22.1.3 Fieldbus versus LAN

The confusion between fieldbus technology and the conventional local area network (LAN) is obvious, as in reality they adopt the same concept of digi-tal and networked communication. LAN provides communication between computer systems in a geographically limited area. A LAN always has one or more servers. Most of the processing occurs on the local computer, not on the servers. The access to the LAN is achieved with a microprocessor-based controller and a medium access unit (MAU). The distributed operating system (DOS) or network operating system (NOS) are the middleware kernels dealing with the management aspects of the entire system. Large quantities of data in the order of up to gigabits per second can be transmitted from one computer to another. However, the timing of messages and their transfers is not critical. Therefore, to some extent, the LAN system is not real-time compliant. A fieldbus on the other hand, interconnects field devices in rugged industrial environments. Digital data in small quantities is transmitted in a time that is predetermined. Fieldbuses conforming to the OSI model have to omit some of the OSI-layer functionalities to attend to the real-time requirements. However, some fieldbuses, for example, LON and Profibus, conform to all the requirements of OSI layers.

22.1.4 Fieldbus versus Contemporary Standards

There exists another set of bus standards, typically originating from a number of sources. As semiconductor companies introduced new microprocessors, they also built a series of board and system products designed mainly to simplify the use and accelerate the acceptance of their devices. These buses were often a direct extension of the microprocessor local bus onto a backplane. This is how Multibus and Vesabus evolved into Versa Module Europe bus (VMEbus).

Another source for buses is the computer industry. In the late 1980s, the inadequacy of the ISA for PC was recognized. It was observed that the architecture became a bottleneck to higher levels of performance. In 1992, Intel proposed a new PC architecture, the peripheral component interface (PCI), as a high bandwidth alternative. Since that time, PCI has not only become the de facto architecture, but it has also established itself as a basic enabling technology in the computer and electronics industries including automation and control sectors. Although the electrical characteristics of the PCI satisfy most applications in the industries, the PCI boards may not be suitable for many applications in terms of their formats and connector concepts. For this reason the CompactPCI specification was developed. CompactPCI is a bus system which is based on the electrical specification of the PCI standard, but is packed into an industrial 3U (100 * 160 cm, 3U) or 6U (233 * 160 cm, 6U) Eurocard format. CompactPCI was developed by PICMG (PCI Industrial Computer Manufacturers Group). The history of the development of fieldbuses also includes the efforts of a number of smaller companies. For example, GESPAC defined the G-64 bus, Pro-Log the STD bus, and, in the recent past, Ampro defined the PC/104 bus. An active but commercially less successful source of buses has been institutions like the Institute of Electrical and Electronic Engineers (IEEE). Notable ventures include the STE bus, and more recently, Futurebus. At present, within the signal-processing and high-speed data-acquisition environment, VME and

PCI architectures are the primary contenders. The most important development has been the adoption of PMC (PCI Mezzanine Card) by the VMEbus community. The industrial and real-time computation community will soon have a wide choice of standardized solutions based upon two important building blocks: VME and PCI.[5]

All such buses were aimed at specific market needs and were relatively independent from any given chip. The buses are designed mostly for signal-processing and intensive data-processing applications such as radar, sonar, imaging, management information system (MIS) servers, transactional banking applications, Ethernet, integrated service digital network (ISDN), industrial control where fast data processing is desired, and so on.[6] The signal-processing architecture is based on digital signal processors (DSP) boards. Although the DSP performance is very high, the resultant architectures are often proprietary. At the same time, it is worth mentioning that most of such bus architectures entertain centralized control implementation. Although some of the buses are capable of being configured onto distributed architecture, the overall installation cost seems to be high when compared with fieldbus-based control implementation.

22.2 Review on Distributed Control

The rapid progress of internationalization following World War II has had an enormous impact on manufacturing industries as it has become the cornerstone of many economic activities. During the 1970s most industrial control systems were relay-based, with nearly a one-to-one I/O ratio. Control systems, consisting of a group of relays, were capable of controlling machine tools, conveyor systems, and other mechanical components such as motors, transportation vehicles, etc. within rugged environments. To increase communication capabilities, implementation simplicity, cost–performance ratio, reliability, and capability, PLC-based control systems emerged in 1969. A PLC is a programmable electronic device that controls machines and processes. It uses a memory to store instructions and execute specific functions that include on/off control, timing, counting, sequencing, and arithmetic and data handling. Functioning as relay replacements, the PLCs are more reliable than relay-based systems, largely because of the robustness of their solid-state components. PLCs save installation, troubleshooting, and labor costs by reducing the wiring and the associated wiring errors. They occupy less space and their reprogrammability features increased flexibility and simplicity with regard to changing control schemes. On the other hand, PLC-based automation was considered a centralized implementation, which has its own loopholes such as central failure. Consequently, a need to break down the control process into "cells" was found as another alternative to improve flexibility (e.g., decentralization of centrally regulated tasks) further. This in turn led to a more decentralized approach to control solutions.

Cell-based control got its name from its architectural design. Process is distributed among a number of processors instead of a single one performing all the control functions. Normally the processors/computers are built into separate modules. Each of them is optimized for a particular function. Often the separate modules are physically distributed around a plant. The modules are connected together via a communication link. The cell-based approach can be considered as the origin of DCS. DCS has quickly become fundamental to most types of automation applications due to its implementation flexibility. The following sections discuss the fundamental aspects of DCS in more detail. They include (i) the architecture and inherent features, (ii) benefits gained from DCS over centralized architecture, (iii) design and implementation aspects, (iv) international standardization initiatives, and (v) advances in DCS-based industrial-control solutions in terms of configurability, design methodology, and diagnostics.

22.3 Fundamental Aspects of DCS

In view of the observation made in the previous sections, control systems can be categorized as centralized control systems (CCS) and DCS. An example of conventional centralized monitoring and control with a central processing unit (CPU) that communicates with a number of field devices, is illustrated in the Figure 22.3. Such systems require individual point-to-point links between each field device and the processing unit.

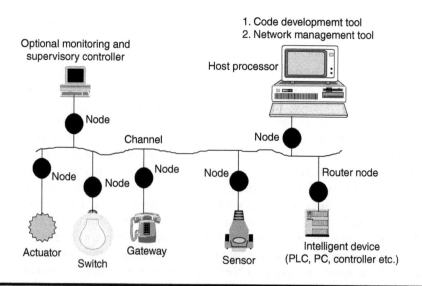

Figure 22.3 A typical fieldbus-based distributed control system (DCS).

The increase in both quantity and quality of I/O information, coupled with increased performance demands, has proved to be a problem for an overloaded centralized control strategy.[7] For a large system the total number of I/O signals can be in thousands, and this may bring problems. Even a powerful computer has difficulty in polling all the inputs within the time limit of the system. Other disadvantages of centralized systems are

- They can be totally paralyzed when the CPU fails.
- They are unable to make the best use of online techniques.
- They are not flexible, i.e., they are slow to take advantage of improvements in technology.
- The installation cost tends to be high.

Hence there has been a move toward distribution of control tasks throughout the system. The emergence of low-cost computing power has enabled the development of *smart distributed systems* for machine, instrumentation, and process control applications.[8] The main difference between DCS and CCS is in the way they access the status information from or to the system. Distributed control is most appropriate when the I/O points are dispersed. A large task is divided among many controllers, resulting in a smooth execution of the whole control process. The possibility of a complete failure is unlikely because one controller will not cause a failure of the whole system. The control problem can be split into levels of control, from basic device-level control to network-level processing. Just as parallel-processing techniques allow individual CPUs to be combined to form supercomputers capable of incredible performances, DCS will allow remote I/O subsystems (sensors/actuators) to communicate with each other to form wide-area automation infrastructures.

The basic elements of DCSs are processing units (local processors, I/O hardware), field devices (sensors, actuators, switch, etc.), memory (for storing control code and intermittent data), communication links, protocols, development tools (compilers and debuggers), and host processors with other supporting software tools that help to download the application codes into the nodes. The host computer supervises and controls the network in run time. The local nodes process input data from sensors or output data to the actuator and perform the control tasks that have been downloaded. A communication link in the system is necessary to enable the nodes to talk to other nodes in the network in run time.

The DCS design process attempts to draw attention to issues like response time, degree of processing power to be allotted at the local level, type of collision-avoidance algorithm, timelines, real-time capability, storage facility, message size, event transfer mechanism, medium type and channel efficiency, topology architecture, communication range, localization through local loops, intrinsic safety, interoperability, interchangeability, scalability, modularity, programmability, and network security. Largely, these attributes can be considered under two domains, namely, "run-time surface" and the "offline surface." The run-time surface embodies the response time, real-time control, intrinsic safety, and operational intelligence, whereas the offline surface embodies aspects such as interoperability, interchangeability, and scalability.[9]

22.4 Standards, Frequency Bands, and Issues

RF technology uses radio waves to carry control signals. Different wireless protocols and methods arose with different RF technologies. These include traditional radio transceivers, cellular telephones, and laptop PCs. Some wireless protocols are defined and administered by the IEEE as part of its 802 standards, which are being developed to achieve increased speeds and new functions. The primary benefits of wireless transmission have led to a number of solutions. These solutions range from voice-oriented, large-scale cellular networks, to data-oriented solutions like wireless LAN (WLAN), wireless personal area networks (WPAN), and wireless sensor networks (WSN). WLAN systems, like the IEEE 802.11 family of standards, are designed with high data rates in terms of tens of Mbit/s over ranges of a kilometer. These parameters provide the user with access to Ethernet. WPAN systems, such as Bluetooth and IEEE 802.15.4 can connect devices wirelessly and are considered energy-efficient.[10] They also support medium data rates in the order of hundreds of Kbit/s to a few Mbit/s and have ranges in the order of a few meters.

Bluetooth is very popular and is a radio standard mainly designed for low power consumption, short range, and with a low-cost transceiver microchip in each device. Bluetooth-based network, called piconet, has already gained popularity. A piconet is a network of devices connected in an ad hoc fashion. A piconet is formed when a portable PC and cellular phone connect each other, and it supports up to eight devices. Within the net, one device acts as the master while the others act as slaves for the duration of the piconet connection. A piconet is also referred to as PAN. Although, the main applications of Bluetooth technology are desktops and laptops, printers, mice and keyboards, cell phones, personal digital assistants (PDAs), and other handsfree devices, it can support other wireless networks. The standard also includes support for more powerful, but certain data logging equipment that can transmit data to a computer via Bluetooth and can also be used for remote controls in the place of infrared which was traditionally used.

Wireless communication takes place in different frequency bands. The fundamental physical laws restrict the transmission power. The 2.4 GHz Industrial Scientific and Medical (ISM) band is a common example. Some proprietary protocols use 900 MHz band. Although this band inherits low throughput it also inherits a better range and wall penetration capability. Further, the band is available in few countries. Above all the 5.8 GHz band has potential advantages in terms of improved throughput, higher noise immunity, and most importantly, smaller antennas. Note that antenna length is inversely proportional to the frequency of operation. Thus, the higher the frequencies of transmission are, lower will be the antenna length. Transmissions at higher frequency band can avoid the effects of the so-called "narrowbanding." Table 22.2 compares three main wireless transmission technologies. These technologies are complementary rather than competitive with each other as they address different needs and have different strengths.[11] In many applications, hybrid networks are mandatory (Figure 22.4). The reason is that any installation to connect the wireless to the wired system has to first retrofit the existing system. This configuration is known as a hybrid system and is preferred in many scenarios. Mesh network is also possible in a hybrid system. One can note that mesh networks may not be the best answer to many of the industrial communication problems because of the lack of deterministic modes with low latency and jitter,

Table 22.2   A Comparison of Major Wireless Standards Using the ISM Band

Standard (Market Name)	802.15.1 (Bluetooth)	802.11b (Wi-Fi)	802.15.4 (ZigBee)
Application focus	Cable replacement	Web, e-mail, video	Control & monitoring
Bandwidth (kBps)	1000–3000	11000	20–250
Transmission range (m)	20 (Class 2) 100+ (Class 1)	100+	20–70, 100+ (ext amplifier)
Nodes supported	7	32	2*64
Battery life (days)	1–7	0.5–5	100–1000+
Power consumption (transmitting)	45 mA (Class 2) <150 mA (Class 1)	300 mA	30 mA
Suitability for low duty-cycle applications	Poor (slow connection time)	Poor (slow connection time)	Good
Spread spectrum technology	Frequency hopping spread spectrum	Direct sequence spread	Direct sequence spread
Memory footprint (kB)	50+	70+	40
Success metrics	Cost, convenience	Speed, flexibility	Power, cost

Source: From Koumpis K. et al., A review and roadmap of wireless industrial control, Available at http://wireless.industrial-networking.com/articles/articledisplay.asp?id=913.

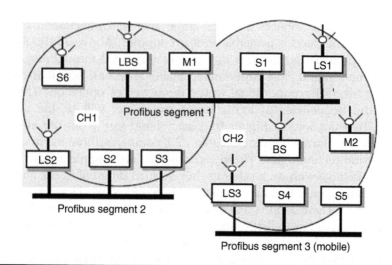

Figure 22.4 A typical hybrid network based on PROFIBUS fieldbus. M, master; S, slaves; LS, link stations; BS, base stations; CH, channel.

but they inherit many other advantages in terms of allowing applications to grow based on demand, as the addition of new nodes is simple. They also adopt short-range transmission, providing good utilization of available bandwidth. The following points should be considered while dealing with wireless links within the industrial environment[11]:

- Technical issues: interfaces
- Security
- Robustness: fail-safe/fail-soft operation
- Interference immunity factor
- Power availability
- Interoperability

22.4.1 An Example

One typical application of a wireless control network is real-time public transport passenger information system (RPTPIS). Wireless network has been brought in as a new feature to increase the attractiveness and usefulness of public transport. The other advantages include vehicles that are more comfortable, low-ticket prices, shorter journey times, and punctuality of timetables. However, during the last decade, the real-time public transport information systems have strongly promoted public transport as opposed to daily use of private cars. The data transmission between vehicles, traffic signal controllers, stop displays, and the central computer is based on radio messages and is handled with radio modems. In the future, the systems can may include portable information using mobile telephones. The real-time passenger information system includes[12] the following:

- Dynamic visual display at bus/tram stops
- Onboard displays and audible information on buses/trams
- Information at home/office via Internet

22.5 Some of the Major Wireless Fieldbuses

This section describes some of the important wireless fieldbuses that have already been developed or are under development. Detail technical information with regard to their services is available in Ref. [13]. The readers are also encouraged to refer to the respective licensed documents for further study.

22.5.1 *Wireless Profibus DP (ELPRO 905U-G-PR)*

Process field bus (PROFIBUS) is one of the most popular fieldbuses, and has been a German national standard since 1991. It has been granted the status of an international standard (IEC 61158) by CENELEC.[14,15] PROFIBUS is used for factory–floor communications. It is designed to deliver real-time services in harsh industrial environments. The MAC protocol of PROFIBUS is based on a token-passing procedure, which is considered as the simplified version of the timed-token protocol,[16] used by masters to grant the bus. While it is very similar to other timed-token MAC protocols, such as IEEE802.4[17] or FDDI,[18] it also supports two categories of messages: synchronous (high priority) and asynchronous (low priority). It offers different profiles, i.e., different sets of protocols and application-layer services. The communication profiles denote different sets of protocols, the application profiles distinguish between different sets of application-layer services, and the physical profiles distinguish between different transmission technologies.[19]

A product, the 905U-G-PR wireless Profibus, provides wireless extension of Profibus DP LANs, including secure firewall isolation and security data encryption. The wireless Profibus extenders are available with both Profibus Master and Slave functionality, and can connect to a Profibus DP LAN at full bus speed of 12 Mb/s. The wireless Profibus has an internal 900 MHz license-free radio band. The unit can transmit over more than 20 miles line-of-sight or up to one mile. The 905U-G-PR can be used to extend a central Profibus LAN to multiple remote devices, link separate LANs, and to form multiple slave devices into a wireless LAN without the need for a master. It can also provide firewall and encryption security. The wireless Profibus units can also communicate directly with other data buses.

Covering up to 20 miles line-of-sight, 3000 ft in congested industrial environments, it works at 450 MHz 5 W licensed radio, 30+ miles and can act as a slave to extend existing physical networks. The typical response time is 30 ms for transfer of 64 digital values. Wireless Profibus DP together with Satel radio modems can easily be connected to a remote area or to a mobile substation, where cabling is difficult or even impossible, by using SATELLINE-3AS radio modem. As claimed, transmission over distances up to 50 km is possible. There is an integrated RS-485 interface in SATELLINE-3AS; so an external interface adapter may not be required. RS-232, RS-422, and RS-485 interfaces are also supported up to 38,400 bps. Data speeds of 9,600 and 19,200 bps are also compatible with this A pair of SATELLINE-3AS radio modems forms a transparent communication link. A SATELLINE 3AS(d) modem works with frequencies between 400 and 470 MHz. The power supply is between 10–30 and 6.6 V. It is also possible to have a SATELLINE-3AS(d) with LCD display. The summary of the important features are[14,20,21] as follows:

- It can act as a Profibus Slave to extend existing Profibus networks
- It provides secure Firewall isolation
- The typical response time is 30 ms (for the transfer of 64 digital values)
- The wireless unit can also act as Profibus Master, connecting to up to 125 Profibus Slaves
- Up to 95 Wireless units can connect in a peer-to-peer network
- It can connect separate Profibus LANs using a secure Firewall
- The wireless transmissions are protected by security encryption
- It can connect remote Profibus sensors and I/O devices
- It can interconnect other data buses and PLCs
- It can facilitate wireless protocol conversion—connect to Modbus, DF1, EtherNet IP, Modbus TCP, DeviceNet, and Modbus Plus
- The internet is enabled via an internal 1.4 MB embedded web system (dynamic HTTP) with event-activated email functionality
- It is used for SCADA applications
- The wireless units can act as repeaters to extend the radio distance
- Configuration and diagnostic softwares are available
- It is easy to use, with drag and click configuration

22.5.2 DeviceNet

The Open DeviceNet Vendor Association (ODVA) announced the launch of wireless DeviceNet™ products that can allow the connecting of compatible products on a wireless fieldbus, creating possibilities for reaping the benefits from challenging factory floor applications. Wireless DeviceNet is good for areas that are difficult to access or that require large amounts of cabling. Designated in the WD30 range and developed by Omron Electronics, the wireless products can reduce expensive installation and wiring costs. A single WD30 master modem allows flexibility by addressing multiple slave modems, each of which can introduce a subnetwork. In addition, each DeviceNet network can accommodate multiple wireless masters, thus making multiple, flexible topologies. A network based on DeviceNet can accommodate up to 32 wireless masters, each with 32 slaves. The low output power of the WD30 transceivers means that interference with other devices is not an issue.[22]

22.5.3 R-Fieldbus

A high-performance radio fieldbus (R-FIELDBUS) that is able to support extended application services within the industrial and manufacturing sectors has been proposed (project no. IST-1999-11316: High Performance Wireless Fieldbus in Industrial Multimedia-Related Environment). As reported, the architecture is based on the integration of emerging wireless technologies for broadband networks with existing industrial communication protocols such as the European Standard EN50170. The objective of this fieldbus architecture is to ease the resolution of problems found in manufacturing plants such as the need for recabling or the need to install new, and moveable, sensors and control units. The wireless capabilities of R-FIELDBUS can bring important improvements in terms of flexibility. The objective also includes provision of full transparent access to any information needed on-site, such as data concerning real-time control and status information, or transparent access to specification drawings and other industrial-type multimedia information (real-time voice and low-resolution digital video sequences). The fieldbus supports user-defined quality of service (QoS) for multimedia communications. The R-FIELDBUS system additionally supports mobile industrial devices and its interoperability with existing devices can be achieved through wired industrial networks. Its primary focus is the development of a state-of-the-art radio-based physical layer and the implementation of the required real-time and multimedia functionalities through the necessary EN 50170 protocol extensions. The main deliverable components of the R-FIELDBUS are[23,24] as follows:

- Requirements and system planning specifications for real-time wireless fieldbuses
- Assessment and selection of a suitable radio technology
- General system architecture
- Specifications of the physical layer
- Specifications of the data link layer extensions

- Real-time performances of the lower-layer subsystems
- Specifications of the higher-layer architecture
- Specifications of the system management
- Field trials and documentation

22.5.4 HART

Highway addressable remote transducer (HART) is a wireless standard aimed at providing wireless capabilities. The wireless HART working group, an activity of the HART Communication Foundation (HCF) had set the goal of producing draft specifications for a wireless standard in early 2006.[25] The HCF is an independent, nonprofit organization providing worldwide support for applications of the HART protocol. The working group plans to coordinate activities with wireless organizations, in the industry such as the ISA SP100 Wireless Committee, to ensure continuity and uniformity with standardization efforts. HCF member companies include major automation suppliers and leaders: ABB, Adaptive Instruments, Elpro Technologies, Emerson, Endress+Hauser, Honeywell, Omnex Controls, Phoenix Contact, Siemens, Smar, and Yokogawa.

22.5.5 Fieldbus Foundation

To achieve plant optimization, the number of different networks, gateways, and subsystems are reduced while information integration between automation systems and MIS is increased. The Fieldbus™ Foundation, a nonprofit organization of over 180 manufacturers and users of automation equipment that was formed to develop the open, integrated fieldbus architecture has the intention of adhering to the above concept. The specification developed by the foundation is FOUNDATION fieldbus™ (FF).[26] FF is also based on the OSI RM. The bottom physical layer provides the connection to the communication media such as wire, fiber optics, and wireless, while the top application layer defines the services needed to send and receive the messages to and from the devices, respectively. To achieve interoperability, FF uses a common language above the application layer that is known as the user layer (UL). The UL defines the structure for performing control functions in the devices. It specifies how distributed functions can be linked across the control network. The UL defines the important application process in terms of resource blocks (RB), function blocks (FB), transducer blocks (TB), and device description (DD) technology. The integrated architecture is shown in Figure 22.5.

To implement the distributed control strategy, functions such as an analog input (AI) in a flow transmitter or an analog O/P (AO) in a valve, for example, are included in FBs. FBs encapsulate control functions

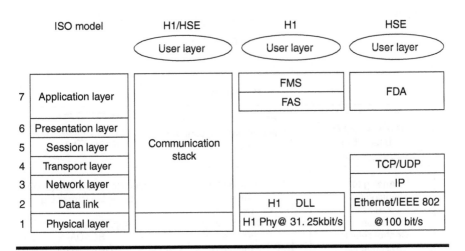

	ISO model	H1/HSE	H1	HSE
		User layer	User layer	User layer
7	Application layer		FMS	FDA
			FAS	
6	Presentation layer	Communication stack		
5	Session layer			
4	Transport layer			TCP/UDP
3	Network layer			IP
2	Data link		H1 DLL	Ethernet/IEEE 802
1	Physical layer		H1 Phy@ 31. 25kbit/s	@100 bit/s

Figure 22.5   The integrated architecture of H1 and HSE: H1 runs at 31.25 kbit/s; HSE runs at 100 Mbit/s; H1 physical layer is an approved IEC 61158 standard; and the HSE physical layer is an approved Ethernet/IEEE 802 standard.

(e.g., PID controller, AI, etc.). RBs define parameters that pertain to the entire application process (e.g., manufacturing ID, device type, etc.). TBs represent interfaces for sensors such as temperature, pressure, and flow. For more complex system-level functions such as batch control, coordinated drive control, I/O gateways, synchronization, and sequencing, the flexible function blocks (FFBs) are useful. The inputs and outputs and the control algorithms of the FFB are modifiable, configurable, and extendable.

The development procedure is as follows: First, the location of FB for a particular device is determined. Once this is done, an FB schedule is created, which is subsequently downloaded to each field device. During runtime, FBs in the device execute at offsets determined by the configuration. Note that FBs contain the information needed for the online control functions, but at this stage of the designing of control networks, additional configuration and display-related information are needed. The additional information is supplied by DD and capability files (CF).

The range of applications covered by FF is shown in Figure 22.6. FF specifies 21 standard FBs to address basic and advanced process control applications. The FFB are dedicated and application-specific and are designed to implement strategies such as supervisory control and data acquisition (SCADA), batch control, PLC sequencing, burner management, coordinated drives, and state-of-the-art I/O interfacing. FFBs are created using programming tools based on standards such as IEC 61131-3. FFBs are also used to map information from other networks such as sensor buses into the FF open architecture. It can be summarized that the main

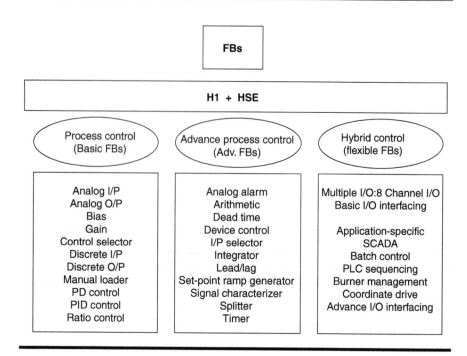

Figure 22.6   Range of application of Foundation Fieldbus.

objective of FF is to facilitate interoperability, field-device integration, subsystem integration, and dataserver integration. Modern data servers are built on high-performance PC platforms with Ethernet access. Data servers can connect to the high-speed ethernet (HSE) backbone and provide a common platform for application softwares. A typical application software includes human machine interface (HMI), enterprise resource planning (ERP), configuration and diagnostic tools, and MIS databases.

FF provides two profiles that are selected based on application needs: H1 and HSE. The physical layer guarantees that the devices will send and receive data bits correctly on the preferred media. H1 and HSE run at 31.25 kbit/s and 100 Mbit/s, respectively. They are designed as a high-performance control backbone for the integration of high-density data generators such as PLCs and analyzers, and data servers. HSE can use Ethernet networking equipment. It is stated that the physical layers of H1 and HSE are the approved IEC 61158 standard and the Ethernet/IEEE 802 standard, respectively.

22.5.6   LonWorks™ Technology

Local operating network (LON) technology, or MonWorks™ from Echelon, USA, makes it possible to design, deploy, and configure low-cost smart devices that communicate with one another. The application domains are consumer electronics, factory automation, commercial building controls, vehicle controls, and home automation.[27] A LON consists of intelligent devices, or nodes powered by the Neuron® Chip processor. Each Neuron chip has three 8-bit processors. Neuron chips are programmed using Neuron C language, a derivative of ANSI C. It supports developing codes for distributed control applications. The nodes are thought of as objects that respond to various inputs and produce the desired outputs. Although the function of a particular node is quite simple, the interaction among nodes enables the network to perform complex tasks in a distributed manner. Nodes are pre-programmed for runtime executions and are capable of sending messages to one another in response to changes in various conditions, and can take action in response to the messages they receive.[28]

LonTalk® is the protocol used by LON. The protocol supports end-to-end acknowledgments with automatic retries. Under this service, a node sending a message expects an acknowledgment from all intended receivers thereby automatically retransmitting the message until all the intended receivers respond. Communication between nodes on a network takes place using the *network variables* (NV) that are defined in each node. The protocol supports twisted pair, power line, coaxial cable, fiber optics and RF. A channel is a physical transport medium. The protocol defines a hierarchical form of addressing user domain, subnet, and node. Multiple dispersed nodes can be addressed using domain and group addresses. A domain is a logical collection of nodes on one or more channels. A subnet is a logical collection of up to 127 nodes within a domain. Up to 255 subnets can be defined within a single domain.

The protocol also supports different types of communication services such as unacknowledged, acknowledged, unacknowledged/repeated, and priority communication services. Under unacknowledged services the node sends out messages on the network whenever the local application decides on it, and it does not wait for responses from receiving nodes. This service provides the widest network bandwidth. Acknowledged service is used when it is important that a message be received at its intended destination. The unacknowledged/repeated service is same as that of acknowledged service, but the message is sent a number of times, determined at the time of node installation. Lastly, priority time slots on a channel can be allocated to improve the response times of critical packets. Only one node is assigned to a particular priority slot. Network management and application software facilitate the design, development, and installation processes. The installation process requires logical connections between devices, which need to communicate with each other using installation tools. One of the important elements of the LON is the transceiver. There are various types of transceivers as listed in Table 22.3.

Table 22.3   LON Transceiver Capabilities

Transceiver (TX) Types	Capabilities
78 kBps twisted pair TX	4600 ft or 1400 m (worst case)
1.25 MBps twisted pair TX	Up to 430 ft or 130 m (worst case)
Power line TX	These transceivers communicates with either a proprietary spread spectrum or a narrow band technology that provides reliable communications for up to 2000 m on a clear line
78 kBps twisted pair, free topology TX	Up to 500 m (1600 ft) with no repeater and up to 1000 m (3200 ft) with one repeater
Radio frequency TX	Both licensed and nonlicensed versions are available in the 400–470 MHz and 900 MHz bands, respectively

Source: Adapted from IST-1999-11316. PROFIBUS Standard DIN 19245 Part I and II, translated from German, *PROFIBUS Nutzer organisation e.V.*, 1992.

22.6   Selecting a Fieldbus

The devolution of intelligence from the central system to the field devices will raise a number of questions. Many criteria are relevant for selecting the appropriate bus system for control applications. Research on as to why customers choose one product or service over competing alternatives gives us the following list of competitive advantages.

22.6.1   Commercial Aspects

The commercial aspects include (i) price advantages, (ii) higher quality, (iii) availability of fieldbus elements, (iv) higher awareness, and (v) better customer services.

22.6.2   Technical Aspects

The technical aspects include: (i) where should the control decisions be taken? (ii) where should the data be stored? (iii) how should the data be accessed? (iv) how much processing power can be incorporated in the field devices? (v) how will short to medium distances to remote I/O units be covered? (vi) what is the throughput and update time? (vii) what is its latency? (viii) what is the link cost? (ix) how can task be partitioned? (x) what type of protocol should be used? (xi) how much is the power consumption?

(xii) what is the error response? and (xiii) how can we configure a new node while the system is running?

22.6.3 Communication Requirements versus Applications

The selection of a fieldbus for a particular type of application is also important. It is anticipated that in the near future, no single fieldbus protocol will predominate and a single fieldbus is unlikely to provide the desired solution for all control applications For example, a packaging plant's control system is very different from a building lighting control system. To understand the requirements for installing a smart device on a control network it is essential to look at the type of the system.

22.7 Discussion and Conclusions

In recent years, the restriction on physical connectivity has caused a parabolic growth in wireless systems. As implied, wireless communication denotes the elimination of wires as physical layers to carry data without requiring any physical changes to target systems. Several forms of data acquisition, monitoring, networking, and control are making inroads into control and automation. Wireless networks are generally characterized by fast deployment and cost effectiveness. The applicability in industrial environments, though limited, is growing steadily. This is due to the fact that the protocols for wireless networks emphasize strict reliability, which is unsuitable for real-time mission-critical industrial control and monitoring applications. Further, although the throughput requirements are comfortable, the operating conditions are harsher. Hence, the requirements for reliability, adaptability, and scalability outweigh speed. Conversely, new suggestions and techniques are constantly being developed to cope with the demanding requirements.

An advantage of RF signaling is that the waves propagate widely (depending on the characteristics of the antenna). They are reliable and provide competitive alternatives to wiring technology. The allowable frequency of the radio signals is determined by each country, although some frequency bands such as the ISM band (2040 Hz) are available for use worldwide. The cost of RF technology has dropped in recent years, especially for short distances. The downside of RF is that it is hard to penetrate metallic building materials; the allowable frequency bands are increasingly crowded and therefore subject to interference; and the short range of low-cost, bidirectional RF devices require either multiple receivers or repeaters to propagate a reasonable distance. The current limitations of RF technology should raise tremendous concern among control system designers. Adopting this immature technology poses critically high risks, life-cycle costs, and warranty exposure. Demonstrating a working RF mesh in test rigs yields very different results from a trial in situ in a real environment. The RF mesh sometimes is not simple for wide-scale deployment. The first step is to carefully evaluate real-world communication range, battery life, and life-cycle costs.

The data and information to be communicated within industrial environments is usually the state information based on small packets associated with insensitive environments and timing requirements. The data are usually temperatures, humidity levels, vibrations, atmospheric pollution levels, chemicals, and electromagnetic noise. The desired throughput of the network is relatively low in contrast to the reliability requirements. Note that wireless communication should be avoided if deterministic transmission with latency under 10 ms is desirable.

In many situations, power-line signaling offers a viable alternative. Most of the control devices in these applications require a connection to electric power, be it AC or DC voltage, making power-line signaling an ideal solution. For free topology, twisted pair wiring even offers proven reliability, no without battery replacement issues, and with very low maintenance costs. For overhead and emergency lighting devices, power line is a better and more reliable alternative to RF.[29]

We have no doubt that RF technology will improve over time; however, we do not believe that manufacturers should be test vehicles for untried technology. RF technology suppliers should continue developing it until they can produce robust field-proven solutions. The value of RF technology—if it can be made to work well at low cost—appears clear for a range of battery-powered sensor-monitoring applications. The value of mesh networking, on the other hand, has yet to be field-proven in control applications in which its sole purpose today is to compensate for poor-quality radios.[30]

Sensor network influences many factors including fault tolerance, scalability, production costs, operating environment, sensor network topology, hardware constraints, transmission media, and power consumption.[31]

References

1. Lawrenz, W., Worldwide status of CAN—Present and future, *2nd International CAN Conference, ICC-95*, London, 3–4 October 1995.
2. Foster, G.T., Glover, J.P., and Warwick, K., Successful LonWorks based machines for PCB manufacture, *Proceedings of the LonUsers International Fall Conference*, Germany, 17 June, 1996.
3. Ericssion, G., A comparison of three fieldbus protocol: FIP, Profibus, and Masterfieldbus, *Proceedings of the International Workshop on Mechatronical Computer Systems for Perception and Action*, Paris, 3–4 June, 1993.

4. Pu, J. and Moore, P.R., Component/image based design of distributed manufacturing machine control system, *Proceedings of the International Conference on Recent Advances in Mechatronics*, Turkey, Vol. 2, ICRAM-95, 1995.

5. Gien, M., Embedded real-time operating systems move to the open world, *Real-Time Magazine*, No. 3 pp. 97–101, July–September 1996.

6. Marshall P.S., *Industrial ethernet*, ISA Press, USA, 2002.

7. Reza, S.R., Smart network for control, *IEEE Spectrum*, Data Communications, Special Report, June 1994.

8. Moore, P.R., Pu, J.S., and Lundgren, J.O., Volvo AGVs open new horizons, Technology Report, Volvo, No. 1, 1997.

9. Mahalik N.P., Trends in industrial control systems; A review, *Journal of the Institutions of Engineers*, vol. 59, no. 4, pp. 75–89, 1998.

10. Andersson, M., Bluetooth for industry, in *The Industrial Ethernet Book*, CA, USA, pp. 10–12, September 2002.

11. Koumpis, K., Hanna, L., Andersson, M., and Johansson, M., A review and roadmap of wireless industrial control. Available at: http://wireless. industrial-networking.com/articles/articledisplay.asp?id=913

12. http://www.satel.fi/bb/home.nsf/pages/Real-Time+Passanger+Information

13. Mahalik, N.P., (Ed.), *Fieldbus Technology; Industrial Network Standard for Real-Time Distributed Control*, Springer Verlag, Germany, 2003.

14. Alves, M., Tovar, E., Vasques, F., Hammer, G., and Röther, K., Real-time communications over hybrid wired/wireless PROFIBUS-based networks, *14th Euromicro Conference on Real-Time Systems, ECRTS' 02*, pp. 142–148, 2002.

15. Project EN 50170 – General purpose field communication system, Technical Report, CENELEC, 1996.

16. Grow, R., A timed token protocol for LAN, Proceedings of *Electro'82, Token Access Protocols*, 17/3, 1982.

17. IEEE Standard 802.4: Token passing bus access method and physical layer specification, IEEE, 1985.

18. ISO 9314-2, Information processing systems—fibre distributed data interface (FDDI)—Part 2: Token ring media access control, ISO, 1989.

19. Andreas, W., Investigations on MAC and link layer for a wireless PROFIBUS over IEEE 802.11, Dr. Eng. Thesis, der Technischen Universit, October 2001.

20. IST-1999-11316. PROFIBUS Standard DIN 19245 Part I and II, translated from German, *PROFIBUS Nutzerorganisation e.V.*, 1992.

21. http://www.controlglobal.com/vendors/products/2005/219.html

22. http://www.industrialnetworking.co.uk/mag/v8-4/p2.html

23. http://www.hurray.isep.ipp.pt/rfieldbus/

24. Koulamas, C., Koubias, S., and Papadopoulos, G., Using cut-through forwarding to retain the real-time properties of Profibus over hybrid wired/wireless architectures, *IEEE Transactions on Industrial Electronics*, vol. 51, No. 6, pp. 1208–1217, 2004.

25. http://www.sensorsmag.com/articles/0805/wireless/signals_main.shtml

26. http://www.knowthebus.org/fieldbus/foundation.asp

27. Markie, T. and J. Jones, LonWorks and PC/104: A winning combination, Engenuity Systems, *International Conference on Fieldbus Communication*, Hinkely, England, 1995.
28. http://www.ieclon.com/LonWorks/LonWorksTutorial.html
29. Montague, J., Why not wireless? *Control Engineering,* October 2001.
30. Gupta, R. and Tennefoss, M., Radio frequency control networking: A technology assessment, Echelon Corporation, 005-0171A_RF_White_Paper.pdf, 2005.
31. Chong, C.Y. and Kumar, S.P., Sensor networks: Evolution, opportunities, and challenges, *Proceedings of IEEE,* pp. 1247–1256, August 2003.

Chapter 23

Supporting Multimedia Communication in the Integrated WCDMA/ WLAN/Ad Hoc Networks

Ju Wang and Jonathan C.L. Liu

23.1 Introduction

Many wireless networks exist and some new ones are expected to be developed soon. The most popular wireless networks today are cellular networks and wireless local area networks (WLAN). The new ones emerging rapidly are the ad hoc and personal area networks. To efficiently utilize the network bandwidth, effective schemes for the call admission control (CAC) and media access control (MAC) are critical. Furthermore, the CAC/MAC design for wireless networks boils down to the optimum resource management and traffic scheduling for both downlinks and uplinks.

These protocols must also be implemented with reasonable complexity to provide the real-time response to the highly dynamic environments, and moreover, be able to support multimedia applications. In general, the following design goals should be achieved: (1) *Short delay*: the network system needs to minimize latency as much as possible when delivering the video frames. (2) *High throughput*: to support these concurrent accesses, the network system needs to sustain high throughput. (3) *High reliability*: the network should not fail easily because of the sustained high throughput.

Given the limited mobile-side transmission power and different traffic quality of service (QoS) requirements, the optimum assignment of time slot, code, and power resources has been the focus of many works.[1–7] However, with the increasing bandwidth demand from mobile applications and the need to support different traffic types, wireless networks need to be carefully integrated. In this chapter, our focus is on how data can be efficiently exchanged between mobile stations and base stations (i.e., uplinks and downlinks) in wireless code division multiple access (WCDMA) networks. Then we extend the discussion to WLAN and ad hoc networks.

CDMA as an efficient means of multiple access cellular technology became the reality in the early 1990s, with the leading paper by Gilhousen[8] et al. and first commercial product from Qualcomm in 1994. CDMA systems are unique such that the technology allows concurrent transmission of all mobile stations in the same frequency bandwidth, and 1/1 frequency reuse among neighboring cells.

Recently, researchers and engineers tried to integrate data service, and even streaming multimedia service, into the cellular network.[1,9–13] Unlike the homogeneous traffic characteristic in traditional voice network, data traffic is very stringent in terms of bit error rate, but might be flexible in terms of delay. For example, e-mail traffic can be delayed without having significant impact to the user experience, while streaming multimedia traffic has a comparable delay requirement as that of a voice call.

Data service can be supported via TDMA, CDMA, or TD-CDMA. Time division multiplexed (TDM) Packet Radio support data service through token circulation.[14] Raw CDMA Data Service supports data traffic in the presence of voice traffic, and is discussed in Refs. [15–18]. In integrated voice/data CDMA systems, the capacity for data calls can be increased by scheduling data transmissions in periods of low voice load and curtailing data transmissions when the voice load is heavy.[17] In Ref. [18], data users followed the ALOHA protocol with retransmission control and contend for the remaining (if any) multiple-access capability of the CDMA channel. In a TD-CDMA network, the uplink and downlink share the same frequency. A data packet from an MS must go through a request-transmission signaling sequence with the BS.[19] Similar to the TDMA data service in Ref. [14], after the BS collects the transmission requests from all mobile terminals during an entire frame length, it uses a packet-scheduling scheme to determine how the packets of multimedia applications are accommodated.

The three main competing CDMA-based 3G systems are CDMA2000,[20] W-CDMA,[21] and TD-SCDMA,[22] all based on direct-sequence CDMA (DS-CDMA). Commonalities among these systems are close-loop power control, high data rate in downlink, link-level adaptation, TDMA fashion transmission, fair queue scheduling, soft handoff, etc. The main differences reside in the frequency bandwidth of carrier, link-adaptation methods, and different implementation of the signaling protocol. TD-SCDMA uses TDD separation between uplink and downlink.

23.1.1 Basic Resources to Manage in CDMA Networks

23.1.1.1 Power Control

Power control[23] is essential in a CDMA system to solve the near–far problem and support multiple users with guaranteed QoS. Consider the uplink as an example. A simplification of the received signal strength (at the BS) for the MS i at time t is

$$r_i(t) = p_i s_i(t) + \sum_{j \neq i} p_j s_j(t) + I_0$$

where $s_i(.)$ is the transmitted signal for the desired user and I_0 the summation of other-cell-interference-plus-noise. Therefore, if the transmitting power of one particular user is much higher than that of others, it will dominant the overall received power at the BS. This will result in high transmission error for the other users. In IS-95, the BS constantly measures the observed signal to interference ratio (SIR) level for all connected MSs, and piggybacks a single-bit power command (increase or decrease transmission power by a fixed amount) every 5 ms.

The power in downlink must coordinate the transmission power of neighboring BSs. In many downlink schemes, orthogonal codes are used to eliminate interference among mobile stations in the same cell. However, orthogonal codes have much higher autocorrelations than the pseudo-noise (PN) code used in the uplink. Thus, downlink transmission in neighboring cells and multipath propagation of signals can still cause interference.

23.1.1.2 Code Control

CDMA systems distinguish channels for different MSs by using different codes. In the forward link, orthogonal codes, such as Walsh codes, can be used with almost zero cross-correlation. For the uplink, since the mobile stations are often not synchronized, orthogonal codes cannot be used. Instead, PN codes with partial orthogonality must be used.

Code assignment has been used as a means to provide different data rates for downlink users in IS-95 early 1990s. VSG-CDMA,[24,25] or varying-gain CDMA systems, allows in the change of the spreading factor to support various data rates in the IS-95 downlink. For uplink, the use of VSG is more difficult since the receiving SIR are directly related to the length of spreading code. Thus the control must also take the SIR requirement into account.[6]

MC-CDMA, or multicode CDMA, is first discussed in Refs. [26, 27]. MC-CDMA assumes fixed spreading factor and allows mobiles to use more-than-one code for different data rates. Again, MC-CDMA is relatively easy to deploy at the CDMA downlink. For the uplink, the use of multiple codes splits the power budget into different substreams, and thus will decrease the perceived SIR level at the base station, assuming that the total transmitting power from the mobile remains unchanged. Therefore, both variable gain spreading (VGS) and MC-CDMA must be administrated at the BS to guarantee QoS.

23.1.1.3 Link-Adaptation and Rate-Power Adaptation

The problem of joint-rate and power adaptation under constrained SIR and power limitation for downlink data transmission in a multicell and VSG-CDMA system is described in Ref. [28]. Similar problems for uplink are also discussed by other authors.[29,30] The optimum joint-rate and power control are often formulated as an SIR-constrained problem. The search for the optimum solution requires considerable computation power for large mobile populations.

For supporting multimedia communication, the common trade off here is to maximize system throughput versus meeting delay guarantees among all users. Owing to user diversity, the maximum data rate at which the BS can send/receive to/from different MSs will not be the same. Thus, a greedy scheduling policy is most likely unfair: some users (e.g., those physically close to the BS) receive excessive bandwidth, while others receive little or no service.

Based on the architecture of CDMA networks, our discussion naturally splits into three categories: (1) schemes on the CDMA uplink, (2) schemes on downlink, and (3) mobility-related issues. Inside each of the three categories, we will show protocol and algorithm designs targeting the support for different QoS and performance metrics.

For uplinks, we address the fundamental knowledge of "soft capacity" and extend the concept to multiple-cell design. While tracing the development of supporting different data types, power-control and rate-adaptation issues emerged. For downlinks, we follow a similar path from single-cell scheduling to multiple-cell models, then extend the (computation-intensive) centralized algorithms to the latest decentralized ones. There are many alternative schemes to design the downlink scheduling, beyond the (TDMA-based) CDMA-2000 standards. We devote a subsection to address the leading schemes.

With the solid understanding of the scheduling issues for uplinks and downlinks, we focus on the mobility management issues next. The unique characteristics of "soft handoff" in CDMA networks are introduced, and the leading methods are explained. Then the power-control issues to support multimedia handoff is explained, focusing on the drawback of the conventional "OR-ON-DOWN" scheme. We have found that it does not support the handoff of multimedia traffic well, and we described a suboptimal solution.

Note that many issues can be applied to two or all three categories; thus some issues will be referred few times during our discussion. For example, fairness issues are applicable in both downlink and uplink, and is indeed proposed for both links (by independent studies). Similar situations hold when we discuss power-control issues and interference modeling.

23.2 Multiple Accesses in CDMA Uplink
23.2.1 Admission Control Models

Admission control algorithms for CDMA cellular networks have been discussed extensively.[18,31–33] The evolution of admission control started with the support of voice traffic only within a single cell. Then the research expanded to support the environment with multiple cells. Eventually the data types supported were further enriched with multimedia content.

23.2.1.1 Single Cell with Voice Traffic

The admission control of the uplink at a single-cell-all-voice system must decide the number of voice connection allowed in a cell. The earliest work can be traced back to Ref. [8] in 1991. When a BS receives a new connection request from the mobile station, a general scheme performs one of the following checks:

- The protocol estimates the potential SIR degradation of all existing connections. If some of the estimated SIRs are below a threshold, then the new call is rejected, otherwise it is accepted. This will require an accurate SIR model, and often the BS-allowable capacity is conservatively estimated.
- Alternately, the BS simply accepts the new call temporarily, and then measures the actual SIR for the existing connections. If the measurement does not meet the SIR requirement, then the new call will be disconnected.

As mentioned earlier, the capacity in CDMA is soft and depends on the level of interference and the desired communication quality (in terms of bit-error-rate (BER)). In a single-cell scenario, the interference is from the transmission of other mobile stations within the same cell. Thus the SIR is often modeled by

$$\frac{P_i * W/R}{\sum_{j \neq i} P_j + \eta},$$

where P_i is the received power for the ith MS, W the total bandwidth, R the data rate, and η the noise level. W/R is often referred to as spreading gain or spreading factor.

Let N be the number of active voice calls and E_b/I_0 the required SIR level. Ref. [8] shows that

$$N \le 1 + \frac{W/R}{E_b/I_0} - \frac{\eta}{S} \qquad (23.1)$$

where a uniform received power S is assumed at the base station. The above equation can be used as a basic admission criteria in a system with perfect power control (such that the received power at the BS is the same for all mobile stations). The capacity determined by Eq. (23.1) is very pessimistic: assuming $W = 1.25\,\text{MHz}$, $R = 8\,\text{kbps}$, and an SIR threshold of 5 db, less than 40 mobiles could be supported per cell. To rectify this, Ref. [8] showed that taking into consideration the voice activity effect and using $120°$ cell-sectorization can improve the capacity by factor of 8.

According to information theory, the maximum data rate for mobile station i is

$$R_i(P_i, P_2, \ldots, P_N) = \log_2\left(1 + \frac{P_i}{\sum_{i \ne j} P_j + \eta}\right) \text{ bits/chip}$$

The system throughput is thus

$$C_t = \sum R_i(P_1, P_2, \ldots, P_N)$$

for a given nonnegative power vector. It is easy to show that for any given N, the maximum throughput is obtained at equal power assignments. As N becomes ∞, optimum C_t converge to $\log_2(e)$. Achieving this boundary condition might require very long block coding and an unrealistic delay. Almost all practical systems require some minimum SIR levels as additional constraints.

This line of work has been pursued intensively to maximize the total capacity of a CDMA cell. However, one recent publication[34] in 2004 did point out that the *fairness* issue should be jointly considered in addition to maximizing the uplink throughput. The authors pointed out that the total uplink throughput may have to be sacrificed for fairness, by the adaptation of the well-known generalized processor sharing (GPS) scheduling principle. Their proposed code-division GPS scheme adjusts the channel rate (service rate) of each traffic flow on a time-slot basis by using multiple orthogonal code channels.

23.2.1.2 Multiple-Cell Interference

Starting in the mid-1990s, the interferences among neighboring cells became the main-stream of research investigation.[9,35–40] Unlike FDMA and TDMA that are clear cut on the frequency and time slots, CDMA in general tries to absorb/reduce the interferences since they are unavoidable. For analysis of system capacity from information theory point of view, Hanly[41,42] provide a thought discussion assuming different MS-BS receiving models. These have resulted in important physical layer innovations, such as signal-combining techniques among neighboring base stations,[43] RAKE receiver to exploits multiple signal path,[44] and multiple-input multiple-output (MIMO) structure[45] in late 1990s.

In general, many MAC layer works follow similar paths to discuss the relationship of multiple-access-interference (MAI) and achieved throughput. Without attacking the complex characteristics of multimedia traffic, Ref. [46] in 2004 addressed the dynamic rate control via varying spreading factor under the SIR constraints, and identified this problem as NP-complete. Thus, heuristic approaches should be sought instead of an optimal one. The first heuristic minimized the *peak interference* to other neighboring cells that is caused by any higher rate allocation in one cell. The second heuristic minimized the composite of the other-cell interferences caused by the sum of interferences generated by all the MSs in a cell to the adjacent cells.

Both of these heuristics are considered more computation-intensive, and thus the authors advocated to have a third heuristic based on the mean-sense approximation of the other-cell interference. The design is to reduce the (computation) overhead resulting from the estimation of the other-cell interference factors. We believe Ref. [46] pointed out a good direction for many computer scientists to pursue further algorithm optimization.

23.2.2 Admission Control with Multimedia Traffic

Extending the CDMA networks to support multimedia traffic is a challenging task since different data types impose their own performance requirements. Researchers in mid-1990s started to address these issues.[9,10,47]

23.2.2.1 Problem Formulation

Though purely theoretically, Sampath[48] considered the admission control when different traffic types are to be supported. The QoS requirement for different traffic types is their minimum SIR γ_i and the required data rate R_i. Assuming perfect knowledge of the channel condition (the MS-to-BS path loss), the admission control problem becomes the sufficient condition to have nonnegative power solution for the following linear system:

$$\frac{W}{R_i} P_i h_i >= \gamma_i \left(\sum_{i \neq j} P_j h_j + \eta W \right), \quad \text{for } i = 1, \dots, N \qquad (23.2)$$

Ref. [48] shows that if the condition

$$\sum_{i=1}^{N} \frac{1}{W/\gamma_i + 1} < 1 \qquad (23.3)$$

holds, then there must exist nonnegative power assignment P_i for the N mobile stations such that all data rate requirements and SIR requirements are satisfied.

A similar condition is also reported by Hanly[41] and Zandler.[49] It is easily seen that the feasible power solution region is convex. Sampath[48] showed that total transmission power can be minimized at the exact solution where the inequality in the linear system (Eq. [23.2]) becomes an equality, which is the boundary point of the feasible region.

The achievable data capacity assuming the above static policy is usually conservative and low. In actual systems, the traffic is not always active for both voice and data traffics. Thus, the left side of Eq. (23.3), called system load, might fluctuate due to the active–silent voice traffic. This makes it beneficial to predict the voice load in the next time slot and determine the amount of data traffic to control the outrage probability and improve the good-put. An analysis of the prediction scheme based on Possion voice activity is discussed by Sampath and Holtzman.[17] Ref. [17] also evaluated a distributed access control method for data traffic, where all mobile stations calculated a transmission probability based on a BS-broadcast load indication, and decided independently whether or not to transmit data packets during the next slot.

Other recent attempts include:

1. In Ref. [50], multicode assignment is considered to provide various QoS for multitype-traffic in a cellular CDMA system. Three types of traffic are defined and treated in different manners, according to their unique traffic nature.
2. Ref. [51] proposed a MAC protocol to utilize the silent period of voice to transmit data traffic. The data terminal compete for the available PN code from the relinquished voice terminal by the slotted ALOHA protocol.
3. In Ref. [52] an admission control protocol for multiservice CDMA is developed based on interference estimation. Instead of using one abstract *DATA* type, multiple data services with different QoS requirements are considered in admission control. They used log-normal distribution to approximate the effect of random user location, shadowing, and imperfect power control.

23.2.2.2 *Optimum Power Assignment and Rate Adaptation*

Soon researchers realized that to maximize the overall cell throughput, one must search an optimum (power, rate) allocation in the feasible space. This is a joint-power-rate optimization problem. The decision variables now are the (power, rate) pair for each mobile station, a total of $2N$ variables. Unfortunately, this problem does not have a close-form solution since the constraints in system (Eq. [23.2]) become nonlinear when both R_i and P_i are variables. A convex searching method is often used to find the optimum solution.

Several uplink power control and scheduling algorithms had been proposed.[1,3,5,30] These methods usually boil down to the control of received SINR for active mobiles to achieve a desired packet error ratio. To eliminate the nonlinearity in the constraints, the maximum rate R_i is treated as a function of the SIR and used in the objective function. Assuming that the maximum transmission power is P_i^{max}, and that the SIR is given as γ_i, the optimum uplink power allocation problem is formulated as an joint-convex optimization problem of the following form:

$$max \sum R_i(P_i) \qquad (23.4)$$

$$\frac{h_i P_i}{\sum_{i \neq j} h_j P_j + \eta} > \gamma_i \qquad (23.5)$$

$$P_i^{max} > P_i > 0 \qquad (23.6)$$

Kumaran[3] shows that the optimum must be achieved at a *corner point* defined by the $2N$ linear constraints. At such a corner point, the N mobile stations either do not transmit, or transmit at their full power. It is also shown that a good scheduler tends to schedule one strong user, or a relatively large number of weak users, into the same time slot. The optimum can be solved by a simplex method efficiently. In Ref. [3], a greedy $O(N^3 \log(N))$ searching method is illustrated to produce an optimum solution. Note that this is considered very computation-intensive, and further complexity reduction should be addressed.

Nevertheless, their observation matches the independent investigations from other researches, such as Refs. [1,7]. In Ref. [1], it is shown that users with similar QoS requirements should be scheduled together (they assumed the same received power for different users), and in Ref. [7], weak users were compensated with high transmission power during the handoff area.

However, one particular publication[53] that appeared in 2003 pointed out that when a compressed video stream is involved, a session-long fixed BER might not result in the best performance. A better power management scheme should adjust the BER of each video packet according to their importance of the contained video data. Ref. [53] further considered the end-to-end video distortion with interframe error propagation. They then propose a revised power management scheme for wireless video service that achieves optimum target BER of each video frame. Though interesting and unique, our experience with video streaming tells us that this approach is difficult to implement. Since high-quality video streams always produce a few number of IP packets, it is quite a challenge to schedule the transmission per packet.

23.2.3 Dynamic Spreading Factors

Though many studies have an excellent work in modeling the PHY and MAC layers, almost no publication in the past has addressed the overall system design aspects until our work reported in Ref. [6]. By adopting the Gold PN (Pseudo Number) sequence from IS-95 and CDMA 2000, we investigated BER performance for different numbers of active users and spreading factors. The performance results clearly indicate that the increase of spreading factors can effectively decrease the BER for a given number of users. For example, with 10 users, increasing the spreading factors from 64 to 96 will reduce BER from 0.0008 to 0.0003 (e.g., 62.5% reduction). Increasing spreading factors increase further to 128 results in a BER of 0.000004, which is 75 times less.*

If the mobile decides to change the traffic carried on a traffic channel, it will send an UPDATE message to the base station. The message has to specify the requested new traffic type (data rate, BER threshold, etc.). The base station will again follow the same steps as it did for an OPEN request, but this time it will not consider an increase in the number of users while doing its computations. By using our proposed admission control protocol, the average BER was always maintained below a 10^{-2} threshold for voice even as the number of users increases. Our system reacted to the increasing demand of users, and new spreading factors have been adopted.

However, our protocol design did reveal design trade-offs in admission delay. We have analyzed the timing components and identified the performance bottlenecks. By examining the CDMA characteristics, our schemes reduce the *admission processing time* with a range of 22–27%. Another unique finding was that multiple access (also uplink) channels for acknowledging the successful change of spreading factors are needed to reduce the *update time* component.

We believe our proposed schemes reflect a good match between the current capability requirement of CDMA and the future demand of multimedia applications.

---

* As the current industrial trend indicates, many hardware manufacturers are considering bringing in the dynamic capability of changing spreading factors in next-generation mobile and base stations.

23.2.4 Uplink Scheduling

Certainly, broader attempts to revamp existing CDMA networks are also possible. For instance, the integrated data service can also be provided in a packet-switching format, where data packets from mobile stations must be scheduled on the frame-by-frame (or slot-by-slot) manner.[2,50,54-58]

An uplink scheduler operates as follows:

■ The BS receives the transmission requests from mobile stations through common access channels or piggy-backed in a data packet. A typical request contains the amount of data pending, the maximum delay tolerance, and the required BER.
■ Based on some scheduling policy, the BS sends out the scheduling decision through its broadcasting channel. Those mobile stations that are notified will be able to transmit in the next time slots, which are typically of fixed length.
■ The BS has to calculate the scheduling decision for each time slot. The scheduler must at least make sure that the BER requirements of all transmitting mobiles are satisfied.

The scheduler should often seek the optimization for the uplink throughput. MAC protocols with packet-delay concerns for MC-CDMA[26] multimedia communication systems have received considerable attention.[1,11,18] Ref. [1] shows a slotted MC-CDMA protocol based on BER scheduling called WISPER. Their results indicate that packets of similar BER requirement should be scheduled together to improve capacity of MAC protocols in MC-CDMA systems.

The recent work in Ref. [11] in 2003 considered a similar problem with the consideration of delay guarantee. They extend the packet scheduling into the MC-CDMA scenario. They also adopted a mini slot, which is only a few bits long, to reduce the slot-time waste when variable packet size is to be supported.

23.3 Multiple Accesses in CDMA Downlink

23.3.1 Single-Cell Capacity

Multiple user support is achieved by orthogonal spreading in the CDMA downlink. In a single-cell scenario and assuming no frequency-selective, the SIR observed at the mobile stations are affected by the background noise only. Thus, when the transmission power at the cell site is not a concern, the number of downlink users equals the number of orthogonal codes available. It is easily seen that the number of orthogonal codes is indeed the length of the code. For example, when the spreading factor is fixed to 64, there are exactly 64 Walsh codes. Multiple data rates are possible by allocating more than one code to a particular user.[26,27,35] An alternative is to allow a VGS factor[59] where spreading codes of different length (still orthogonal to each other) are used to satisfy the different data rates.

When the total transmission power is limited at the cell site, it is necessary to allocate sufficient power to mobile stations and maximize the throughput. The admission problem becomes the feasible power solution to the following problem:

$$\frac{b_i P_i}{\eta} > \gamma_i, \quad \text{for all } i \tag{23.7}$$

$$\sum P_i = P_{budget} \tag{23.8}$$

where P_{budget} is the power budget at the cell site. The necessary and sufficient condition of admission criteria is

$$\sum \eta \frac{\gamma_i}{b_i} < P_{budget}$$

Recent works,[15,59] are particularly interesting since the problem is modeled to address multimedia QoS concerns. Ref. [15] in 2002 advocated that the policy can only accept a call when the quality of existing ones can be guaranteed. Ref. [59] in 2004 demonstrated that the exhaustive search for an optimal combination of dynamic rates and power adaptation is quite difficult due to exponential time complexities. They proposed suboptimal and near-optimal approaches.

23.3.2 Multicell Capacity

The capacity of the downlink in multicell environment must take into consideration the transmission of all base stations that share the same frequency bandwidth. This problem needs be tackled with a different formulation.

23.3.2.1 One-Cell-One-Mobile Model

Zander[49] considers a multicell situation where each cell serves exactly one mobile station. Although this abstract model is naturally suitable for a TDMA system, it is also a close approximation for the CDMA downlink cochannel mobiles since each channel can serve one mobile. Assuming M cells in the system, they describe the SINR at the mobile station i as

$$\Gamma_i = \frac{P_i}{\sum_{j=1}^{M} P_j Z_{i,j} - P_i} \tag{23.9}$$

where P_i is the transmission power at cell i (which mobile i is associated to), and $Z_{i,j} = h_{i,j}/h_{i,i}$ is the normalized link gain matrix. In this formulation, we ask for the condition of feasibility when a set of SINR goals are given.

When a common SINR goal γ is applied to all mobile stations, the above SINR inequality translates to the following matrix notation:

$$ZP \geq \frac{1+\gamma}{\gamma}P$$

Zendar shows an elegant result, which says that a positive power vector **P** exists and satisfies all inequalities in Eq. (23.9) if the matrix **Z** has a nonnegative eigenvalue λ^*, and the eigenvector $\mathbf{P}^*$ corresponding to λ^* is a feasible power solution. Thus, finding the feasible power vector for such a model is reduced to finding the maximum nonnegative eigenvalue and the corresponding eigenvector.

23.3.2.2 Intracell Power Sharing with Multicell Interference

In Ref. [8], the problem is slightly different: the set of K BSs have the maximum transmitting powers S_i, $i = 1, \ldots, K$, and each base station serves N_i mobile stations. Each mobile has its data rate requirement $R_{i,j}$ and SINR requirement (E_b/N_{0ij}), for the jth mobile station at cell i. The goal is to determine whether or not their is a feasible power allocation such that data rate, SINR, and power budget are all satisfied. Notice that one mobile station only receives a fraction of transmission power from one BS. Their is no signal-combining technique used here. For a given mobile station, transmission of all other $(K - 1)$ BSs is external noise.

The sufficient condition given Ref. [8] is

$$\sum_{j=1}^{N_i} \left(\frac{E_b}{N_0}\right)_{i,j} \frac{R}{\beta W} \left(1 + \sum_{k=2}^{K} \frac{h_{k,j}S_k}{h_{i,j}S_i} + \frac{\eta}{h_{i,j}S_i}\right) \leq 1 \quad \text{for } i = 1, \ldots, K \quad (23.10)$$

Based on the above condition, a centralized admission control procedure works as follows:

- Collect all path loss information, data rate, and SINR requirements from mobile stations.
- Verify the K conditions in Eq. (23.10), if all the conditions are satisfied, report the problem is feasible.

Thus, to add one mobile station, all K involved BSs must be checked. Similar to the previous case, a centralized algorithm will need to collect all instantaneous cell–mobile link loss parameters for decision making. Eq. (23.10) is somewhat conservative because it assumes that the total transmission power at each cell is fixed to its maximum power S_k. In reality, the forward transmission is also related to the number of mobiles in that cell and could vary from time to time due to mobile movement.

Following Zendar's modeling, Baccelli[60] in 2003 specifically expressed the intracell interference and intercell interference in a set of linear inequalities

$$\overline{\mathbf{P}} \geq \overline{\mathbf{A}}\overline{\mathbf{P}} + \overline{\mathbf{b}}$$

where the dimension of the power vector $\overline{\mathbf{P}}$ is the sum of all mobiles over all cells involved. Matrix $\overline{\mathbf{A}}$ is obtained after rearrange the SIR constraints, and can be treated as an effective interference matrix, where the rows of $\overline{\mathbf{A}}$ are normalized by the SIR requirement of individual MS. To efficiently obtain the sufficient and necessary feasible power solution, the problem is broken into two linear problems with significantly reduced dimensions:

1. A global power allocation problem of the form

$$\mathbf{P} \geq \mathbf{A}\mathbf{P} + \mathbf{b}$$

The dimension of the power vector $\mathbf{P}$ is the number of cells. The construction of the matrix $\mathbf{A}$ and $\mathbf{b}$ from the matrix $\overline{\mathbf{A}}$ and vector $\overline{\mathbf{b}}$ at the original problem is detailed in Ref. [60]. The main result from Ref. [60] shows that the global power allocation is feasible if the spectral radius (or maximum eigenvalue) of matrix $\mathbf{A}$ is greater than or equal to 1.

2. A local power allocation problem of the form

$$\overline{\mathbf{P}_j} \geq \mathbf{A}_j\overline{\mathbf{P}_j} + \overline{\mathbf{b}_j}$$

which deals with the SNIR in cell j.

The main result is that if the global power allocation can be satisfied and $\mathbf{P} = (P_1, P_2, \ldots, P_K)$ is a feasible power vector, then the feasible condition for local power allocation problem at cell i is

$$\alpha_i \sum_{j=1}^{N_i} H_{i,j} < 1$$

where $H_{i,j} = \gamma_{i,j}/(1 + \alpha_i \gamma_{i,j})$, α_i is the intracell orthogonal degree, and $\gamma_{i,j}$ represents the SINR requirement for the jth mobile in cell i.

23.3.2.3 Decentralized Downlink Power Control

The centralized admission check above could be unrealistic to implement, since all BSs nearby must check the admission condition. Owing to the mobility of mobile stations, the channel gain $b_{i,j}$ will change from time to time, which makes the centralized admission algorithm very expensive. Many distributed forward-link power control algorithms are discussed in the literature.[42,60,61]

The decentralized power control algorithm in Ref. [61] performs distributed power control at all BSs as follows:

■ Each BS chooses a starting power level arbitrarily.
■ The mobile station will measure the SINR level and report $\gamma_i(n)$ to the associated BS.
■ BS adjusts the transmission power by

$$P_i(n+1) = \beta P_i(n) \left(1 + \frac{1}{\gamma_i(n)}\right)$$

where $\beta > 0$.

It can be shown that the above algorithm will approach a uniform SINR level at all cells and remain stable. Hanly[42], extended this algorithm to handle nonuniform SINR requirements.

In Ref. [60], Baccelli proposes a decentralized downlink admission control protocol based on the two-step condition check. The admission control can be done locally at each BS station without considering the external cells.

23.3.3 Scheduling Issues for Downlinks

Downlink bandwidth allocation can also be designed by allocating orthogonal codes to mobile stations. Depending on the user's bandwidth requirement, multiple codes can be assigned to one user in an MC-CDMA downlink. One particular problem is that some traffic is inherently bursty. Thus the mobile will often over-book channel bandwidth to guarantee the busy-time requirement. Over-book, however, might cause bandwidth underutilization at the idle times. This is solved by Kam[62] by permitting many users to share a common orthogonal variable spreading factor (OVSF) code using a slot scheduler at the base station.

23.3.3.1 TDM-based Scheduling

The CDMA2000 1xEVDO[20], also known as IS-856 or CDMA-HDR, is one of the recent efforts to provide high-speed data service in the 3G system through link-adaptive transmission. IS-856 uses TDM for its forward link. Full power and all forward data channels (codes) are assigned to one active user at a time. This is in contrast to the code-division-multiplexed IS-95 forward link, where data for multiple active users are synchronized in transmission. In IS-856, with full forward transmitting power allocated to one user at a time, no forward link power control is used. Instead, IS-856 uses rate-adaptation based on the mobile side SINR measurement, which is feedback from the mobile station via a reverse link. The link-adaptation significantly improves the forward link.

23.3.3.2 Wireless Fair Queuing

Scheduling fairness is usually obtained via GPS[63] and it variation. These wireless-fair-queuing-based schemes[54,58,63,64] guarantee a delay upper bound to a flow by allocating to it a deterministic share of the transmission capacity. The wireless fair scheduling takes as input the weight (or bandwidth share) for connected mobile stations. The goal is to schedule downlink transmission such that the actual data rate received at mobiles are proportional to their weight.

Stamoulis[64] showed that the single-parameter controlled scheduling in GPS is not sufficient to support multirate multimedia services with widely diverse delay and bandwidth specifications. For example, video and audio have delay requirements of the same order, but video has an order of magnitude greater bandwidth requirement than audio. Therefore, delay-bandwidth coupling could lead to bandwidth underutilization. Ref. [64] uses a fair queuing method for bandwidth allocation and multicode CDMA to minimize queuing delays in wireless networks. In their method, the queuing delay is minimized by finding the optimum time-varying weight, which is translated to the number of CDMA codes to be used. This is solved by a dynamic programming algorithm. Once the weight vector is obtained, a FQ scheduler is used for actual scheduling.

One difficulty of wireless fair queuing algorithm is related to the transmission compensation when the channel is erroneous. Consider two consecutive time windows t_1 and t_2, both equally shared by M flows (mobile connections). Assume that one mobile lost its connection to the BS during t_1 due to channel error, the wire-line GPS scheduler will split bandwidth equally among $M - 1$ good mobiles during t_1, and resume to the M-way split for time window t_2. That is, no compensation is made across different time windows. If a mobile station constantly experiences bad channel quality, its actual data rate will be much lower than in the ideal case with an error-free channel. Notice that such is not the case in the wire-line network, where backlogged packets are always proportionally scheduled according to their weights.

The wireless fair queuing algorithms[54] thus differentiate between a nonback-logged flow and a backlogged flow that perceives channel error. With the wireless fair queuing scheduler, a mobile station backlogged due to transmission error is compensated in the future when the channel becomes usable.

However, compensation should be controlled within a reasonable time period to prevent prolonged delay for the *normal* flows, especially when delay-sensitive traffic is presented. Wireless fair queuing thus consists of several components:

■ Error-free service model, which is a referencing service model when channels are always good.
■ Lead and lag models, determining which mobiles are ahead of the error-free service and which are lagged behind due to channel fluctuations. The amount of lead or lagged data is usually the difference between the actual received service and that from the error-free service model.
■ Compensation model, determining which mobiles should be compensated in the upcoming time frame.
■ Additional policy to support delay-sensitive mobiles.

In terms of complexity, the GPS is on the order of $O(\log(V))$, where V is the number of queued packets. Wireless fair queuing require $O(\log(U))$ in terms of the number of mobile stations in system.

23.3.3.3 Packet Dropping Control

Kong et al.[58] discussed packet scheduling in an MC-CDMA network with a guaranteed delay bound for time-sensitive traffic. They assume a time-division MAC framework, where a jumbo MAC frame is used for access control, uplink traffic, and downlink traffic. The transmission at uplink follows the request (mobile)-scheduling (BS) transmit stage for each MAC frame period. For time-sensitive traffic, delay guarantee must be strictly enforced, i.e., packets past their delivery deadline are of little value and might be disposed. In the case of a wireless CDMA network, the capacity of a cell might vary significantly due to the link condition and external interference. Thus a hard delay guarantee often results in a conservative admission policy and low radio utilization. An alternate point of view is to admit connections over optimistically, and achieve delay guarantee by dropping packets. In other words, the cost of delay guarantee is shifted over to the packet-dropping ratio. This raises the issue of control packet dropping such that the average dropping ratios are proportionally distributed according to traffic weights, and with reasonable fairness.

In Ref. [58], Kong et al. showed that a simple scheduler can be designed to provide fairness in dropping ratio, in a similar manner as in the wireless fair queuing. In their proposed scheme, the transmission requests are queued FIFO for individual mobile station. The scheduler at the BS adopt a water-filling approach to decide which head-of-line request should be scheduled by comparing the expired-to-eligible fragment ratios (EERs) among mobile stations. The EER for mobile i is define as

$$\frac{\text{(No. of eligible packets)} - \text{(No. of expired packets)}}{\phi_i * \text{(No. of eligible packets)}}$$

where ϕ_i is the weight for mobile i. When the time slot in a MAC frame is filled out, the unscheduled packets are discarded if the deadline is about to be passed. The number of dropping/expired packets is updated every MAC frame. As with the GPS and wireless queuing, the complexity of this scheduler is low enough $(\log(M))$ to support frame-by-frame operation. Simulation results in Ref. [58] confirm that the dropping ratio is distributed fairly according to the mobile weights.

23.3.3.4 Multicode Rate with Fairness

Many recent studies[3,30] show that, a throughput-maximizing transmission in uplink should allocate the maximum allowable rate to only one user with the highest channel gain. A similar transmission scheme is believed to be optimum for the downlink channel. Recent developments of the 3G high-speed downlink tend to use a dedicated frequency band for high-rate data channel, such as the CDMA/HDR system in CDMA2000.[20]

Just recently in 2004, Kim[57] investigated the fairness problem in a slot-based MC-CDMA downlink protocol. It is argued that the dedicated CDMA/HDR downlink, with its link-adaptation in transmission, can cause significant unfairness in the sense that only a very small number of users receive near the maximum allowable rate. The fairness is measured by the packet latency, defined as waiting time before a fixed-length packet is received at the mobile station. Similar to other fairness protocols, the control algorithm in Ref. [57] adjusts the allocation of slot time based on the estimated data rate between BS and individual mobiles. The estimation of data rates is done at the mobile station, where the BS-MS SINR is monitored, as well as external interference. The MS will then send a data-rate request to the BS and wait for service.

The decision in the BS is to decide the number of time slots for the next scheduling period (a scheduling period is believed to be short enough compared to the slow channel fading): the number of slots is inversely proportional to the estimated data rate for each mobile. Since the channel condition is assumed to be stable during each scheduling period, the order of transmission is not specified. Nevertheless, a deadline monotonic scheduler within the scheduling period might be useful to minimize the packet delay.

It is also worth pointing out that Ref. [57] uses quasi-orthogonal codes to achieve over-saturated signal formation. Such a coding scheme is shown to outperform the conventional multicode CDMA system in a multicell fading environment in term of throughput. However, advanced mobile-side signal-processing techniques must be used, such as iterative decoding, maximum ratio signal combining, and self-interference cancellation.

23.4 Mobility Management

Compared with wire-line networks, one of the unique functions that wireless networks need to support is mobility management. Since the MSs can potentially move all the time, the system needs to support multimedia application in a continuous manner without any interruption. Therefore, handoff schemes are particularly important, and the supporting schemes for location update and tracking serve as the enabling techniques for reaching the MSs.

23.4.1 Location Update and Tracking

The schemes to support location update and tracking can be viewed as a signaling protocol instead of data protocol. Thus, we will only touch on the latest developments for signaling protocols and focus on the data protocol developments for the rest of this section. Two basic operations are involved in mobility management: location update and paging. Location update is the process through which a system tracks the location of mobile users that are not in conversation. The up-to-date location information of a mobile user is reported by the mobile user dynamically. A location area may include one or more cells. When an incoming call arrives, the system searches for the mobile user by sending polling signals to cells in the location area. This searching process is referred to as paging. The paging cost is defined as the number of cells to be searched to locate a mobile unit.

It is widely accepted that a large location area will result in a decrease in the number of location updates, but will increase the paging cost.[65] The trade-off between location update and paging cost is controlled by the determination of location area, location update scheme, and paging search strategy. Most commercial systems adopt a static location update scheme, where a mobile user will perform a location update when it switches to a new cell. This might cause a significant amount of updating traffic when the mobile is at the boundary of multiple cells, similar to the ping-pong phenomena in hard handoff. To address this problem, Ref. [65] discussed three dynamic location update schemes:

- distance-based method, where mobile location is updated when the distance between the current cell and the previously updated cell exceeds a threshold d,

■ movement-based method, where update will be triggered when a mobile crosses the cell's boundary *m* times, and

■ time-based methods, where location is updated every *t* units of time.

To reduce the costs associated with location update and paging, the ability to predict the mobile location has been investigated. Ref. [66] estimated the location probability by assuming that the MSs take the shortest path's movement pattern on four possible directions. In Ref. [67], a two-level movement model is used to predict the next cell; both mobile speed and direction are considered.

The recent work in Ref. [68] further discussed how to decide the optimum *m* threshold for dynamic movement-based location update. They also argued that different mobile stations have diverse mobility patterns and call arrival parameters. Thus the value of the movement threshold should be determined on a per-user basis. Under the assumption of equal future cell probability, the author proves that the cost function is convex and an unique movement threshold exists with optimal cost. Their analysis shows that the optimal threshold decreases as the call-to-mobility increases. Their result suggests that a high mobility should use a small movement threshold. Also, the variance of mobile residence time seems to have little effect on the optimal threshold.

23.4.2 Soft Handoff in CDMA Networks

Handoff[7,38,69–74] is a phenomena related to the mobile movement and the cell structure of cellular networks. When a mobile is about to move out of the range of current cell (BS), it must be able to connect to another cell (BS) for a continuous connection. Associated with this process is additional signaling overhead, re-assignment of resources, the power control problem, and other issues.

Traditional handoff algorithms are based on the voice-only network, carrying homogeneous traffic. The major performance focus of these algorithms is to guarantee the connectivity of live calls and to keep the failure rate as small as possible. Therefore some important metrics and criteria used include received signal strength (RSS), SIR distance, and traffic load.

An important factor that makes the handoff problem nontrivial is signal strength observed at the mobile and base stations is subject to shadowing and fading effects. Therefore, a scheme completely relying on one fixed SINR observation will inevitably result in (1) a relatively high SINR value and frequent handoffs, referred to as ping-pong effects in literature, or (2) a relative low SINR threshold that could cause significant degradation of call quality.

The CDMA system supports a unique handoff scheme—soft handoff.[1,73] Soft handoff allows more than one base station communicating with the mobile station during the hand-over process, thus there is no break period when the mobile moves to another cell. With soft handoff, mobile stations maintain an active set of base stations. All information to the mobile station will be transmitted (redundantly) by all BSs in its active BS list. In the uplink, the transmission of the mobile station will be joint-decoded by all base stations in the active cell. The joint-decoding is usually archived by selecting the cell with maximum SINR (easy to implement and widely used in commercial systems), or the maximum ratio combining (MRC) technique. The latter one is known to produce the best SINR that is possible from the multiple signal branches, but requires complete knowledge of time-varying path loss at the BSs and is much more expensive to implement. It had been shown that soft handoff can reduce the transmission power of the mobile, and increase the system capacity.[38,73]

Note that even having the parameters optimized can be a design factor. Handoff parameter optimization is very important for soft handoff control to capture the trade-offs involved in performance indicators such as average uplink and downlink interference, expected number of BSs in the active set, expected number of changes in the active set, trunking resource efficiency, carried traffic, and call quality. Owing to the complexity of the problem, much of the prior work on soft handoff design has been heuristic.

In Ref. [71], Asawa and Stark model the MS-BS signal as an log-normal stochastic process and seek the optimal switching policy to maximize the expectation of a reward-minus-cost function. The reward term monotonically increases as the signal quality, while the cost term is proportional to the number of handoffs (see Eq. (1) in Ref. [71]). By selecting different rewarding functions and cost values, different optimization goals can be achieved. By analysis of the instantaneous switching gain and cost, the necessary and sufficient condition of the optimum handoff policy is derived.

Channel/bandwidth reservation strategies are another important part in handoff algorithms.[75] In Ref. [76], the BS considers incoming and outgoing handoffs from/to adjacent cells. The BS then calculates the total required bandwidth in its cell for both handed-off and existing connections. This scheme assumes that mobile dwell time follows an exponential distribution and the average handoff rate is known. The estimation of handoff rate and the new call request is based on a fixed stochastic model and the number of mobiles in adjacent cells.

Levin[77] discusses the *shadow cluster* concept, which is a number of cells close to a mobile station, to estimate future resource requirements. Their idea is that if the mobility of each mobile is predictable and accurate, then it is possible to determine the number of mobiles in each cell precisely and thus the number of channels needed in the future. This will allow the optimum channel reservation with minimum handoff drop probability and maximizing channel utilization. Such a scheme inevitably requires excessive message exchanging among neighboring cells and requires global optimization, and thus might be too expensive for a large cluster size.

Ref. [69] further provides a discussion of trade-offs in soft handoff, in handling mobility and shadow fading. A nice feature of this algorithm is that it incorporates the effects of mobility and shadow fading in the handoff decision. The authors also investigated the effects of different combining techniques, including select combining (SC), equal-gain combining (EGC), and optimum combining (OC).

23.4.3 Leading Handoff Methods

Nevertheless, the leading standard handoff schemes of CDMA networks are still based on SINR alone. The decision of handoff could be made by the mobile stations, or initiated from the BS. A widely accepted model is called mobile-assisted approach, which is used in 3G systems. In such a model, mobile stations constantly monitor the SINR of the pilot signal of the nearby BSs. If the pilot signal of a BS continuously increases and above a certain threshold, it will be added into an active set. The mobile station will then acquire the channel information and try to establish a connection with the new BS. If successful, a redundant communication link is established. However, if the pilot signal of a BS decreases below a certain threshold, the mobile might disconnect from that BS.

The pilot signal strength follows a power-degraded log-normal distribution

$$S = Gd^n 10^\zeta$$

where G is a constant representing the fixed pilot transmitting power and overall antenna gain and d the distance between the mobile station and the BS of interest. The power attenuation factor n ranges from 2 to 6, but is typically fixed in a given environment. ζ represents the random fading and shadowing factor, usually assumed Gaussian.

The three control parameters used in the IS-95[24,78] handoff algorithm are (1) the signal strength threshold T_{add} to add a BS, (2) the signal strength threshold T_{drop} to drop a BS, and (3) an additional timer threshold to drop a BS. Introducing T_{drop} has been shown to be very effective in reducing the number of unnecessary handoffs.

Along this line of work, handoff schemes for conventional (i.e., voice-only) traffic type has been investigated extensively. For instance, resource allocation in general has been discussed in many works[76,79,80] starting in the late 1990s. In Ref. [76], a distributed call admission scheme with the consideration of neighboring cells are reported. The movement of mobiles as well as the system load in nearby cells is used to make the admission decision. More recently, Ref. [81] discussed channel reservation schemes when balancing the trade-off of call blocking and dropping probability. Such a scheme must estimate the handoff rate, based on the mobile movement pattern and load in neighboring cells. However, these schemes were not specifically designed for supporting the handoff of multimedia traffic. Recent development in this direction tries to address resource reservation for CBR and VBR traffic,[82] based on user mobility information and a tighter jitter and delay bound.

23.4.4 Power Control with Spreading Factors for Multimedia Handoff

To support the handoff for multimedia traffic, power control methods emerged as an effective means to balance the overall consideration from various aspects. Lee[80] in 2000 presented a power-assignment problem for mobiles supporting different QoS requirements. They formulated the admission problem by a set of inequalities of desired SIR for all of the active mobiles and provided a method to drive a feasible solution. However, in this study, the processing gain is fixed to a predefined value for each traffic type. Thus the admissible region will be strictly limited.

In 2001, Zhou et al.[79] studied a forward link power control scheme in a two-cell CDMA network supporting different QoS requirements. The scheme is optimized in the sense of minimizing base station transmitting power, and maximizing system throughput. However, the study is based on the assumption of a fixed spreading factor, and did not address the resource assignment for the reverse link. Power control in handoff is also discussed just recently in Ref. [70] in 2003.

To support multimedia traffic, handoff mobiles should be power-controlled in conjunction with the ability of changing the spreading factors to maintain an acceptable BER quality. Therefore, the performance goals of the handoff algorithm with dynamic spreading factor are summarized as follows: (1) all handoff requests should be processed within a given time period, which is important when the system in under heavy load and (2) the handoff algorithm should seek the highest throughput while maintaining the BER performance of the handoff mobile during the handoff period.

23.4.4.1 Performance of OR-ON-DOWN Power Control

We have found that the conventional OR-ON-DOWN power-control scheme did not perform well in supporting multimedia traffic during the hand-over period. With the dynamic spreading factor enabled system, the mobile might not be able to communicate with all the cells in the active set well, particularly with the highly loaded cell. For instance, assume a mobile handoffs from a *cell*0 with load of 10 (desired SF = 32) toward *cell*1 with load of 20 (desired SF = 64). The SF of the mobile is initially 32. Now if *cell*1 issues a DOWN command to the mobile (since the mobile is moving toward *cell*1 and finds that the signal is becoming too strong), the mobile will follow the command as required by the OR-ON-DOWN rule. This will in turn reduce the received signal strength in *cell*0 and thus may increase the link BER. For the reverse link between the handovering mobile and *cell*1, the BER is also high since the assigned spreading factor of 64 is only good for *cell*0, but is not long enough to guarantee the desired BER level in the heavily loaded *cell*1. Therefore the overall link quality will be below the expected since neither link can provide good performance.

Using the above example, when the mobile moves from *cell*1 to *cell*0, we observed that the BER requirement of the handoff mobile is also violated (see Figure 3 in Ref. [7]). At a range of 800 m, when the mobile enters the handoff area, there is a sudden increase of BER (the spreading factor is adjusted to 32). Meanwhile the OR-ON-DOWN power control will limit the signal power such that *cell*1, which is closer to the mobile, receive the signal necessary strong. Though the received signal level is maintained in *cell*1, the change of SF from 64 to 32 significantly increases the BER in the *cell*1. *cell*0, which is receiving a weaker than expected signal due to the DOWN power command issued to the mobile from *cell*1, also cannot provide the expected link quality. Overall, the BER in this period exceeds the allowed value.

23.4.4.2 BER Model in Handoff Area

To provide a joint-power-spreading-factor control for the handoff mobiles, it is necessary to precisely model the BER of handoff mobiles. Let $SIR_{j,i}$ be the SIR of the jth handoff mobile at the ith cell, we have

$$SIR_{j,i} = \frac{E_j}{\eta_i} = \frac{p_j SF_j}{m_i + I_b + fI_o} = \frac{(P_j r_{j,i}^{-\mu} 10^{\frac{\eta}{10}}) SF_j}{m_i + fI_o + I_b} \quad (23.11)$$

$$I_o = \frac{1}{|A|} \sum_{u \in A} m_u \quad (23.12)$$

$$I_{b,j} = \sum_{k \neq j} (P_k r_{k,i}^{-\mu} 10^{\frac{\eta_k}{10}}) \quad (23.13)$$

where p_j is the received signal power at cell i for mobile j, and SF_j the spreading factor of the jth mobile. m_u the system load of cell i. f represents the factor for other cells,[73] and A is the set of active cells. In estimating the interference from other cells (I_o), we use the average number of mobile users in the surrounding cells as the mobile density (number of mobiles/cell). However, the interference caused by the handoff mobiles I_b needs to be considered as a separate source from the general other-cell-interference term, since they might transmit stronger signals than expected (as shown in Section 23.4).

The received signal strength is modeled by a path loss model: $p_{j,i} = P_j r_{j,i}^{-\mu} 10^{\frac{\eta}{10}}$, where P_j is the mobile's transmitting power, $r_{j,i}$ the distance between the mobile and base stations, $\mu\ (=3.5)$ the path loss order, and η the zero-mean random power fluctuation. We also assume that the nonhandoff mobile is under perfect power control of its resident cell, with a normalized received power level of $1^{\dagger}$.

Using the diversity decoding for uplink channel, the best link among the active set is chosen for the jth handoff mobile. If A_j is the active set of the jth mobile, we have

$$SIR_j = \min_{i \in A_j} SIR_{j,i} \qquad (23.14)$$

The SIR for the in-cell mobiles in each active set could be similarly obtained. Note that the interference from all of the handoff mobiles is now regarded as noise, and the regular in-cell mobile users have unit received signal strength. For traffic type t in the ith active cell with spreading factor SF_t, the SIR is approximated by

$$SIR(i, t) = \frac{SF_t}{(m_i + \sum_j (P_j r_{j,i}^{-\mu} 10^{\frac{\eta}{10}}) + fI_o)} \qquad (23.15)$$

Let H denote the set of handoff mobiles at any time. The optimal SF/power assignments that maximum throughput for the handoff mobile can be obtained by solving the following problem:

$$\max z = \sum_{j \in H} \frac{W}{SF_j} \qquad (23.16)$$

$$SIR_j \geq b_j, j \in H \qquad (23.17)$$

$$SIR_{i,t} \geq b_t, i \in A, \text{ and } t \in \{\text{traffic type set}\} \qquad (23.18)$$

23.4.4.3 Suboptimal SF/POWER Admission Algorithms

Without nonnegative power constraints, the optimal solution for Eq. (23.18) can be obtained by Lagrange's method after relaxing the SINR constraints into linear form (see Ref. [7] for details). However, it might not result in a feasible solution for the original problem due to the possible zero-value

---

† In practice, the received powers of different traffic are not uniformly assigned, thus they cause a different level of interference.

power assignment. Furthermore, the allowed spreading factor in reality can only be taken from a limited set of positive integers. This necessitates a postprocessing to convert the optimal vector into a feasible solution. For example, if the spreading factor has a value of 35.5, we should replace it with the closest greater integer value that is in the feasible solution space.

The SINR constraints in Eq. (23.16)–Eq. (23.18) is linearized first (the actual reduced linear format is given in Ref. [7]). The heuristic algorithm then goes through a series of steps: (1) we find the optimal power vector for problem in Eq. (23.18), using a standard simplex method, (2) this is followed by a check to see all P_j in the solution vector is greater than their desired threshold.

If true, the continuous SF_j solution are

$$SF_j = \frac{b_j\left(m_j + \frac{f}{|N_j|}\sum_{u \in N_j} m_u + \sum_{k \neq j}\left(P_k r_{k,j}^{-\mu} 10^{\frac{n_k}{10}}\right)\right)}{\left(P_j r_{j,j}^{-\mu} 10^{\frac{n_j}{10}}\right)} \quad (23.19)$$

After round up to the closest integer, the SF_j values are substituted in Eq. (23.19) to obtain the target P_j values.

If not, a transformation should be done which will further simplify the linear problem in Eq. (23.18). We then fix the problematic decision variables to the minimal power level and replace them back into Eq. (23.18). This will generate a new linear programming problem with a reduced number of decision variables. The revised problem will be processed again using the simplex method, and the same check process will be applied to the new optimal vector. It can be seen that the algorithm will eventually terminate, and the maximum loop time will be less than the number of handoff mobiles.

23.5   Design Integration with Ad Hoc Networks

The main disadvantage of WCDMA networks is perhaps on the limited bandwidth available to the users. The emerging multimedia applications (e.g., streaming for high-quality HDTV video) demands high throughput. Therefore, further integration with WLAN and ad hoc networks provides the unique opportunity to connect every user/device with high bandwidth. The primary advantage is the 11-/45-Mbps bandwidth offered by 802.11b/a/g WLAN (infrastructure mode) and perhaps even higher bandwidth for the ad hoc and personal area networks (PAN) networks.

How to integrate these three networks together into a highly efficient and seamless network requires systematic investigation in the future. One typical approach is to have a hierarchical design with the combined WCDMA/WLAN/ad hoc serves as the top–down structure. The integration of WCDMA/WLAN requires the intelligent selection of gateway points in either the WLAN portion or the 3G network to connect the users to the WCDMA core network anywhere.

There are a few schemes[83–86] proposed in recent years for such an effort. However, the majority of these schemes assumed that the bandwidth for the 3G core networks will be increased significantly in the near future. We believe it will take a longer time for WAN/MAN such as WCDMA to deliver high-bandwidth throughput. On the other hand, WLAN and ad hoc networks will have much faster development on delivering high-throughput products.

Thus, our approach[87] uses WLAN to cluster mobile users and reduce the 3G radio activity. The philosophy is that when the 3G radio link is less crowded, it most likely provides higher efficiency for the users. Based on the relative BS/AP positions of the WCDMA and WLAN networks, we analyzed six cases of configurations. These six cases cover the majority of scenarios when WCDMA's BS interacts with WLAN's APs. We have formulated the problem, and produced the suboptimal solutions to reduce the overall interference between these devices.

When many users connect to a single AP of WLAN, the load imbalance becomes apparent. Ad hoc networks can be jointly integrated between WLAN/ad hoc as the relay points to achieve better load balance. Since ad hoc networks mostly work within a limited distance (e.g., within tens of meters), it is natural to have PANs connected with the combined WCDMA/WLAN networks to extend the global connectivity.

One approach to integrate ad hoc networks into the combined WCDMA/WLAN networks is to follow the top–down structure, which only allows the ad hoc networks connect to WLAN only (instead of providing connectivity to the WCDMA core networks, though it is possible). However, even with this simplified structure division, the overall design task still remains to be a challenge.

The key factor is, with the ad hoc networks, relay can be mobile. Though the AP's location is fixed, it is open to decide which mobile station should serve as the relay point to connect to the AP on behalf of other mobile stations of the same ad hoc networks. Therefore, a higher complexity of overall system design should be addressed. These issues include what media access methods should the system provide to support different traffic types and what relay structure should be determined with the goal to maximize the overall throughput.

References

1. I. Akyildiz, D.A. Levine and I. Joe, "A slotted CDMA protocol with BER scheduling for wireless multimedia networks," *IEEE/ACM Transactions on Networking*, vol. 7, pp. 146–158, 1999.

2. E. Esteves, "On the reverse link capacity of CDMA2000 high rate packet data systems," *Proc. IEEE ICC*, vol. 3, pp. 1823–1828, 2002.
3. K. Kumaran and L. Qian, "Uplink scheduling in CDMA packet data systems," *Proc. IEEE INFOCOM*, July 2003.
4. B. Li, L. Li, B. Li, K. Sivalingam, and X. Cao, "Call admission control for voice/data integrated cellular networks: performance analysis and comparative study," *IEEE Journal on Selected Areas in Communications*, vol. 22, no. 4, pp. 706–718, 2004.
5. C.W. Sung and W.S. Wong, "Power control and rate management for wireless multimedia CDMA systems," *IEEE Transactions on Communication*, vol. 49, no. 7, pp. 1215–1226, 2001.
6. J. Wang and J. Liu "Multimedia support for wireless W-CDMA with dynamic spreading," *ACM Journal of Wireless Network*, vol. 8, 355–370, 2002.
7. J. Wang and J. Liu, "Handoff algorithms in dynamic spreading WCDMA system supporting multimedia traffic," *IEEE Journal on Selected Areas in Communications*, vol. 21, no. 10, pp. 1652–1662, 2003.
8. K. Gilhousen, I. Jacobs, R. Padovani, A. Viterbi, L. Weaver, and C. Wheatley, "On the capacity of a cellular CDMA system," *IEEE Transactions on Vehicular Technology*, vol. 40, no. 2, pp. 303–312, 1991.
9. D. Ayyagari and A. Ephremides, "Optimal admission control in cellular DS-CDMA systems with multimedia traffic," *IEEE Transactions on Wireless Communications*, vol. 2, no. 1, pp.195–202, 2003.
10. B. Bing and R. Subramanian, "Enhanced reserved polling multi-access technique for multimedia personal communication systems," *ACM Journal of Wireless Networks*, vol. 5, no. 3, pp. 221–230, 1999.
11. C. Chang and K. Chen, "Medium access protocol design for delay-guaranteed multi-code CDMA multimedia networks," *IEEE Transactions on Wireless Communications*, vol. 2, no. 6, pp. 1159–1167, 2003.
12. C.L.I, C.A. Webb III, H.C. Huang, S. Brink, S. Nanda, and R.D. Gitlin, "IS-95 enhancements for multimedia services," *Bell Labs Technical Journal*, pp. 60–80, autumn 1996.
13. T. Kwon, Y. Choi, C. Bisdikian, and M. Naghshineh, "QoS provisioning in wireless/mobile multimedia networks using an adaptive framework", *ACM Journal of Wireless Networks*, vol. 9, no. 1, pp. 51–59, 2003.
14. A. Acampora and S. Krishnamurthy, "A new adaptive MAC layer protocol for broadband packet wireless networks in harsh fading and interference environments," *IEEE/ACM Transactions on Networking*, vol. 8, no. 3, pp. 328–336, 2000.
15. W. Jeon and D. Jeong, "Call admission control for CDMA mobile communications systems supporting multimedia services," *IEEE Transactions on Wireless Communications*, vol. 1, no. 4, pp. 649–659, 2002.
16. N.B. Mandayam and J.M. Holtzman, "Analysis of a simple protocol for short message data service in an integrated voice/data CDMA system," *Proc. IEEE MILCOM'95*, San Diego, CA, November 1995.
17. A. Sampath and J. Holtzman, "Access control of data in integrated voice/data CDMA systems: Benefits and tradeoffs," *IEEE Journal on Selected Areas in Communications*, vol. 15, no. 8, 1997.

18. M. Soroushnejad and E. Geraniotis, "Multi-access strategies for an integrated voice/data CDMA packet radio network," *IEEE Transactions on Communication*, vol. 43, pp. 934–945, 1995.

19. X. Wang, "Wide-band TD-CDMA MAC with minimum-power allocation and rate- and BER-scheduling for wireless multimedia networks," *IEEE/ACM Transactions on Networking*, vol. 12, no. 1, pp. 103–116, 2004.

20. Q. Wu and E. Esteves, "The CDMA2000 high rate packet data systems," Qualcomm Technical Report, 80-H0593-1.

21. H. Holma and A. Toskala, *"WCDMA for UMTS: Radio Access for Third Generation Mobile Communications,"* John Wiley, New York, 2001.

22. Third generation partnership project-2 website: http://www.3gpp2.org.

23. J. Holtzmann, S. Nanda, and D. Goodman, "CDMA power control for wireless networks," *Third Generation Wireless Information Networks*, Kluwer Academic Publishers, Norwell, MA, USA, pp. 299–311, 1992.

24. EIA/TIA/IS-95 Interim Standard, *Mobile Station-Base Station Compatibility Standard for Dual-Mode Wide-Band Spread Spectrum Cellular System*, Telecommunication Industry Association, July 1993.

25. T. Minn and K. Siu, "Dynamic assignment of orthogonal variable-spreading-factor codes in W-CDMA," *IEEE Journal on Selected Areas in Communications*, vol. 18, no. 8, pp. 1429–1440, 2000.

26. C.L. I. and R.D. Gitlin, "Multi-code CDMA wireless personal communication networks," in *Proc. IEEE ICC'95*, Seattle, WA, pp. 1060–1064, June 1995.

27. T. Nagaosa and T. Hasegawa, "Code assignment and the multi-code sense scheme in an inter-vehicle CDMA communication network," *IEICE Transactions Fundamentals*, vol. E81-A, no. 11, pp. 2327–2333, 1998.

28. D. Kim, E. Hossain, and V. Bhargava, "Downlink joint rate and power allocation in cellular multirate WCDMA systems," *IEEE Transactions on Wireless Communications*, vol. 2, no. 1, pp. 69–80, 2003.

29. S. Oh and K.M. Wasserman, "Adaptive resource allocation in power constrained CDMA mobile networks," *IEEE Proc. WCNC'99*, pp. 510–514, 1999.

30. S. Ulukus and L.J. Greenstein, "Throughput maximization in CDMA uplinks using adaptive spreading and power control," *IEEE Proc. ISSSTA'00*, pp. 565–569, September 2000.

31. J. Peha and A. Sutivong, "Admission control algorithms for cellular systems," *ACM Journal of Wireless Networks*, vol. 7, no. 2, pp. 117–125, 2001.

32. S. Ramakrishna and J. Holtzman, "A scheme for throughput maximization in a dual-class CDMA system," *IEEE Journal on Selected Areas in Communications*, vol. 16, no. 6, 1998.

33. S. Shen, C. Chang, C. Huang, and Q. Bi, "Intelligent call admission control for wideband CDMA cellular systems," *IEEE Transactions on Wireless Communications*, vol. 3, no. 5, pp. 1810–1821, 2004.

34. L. Xu, X. Shen, and J. Mark, "Dynamic fair scheduling with QoS constraints in multimedia wideband CDMA cellular networks," *IEEE Transactions on Wireless Communications*, vol. 3, no. 1, pp. 60–73, 2004.

35. S. Choi and D. Cho, "Capacity evaluation of forward link in a CDMA system supporting high data-rate service," *Proc. IEEE GLOBECOM*, pp. 123–127, 2000.

36. P.K. Frenger, P. Orten, and T. Ottosson, "Code-spread CDMA with interference cancellation," *IEEE Journal on Selected Areas in Communications*, vol. 17, no. 12, pp. 2090–2095, 1999.

37. H. Fukumasa, R. Kohno, and H. Imai, "Design of pseudo-noise sequences with good odd and even correlation properties for DS/CDMA," *IEEE Journal on Selected Areas in Communications*, vol. 12, no. 5, pp. 828–836, 1994.

38. C.C. Lee and R. Steele, "Effect of soft and softer handoffs on CDMA system capacity," *IEEE Transactions on Vehicular Technology*, vol. 47, no. 3, pp. 830–841, 1998.

39. M. Lops, G. Ricci, and A.M. Tulino, "Narrow-band-interference suppression in multiuser CDMA systems," *IEEE Transactions in Communications*, vol. 46, no. 9, pp. 1163–1175, 1998.

40. S. Oh and K.M. Wasserman, "Dynamic spreading gain control in multiservice CDMA networks," *IEEE Journal on Selected Areas in Communications*, vol. 17, no. 5, pp. 918–927, 1999.

41. S. Hanly, *Information Capacity of Radio Network*, Ph.D. thesis, University of Cambridge, 1993.

42. S. Hanly, "Capacity and power control in spread spectrum macro diversity radio networks," *IEEE Transactions on Communications*, vol. 44, pp. 247–256, 1996.

43. M. Win and J. Winters, "Exact error probability expressions for MRC in correlated nakagami channels with unequal fading wireless parameters and branch powers," *Proc. IEEE GLOBECOM'99*, pp. 2331–2335, 1999.

44. H. Liu and K. Li, "A decorrelating RAKE receiver for CDMA communications over frequency-selective fading channels," *IEEE Transactions on Communications*, vol. 47, no. 7, pp. 1036–1045, 1999.

45. G.J. Foschini and M.J. Gans, "On limits of wireless communications in a fading environment when using multiple antennas," *Wireless Personal Communications*, vol. 6, pp. 311–335, 1998.

46. D. Kim, E. Hossain, and V. Bhargava, "Dynamic rate adaptation and integrated rate and error control in cellular WCDMA networks," *IEEE Transactions on Wireless Communications*, vol. 3, no. 1, pp. 35–49, 2004.

47. C. Ho and C. Lea, "Improving call admission policies in wireless networks," *ACM Journal of Wireless Networks*, vol. 5, no. 4, pp. 257–265, 1999.

48. A. Sampath, P.S. Kumar, and J.M. Holtzman, "Power control and resource management for a multimedia wireless CDMA system," *Proc. PIMRC95*, Toronto, Canada, September 1995.

49. J. Zander, "Performance of optimum transmitter power control in cellular radio systems," *IEEE Transactions on Vehicular Technology*, vol. 41, pp. 57–62, 1992.

50. S. Choi and K.G. Shin, "Uplink CDMA systems with diverse QoS guarantees for heterogeneous traffic," *Proc. ACM MOBICOM*, pp. 120–130, 1997.

51. M. Naraghi-Pour and H. Liu, "Integrated voice-data transmission in CDMA packet PCN's," *Proc. IEEE ICC*, pp. 1085–1089, 2000.
52. L. Zhuge and V. Li, "Interference estimation for admission control in multiservice DS-CDMA cellular systems," *Proc. IEEE GLOBECOM*, pp. 1509–1514, 2000.
53. I. Kim and H. Kim, "An optimum power management scheme for wireless video service in CDMA systems," *IEEE/ACM Transactions on Wireless Communications*, vol. 2, no. 1, pp. 81–91, 2003.
54. V. Bharghavan, S. Lu, and T. Nandagopal, "Fair queuing in wireless networks," *IEEE Personal Communication*, vol. 6, pp. 44–53, 1999.
55. S. Borst, "User-level performance of channel-aware scheduling algorithms in wireless data networks," *Proc. IEEE INFOCOM*, July 2003.
56. R. Jantti and S. Kim, "Transmission rate scheduling for the non-real-time data in a cellular CDMA system," *IEEE Communications Letters*, vol. 5, no. 5, May 2001.
57. D. Kim, "Optimum packet data transmission in cellular multirate CDMA systems with rate-based slot allocation," *IEEE Transactions on Wireless Communications*, vol. 3, no. 1, pp. 165–175, 2004.
58. P. Kong, K. Chua, and B. Bensaou, "A novel scheduling scheme to share dropping ratio while guaranteeing a delay bound in a multicode-CDMA network, *IEEE/ACM Transactions on Networking*, vol. 11, no. 6, pp. 994–1006, 2003.
59. D. Kim, E. Hossain, and V. Bhargava, "Dynamic rate and power adaptation for provisioning class-based QoS in cellular multirate WCDMA systems," *IEEE Transactions on Wireless Communications*, vol. 3, no. 5, pp. 1590–1601, 2004.
60. F. Baccelli and B. Blaszczyszyn (INRIA-ENS), "Downlink admission/congestion control and maximal load in CDMA networks," *Proc. IEEE INFOCOM*, July 2003.
61. J. Zander, "Distributed co-channel interference control in cellular radio systems, *IEEE Transactions on Vehicular Technology*, vol. 41, pp. 305–311, 1992.
62. A. Kam, T. Minn, and K. Siu, "Supporting rate guarantee and fair access for bursty data traffic in W-CDMA," *IEEE Journal on Selected Areas in Communications*, vol. 19, no. 11, pp. 2121–2130, 2001.
63. A. Parekh and R. Gallager, "A generalized processor sharing approach to flow control — The single node case," *IEEE/ACM Transactions on Networking*, vol. 1, no. 3, pp. 344–357, 1993.
64. A. Stamoulis, N. Sidiropoulos, and G.B. Giannakis, "Time-varying fair queuing scheduling multicode CDMA based on dynamic programming," *IEEE GLOBECOM01*, pp. 3504–3508, 2001.
65. A. Bar-Noy, I. Kessler, and M. Sidi, "Mobile users: To update or not to update?" *ACM-Baltzer Journal of Wireless Networks*, vol. 1, no. 2, pp. 175–186, 1995.
66. A. Abutaleb and V.O.K. Li, "Paging strategy optimization in personal communication system," *ACM-Baltzer J. Wireless Networks (WINET)*, vol. 3, pp. 195–204, 1997.

67. T. Liu, P. Bahl, and I. Chlamtac, "Mobility modeling, location tracking, and trajectory prediction in wireless ATM networks," *IEEE Journal on Selected Areas in Communications*, vol. 16, pp. 922–936, 1998.

68. L. Jie, H. Kameda and K. Li, "Optimal dynamic mobility management for PCS networks," *IEEE/ACM Transactions on Networking*, vol. 8, no. 3, pp. 319–327, 2000.

69. M. Akar and U. Mitra, "Soft handoff algorithms for CDMA cellular networks," *IEEE Transactions on Wireless Communications*, vol. 2, no. 6, pp. 1259–1274, 2003.

70. M. Akar and U. Mitra, "Joint power and handoff control using a hybrid systems framework," *Proc. IEEE INFOCOM*, July 2003.

71. M. Asawa and W.E. Stark, "Optimal scheduling of handoffs in cellular networks," *IEEE/ACM Transactions on Networking*, vol. 4, pp. 428–441, 1996.

72. B. Hamdaoui and P. Ramanathan, "A network-layer soft handoff approach for mobile wireless IP-based systems," *IEEE Journal on Selected Areas in Communications*, vol. 22, no. 4, pp. 630–642, 2004.

73. A.J. Viterbi, A.M. Viterbi, K.S. Gilhousen, and E. Zehavi, "Soft handoff extends CDMA cell coverage and increases reverse link capacity," *IEEE Journal on Selected Areas in Communications*, vol. 12, pp. 1281–1288, 1994.

74. N. Zhang and J.M. Holtzman, "Analysis of a CDMA soft handoff algorithm," *IEEE Transactions on Vehicular Technology*, vol. 47, pp. 710–714, 1998.

75. D. Hong and S.S. Rappaport, "Traffic model and performance analysis for cellular mobile radio telephone systems with prioritized and non-prioritized procedures," *IEEE Transactions on Vehicular Technology*, vol. 35, no. 3, p. 7792, 1986.

76. M. Naghshineh and M. Schwartz, "Distributed call admission control in mobile/wireless networks," *IEEE Journal on Selected Areas in Communications*, vol. 14, no. 4, pp. 711–717, 1996.

77. D.A. Levine, I.F. Akyildiz, and M. Naghshineh, "A resource estimation and call admission algorithm for wireless multimedia networks using the shadow cluster concept," *IEEE/ACM Transactions on Networking*, vol. 5, no. 1, pp. 1–12, 1997.

78. L. Harte, *CDMA IS-95 for Cellular and PCS*, McGraw-Hill, New York, 1999.

79. C. Zhou, M. Honig, and S. Jordan, "Two-cell utility-based resource allocation for a CDMA voice service," *IEEE Proceeding of 54th Vehicular Technology Conference*, vol. 1, pp. 27–31, 2001.

80. T. Lee and J. Wang "Admission control for variable spreading gain CDMA wireless packet networks," *IEEE Transactions on Vehicular Technology*, vol. 49, no. 2, pp. 565–575, 2000.

81. S. Choi and K. Shin, "A comparative study of bandwidth reservation and admission control schemes in QoS-sensitive cellular networks," *ACM Journal of Wireless Networks*, vol. 6, no. 4, pp. 289–305, 2000.

82. D. Zhao, X. Shen, and J. Mark, "QoS performance bounds and efficient connection admission control for heterogeneous services in wireless cellular networks," *ACM Journal of Wireless Networks*, vol. 8, no. 1, pp. 85–95, 2002.

83. M. Buddhikot, G. Chandranmenon, S. Han, Y.-W. Lee, S. Miller, and L. Salgarelli, "Integration of 802.11 and third-generation wireless data networks," *Proc. IEEE INFOCOM*, March 2003.

84. E. Uysal-Biyikoglu, B. Prabhakar, and A. El Gamal, "Energy-efficient packet transmission over a wireless link," *IEEE/ACM Transactions on Networking*, vol. 10, no. 4, pp. 487–499, 2002.

85. H. Wu, S. De, C. Qiao, E. Yanmaz, and O. Tonguz, "Handoff performance of the integrated cellular and ad hoc relaying (iCAR) system," *ACM Wireless Networks (WINET)*.

86. H. Wu, C. Qiao, S. De, and O. Tonguz, "An integrated cellular and ad hoc relaying system: iCAR," *IEEE Journal on Selected Areas in Communications*, vol. 19, no. 10, pp. 2105–2115, 2001.

87. J. Wang and J. Liu, "Optimizing uplink scheduling in an integrated 3G/WLAN network," *International Journal of Wireless and Mobile Computing*, 2005.

Index

Actuators, 107–131. *See also under*
    Energy conservation:
    mobile robots/actuators, 129
Ad hoc networks, 226–250. *See also
    under* Opportunism
Ad hoc on-demand distance vector
    routing (AODV), 50
Ad hoc traffic indication messages
    (ATIM), 372, 464
Adaptive self-configuring sensor
    networks topologies (ASCENT),
    388–392
Adaptive timing synchronization
    procedure (ATSP), 373
Address autoconfiguration, 449–451:
    ad hoc address autoconfiguration,
    449–450
    hardware-based solution, 449
    MANETconf, 451–452
    weak DAD, 450–451
Advanced encryption standard (AES),
    410:
    and the counter-mode with CBC-MAC
        protocol (AES-CCMP), 416–417
Aggregation techniques, 36, 37
Angle of arrival (AoA) technique, 256
Announcement traffic indication message
    (ATIM), 314
Application layer security, 159–161
Asymmetric links, 118–119
Asynchronous link protocols, 72–73,
    375–379

Bayes' rule, 264
Berkeley MAC (B-MAC) protocol, 72
Binary model, 4:
    coverage solutions based on, 4–10
Bit error rate (BER), 119–120
Blind-guidance systems:
    blind alarm surveillance system,
    210–214

design and implementation, 214–218
initialization, 213–214
installation, 212–216: basic elements,
    215; interface architecture, 215
canes, 199–203
    design and implementation, 203–206
    as multifunction aid equipment,
        201–202
    RFID in, 204
dogs, 200
fingerprint identification system, 202
FLASH use in, 216
robots, 199–203
    design and realization of, 206–210
    multirobots, smart cooperation
        strategy for, 203
    omnidirectional images, 208–210
    Robot's main body fabrication, 202
    smart controller units, 203
    vision system for, 202–203
Bluetooth technology, 580
Breach weight, 13

CA-AODV, 477–478
Carrier sense multiple access (CSMA), 72:
    carrier sense multiple access with
        collision avoidance (CSMA/CA),
        326
Cellular aided mobile ad hoc network
    (CAMA), 495–496
Centralized protocols, 110–111
Channel assignment, 277–297, 442–446:
    grids, 285–287
        bidimensional grid, 285
        cellular grid, 285
        optimal $L(\delta_1, \delta_2)$ coloring of grids,
            286
        optimal $L(\delta_1,1,...,1)$ coloring of
            bidimensional grids, 286–287
    interval graphs, 288–291

Channel assignment (*Contd.*)
    approximate $L(\delta_1,...,\delta_t)$ coloring of
        interval graphs, 288–291
    approximate $L(\delta_1,\delta_2)$ coloring of unit
        interval graphs, 291
    preliminaries, 280–282 rings, 282–285
    optimal $L(\delta_1,1,...,1)$ coloring of rings,
        283–285
    trees, 292–296
    approximate $L(\delta_1,..., \delta_t)$ coloring of
        trees, 292–296
Channel hopping, 317–320:
    CHMA, 317–318
    SSCH, 318–320
Channel precedence indicator (CPI), 453
Channel reservation notification (CRN),
    311
Channel usage list (CUL), 469
Channel utilization comparisons, in QoS,
    339–340
Ciphertext, 401
Cluster-based scheme/solution:
    to MMAC, 472–473
        highest connectivity cluster
            algorithm, 472
        MIX algorithm, 472
    in WMN, 455
Cluster-head gateway switch
    routing(CGSR), 47
Coal mine safety application, 98–102:
    location tracking, 100–101
    network topology, 101
    rescue sensor network in, 99
    voice streaming, 102
Co-channel reuse distance, 278
Collaborative path planning algorithm
    (CPPA), 181–184
Communication protocols, 25–59. *See
    also under* Wireless sensor
    networks:
    comparative studies, 53–56
        on communication pattern, 53–54
        on performance evaluation,
            54–56:evaluation metrics, 54–55;
            overhead to maintain the routes,
            55; reliability and fault tolerance,
            55
    dual-operation communications, 38
        sensor-to-sink type, 38
        sink-to-sensor type, 38
        types, 38
    global IDs, 27
    ID establishment, 31–33
        field ID assignment, 32

        field ID, 32
        local ID assignment, 33
        node ID, 31
    overhead of protocols, comparison,
        57–58
    power issues in, 26–27
    processing, storage, and transmitting,
        27
    routing protocols, 39–53. *See also
        under* Routing protocols
    sensor to sink communication, 35–37
        all-to-sink communication, 37
        one-to-sink communication, 36
        region-to-sink communication, 36:
            combination, 36; concatenation,
            36; reduction, 36–37
        subset-to-sink communication, 37
    sensor-to-sensor communications,
        37–38
    sink-to-sensor communications, 33–35
        sink-to-all communication, 33–34
        sink-to-one communication, 34
        sink-to-one communication, 34–35
        sink-to-subset communication, 35
Concatenation, 36
Connected dominating set (CDS), 22
Connectivity-preserving channel
    assignment, 445–446
Contentionaware admission control
    protocol (CACP), 346
Core-extraction distributed routing
    algorithm (CEDAR), 345
Coverage and connectivity, 3–23:
    computing coverage, 4–14
        binary model, solutions based on,
            4–10
        probabilistic model, solutions based
            on, 10–12. *See also under*
            Probabilistic model
        solutions based on exposure, 12–14
    coverage and connectivity, 20–23
    coverage and scheduling, 14–20
        distributed scheduling solutions,
            17–20
        set covering solutions, 15–17
    coverage problem, definitions, 4
Coverage configuration protocol (CCP), 5
1-Coverage-preserving (1-CP) protocol,
    19:
    energy-based, 19

Data confidentiality protocols in RSN,
    414–417
802.11 DCF protocol, 327–328

Decryption process, 401:
  TKIP, 424–425
Dedicated control channel, 303–313:
  Bi-MCMAC, 311–313
  GRID and GRID-B, 309–311
  on-demand channel assignment,
    304–307 DCA-PC, 307–309
  channel selection manner, 308
  operation, 307
Delay rule, 525
Depth-first search (DFS) based routing,
  112
Digital signal processing (DSP)
  architecture, 192
Digital subscriber line (DSL), 441
Directed local spanning subgraph
  (DLSS), 392, 395–397
Directed relative neighborhood graph
  (DRNG), 392–395
Direction of arrival (DoA) approach, 256
Distributed control system (DCS),
  fieldbus technology for, 571–592.
  See also under Fieldbus
  technology
Distributed coordination function (DCF),
  326, 384
Distributed intrusion detection, 509–511
Dominating-awake-interval protocol,
  375–376
Dual-operation communications, 38
Duplicate address detection (DAD), 449
Dynamic channel assignment (DCA),
  469–470:
  operation, 470
Dynamic host configuration protocol
  (DHCP), 449
Dynamic source routing (DSR), 50, 345

EAP authentication process, 422–423
EDCF-DM protocol, 333–334
Elliptic curve digital signature algorithm
  (ECDSA), 151
Elliptic-curve cryptography (ECC),
  150–151
Encryption, 400–404:
  TKIP, 423–424
Energy conservation in sensor and
  sensor–actuator networks,
  107–131:
  fault tolerant sensor and sensor
    actuator networks, 128–130
  for asymmetric links, 118–119
  heterogeneous sensor networks,
    108–110

homogeneous sensor networks,
    107–108
  localized algorithms in, 110–112
  localized coordination framework for,
    121–128
    actuator–actuator coordination, 121,
      127–128
    anycasting with guaranteed delivery,
      126–127
    sensor–actuator coordination, 121,
      122–125: anycasting in, 122, 125
  minimum-energy broadcasting and
    multicasting, 113–115
    broadcasting, 113–115
    multicasting, 114–115
  power-aware routing with a realistic
    physical layer, challenges of,
    119–121
  power-aware routing, 115–116
    controlled mobility for, 116–118
Energy conservation protocols for
    wireless ad hoc networks,
    371–397:
  power control, 382–387
    BASIC protocol, 382
    by busy tones, 385–387
    topology control by, 392–397:
      DRNG, 392–395 DLSS, 395–397
    with periodical pulses, 382–385:
      carrier sensing range, 382; carrier
      sensing zone, 382; transmission
      range, 382
  power management, 371–382
    ad hoc networks, 372–374
    asynchronous power-saving
      protocols, 375–379:
      dominating-awake-interval
      protocol, 375–376;
      periodically-fully-awake-interval
      protocol, 375–376; quorum-based,
      375
    of IEEE 802.11
    semiasynchronous power-saving
      protocols, 379–382:
      location-based schemes, 380–381;
      SNR-probability-based, 380–381
    topology control by, 388–392:
      ASCENT, 388–390; span, 390–392
    topology control protocols, 387–397.
      See also separate entry
Extended interframe space (EIFS), 384
Extensible authentication protocol over
    LANs (EAPOL), 405, 406–408

Fault tolerant sensor and sensor actuator networks, 128–130
Fieldbus technology for distributed control applications, 571–592:
  DeviceNet, 584–585
  distributed control, review on 577–578
  fieldbus foundation, 586–588
    range of applications covered by, 587
  fundamental aspects of, 578–580
  Highway addressable remote transducer (HART), 586
  LonWorks™ technology, 588–590
  origin, 572–573
  OSI model and fieldbus layers, 573–575
  R-Fieldbus, 585–586
  selecting a, 590–591
    commercial aspects, 590
    communication requirements versus applications, 591
    technical aspects, 590–591
  services claimed for, 574–575
    application layer, 574–575
    datalink layer, 574
    management layer, 575
    physical layer, 574
  standards, frequency bands, and issues, 580–583
  versus contemporary standards, 576–577
  versus LAN, 575–576
  wireless profibus DP (ELPRO 905U-G-PR), 583–584
Firefly, 65–104:
  coal mine safety application, 98–102. *See also separate entry*
  FireFly Jr, 67
  fireFly sensor nodes (SNs), 65–71
  hardware-assisted time synchronization, 68–71. *See also under* Time synchronization
  multihop wireless networks, TDMA link layer protocol for, 71–87. *See also separate entry*
  nano-RK, 87–98. *See also separate entry*
Fixed and switchable channels, 476–477
FLASH use, in blind-guidance systems, 216
Foraging and gathering, in swarm-bot systems, 172–177:
  direction measurement, 175–177
  inter-distance maintenance, 177
  position measurement, 173–175

process of electing a coordinator, 172–173
Frame check sequence (FCS) suffix, 401
Free channel list (FCL), 303, 470
Fully synchronized link protocols, 73

Gateway discovery, 455–456
Gateway selection, 457–458:
  link-quality-based scheme, 457
  traffic-load-based scheme, 457–458
Gateway-based scheme, in WMN, 453–454
Gathering, 172–177. *See also under* Foraging
Gaussian fit sensor model, 265
Geographic adaptive fidelity (GAF), 49
Geographic hash table (GHT), 43
Geographical and energy aware routing (GEAR), 41–42
Glacier bed deformation monitoring system, 31
Global positioning system (GPS), 309
3GPP/WLAN interworks, 487–490:
  interworking architecture, 487–490
  simplified integration view, 488
Greedy perimeter stateless routing (GPSR), 40
Greedy-face-greedy (GFG) routing algorithm, 126
Greedy-MSC Heuristic algorithm, 17
Guide, alarm, recovery and detection (GUARD) guide system, 192

Hardware-assisted time synchronization, 68–71
Heterogeneous sensor networks, 108–110
Heterogeneouswireless multihop networks, 492–499:
  architectures and integrated routing for, 494–495
  CAMA, 495–496
  HCMN, 497
  two-tier heterogeneous MANET architecture, 496–497
Hierarchical cooperative intrusion detection, 511–512
Highway addressable remote transducer (HART), 586
Home band, 247
Homogeneous sensor networks, 107–108
Hot spot depletion problem, 35
Hybrid coordination function (HCF), 330
Hybrid methods, 267–270:
  model tree, 269–270

structure, 270
multiple regression, 269

Ideal routing protocol, 116
IEEE 802.11i security, 410–417:
    802.11i operational phases, 412–414
    802.1X authentication, 413
    RADIUS-based key distribution,
        413–414
    security capabilities discovery, 413
    data confidentiality protocols in RSN,
        414–417
    AES-CCMP, 416–417
    temporal key integrity protocol,
        415–416
    key hierarchy of 802.11i, 411–412
    temporal key integrity protocol and,
        419–434. *See also under* Temporal
        key integrity protocol
IEEE 802.11s mesh networks, 478–480
IEEE 802.11u integration with external
    networks, 491–492
IEEE 802.15 mesh networks, 480–481
IEEE 802.16 mesh networks, 481
IEEE 802.21 MIH function, 490–491
Independent basic service set (IBSS), 373
Indicator-based scheme, in WMN, 453
Integrated cellular network, security
    issues in, 535–566:
    architecture, 537–540
    communication and mobility,
        538–539
    multihop routing protocols, 539–540
    investigation and analysis of, 554–563
    charging and rewarding, 558–562
    integrated internet and MANET,
        562–563
    multihop cellular IP, 560
    two-hop relay, 557–558: security
        authentication in, 558
    unified cellular and ad hoc network,
        555–557: UCAN secure crediting,
        556
    new security issues and challenges,
        564–566
    distributed trusted relationship,
        564–565
    multihop routing security, 565
    potential security threats, 542–554
    internet connectivity attacks,
        543–547. *See also separate entry*
    MS mobility attacks, 547–552. *See*
        also separate entry
    multihop routing attacks, 552–553

packet forwarding attacks, 553–554
specific wireless protocol attacks,
    554
security impacts from, 540–542
    internet and infrastructure support,
        540–541
    multimode with multiple radio
        interfaces, 542
    wireless multihop communication,
        541–542
Integrated heterogeneous wireless
    networks, 483–502:
    3GPP/WLAN interworks, 487–490. *See*
        also separate entry
    characteristics, 484
    heterogeneouswireless multihop
        networks, 492–499. *See also*
        separate entry
    IEEE 802.11u integration with external
        networks, 491–492
    IEEE 802.21 MIH function, 490–491
    infrastructure-based, 485–492
    integrated routing protocol, 497–499
    research issues for, 500–502
Integrated routing protocol, 497–499
Integrated WCDMA/ WLAN/ad hoc
    networks, 595–621. *See also under*
    Supporting multimedia
    communication
Integrity rule, 525
Interference phenomena, 278:
    hidden interferences, 278
Interference-free node scheduling, 76–79
Internet connectivity attacks, 543–547:
    eavesdropping and traffic analysis,
        543–544
    internet–wireless DDoS, 545–546
    unauthorized internet access and
        attacks, 544–545
    wireless–internet DDoS, 546–547
Interval rule, 525
Interval-coloring, 289
Intrusion detection system (IDS),
    505–530:
    background, 505–506
    cross-feature analysis for, 513–514
    for mobile ad hoc networks, 506–519.
        See also under Mobile ad hoc
        networks requirements for,
        507–508 security vulnerabilities,
        508
    for wireless sensor networks, 519–530.
        See also separate entry
    game theory approach to, 515–516

Intrusion detection system (IDS) (*Contd.*)
  misuse detection and anomaly
    detection, 516–517
  SVM-based, 514–515

Jamming:
  attack, 137–138
  rule, 525

Key distribution center (KDC), 143–144
Key handshake procedure, 433–434
Key management, in security, 140–155:
  concepts, 140–141
  group key management, 153–154
  group-based key predistribution and
    management, 146–147
    cross function-based, 147
    grid-based, 146–147
  key revocation, 154–155
  key space-based redistribution,
    141–142
  public key-based key establishment,
    149–153
  random key predistribution, 142–146
  time-dependent key predistribution,
    147–149
Keystream, 401

Load-aware channel assignment, 443–445
Local shortest path tree (LSPT), 113
Localization anomaly detection, 523–524
Localization techniques, 255–276:
  location tracking, 272–274. *See also
    separate entry*
  nondedicated systems, 255, 256–272.
    See also Nondedicated localization
    techniques
Localized encryption and authentication
    protocol (LEAP), 147–148
Localized protocols, 110–111
Location tracking, 272–274:
  Markov-based location tracking,
    273–274
  viterbi-like location tracking, 273
Location-aided routing (LAR), 39:
  request zone in, 40
Log-linear scheduler, 230
Log-normal shadowing model, 11
LonWorks™ technology, 588–590
Loosely synchronous link protocols, 73
Low energy adaptive clustering hierarchy
    (LEACH), 48

Mantis OS, 89
Markov decision process (MDP), 527
Markov-based location tracking, 273–274
Media access control protocol (MAC),
    197
Michael function, fragility of, 430–431:
  inverse function of Michael, 430
Minimap integration, 178–181
Minimum-energy broadcasting and
    multicasting, 113–115
MMAC, 314–317
MOAR protocol, 247–249:
  attacks against, 508
  IDS for, 508–519
    distributed intrusion detection,
      509–511
    hierarchical cooperative intrusion
      detection, 511–512
    mobile agent-based intrusion
      detection systems, 512–513
    watchdog and pathrater, 518
  intrusion prevention for, 508–509
Mobile ad hoc network (MANET):
  multichannel MAC protocols for,
    301–322. *See also under*
    Multichannel MAC protocols
  performance study, 249–250
  versus WSNs, 28–30. *See also under*
    WSNs
MS mobility attacks, 547–552:
  advertisement with CoA, 547
  avoiding tracing
    by changing identity, 552
    by mobility, 551–552
  bogus registration, 549
  BS cache poisoning, 551
  BS/FA discovery, 547
  forged BS/FA, 550
  mobility binding, 548
  registration poisoning, 548–549
  registration replay attack, 549–550
  registration reply, 548
  registration request, 548
Multichannel MAC protocols for MANET,
    301–322:
  comparison, 320–321
  design issues, 302–303
  open issues, 321–322
    losing channel information, 321
    disordered channel reuse, 321–322
    multiChannel MANET, scheduling in,
      322
  types, 303–320

channel hopping, 303, 317–320. *See also separate entry*
dedicated control channel, 303–313. *See also separate entry*
split phase, 303, 313–317. *See also separate entry*
Multichannel protocols and standard activities, of WMN, 461–481:
multichannel MAC protocols, 462–473
cluster-based solution, 472–473
DCA, 469–470 MUP, 470–472
IEEE 802.11s mesh networks, 478–480: channel selection, 479; link establishment, 479; neighbor discovery, 479
IEEE 802.15 mesh networks, 480–481
multichannel routing protocols, 473–478. *See also separate entry*
standard activities of, 478–481
with a single transceiver, 463–469: channel selection/scheduling problem, 463; channel switching problem in, 463; MMAC, 464–466; multichannel hidden terminal problem, 463–464; SSCH, 467–469
Multichannel routing protocols, 473–478:
CA-AODV, 477–478
fixed and switchable channels, 476–477
Hyacinth, 473–475
routing procedure proposed in, 474
WCETT metric, 475–476
Multihop traffic indication message (MTIM) window, 375
Multihop wireless networks, TDMA link layer protocol for, 71–87:
current MAC protocols, 72–75
asynchronous link protocols, 72–73
fully synchronized link protocols, 73
loosely synchronous link protocols, 73
end-to-end latency, 86–87
explicit rate control, 81
interference-free node scheduling, effectiveness of, 76–79
lifetime, 83–86
logical topology control, 76
network scheduling, 79–81
operation procedures, 75–76
TDMA slot mechanics, 81–83
Multiple accesses in CDMA uplink, 599–613:
admission control models, 599–601

multiple-cell interference, 600–601
single cell with voice traffic, 599–600
admission control with multimedia traffic, 601–604
optimum power assignment and rate adaptation, 603–604
problem formulation, 601–602
downlinks, scheduling issues for, 609–613
multicode rate with fairness, 612–613
packet dropping control, 611–612
TDM-based scheduling, 609–610
wireless fair queuing, 610–611
dynamic spreading factors, 604
multicell capacity, 606–609
decentralized downlink power control, 608–609
intracell power sharing with multicell interference, 607–608
one-cell-one-mobile model, 606–607
single-cell capacity, 605–606
uplink scheduling, 605
Multi-radio link-quality source routing (MR-LQSR), 475
Multiradio unification protocol (MUP), 470–472

Nano-RK, 87–98:
architecture, 92–98
energy management support, 94
reservation paradigm, 93–96
static approach, 93
current sensor network operating systems, 89–90
Mantis OS, 89–90
TinyOS, 89
integration with RT-link, 97–98
sensor networking RTOS, design goals, 90–92
task management and scheduling, 96–97
Narrowbanding, 581
Nearest neighbor in signal space (NNSS) technique, 261
Neighbor elimination scheme (NES), 113
Network address translation (NAT), 449
Network allocation vector (NAV), 304, 383
Network layer security, 157–159
Network scheduling, 79–81
Network-planning techniques, 442–448:
channel assignment, 442–446
classes in, 443

Network-planning techniques (*Contd.*)
connectivity-preserving channel
assignment, 445–446
gateway placement, 446–448
load-aware channel assignment,
443–445
Nondedicated localization techniques,
256–272:
fundamental principles, 256–259
triangulation, 256–257
trilateration, 256
hybrid, 267–270. *See also separate
entry*
path loss model, 270–272
radio map, 259–266
NNSS technique, 261
probabilistic techniques, 263–266
smallest polygon, 262–263
reference point (RP), 266–267
Nonquorum intervals, 378

OAR protocol, 244–247
Opaque transitory master key (OTMK),
147–148
Opportunism in wireless networks,
223–253:
avenues and basic principles, 223–227
degrees of freedom, 225–226: source
"degrees of freedom", 226; spatial
dimensions, 225; temporal and
spectral dimensions, 225
roadmap, 226–227
source opportunism, 226, 227–234
fully informed schedulers, 228–232
problem formulation, 227–228
uninformed robust schedulers,
232–234
spatiotemporal-spectral opportunism
in ad hoc networks, 247–250: MOAR
protocol, 247–249. *See also
separate entry*
temporal–spatial opportunism
in ad hoc networks, 227, 241–247:
multi-rate IEEE 802.11, 242; OAR
protocol, 244–247; review of IEEE
802.11, 241–242
over a single link, 226, 234–241:
channel models and problem
formulation, 236–238; multiple
antenna representation, 238–239;
throughput maximizing resource
allocation, 239–241
Out-of-band time synchronization in
firefly nodes, 69–71

Packet delay comparisons, in QoS,
337–339
Packet leashes, 523
Pairwise master key (PMK), 411
Pairwise transient key (PTK), 411
Partial synchronization technique,
468–469
Path loss model, 270–272
Perimeter coverage, 8
Periodically-fully-awake-interval
protocol, 375–376
Peripheral component interface (PCI),
576
Physical layer security, 136–140
Portable operating system interface
(POSIX), 95
Power management, for wireless ad hoc
networks, 371–382. *See also under*
Energy conservation protocols
Power-saving mechanism (PSM), 464
Preferable channel list (PCL), 314, 465
Priority ceiling protocol emulation
protocol, 97
Probabilistic model, 4, 263–266:
coverage solutions based on, 10–12
log-normal shadowing model, 11
signal decay model, 10
data structures, 351–352
route reply (RREP) packets, 351
route request (RREQ) packets, 351
fundamental rules, 352–354
novel QoS routing protocol, 350–351
phases, 344
bandwidth reservation phase, 344,
350, 358–359
route-discovery phase, 344, 350,
354–358

QoS routing protocols for MANET,
343–368:
reviews of, 345–348
reservation-based, 346–348
route maintenance, 359–360
simulation results, 360–367
bandwidth requirement impact,
361–363
host density impact, 365
mobility impact, 365–367
traffic load impact, 363–365
Quality of service (QoS) in wireless ad
hoc networks, 325–340:
background, 327–331
802.11 DCF potocol, 327–328
802.11 EDCF, issues of, 331

enhanced DCF of IEEE 802.11e,
329–330
for mobile ad hoc networks, 343–368.
See also under QoS routing
protocols
performance evaluation, 334–340
channel utilization comparisons,
339–340
packet delay comparisons, 337–339
simulation setup, 334–335
throughput comparisons, 335–337
proposed EDCF-DM protocol, 331–334
architecture, 332
network condition measurement,
331–332
traffic state measurement, 332–333
Quorum-based protocol, 375–377

Radio frequency identification (RFID),
204
Radio transmission range rule, 525
Random fashion routing, 50–53:
energy-aware routing, 51–52
ReInForM, 52
rumor routing, 50–51
Random key predistribution, 142–146
RC4 algorithm, 402–404:
pseudo-random sequence generation
(Stage 2), 403
S-box initialization (Stage 1), 403
Receiver-based auto rate (RBAR), 243
Reference point (RP) method, 266
Remote authentication dial in user
service (RADIUS), 405
Repetition rule, 525
Response implosion problem, 34
Retransmission rule, 525
Routing protocols in communication,
39–53:
indicator-based approaches, 39–49
cluster-based, 47–49: CGSR, 47;
GAF, 49; LEACH, 48
geography-based, 39–44:
BOUNDHOLE, 42; GEAR, 41–42;
GPSR, 40; TTDD, 43–44
gradient based, 44–47: ARRIVE, 45;
MCFN, 44
indicator-free approaches, 49–53
on-demand fashion, 49–50: AODV,
50; DSR, 50
random fashion, 50–53. *See also
separate entry*
RT-link, 71–87. *See also under* Multihop
wireless networks

Scheduling, 14–20
Secure crediting, 555
Security issues:
in integrated cellular network,
535–566. *See also under*
Integrated cellular network
in WEP, *see under* Wired equivalent
privacy protocol
in wireless sensor networks, 135–161
application layer security, 159–161:
secure localization, 159; secure
time synchronization, 159–161
key management, 140–155. *See also
separate entry*
link layer security, 155–157: TinySec
packet format for, 156
network layer security, 157–159
node compromise, 139–140
physical layer security, 136–140
secure aggregation, 161
topology change, 137–139: jamming
attack, 137–138; physical removal
and relocation, 138–139
802.1X security measures, 405–410:
Cisco-LEAP, 408
EAP-FAST, 408
EAP-MD-5, 407
EAP-PEAP, 407–408
EAP-TLS, 407
EAP-TTLS, 408
IEEE 802.1X authentication operations,
408–410
EAPOL operations on, 409
security problems in, 410:
man-in-the-middle attack, 410;
session hijacking, 410
IEEE 802.1X framework, 405–406
security problems, 404–405
data confidentiality, 404
data integrity, 404
user authentication, 405
Self-configuring techniques, 448–458:
address autoconfiguration, 449–451.
See also separate entry
dynamic channel selection, 451–455
cluster-based scheme, 455
gateway-based scheme, 453–454
indicator-based scheme, 453
tree-based scheme, 454–455
gateway discovery, 455–456
gateway selection, 457–458. *See also
separate entry*
Semiasynchronous power-saving
protocols, 379–382

Sensor and sensor–actuator networks, energy conservation in, 107–131. *See also under* Energy conservation
Sensor–actuator coordination, 121, 122–125
Set cover, 15–16
Shadow cluster concept, 615
Short interframe space (SIFS), 384
Signal decay model, 10
Sixth-generation ant-bot, 168
Slotted seeded channel hopping (SSCH), 467–469
Smart blind alarm surveillance and blind guide network system, 191–218. *See also under* Wireless optical communication:
blind-guidance cane and robot, 199–203. *See also under* Blind-guidance systems
function block diagram, 194
transimpedance amplifier, 196
Span, 390–392
Split phase, 313–317
MMAC, 314–317
Steiner tree approach, 114
Subscription service provider network (SSPN), 491
Supporting multimedia communication, 595–621:
ad hoc networks, design integration with, 620–621
basic resources to manage in, 597–599
code control, 597–598
link-adaptation and rate-power adaptation, 598–599
power control, 597
mobility management, 613–621
leading handoff methods, 616–617
location update and tracking, 613–614
OR-ON-DOWN power control, performance of, 617–618: BER model in Handoff area, 618
power control, 617
soft handoff in, 614–615
suboptimal SF/POWER admission algorithms, 619–620
multiple accesses in CDMA uplink, 599–613. *See also separate entry*
Swarm-bot systems, 167–186:
cooperative localization algorithm, 169–172
absolute coordinate, 170

initial phrase, 171
iterative phrase, 171
relative coordinate, 170
coordinator process algorithm, 185
CPPA, 181–184
foraging and gathering, 172–177. *See also separate entry*
hardware structure, 170
member process algorithm, 186
minimap integration, 178–181
sixth-generation ant-bot, 168
system architecture, 168–169

TDMA link layer protocol, for multihop wireless networks, 71–87
Temporal key integrity protocol (TKIP) and its security issues in IEEE 802.11i, 419–434:
decryption process, 424–425
encryption process, 423–424
fragility of Michael, 430–431. *See also under* Michael function
key handshake procedure, 433–434
TKIP countermeasures, 431–433
for the authenticator, 432
for the supplicant, 433
TKIP message integrity code, 427
TKIP MIC processing format, 426–427
TKIP mixing function, 427–430
phase 1, 428–429
phase 2, 429–430
TKIP MPDU message format, 426
WEP and its weakness, 420–421
wi-fi protected access, 421–423. *See also separate entry*
Throughput comparisons, in QoS, 335–337
Time difference of arrival (TDoA) technique, 256
Time of arrival (ToA) technique, 256, 258–259
Time synchronization:
benefits of, 68–69
group time synchronization, 160
hardware-assisted, 68–71 out-of-band time synchronization, 69–71
multihop time synchronization, 160
network-wide time synchronization, 160
secure time synchronization, 159–161
sender–receiver time synchronization, 160

time-synchronized real-time sensor networking platform, *see under* Firefly

Timing synchronization function (TSF), 373

TinyOS, 89

TinySec packet format, 156

Topology control protocols, 387–397: factors affecting, 387–388

Traffic state measurement, 332–333

Transimpedance amplifier (TIA), 194–195

Transitional security network (TSN), 410

Transmission control protocol (TCP), 311

Transmission opportunity (TXOP), 330

Tree-based scheme, in WMN, 454–455

Tree-coloring algorithm, 292–293

Triangulation, 256–257

Trilateration, 256–259

Two-hop relay system, 557–558

Two-tier data dissemination (TTDD), 43

Uninformed robust schedulers, 232–234

Unit disk graph (UDG) model, 112

Value iteration algorithm (VIA), 230

Virtual clustering, 73

Viterbi-like location tracking, 273

Voronoi diagram, 12–13

Wakeup schedule function (WSF), 379

Watchdog and pathrater, 518

4-Way handshake, 413–414

Weak DAD, 450–451

Weight clustering approach (WCA), 526

Wi-Fi networks, 193–196

Wi-Fi protected access, 421–423: 802.1X EAP-based authentication, 422–423

Wired equivalent privacy (WEP) protocol, 399–405: 802.1X security measures, 405–410. *See also separate entry* and its weakness, 420–421 decryption operation, 402 encryption operation, 401–402 procedure of, 401 goals of, 421 IEEE 802.11 security framework, 400 IEEE 802.11i security, 410–417. *See also separate entry* RC4 algorithm, 402–404. *See also separate entry* security weaknesses, 399–405

WEP cryptographic operations, 400–404

Wireless ad hoc networks: energy conservation protocols for, 371–397. *See also under* Energy conservation protocols quality enhanced service for, 325–340. *See also under* Quality of service (QoS)

Wireless LAN: channel assignment in, 277–297. *See also under* Channel assignment localization techniques for, 255–276. *See also under* Localization techniques security, 399–417. *See also under* Wired equivalent privacy (WEP) protocol security weaknesses, 399–405

Wireless mesh network (WMN), 439–458: design principles, 439–458 generic architecture and basic requirements, 439–442 as a metropolitan-area network, 441 multichannel protocols and standard activities, 461–481. *See also under* Multichannel protocols and standard activities network-planning techniques, 442–448. *See also separate entry* self-configuring techniques, 448–458. *See also separate entry*

Wireless multicast advantage (WMA), 30

Wireless networks: intrusion detection for, 505–530. *See also under* Intrusion detection system opportunism in, 223–253. *See also under* Opportunism

Wireless optical communication: analog front-end circuit, 195 function block diagram, 195 network, design of, 196–199 smart blind alarm surveillance and blind guide network system on, 191–218. *See also under* Smart blind alarm surveillance

Wireless optical transceiver, manufacture of, 193–196

Wireless Profibus DP (ELPRO 905U-G-PR), 583–584

Wireless sensor networks (WSNs). *See also under* Firefly:
autonomous swarm-bot systems for, *see under* Swarm-bot systems
basic components of, 26
communication patterns in, 30–38. *See also under* Communication protocols
coverage and connectivity of, 3–23. *See also under* Coverage and connectivity
k-covered, 5
k-perimeter-covered, 7
paradigm of routing in, 26
security in, 135–161, *see under* Security issues
versus MANETs, 28–30
  differences, 29–30: communication paradigm, 29; density, 30; design objectives, 29; major research challenges, 30; node identifications, 29; protocol design fashion, 30; resources, 29
  similarities, 28: ad hoc mode, 28; power issue, 28; resource constraint, 28; wireless communications, 28
α-covered, 9
Wireless sensor networks, IDS for, 519–530:
attacks against, 520–522
  black hole attack, 521
  defenses against, 523–525: localization anomaly detection, 523–524; localized encryption and authentication protocol, 524; packet leashes, 523; radio

resource testing, 524–525; random key predistribution, 525
  flooding attack, 522
  jamming, 521
  misdirection, 521
  physical attacks, 521
  rushing attack, 522
  selective forwarding attack, 521
  sinkhole attack, 521
  stealthy attack, 522
  Sybil attack, 521
  wormhole attack, 522
challenges on, 519–520
intrusion detection for, 525–528
  decentralized high-level rule-based IDS model, 525–526: delay rule, 525; integrity rule, 525; interval rule, 525; jamming rule, 525; radio transmission range rule, 525; repetition rule, 525; retransmission rule, 525
  failure recovery and, 527–528
  hierarchical-based, 528
  MDP-based intrusion detection scheme, 527
  noncooperative game approach, 526–527
for physical intrusion detection, 528–530
  reconfigurable sensor network for, 528–529
  SOC- and HMM-based intrusion detection, 529
for target detection, 529–530
Wormhole attack, 139

Zigbee platform, 197